G-W
PUBLISHER

THIRD EDITION

FLUID POWER

Hydraulics and Pneumatics

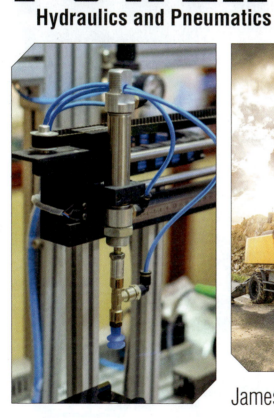

James R. **Daines** | Martha J. **Daines**

Basic principles and applications of hydraulics and pneumatics for technicians.

Be Digital Ready on Day One with EduHub

EduHub provides a solid base of knowledge and instruction for digital and blended classrooms. This easy-to-use learning hub delivers the foundation and tools that improve student retention and facilitate instructor efficiency. For the student, EduHub offers an online collection of eBook content, interactive practice, and test preparation. Additionally, students have the ability to view and submit assessments, track personal performance, and view feedback via the Student Report option. For instructors, EduHub provides a turnkey, fully integrated solution with course management tools to deliver content, assessments, and feedback to students quickly and efficiently. The integrated approach results in improved student outcomes and instructor flexibility.

Michael Jung/Shutterstock.com

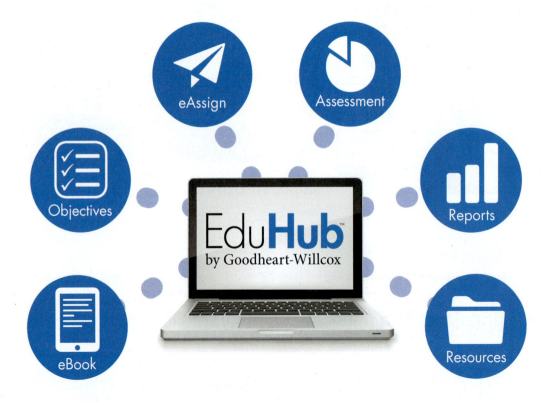

eBook

The EduHub eBook engages students by providing the ability to take notes, access the text-to-speech option to improve comprehension, and highlight key concepts to remember. In addition, the features enable students to customize font and color schemes for personal viewing, while links to links to vocabulary practice activities and e-flash cards provide reinforcement and enrich understanding.

Objectives

Course objectives at the beginning of each eBook chapter help students stay focused and provide benchmarks for instructors to evaluate student progress.

eAssign

eAssign makes it easy for instructors to assign, deliver, and assess student engagement. Coursework can be administered to individual students or the entire class.

Monkey Business Images/Shutterstock.com

Assessment

Self-assessment opportunities enable students to gauge their understanding as they progress through the course. In addition, formative assessment tools for instructor use provide efficient evaluation of student mastery of content.

Reports

Reports, for both students and instructors, provide performance results in an instant. Analytics reveal individual student and class achievements for easy monitoring of success.

Score	Items	
	🖨 Print	⬇ Export
100%	🟢	🟢
80%	🟢	🔴
100%	🟢	🟢
80%	🟢	🔴
100%	🟢	🟢
100%	🟢	🟢

Instructor Resources

Instructors will find all the support they need to make preparation and classroom instruction more efficient and easier than ever. Lesson plans, answer keys, and PowerPoint® presentations provide an organized, proven approach to classroom management.

Learn more about EduHub at www.g-w.com/eduhub

Guided Tour

Chapter Outline provides a preview of the chapter topics and serves as a review tool.

Objectives clearly identify the knowledge and skills to be obtained when the chapter is completed. Objectives are tied to the chapter outline, to the text, and to our end-of-chapter materials.

Key Terms list the important terms to be learned in the chapter.

Examples demonstrate the mathematical concept that has just been presented, showing the math skills required to solve real-world problems.

Caution features point out safety-related issues to help you avoid potentially dangerous materials and practices.

Note features present additional information related to the topic being discussed to deepen understanding.

Illustrations have been designed to clearly and simply communicate the specific topic.

Summary provides an additional review tool for you and reinforces key learning objectives.

Apply and Analyze activities develop higher-order thinking skills, extend your learning, and help you analyze and apply knowledge.

Summary

- Compressed-air units produce pressurized air for the pneumatic system.
- A compressed-air unit consists of a prime mover, a compressor, and components that condition and store the pressurized air used by the system workstations.
- Compressed-air units can be portable or permanently mounted to provide air to fixed workstations from a central supply.
- The basic operation of any compressor design used in a pneumatic system includes three phases: air intake, air compression, and air discharge.
- Compressors are classified by displacement (positive-displacement or nonpositive-displacement designs) and fundamental pumping motion (rotary or reciprocating designs).
- Positive-displacement compressor designs mechanically reduce a compression chamber in size to compress air. Nonpositive-displacement compressors use impellers or vanes rotating at high speed to increase air velocity and pressure.
- Reciprocating-piston compressors use a cylinder and a reciprocating piston to compress air. Rotary designs use continuously rotating vanes, screws, or lobed impellers to move and compress the air.
- Compressor-capacity control systems are used to match the volume of compressed air produced by the compressor to the volume of compressed air consumed by the working pneumatic system. The closer the compressor air output matches system consumption, the more cost effective the system operation.
- Compressor-capacity control systems include bypass, start-stop, inlet valve unloading, speed, and inlet size controls.
- Determining the required output of a compressor can be accomplished by identifying the actuators used, the volume of compressed air needed for the operation of each item, and the percentage of time each functions during system operation.

Internet Resources

The following are some useful resources available on the Internet. Enter a company or organization name into a search engine to access its website. Explore the various areas of the sites to discover useful fluid power resources.

Atlas Copco. Enter "screw compressors" in the product search box to review product information on the various rotary screw compressors manufactured by the company. Includes stationary and portable units.

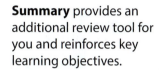

Dresser-Rand. Describes a centrifugal compressor designed for a process application in the oil, gas, or chemical industries.

FS-Elliott Co., LLC. Describes a line of centrifugal compressors. Illustrated with a general description of the operation of a two-stage unit including intercoolers.

Hydraulics & Pneumatics Magazine. Review two articles: *Air Compressors–Part 1* and *Air Compressors–Part 2*. These articles cover the styles of compressors available in the pneumatic field, emphasis on rotary compressors. Also includes information on power and efficiency, lubricated or lubrication-free compressors, cooling methods, and capacity control.

Ingersoll Rand Company. Provides information on reciprocating air compressors manufactured by the company. The brochure linked on the page refers to single- and two-stage air compressors.

Jenny Products, Inc. Guide to selecting an air compressor for a small business operation or home shop. Covers factors such as the need for staging, type of prime mover, and specific questions that should be answered before purchasing a compressor.

Saylor-Beall Manufacturing Company. Manufacturer that produces reciprocating and screw compressor units. Information contained includes product brochures and service manuals with illustrated, informative materials covering general questions and answers.

SENCO Brands, Inc. Product manual for a single-stage, reciprocating compressor. Applications for these compressors range from the home shop to industrial situations.

Chapter Review

Answer the following questions using information in this chapter.

1. A compressed air unit classified as a portable unit gives a(n) _____ air supply.
   A. industrial
   B. prime mover
   C. receiver
   D. central
2. *True or False?* Double-acting reciprocating compressors have compression chambers at each end of the pistons.
3. *True or False?* The crankshaft is the only moving part in a centrifugal compressor.
4. *True or False?* Displacement of a compressor is the volume of air displaced per revolution of the unit.

5. *True or False?* Single-acting compressors use cylinders, pistons, connecting rods, and crankshafts similar to the type found in internal combustion engines.
6. An automatic compressor capacity control method is commonly used with a smaller compressor using _____ as the prime mover.
   A. steam power
   B. electric motors
   C. pneumatics
   D. gasoline engines
7. *True or False?* The housing around the centrifugal compressor forms a volute collector that directs the airflow through the compressor.
8. Lobe-type compressors typically only develop air pressures ranging from 10 to _____ psi.
   A. 20
   B. 35
   C. 60
   D. 100
9. Typically, today's nonpositive-displacement compressors compress air using _____.
   A. impellers
   B. cylinders
   C. chambers
   D. pistons
10. *True or False?* An impeller turning at relatively low speed moves air through a centrifugal air compressor.
11. In the lobe-type compressor, there is no reduction of air volume in the compressor as the impellers turn. The air is simply swept from the inlet port to the _____ port.
12. Using a number of compressor units to increase pressure in small increments is called _____.
    A. start-stop
    B. capacity control
    C. bypass
    D. staging
13. Compressor-air output and system-air demand are matched by using some type of _____ control system.
14. A common method of compressor-capacity control used with reciprocating compressors is inlet valve _____.
15. *True or False?* When selecting a pneumatic system compressor, only one or two factors need to be considered.

Apply and Analyze

1. Which type of compressor would you choose for each of the situations listed below? Why? Identify specific aspects of compressor design or operation that allow the compressor to meet the stated needs.
   A. A steady, continuous supply of air.
   B. Low volumes of high-pressure air.
   C. Very large volumes of oil-free pressurized air.
   D. Minimal vibration of the compressor.
   E. Large volumes of air, but space for installation is limited.
2. For each of the following compressor types, where does energy loss occur during operation?
   A. Reciprocating-piston compressor.
   B. Rotary sliding-vane compressor.
   C. Centrifugal compressor.
3. You are assisting a company that is installing new compressors and updating their pneumatic system. The company is interested in minimizing not only initial investment costs, but also costs of operation.
   A. What specific advice would you give them in their choice of compressors?
   B. What information would you need to determine the best method(s) of compressor-capacity control for improving the efficiency of the system during operation? Explain your answer.

Research and Development

1. Lubrication is critical for the performance and service life of many air compressors. Investigate the different methods of lubrication used for the main types of positive-displacement compressor. Prepare a chart that illustrates the benefits and drawbacks of each lubrication method.
2. Using air compressor manufacturers' product listings found online, find and compare two compressors which could be used for similar applications, but which have different basic designs, such as rotary and reciprocating. Develop a fact sheet to be used to help a potential customer choose between the two compressors.
3. Investigate the historical development of air compressors. Produce a time line that shows design changes as well as the use of new materials or machining methods to improve performance.
4. Research an application that uses dynamic compressors to pressurize air. Write a report that describes the application, discusses the design and capacity of the compressor, and explains why this type of compressor is used.

Research and Development activities provide additional opportunities for you to explore and deepen your understanding of key concepts from each chapter.

Internet Resources help you find websites that extend learning beyond the scope of this text.

Chapter Review questions allow you to demonstrate knowledge, identification, and comprehension of chapter material.

EduHub

EduHub provides a solid base of knowledge and instruction for digital and blended classrooms. This easy-to-use learning hub provides the foundation and tools that improve student retention and facilitate instructor efficiency.

For the student, EduHub offers an online collection of eBook content, interactive practice, and test preparation. Additionally, students have the ability to view and submit assessments, track personal performance, and view feedback via the Student Report option. For the instructor, EduHub provides a turnkey, fully integrated solution with course management tools to deliver content, assessments, and feedback to students quickly and efficiently. The integrated approach results in improved student outcomes and instructor flexibility. Be digital ready on day one with EduHub!

- **eBook content.** EduHub includes the textbook in an online, reflowable format. The eBook is interactive, with highlighting, magnification, note-taking, and text-to-speech features.
- **Assignments.** In EduHub, students can complete the text review questions as online assignments. Instructors can assign key term activities, chapter reviews, and laboratory activities from the Lab Workbook. Many activities are autograded for easy class assessment and management.

Student Tools

Student Text

Fluid Power: Hydraulics and Pneumatics is an introductory text targeted to students pursuing a technician-level career path. It presents the fundamentals of this subject with extensive coverage of both hydraulic and pneumatic systems. Coverage includes details on the design and operation of hydraulic and pneumatic components, circuits, and systems.

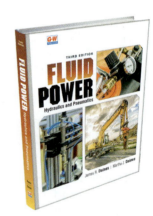

The approach used throughout the book is based on developing an understanding of the operation of fluid power component parts and circuits. Understanding component operation requires comprehension of fundamental principles of physics and basic mathematical formulas. These fundamentals are discussed with topics that require them for comprehension of component operation.

Lab Workbook

The Lab Workbook that accompanies *Fluid Power: Hydraulics and Pneumatics* includes instructor-created activities to help students recall, review, and apply concepts introduced in the book.

Instructor Tools

LMS Integration

Integrate Goodheart-Willcox content in your Learning Management System for a seamless user experience for both you and your students. Contact your G-W Educational Consultant for ordering information or visit www.g-w.com/lms-integration.

Instructor Resources

Instructor Resources provide all the support needed to make preparation and classroom instruction easier than ever. Available in one accessible location, you will find Instructor Resources, Instructor's Presentations for PowerPoint®, and Assessment Software with Question Banks. These resources are available as a subscription and can be accessed at school, at home, or on the go.

Instructor Resources One resource provides instructors with time-saving preparation tools, such as answer keys, chapter outlines, editable lesson plans, and other teaching aids.

Instructor's Presentations for PowerPoint® Instructor's Presentations for PowerPoint® provide a useful teaching tool when presenting the lessons. These fully customizable, richly illustrated slides help you teach and visually reinforce the key concepts from each chapter.

Assessment Software with Question Banks Administer and manage assessments to meet your classroom needs. The following options are available through the Respondus Test Bank Network:

- A Respondus 4.0 license can be purchased directly from Respondus, which enables you to easily create tests that can be printed on paper or published directly to a variety of Learning Management Systems. Once the question files are published to an LMS, exams can be distributed to students with results reported directly to the LMS gradebook.

- Respondus LE is a limited version of Respondus 4.0 and is free with purchase of the Instructor Resources. It allows you to download question banks and create assessments that can be printed or saved as a paper test.

G-W Integrated Learning Solution

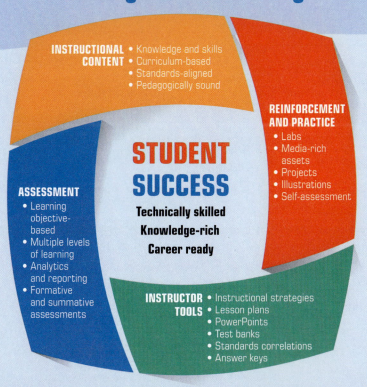

INSTRUCTIONAL CONTENT
- Knowledge and skills
- Curriculum-based
- Standards-aligned
- Pedagogically sound

REINFORCEMENT AND PRACTICE
- Labs
- Media-rich assets
- Projects
- Illustrations
- Self-assessment

STUDENT SUCCESS
Technically skilled
Knowledge-rich
Career ready

ASSESSMENT
- Learning objective-based
- Multiple levels of learning
- Analytics and reporting
- Formative and summative assessments

INSTRUCTOR TOOLS
- Instructional strategies
- Lesson plans
- PowerPoints
- Test banks
- Standards correlations
- Answer keys

The G-W Integrated Learning Solution offers easy-to-use resources that help students and instructors achieve success.

▸ **EXPERT AUTHORS**
▸ **TRUSTED REVIEWERS**
▸ **100 YEARS OF EXPERIENCE**

EMPLOYABILITY SKILLS · TECHNICAL SKILLS · ACADEMIC KNOWLEDGE · INDUSTRY RECOGNIZED STANDARDS

THIRD EDITION

FLUID POWER

Hydraulics and Pneumatics

by

James R. **Daines** | Martha J. **Daines**

Publisher
The Goodheart-Willcox Company, Inc.
Tinley Park, IL
www.g-w.com

Library of Congress Catalog Card Number 2018025320

ISBN 978-1-63563-473-0

4 5 6 7 8 9 – 20 – 23 22

Image credits: Front cover (clockwise from top right): Ambiento/Shutterstock.com, Juan Enrique del Barrio/Shutterstock.com, Dardiamon67/Shutterstock.com. Section 1 opener: Rawi Rochanacipoart/Shutterstock.com. Section 2 opener: Dmitry Kalinovsky/Shutterstock.com. Section 3 opener: DJ Srki/Shutterstock.com.

Library of Congress Cataloging-in-Publication Data
Names: Daines, James R., author. | Daines, Martha J., author.
Title: Fluid Power / by James R. Daines, Martha J. Daines.
Description: Third edition. | Tinley Park, IL : Goodheart-Willcox Company, Inc.,
    [2020] | Includes index.
Identifiers: LCCN 2018025864 | ISBN 9781635634730
Subjects: LCSH: Fluid power technology. | Hydraulic machinery. | Pneumatic machinery.
Classification: LCC TJ840 .D34 2020 | DDC 620.1/06--dc23 LC record available at https://lccn.loc.gov/2018025864

Preface

This book is the outcome of many years of persistent interest in fluid power, beginning with the operation of farm equipment and involvement in a 4H tractor-maintenance project when I was a teenager, continuing throughout my college years and far into my professional life. Partly, the fascination came from the science foundations that are there and partly from the practical applications that brought those abstract concepts to life. There is orderliness to hydraulics and pneumatics that, beyond the work accomplished, has a sense of aesthetics. So the study of fluid power and the challenge of teaching others to use it has been a satisfying journey.

Our lives would be very different today if early civilizations had not recognized the potential of using air and water to do work. From the first rough waterwheels and windmills to the sophisticated applications we see today, fluid power has enabled us to do what was, in many cases, first thought to be impossible. Now, with computer interfaces, new materials, and imaginative technology to help us, we may be just at a new dawn of breakthroughs to come.

The purpose of this book is to introduce you to the field of fluid power. It lays the groundwork for building an understanding of the concepts involved and systematically guides the use of these concepts in designing functional circuits. Purposefully, the questions asked will require problem solving and creative thinking; that is what real life requires. Mastery of the subject matter is not easy, but careful study and thoughtful experimentation will make it happen.

Many individuals, companies, and organizations have contributed to this learning resource. I wish to acknowledge their assistance and express my sincere appreciation to them all. Likewise, I am deeply grateful to the publisher, Goodheart-Willcox, for continued encouragement and support while this was written. I hope the book will be a useful tool for the field.

—Jim Daines

About the Authors

James R. Daines is Professor Emeritus of Technology at the University of Wisconsin—Stout. He holds an EdD degree from the University of Missouri and BS and MS degrees from Stout State College, Menomonie, Wisconsin. Professional experience includes 25 years teaching power- and media-related courses to secondary, postsecondary, and adult students. In addition, 10 years were spent as president of a company specializing in developing courses and instructional materials for power and other related technologies. Those materials included extensive computer-based courses for hydraulics and pneumatics. Other experiences related to fluid power involved working on the design and construction of training benches and instructional materials for a producer of fluid power training equipment.

Martha J. Daines holds a PhD from the University of Chicago and a BS degree from the University of Wisconsin—Madison. Martha's professional experience includes 10 years investigating multiphase flow in high-pressure environments and 15 years teaching technology and power courses to students at the University of Wisconsin—Stout.

New to This Edition

The following changes have been made to the third edition of *Fluid Power: Hydraulics and Pneumatics* to strengthen the integrated learning solution and to keep your class up to date on the latest technology.

- **Integrated Learning Solution (ILS)** has been bolstered by strengthening the relationship between objectives and the chapter outline, the text, and our end-of-chapter materials. The addition of summaries provides another tool for student review. *Apply and Analyze* and *Research and Development* activities have been added, extending learning through higher-order thinking skills and hands-on activities.

- **Examples** have been added throughout the text to show math equations in use and the work needed to solve a problem.

- **Chapter reorganization** has moved the safety chapter forward in the text to emphasize its importance.

- **Chapter 1: Introduction to Fluid Power** has streamlined historical content. The chapter takes a broader approach to teaching students the evolution of fluid power throughout history.

- **Chapter 5: Fluid Power Standards and Symbols** has streamlined standards content to focus more on fluid power symbols.

- **Figures** have been updated to show newer technology and to help students make real-world connections with chapter concepts.

Reviewers

The author and publisher wish to thank the following industry and teaching professionals for their valuable input into the development of *Fluid Power: Hydraulics and Pneumatics*.

Wesley Gubitz
Cape Fear Community College
Castle Hayne, NC

John Harris
Ivy Tech Community College
Shellersburg, IN

Jon W. Hineman
Ivy Tech Community College
Terre Haute, IN

Darin Maltsberger
Mitchell Technical Institute
Mitchell, SD

Charles Millar
Fanshawe College
London, Ontario

Tim Tewalt
Chippewa Valley Technical College
Eau Claire, WI

Acknowledgments

The author and publisher wish to thank the following industry and teaching professionals for their valuable input into the development of *Fluid Power: Hydraulics and Pneumatics*.

A & A Manufacturing Co., Inc.

AGCO Corporation

API Heat Transfer

Atlas Copco

Badger Iron Works, Inc.

Bailey International Corporation

Brand Hydraulics

Brand X

Caterpillar, Inc.

Chicago Pneumatic

CNH America LLC

Continental Hydraulics

Deere & Company

DeVilbiss Air Power Company

Donaldson Company, Inc.

Dresser Instruments; Dresser, Inc.

Eaton Fluid Power Training

Ergodyne

Fairchild Industrial Products

Gates Corporation

Grayling Recreation Authority, Hanson Hills Recreation Area

Handi-Ramp

HYDAC Technology Corporation

IMI Norgren, Inc.

International Organization for Standardization

Kim Hotstart Manufacturing Company

Lab-Volt Systems, Inc.

Manufactured Housing Institute

MDMA Equipment—Menomonie

Miller Electric Mfg. Co.

MTS Systems Corporation

The Oilgear Company

Parker Hannifin

Phelps Industries, Inc.

PSS—Steering and Hydraulics Division, England

Sauer-Danfoss, Ames, IA

Schroeder Industries LLC

Sporlan Division, Parker Hannifin Corporation

Star Hydraulics, Inc.

USDA

Yates Industries, Inc.

Zinga Industries, Inc.

Brief Contents

Contents

FLUID POWER PRINCIPLES

1 Introduction to Fluid Power
The Fluid Power Field

Since the beginning of time, long before written history, people have searched for ways to conveniently transmit energy from its source to where it is needed and then convert the energy into a useful form to do work. This chapter introduces the fluid power field as an approach that provides an effective means of transferring, controlling, and converting energy.

Learning Objectives

After completing this chapter, you will be able to:

- Define the terms fluid power, hydraulic system, and pneumatic system.
- Explain the extent of fluid power use in current society and provide several specific examples.
- List the advantages and disadvantages of fluid power systems.
- Discuss scientific discoveries and applications important to the historical development of the fluid power industry.

Key Terms

actuator	hydraulic	scientific method
compact hydraulic units	hydraulic accumulator	steam engine
cup seal	hydraulic intensifier	water screw
fluid	Industrial Revolution	waterwheel
fluid compressibility	pneumatic	windmills
fluid power	prime mover	

1.1 Definition of Fluid Power

The basis of *fluid power* is pressurized fluids. A *fluid* is a liquid or gas substance, such as air, water, or oil, that easily flows and tends to assume the shape of the container in which it is stored. The fluids are incorporated into physical hardware systems that generate, transmit, and control power in a wide variety of consumer and industrial applications. Today, it would be difficult to identify a product that has not been affected by fluid power at some point along the route from raw material to final installation.

Fluid power systems are versatile contributors to industry. Applications range from brute force needed in heavy industry to the sensitive positioning of parts in precision machining operations, **Figure 1-1**. The systems are generally grouped under the two broad classifications of *pneumatic* and *hydraulic*. Pneumatic systems use gas, usually air, while hydraulic systems use liquids, usually oil. Other fluids can be used in special applications.

Fluid power is one of the three types of power transfer systems commonly used today. The other systems are mechanical and electrical. Each of the systems transfers power from a *prime mover* (source of energy) to an *actuator* that completes the task (work) required of the system.

Fluid power systems use the prime mover to drive a pump that pressurizes a fluid, which is then transferred through pipes and hoses to an actuator.

Mechanical systems transfer power from the prime mover to the point of use by means of shafts, belts, gears, or other devices. Electrical systems transfer power using electrical current flowing through conductors. Typical applications in business, industrial, and consumer products and systems use combinations of fluid, mechanical, and electrical power transfer methods.

1.2 Fluid Power Industry

The fluid power industry is a complex entity. It includes education, design and manufacturing of components, design and assembly of systems using those parts, and troubleshooting and maintenance needed to keep the systems performing efficiently, **Figure 1-2**. In addition, a complex sales and distribution system ensures users access to replacement components and information concerning service, new and improved component designs, and new system applications.

Growth of the fluid power industry has required a parallel growth in the number of people who understand and can work effectively with fluid power systems. These people range from engineers responsible for designing the components to mechanics responsible for maintenance and repair of fluid power equipment. The type of education and training available to prepare these people varies considerably. Formally organized programs exist in two-year technical and community colleges, four-year universities, and in programs offered by component manufacturers. Many individuals seek fluid power training after exposure to the field through their jobs.

The fluid power industry is a broad field and a key contributor to the success of many businesses and industries. Fluid power is used extensively in

Deere & Company

Figure 1-1. Equipment used in construction and street maintenance is an example of a fluid power application commonly encountered in daily life. This backhoe is capable of producing the brute force needed to break and move concrete.

Andrey N. Bannov/Shutterstock.com

Figure 1-2. The service of fluid power systems in business and industry provides employment for many highly trained individuals.

manufacturing, construction, transportation, agriculture, mining, military operations, health, and even recreation. The list is almost endless. Applications vary and components have different appearances in the various applications. System sizes range from miniature to massive, but fluid power principles provide the needed power, force, and control.

Fluid power has been a key contributing factor in the development of current agricultural equipment. Modern farm equipment uses hydraulics extensively. These uses range from simple hydraulic cylinders that raise and lower implements to complex devices that maintain clearances, adjust torque, and provide easy control of speed and direction on tractors and a variety of specialized planting, harvesting, and processing equipment, **Figure 1-3**.

Fluid power is used in some form in all modern transportation systems. These uses range from automobiles to complex, wide-body aircraft found on international flights. Specific examples of the application of fluid power principles include hydraulic and pneumatic braking systems, power-assisted steering found on most forms of wheeled vehicles, hydrostatic transmissions that provide almost unlimited speed and torque control, and suspension systems that use hydraulic and/or pneumatic dampening.

The construction industry is a very diverse industry. Construction activities include the building of residences and all types of commercial structures, roads and highways, irrigation systems, harbor facilities, and

a wide variety of other projects. The industry makes use of many types of earthmoving equipment, material-handling equipment, and specialized fastening and finishing devices. Examples of typical applications that make use of fluid power include: backhoes for excavation; cranes for moving, lifting, and positioning materials; vibrators for consolidating concrete after it has been placed; and nail-driving apparatuses.

Manufacturing organizations rely heavily on fluid power. Applications range from huge presses in automobile body fabrication plants, which form body panels, to packaging equipment for miniature parts in electrical component manufacturing operations. These applications use hydraulics and pneumatics to make the equipment operate as needed. Equipment needs may range from huge forces to draw metal into desired shapes, to a gentle nudge accurately positioning a part for machining, to the deliberate movement of sanders performing a final finish sanding. Fluid power can easily provide each of these varying levels of force. In many instances, the desired results can be obtained using off-the-shelf equipment. In other situations, standard components may be used to assemble circuits and systems to produce the desired result, **Figure 1-4**.

Mining companies use fluid power both in open-pit and underground operations. Spectacular examples of an application in this industry are the huge shovels used in coal strip mining operations. These shovels remove the overburden from veins of coal that are near the surface. Some of these shovels are several

smereka/Shutterstock.com

Figure 1-3. Most modern combine harvesters have a hydrostatic drive system and utilize hydraulics for stability and to raise and lower the header.

Peter Hansen/Shutterstock.com

Figure 1-4. Fluid power applications have been used during the manufacture or processing of most consumer products today. This carnival ride makes extensive use of fluid power.

stories high and they can remove multiple cubic yards of material during each pass of the scoop. The shovels use large numbers of fluid power systems and circuits for movement and control. Other open-pit mining operations use more standard front-end loader and truck designs for loading and moving ore. In addition, many fluid power applications, both pneumatic and hydraulic, can be found in mine drilling, crushing, and material-handling equipment. Fluid power applications are especially desirable in underground mining locations. Accumulations of gas may produce potentially explosive conditions, limiting or preventing the use of electrical devices.

Land, sea, and air defense forces use fluid power to assist in moving personnel, supplies, and equipment to support their operations. The military makes use of a full range of fluid power components and circuits. Many of these parallel civilian commercial applications, but others are highly specialized and are not directly duplicated in commercial applications. Hundreds of applications exist, ranging from power-assisted steering of land vehicles to the precision positioning of rocket launchers for air defense, **Figure 1-5.**

1.3 Fluid Power Systems

Fluid power is a highly versatile power transmission system, as illustrated by the range of applications discussed earlier in this chapter. No system, however, is entirely suitable for all applications. All power-transmission systems have characteristics that are desirable in one application, but turn into disadvantages in

Lone Pine/Shutterstock.com

Figure 1-5. Complex defense systems make extensive use of fluid power.

other situations. A system cannot have every desired advantage without disadvantages. Understanding system characteristics, as well as what is needed for a particular result, will help in producing an effective and efficient application.

The range of applications that use fluid power makes the development of a simple list of advantages and disadvantages difficult, since examples that do not "fit" can easily be found. This problem is further complicated by the inherent differences of the two major divisions of the fluid power field: hydraulics and pneumatics.

1.3.1 System Characteristics

Although hydraulic and pneumatic systems share the characteristics of energy transfer by means of fluid pressure and flow, differences affect how and where they are applied. These differences include:

- Accuracy of actuator movement.
- Operating pressure.
- Actuator speed.
- Component weight.
- Cost.

Accuracy of movement

Fluid compressibility is the inherent characteristic that differs between hydraulic and pneumatic systems. A gas is compressible, while a liquid can be compressed only slightly. Hydraulic systems, therefore, can produce more accurate, easily controlled movement of cylinders and motors than pneumatic systems. Compressibility produces a more "spongy" operation in pneumatic systems that is not suitable where highly accurate movement is required.

Operating pressure

Hydraulic systems can operate at much higher pressures than pneumatic systems. Hydraulic system operating pressure ranges from a few hundred pounds per square inch (psi) to several thousand psi. Pressures of more than 10,000 psi are used in special situations. Pneumatic systems, in contrast, normally operate between 80 to 120 psi. High-pressure pneumatic systems normally are not used.

Actuator speed

Pneumatic systems are commonly used when high-speed movement is required in an application. Rotation speeds of over 20,000 revolutions per minute (rpm) are possible. Rapid-response cylinder operation is also possible with pneumatic systems. These designs are generally found in situations involving lighter loads and lower accuracy requirements.

Component weight

System operating pressure affects the structure of components. Hydraulic systems operate at higher pressures, requiring the use of stronger materials and massive designs to withstand the pressure. Pneumatic systems operate at much lower pressures and, therefore, can be manufactured using lightweight materials and designs that minimize the amount of material.

Hydraulic applications tend to involve equipment that handles heavier weights, requiring both higher system operating pressure and physical strength of machine parts. **Figure 1-6** shows a front-end loader. The cylinders used to operate the bucket must be of a construction that can withstand high system pressure and the heavy load of the bucket and its contents.

Pneumatic systems tend to involve applications where ease of handling and light weight are critical for effective operation of the tool or system. **Figure 1-7** shows a pneumatic grinder being used on a large metal casting. The grinder is lightweight and very portable. The tool is easily manipulated by an individual and constructed to provide a long service life.

Cost

The cost of fluid power systems ranges widely. A variety of situations exist and a number of solutions are available for each one. The solution selected to solve the problem directly affects the cost. Understanding system advancements, basic characteristics of hydraulic and pneumatic systems, and knowing which standard components are available are necessary to produce a system that does the best job at the lowest cost.

Deere & Company

Figure 1-6. Hydraulic systems are commonly used in applications where high system pressures are needed to complete the required work.

Skinfaxi/Shutterstock.com

Figure 1-7. Pneumatics are commonly used where high speeds and lightweight tools are needed.

The cost of system operation is a factor that must be considered. Generally, pneumatic systems are more expensive to operate than hydraulic systems. This cost can be directly associated with the compression, conditioning, and distribution of air. Careful maintenance to eliminate leakage can greatly reduce operating cost.

1.3.2 Advantages and Disadvantages of Fluid Power Systems

Fluid power systems have several advantages and disadvantages when compared with mechanical and electrical power transfer systems. Several of the important advantages and disadvantages of fluid power systems are presented in the next two sections.

Advantages

The following list of advantages applies to both hydraulic and pneumatic systems, except as noted:

- An easy means of multiplying and controlling force and torque.
- Infinitely variable speed control for both linear and rotary motion.
- Overloading the system simply stalls the actuator without damage to the components.
- Provides an easy means of accurately controlling the speed of machines and/or machine parts.
- Provides the ability to instantly stop and reverse linear and rotary actuators with minimal shock to the system.
- Systems easily adapt to accommodate a range of machine sizes and designs.
- Systems readily adapt to external control methods, including mechanical, pneumatic, electrical, and electronic systems, **Figure 1-8**.

Monkey Business Images/Shutterstock.com

Figure 1-8. Both hydraulic and pneumatic systems can be readily adapted to external control systems, including state-of-the-art electronic designs.

- Systems can easily provide component lubrication.
- Large volumes of compressed air may be easily stored in pneumatic systems to provide energy for intermittent, heavy system demand, **Figure 1-9**.
- Pneumatic systems provide clean operation with minimal fire hazard.

Goodheart-Willcox Publisher

Figure 1-9. Pneumatic systems can easily store large volumes of compressed air to meet the intermittent high demand of some systems. The horizontal tank under the compressor serves as the storage area (receiver) for compressed air.

Disadvantages

The following list of disadvantages applies to both hydraulic and pneumatic systems, except as noted.

- Higher safety factors associated with high-pressure oil and compressed air.
- Susceptibility to dirty environments, which can cause extreme component wear without careful filtration.
- Fluid leakage and spills cause a slippery, messy work environment around hydraulic equipment.
- Fire hazard with hydraulic systems using combustible oils.
- Special handling and disposal procedures for hydraulic oil required by environmental regulations.
- High cost of compressing and conditioning air for use in pneumatic systems.
- Reduced accuracy in actuator speed control in pneumatic systems caused by compressibility of air.
- Noise level of pneumatic systems when air is directly exhausted to the atmosphere from components.

1.4 Historical Perspective of Fluid Power

Awareness of the historical background of our civilization, our particular culture, and our nation is generally considered to be an important part of the preparation to be a responsible citizen. This awareness provides an appreciation of our current situation by providing information about the origin and development of cultural, religious, and political ideas, principles, and systems. Likewise, an awareness of the historical development of a technical field of work should provide an appreciation of what we have, some idea of how it was achieved, and an appreciation of the cultural changes associated with the technical field.

This section provides a brief discussion of some aspects of technical development and innovation as viewed by Western culture. Volumes of information are available on developments throughout history involving Islamic and Eastern contributions as well as those reflecting Western thought. Fluid power, as the term is currently used in our society, describes a relatively new field. However, the roots of this field extend far back into history.

1.4.1 Historical Awareness

The use of fluid power has developed along with civilization. The natural movements of air and water were probably the first sources of power used by early humans.

It is speculated that the use of crude sails to reduce the effort to move boats was the first attempt to harness this natural power, **Figure 1-10**, but origins of fluid power are difficult to identify. Recorded evidence does not exist in most cases. Archeology gives us many hints, but speculation still plays a major role in our current thinking about how existing scientific principles and mechanical devices were developed.

It is doubtful whether the physical principles associated with modern hydraulics and pneumatics were fully understood in early history or even at the beginning of the Industrial Revolution. The physical principles and design features that we consider basic to the operation of fluid power systems were developed over several centuries. Many of these developments were the products of observation or experimentation more than due to an understanding of advanced scientific principles. Application of the scientific method played an ever-increasing role in the development of both fluid power theory and machinery design. The *scientific method* dictates accurate measurement, controlled testing, reproducibility of results, and systematic demonstration and reporting of results.

Figure 1-10. People have used the natural movement of both air and water throughout history to reduce work and aid in transportation. Using sails on ships is an example of using fluid power (wind).

1.4.2 Applications of Fluid Power through History

Fluid power developments have been essential to the development of human civilization. Harnessing the power of fluids has most greatly affected the following areas:

- Transportation.
- Movement of water.
- Generation and transmission of power.

These elements have had far-reaching effects on the development of our industrial society.

Antiquity

Artwork from Egyptian tombs that date back to 2500 BCE shows reed boats with bipod masts and sails designed for problems associated with river sailing. Sails for the propulsion of ships continued to be refined through the ages. Many sailing principles were applied to the development of windmills and, eventually, pneumatic components.

Flowing water in rivers and streams was also used in transporting boats and materials. The Egyptians, Persians, and Chinese built elaborate systems of dams, ditches, and gates for water control and irrigation. These various applications eventually led to the development of variations of the waterwheel to lift water for irrigation purposes and turn simple mills. These early uses of fluid power were dependent on vast quantities of low-pressure air and water supplied by nature. They also were subject to variations in the weather, which made them somewhat unpredictable and only partially under the control of the operators.

Movement and control of water for irrigation, flood control, and municipal water systems also produced ideas that eventually were used in fluid power systems. Controlling the annual flood of the Nile River and moving water for irrigation purposes during the remainder of the year produced a better understanding of water flow in channels. The *water screw*, **Figure 1-11**, is still used in parts of Egypt and other areas of the world for moving or lifting small volumes of water. The Roman Empire constructed great aqueducts for the movement of water. Ruins of these systems can still be found throughout the territories the Romans controlled, **Figure 1-12**.

It is generally believed that the energy of running water was first effectively harnessed during the first century BCE. Early *waterwheels* used a wheel rotated by a running stream, **Figure 1-13**. These designs can produce up to three horsepower. These wheels used both undershot and overshot designs. Undershot wheels were placed directly in a stream. Overshot wheels used water directed on the top of the wheel through sluice channels from dams or natural waterfalls. Waterwheels revolutionized the grinding of grain and were gradually adapted to other purposes.

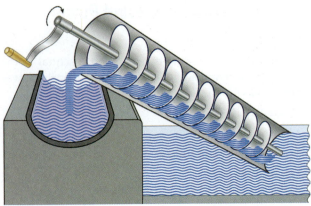

Fouad A. Saad/Shutterstock.com

Figure 1-11. Archimedes is credited with the design of the water screw, which has been used to move and lift water since the third century BCE. The basic principles found in this device are still used in modern fluid power components.

Goodheart-Willcox Publisher

Figure 1-12. An understanding of basic scientific principles has developed slowly throughout civilization. Ruins of the aqueduct system of the Roman Empire illustrate an early need for a practical understanding of fluid flow.

Ahn Eun sil/Shutterstock.com

Figure 1-13. Waterwheels use the weight or forces of water to turn a wheel. Energy is transferred from the wheel by gears and belts to power a mill.

The use of *windmills* for doing work did not appear until after the decline of the Roman Empire. The earliest record of a windmill-like device is from central Asia. These wind-driven devices were used to turn prayer wheels. Historians believe that the first "real" windmills were developed and used in central Asia about 400 CE.

Middle Ages

During the Middle Ages following the fall of the Roman Empire, the development of new technology was extremely slow in Western civilization. In the Western countries of Europe, existing Roman structures were allowed to gradually deteriorate as major emphasis was placed on defense, conquest, development of monastic (religious) orders, and support of the Crusades. The Islamic world, however, continued to make progress on public water supply systems, public baths, and power-driven mills. Many of these ideas were introduced in Europe during the twelfth and thirteenth centuries by returning Crusaders.

At first, waterwheels were just a means of turning millstones for grinding grain for local consumption. Waterwheels began to be recognized as an important source of power for other uses during the fourth and fifth centuries. They were slowly adapted for use in sawmills, paper mills, iron mills, and mining operations. An example of the slow, but pervasive, growth of waterpower as a prime mover is that the earliest recorded use in England was in the eighth century with over 5,000 in use by the eleventh century. Water mills were profitable operations and found in every community that had a suitable stream.

Windmills continued to be developed in several forms in the Near East, where easy access to water was limited. Records indicate that windmills were extensively used in Afghanistan and Persia, with a number of designs developed to control speed. Speed control was required because of strong winds in the area. Use of wind power spread throughout the Islamic world during the period.

The first recorded use of windmills in Western Europe was during the twelfth century, **Figure 1-14**. These mills were initially used for grinding grain, but were gradually modified for use in other applications, such as sawing lumber, pumping water, and manufacturing. Holland and the other "low countries" of Europe were prime users of these mills because of their flat terrain and consistent winds.

Windmills gradually developed with improvements in design and increases in size. Power output ranged from 4–5 horsepower for the average mill to over 15 horsepower for larger mills. At the peak of windmill use, Holland had over 8,000 mills, with many of them having sail spans as large as 100 feet.

fokke baarssen/Shutterstock.com

Figure 1-14. Windmills have played an important part in the development of civilization. Their use is limited to coastal and plains areas with consistent winds.

Many mechanical improvements were made as water and windmills became prime sources of power. These improvements included methods of power transmission through shafts, cogged wheels, crude gears, and cams. These devices were needed to transmit the power and transform it into the type of motion needed to do the desired work. Many ingenious devices were developed that have contributed to the body of knowledge as the world moved toward the Industrial Revolution.

Industrial Revolution

The eighteenth and nineteenth centuries are often designated as the period in which an "industrial revolution" occurred. That period of history produced tremendous changes in world society, particularly in Great Britain. Great Britain had established extensive world trade, a strong financial base, large local reserves of iron ore and coal, and also had key individuals interested in the practical development of scientific principles. These combined factors led to the *Industrial Revolution*.

Changes during the Industrial Revolution occurred much more quickly than those during the Middle Ages. However, identification of specific dates or events that started the "revolution" is not possible. It must also be emphasized that the industrial change was not just tied to mechanical devices, but involved changes as diverse as world trade and application of scientific methods. Nevertheless, development of new prime movers must be considered to be the key to the Industrial Revolution, **Figure 1-15**. The *steam engine*, internal combustion engine, and gas turbine all were developed during that period. Unlike water and windmills, these devices were mobile and not dependent on local weather or terrain conditions. Even with those new power-generating devices, water and windmills continued to play important roles in industrial power generation throughout the middle 1800s and much later in smaller, local and rural applications.

The use of machines instead of hand tools was another key to the changes of the period. These machines exceeded the capabilities of skilled craftsmen to develop products. They allowed production rates that met demands resulting from expanded trade areas.

The development of manufacturing increased the need for methods and devices to move water and transmit power. The procedures and devices developed to satisfy these needs are where fluid power, as we think of it today, began to emerge. Many of the early fluid power elements involved pumping water from mines or for manufacturing uses. James Watt (who is usually identified as the inventor of the steam engine but was also a prolific developer of new manufacturing techniques) designed steam-driven, reciprocating pumps that were used in the late 1700s and early 1800s. By the mid-1800s, these pumps were widely used. The East London Water Company operated a unit with a 100-inch diameter cylinder and a stroke of 11 feet. By the 1870s, patents were

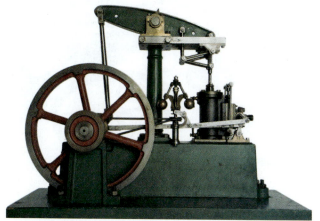

Baptist/Shutterstock.com

Figure 1-15. The steam engine provided a reliable, portable source of power. Its development is considered one of the key factors in the development of the Industrial Revolution.

issued for variable-stroke, piston pumps. The centrifugal pump was invented in the late 1600s, but was not generally used until the mid-1800s. The jet pump was also produced during this same period.

A key factor that allowed the development of many practical fluid power components was the *cup seal*. This type of seal uses the pressure within the system to force a sealing collar against a shaft or ram to prevent fluid loss, **Figure 1-16**. Invented in 1795 by English engineer Joseph Bramah and his assistant Henry Maudslay, the cup seal allowed hydraulic presses to be built that both did not leak and could be continuously used.

Two other devices that were especially important to the development of fluid power during the 1800s were the *hydraulic accumulator* and the hydraulic intensifier. Large, weighted accumulators were used to store pressurized fluid from a pump during the idle time of a press or other machine. During peak operating time, the charged accumulator assisted the pump by supplying high volumes of fluid needed for operation.

The *hydraulic intensifier* was used to more easily obtain high system pressures for use in bailing, metal forming, forging, or other applications. The intensifier, which usually consisted of some combination of large and small diameter rams, used the area differences of the rams to boost system pressure above the capability of the pump. This principle allowed the use of higher pressure even when the technology of the time did not allow pumps to operate at extremely high pressure.

Pressurized water was extensively used to distribute power to businesses and manufacturers in several cities in Great Britain. By 1900, it was considered economical to transmit power up to 15 miles from a centralized pumping station. These systems provided

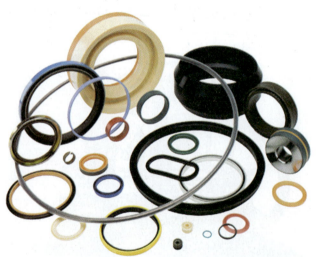

Figure 1-16. Modern sealing devices are often taken for granted. However, their development in the late 1700s by Bramah and Maudslay was instrumental to the practical application of fluid power principles.

water up to several hundred pounds per square inch of pressure that could be directly used for the operation of presses, hoists, and water motors. These systems continued in use until electrical generating and distribution systems were perfected.

Fluid power development and use during the nineteenth century was very extensive. This use involved the generation of power through the design of effective water turbines; the transmission of power using central power stations and elaborate distribution lines; and the use of fluid power in construction, manufacturing, and material distribution systems.

Recent history

The emphasis on the development of fluid power applications decreased as the use of electricity grew in the late 1800s and early 1900s. Development concentrated more in the heavy industrial and mobile areas where fluid power applications appeared to be most practical. Generally, these factors have promoted progress in fluid power applications in recent years:

- Development of new materials.
- Miniaturization of components.
- Effective electrical/electronic control.

Improvements were made in sealing devices and machining techniques that reduced both internal and external leaks. This, in turn, improved system efficiency. Reducing external leaks also allowed use in applications where cleanliness was an important factor. Water was replaced in hydraulic systems by petroleum-based fluids, which improved lubrication and eliminated the danger of freezing in cold climates.

Military applications contributed to the development of fluid power. A milestone occurred in 1906 when hydraulic systems first appeared on a warship, the battleship USS Virginia. These hydraulic systems replaced many mechanical and electrical systems. Since that time, all branches of the military have incorporated fluid power on numerous devices to solve problems in gunnery, materials handling, navigation, and support services.

The development and refinement of the concept of *compact hydraulic units* in the 1920s had a far-reaching effect, extending to today's fluid power applications, **Figure 1-17**. These self-contained systems include the power source, pump, and reservoir. The units have been applied to a full range of industrial and consumer applications. Central hydraulic and pneumatic power systems have remained in large industrial applications, **Figure 1-18**, but the more compact direct system adds flexibility. It has been the key to the development of today's large mobile hydraulics field.

The development of new materials and manufacturing techniques has promoted the design of new fluid power concepts and allowed practical application of

Continental Hydraulics

Figure 1-17. Compact power sources such as these hydraulic units have added flexibility to the application of fluid power.

old ideas. The miniaturization of components has produced new applications, while the combining of electrical/electronic control and fluid power systems has produced more effective machines. Two components that illustrate these factors are hydrostatic transmissions and servo systems. These units are heavily used in both industrial and consumer applications.

Today's automobile can be used as an example of the use of fluid power. The body of the vehicle is formed by huge hydraulic-powered presses. Hydraulically-controlled resistance welding equipment assembles those parts, while untold numbers of other hydraulic

and pneumatic tools are used in the production of the additional parts and during the final assembly process, **Figure 1-19**. Intricate fluid power systems are also used in the steering, braking, and ride control systems of the vehicle, which promote safety and comfort of the driver and passengers.

Fluid power has grown tremendously during the 1900s and into the 2000s. Current space technology, manufacturing industry demand, and consumer interest indicate that this growth will continue into the twenty-first century, **Figure 1-20**.

xieyuliang/Shutterstock.com

Figure 1-19. An automobile assembly line illustrates the diverse use of fluid power in industry today. A full range of both pneumatic and hydraulic applications can be found in such operations.

Atlas Copco

Figure 1-18. Central power systems are often used in large manufacturing facilities to produce compressed air for pneumatic applications.

ssuaphotos/Shutterstock.com

Figure 1-20. Windmills have experienced a resurgence as prime movers in the windmill farms currently used for electrical generation. As energy costs rise, so does interest in harvesting this "free" energy.

Summary

- Fluid power systems use pressurized fluids to transfer energy from a prime mover to an actuator that performs work.
- Fluid power systems are generally grouped under two broad classifications: hydraulic and pneumatic.
- Hydraulic systems generally use oil as the system fluid, while pneumatic systems use air.
- The fluid power industry is a broad field that includes education, design and manufacturing of components, design and assembly of systems using those parts, and troubleshooting and maintenance needed to keep the systems performing efficiently.
- Fluid power is used extensively in manufacturing, construction, transportation, agriculture, mining, military operations, health, and even recreation.
- Advantages of both hydraulic and pneumatic systems include easy control of force, torque, speed, and direction of actuators.
- The natural movement of air and water was used in the earliest applications of fluid power; wind and water mills were early prime movers that harnessed this natural movement to provide power until well into the Industrial Revolution.
- Many early fluid power devices were developed through observation or experimentation rather than scientific theory.
- Compact, self-contained power units which contained the prime mover, pump, and reservoir were invented in the early 1900s and had considerable influence on the development of fluid power as we know it today.

Internet Resources

The following are some useful resources available on the Internet. Enter a company or organization name into a search engine to access its website. Explore the various areas of the sites to discover useful fluid power resources.

National Fluid Power Association (NFPA). Summarizes the basic aspects of fluid power systems as well as information on applications.

The Great Idea Finder. Provides information on James Watt and other inventors who made major contributions to industrial development during the Industrial Revolution.

Wikipedia: The Free Encyclopedia. Offers information on the history and operation of hydraulic and pneumatic power-transmission systems and articles on related topics.

Chapter Review

Answer the following questions using information in this chapter.

1. Fluid power systems use _____ fluids to transmit power.
2. The physical components in a fluid power system are used to generate, transmit, and _____ power to produce the desired results in an application.
3. *True or False?* Fluid, mechanical, and electrical power are transfer systems commonly used in industry today.
4. *True or False?* Fluid power is used exclusively in the agriculture industry.
5. _____ power systems can provide lightweight, easily handled tool applications.
6. The cost of operating a pneumatic fluid power system is affected by _____.
   A. conditioning the air
   B. compressing the air
   C. distributing the air
   D. All of the above.
7. Clean operation with minimum fire hazards is a characteristic of _____ systems.
8. Which of the following is a disadvantage of using a fluid power system?
   A. No speed control for linear and rotary motion.
   B. Higher safety factors associated with high-pressure oil and compressed air.
   C. Cannot easily be adapted to accommodate a range of machine sizes and designs.
   D. All of the above.
9. *True or False?* Many early fluid power developments were the products of observation and experimentation rather than understanding of scientific principles.
10. Artwork from Egyptian tombs indicates that sails were used to assist in the propulsion of boats as early as _____ BCE.
    A. 200
    B. 1000
    C. 2500
    D. 3000

11. The use of _____ for doing work did not appear until after the decline of the Roman Empire.

12. The period of history during the eighteenth and nineteenth centuries, known as the _____, produced tremendous changes in industry, including the development of many fluid power concepts and components.

13. The invention of new _____ is usually considered to be the key that allowed the development of the Industrial Revolution.

14. *True or False?* The invention of the cup seal led to the development of the first functional pneumatic system.

15. The development of new _____ and manufacturing techniques has promoted the design of new fluid power concepts and allowed practical application of old ideas.

Apply and Analyze

1. Skid-steer loaders typically employ hydraulic systems for operations.
   A. What impact would switching to a mechanical power transmission system have on loader operation? Why?
   B. What impact would switching to a pneumatic system for lifting have on how this loader could be used? Why?

2. You are designing an automated system to hold a part in place during drilling. What information would you need to allow you to choose between a hydraulic and a pneumatic system? Why?

Research and Development

1. Examine your home or other residence. List and describe at least three systems or appliances that use component parts or basic concepts associated with fluid power.

2. Archimedes, Blaise Pascal, Robert Boyle, Jacques Charles, and Daniel Bernoulli are among the many individuals who established the basic principles on which fluid power applications are based. Research one of these individuals and prepare a short report describing the principle identified by the individual, the process by which it was identified, and the specific importance of the discovery for fluid power applications.

3. The first cup seal developed by Joseph Bramah and Henry Maudslay was made of leather. The design and materials used in current hydraulic seals vary depending on the application. Investigate and describe at least one type of hydraulic seal currently used. Prepare a written report discussing the design, composition, and an application in which the seal is used.

4. In the fluid power industry, trends are leaning toward miniaturization, increased energy efficiency, reduced environmental impact, and use of electronic controls. Find a current example that illustrates one of these trends. Create a short presentation for the class about your example.

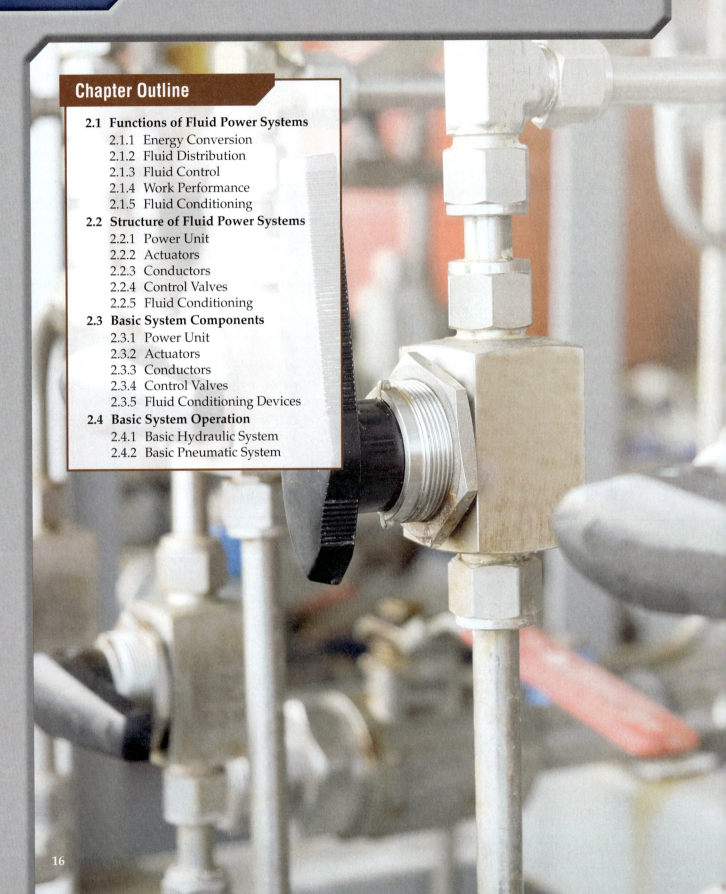

2

Fluid Power Systems
The Basic System

The general definition of a system is a set or group of principles, beliefs, things, or parts that form a whole, such as a system of government, an accounting system, a mountain system, or the digestive system. The parts of a system work together to produce a desired end result. This chapter introduces the basic design, structure, and operation of fluid power systems.

Learning Objectives

After completing this chapter, you will be able to:

- Explain the functions of fluid power systems.
- Identify the basic structure of fluid power systems.
- List the basic component groups involved in the structure of fluid power systems.
- Describe the function of the components involved in basic fluid power systems.
- Describe the similarities and differences of hydraulic and pneumatic systems.
- Explain the operation of basic hydraulic and pneumatic systems.

Key Terms

adapters
component group
compressor
conductor
control valve
cylinder
directional control valve
filter
fitting

flow control valve
fluid conductor
heat exchanger
hoses
mechanical coupler
motor
pipe
power unit
pressure control valve

pressure regulator
prime mover
pump
receiver
reservoir
separator
system function
tube

2.1 Functions of Fluid Power Systems

Fluid power systems are made up of *component groups* containing parts designed to perform specific tasks. These component groups act together to perform the work desired by the system designer. The work may involve simple or complex tasks, but the component groups perform specific *system functions* that are basic to all fluid power systems. There are five functions that are basic to system operation of any fluid power system, **Figure 2-1**:

- Energy conversion.
- Fluid distribution.
- Fluid control.
- Work performance.
- Fluid conditioning.

Each of these functions must be performed by a fluid power component if the system is to operate efficiently and provide a reasonable service life. The operating environment, power output, and complexity of the system establish the number of components required to perform a particular function.

2.1.1 Energy Conversion

Fluid power systems do not *generate* energy, but *transform* it into a form that can be used to complete a task. The process begins with a prime mover pressurizing a fluid. It ends with an actuator using the energy stored in the pressurized fluid to perform work.

2.1.2 Fluid Distribution

The operation of fluid power systems requires the distribution of fluid to the components in the system. Various types of lines are involved in this function. Valves and other components also serve to assist in fluid distribution.

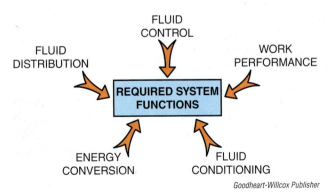

Goodheart-Willcox Publisher

Figure 2-1. The basic functions of a fluid power system.

2.1.3 Fluid Control

Fluid power systems require the control and regulation of the fluid in the system to perform the tasks desired by the system designer. A number of different components are used to control fluid flow rate, direction, and pressure in a system. Control of these three elements allows the system to provide the desired operating characteristics.

2.1.4 Work Performance

Making use of the energy stored in the pressurized fluid of the system is the primary function of a fluid power system. This process involves actuators that convert the energy stored in the pressurized fluid to linear or rotary motion to perform the desired work.

2.1.5 Fluid Conditioning

Fluid power system performance and service life require a fluid that is clean and provides lubrication to system components. This function involves storing fluid, removing dirt and other contaminants, and maintaining proper system operating temperature.

2.2 Structure of Fluid Power Systems

The physical appearance of fluid power systems varies considerably, depending on the type of fluid used, application, and power output. However, each system is structured using component groups that perform the various system functions. The structure of typical fluid power systems involves five component groups, **Figure 2-2**:

- Power unit.
- Actuators.
- Conductors.
- Control valves.
- Fluid conditioning.

Each component group has primary tasks to perform that relate to one of the five system functions. However, each group also performs secondary tasks relating to one or more of the other functions.

2.2.1 Power Unit

The power unit group of components deals primarily with the energy-conversion function of the system. The unit consists of a prime mover, pump, and reservoir. The prime mover is the source of energy for the system. The energy produced by the prime mover turns the pump, which produces fluid flow that transmits energy through the system. The reservoir serves as a storage unit for system fluid. It also performs fluid maintenance functions.

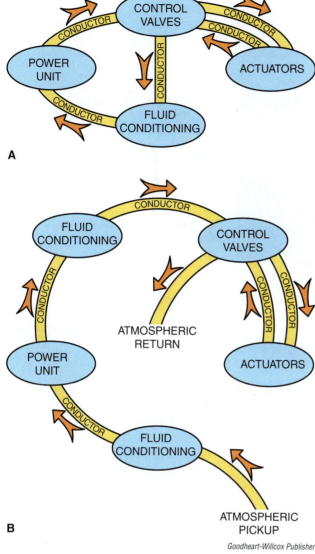

Conductors perform such tasks as the intake of fluid for the pump, distribution of fluid to and from control valves and actuators, transmission of sensing and control pressures, and the draining of liquids that have accumulated in components.

2.2.4 Control Valves

Three different types of valves are required to perform the fluid control function in a fluid power system. The valves in the control valves group are:

- Directional control valves.
- Pressure control valves.
- Flow control valves.

Directional control valves provide control over fluid flow direction in sections of a system to start, stop, and change the direction of actuator movement. Pressure control valves are used to limit the maximum pressure of the system or in a section of the system. Flow control valves provide control over fluid flow rate in a section of a system to control the rate of movement of an actuator.

2.2.5 Fluid Conditioning

The fluid conditioning group involves maintaining and conditioning system fluid. This requires removal of dirt and moisture from the fluid and ensuring proper operating temperature. Specially designed components are available to perform these tasks. However, basic systems often maintain the fluid using other system components that perform the task as a secondary function. *Filters* are used to remove dirt and moisture from systems, although a properly designed reservoir can perform this task under certain conditions. Maintaining the proper system operating temperature can require the use of a *heat exchanger*. However, this task is usually performed by dispensing heat through the reservoir, system lines, and other components.

2.3 Basic System Components

Fluid power systems are constructed of various components that perform the specific system functions. The number and appearance of components required to perform these functions varies considerably depending on the complexity and accuracy of the work performed, the environment in which the system operates, and the manufacturer of the component. Understanding the operation of fluid power systems requires an understanding of the construction and operation of the components that form the structural groups of the system. Refer to **Figures 2-3** and **2-4**.

Goodheart-Willcox Publisher

Figure 2-2. The structure of fluid power systems. A—Hydraulic system. B—Pneumatic system.

2.2.2 Actuators

The actuators group of components performs the work done by the system. These components convert the energy in the system fluid to linear or rotary motion. The basic actuators are cylinders for linear motion and motors for rotary motion. A variety of cylinder and motor designs are used to produce the specific motion needed to complete the work required of the system.

2.2.3 Conductors

Fluid distribution is the primary function of the conductors group of components. Pipes, hoses, and tubes serve as the conductors that confine and carry system fluid between the pump and other system components.

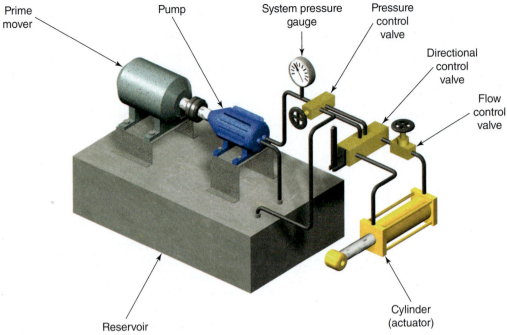

Figure 2-3. A basic hydraulic fluid power system. Note the components and their relationships to each other.

2.3.1 Power Unit

The appearance and structure of the power unit in a fluid power system can vary considerably, depending on the application of the system. In basic systems, the *power unit* consists of a prime mover, pump, reservoir or receiver, mechanical coupler, and the fluid conductors required to make the unit operational. Terminology and design vary somewhat between pneumatic and hydraulic systems, but direct comparisons can be made between the two systems. For example, in pneumatic systems, the power unit is generally referred to as the *compressor*, **Figure 2-5**. In hydraulic systems, the power unit is called a *pump*, **Figure 2-6**.

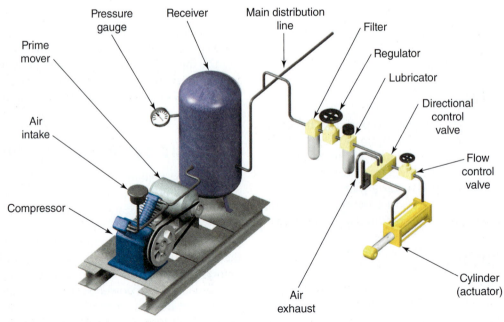

Figure 2-4. A basic pneumatic fluid power system. Note the components and their relationships to each other.

Goodheart-Willcox Publisher

Figure 2-5. A typical pneumatic system compressor (power unit). The prime mover in this application is a gasoline engine.

The ***prime mover*** (source of energy) of a system is usually an electric motor or an internal combustion engine. However, gas or steam turbines may be used to power large industrial operations. System size, operating environment, and mobility are factors that determine the device selected as the prime mover. Prime movers range from fractional horsepower electric motors and small single cylinder gasoline engines to gas turbines producing thousands of horsepower.

The pump or compressor is the heart of a fluid power system. These components produce the fluid flow used to transmit energy from the prime mover throughout the system. As the prime mover turns the pump or compressor, varying pressure conditions are created within the units that cause fluid to enter, move through, and then exit into the system. One of several different designs may be used in a fluid power system, **Figure 2-7.** However, each design involves an intake

Continental Hydraulics

Figure 2-6. A typical basic hydraulic system pump (power unit).

Continental Hydraulics

Figure 2-7. The internal parts of a vane-type pump. The vane design is used in both hydraulic pumps and pneumatic compressors.

area in which a lower-than-atmospheric pressure is created and an outlet area in which a higher-than-atmospheric pressure is generated.

The *reservoir* is the storage area for oil in a hydraulic system. The unit may be a simple box-like container or it may be a cavity in the base of a machine that serves to store system fluid. A pneumatic system uses a *receiver* to store compressed air. This unit is usually a cylindrical tank. The reservoir or receiver also plays other important roles in system operation, such as contributing to system temperature control and fluid cleaning.

The other components that complete the basic power unit are a mechanical coupler and fluid conductors. The *mechanical coupler* is used to connect and align the power output shaft of the prime mover and the pump or compressor shaft to ensure smooth transmission of energy between the components. A *fluid conductor* is used to move fluid from the reservoir to the pump and from the compressor to the receiver.

2.3.2 Actuators

Actuators are the components that convert the energy of the pressurized fluid to mechanical movement to perform the work for which the system was designed. Cylinders and motors are the two basic types of actuators used in fluid power systems. *Cylinders* provide linear motion, while *motors* provide rotary movement. A variety of specially designed actuators provide combinations of these motions for specific applications.

Cylinders

A basic cylinder consists of a cylinder body, cylinder end caps, piston, piston rod, ports, and seals, **Figure 2-8.** A closed chamber is produced in the cylinder body when the end caps are attached to the body. The piston rod is attached to the piston and this assembly is located in the chamber with the rod extending

through one of the end caps. The closed chamber is divided into two parts by the piston. Ports are located in each of the end caps. Appropriate seals are located between the end caps and cylinder, piston and cylinder, and piston rod and end cap.

Forcing fluid into the cylinder through a port causes the piston rod to move out of the cylinder on the extension stroke and back into the cylinder on the retraction stroke. This produces linear motion for use in a machine. **Figure 2-9** shows a cylinder used in the steering system of a lawn tractor.

Motors

Fluid power motors have many of the same basic design characteristics as pumps. There are several different motor designs, ranging from simple to very complex. A basic motor consists of a housing, rotating elements, power output shaft, ports, and seals. The housing provides a closed chamber to retain system fluid. The rotating elements divide the housing chamber into intake

Goodheart-Willcox Publisher

Figure 2-9. A typical fluid power cylinder. This cylinder is used to assist the steering of a lawn tractor.

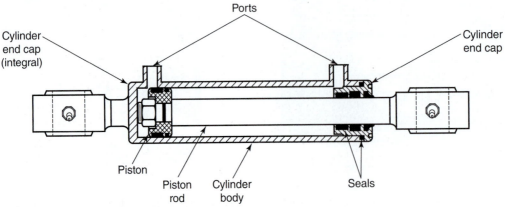

Used with permission of CNH America LLC

Figure 2-8. A cross section of a fluid power cylinder.

and outlet sections. The power output shaft is attached to the rotating elements. Ports are provided in both the intake and outlet sections of the housing chamber. Seals are provided to prevent leakage of fluid between the housing and rotating elements and between the housing and the power output shaft.

Forcing fluid into the motor causes the rotating elements to turn the power output shaft. This produces rotary motion for use in a machine. **Figure 2-10** shows a fluid power motor in an agricultural application.

2.3.3 Conductors

Conductors confine the system fluid as it is distributed throughout the system. Pipes, tubes, and hoses are the three general types of conductors used in fluid power systems. Special manifolds (larger channels) consisting of multiple passageways are used in systems where space and weight are important factors. The type of conductor used depends on the type of fluid, system pressure, required component movement, and the environment in which the system operates. This group of components also includes a variety of fittings and adapters to allow easy assembly while ensuring a system free of leaks.

Pipes are rigid conductors commonly used in stationary fluid power installations. Piping is lower in cost than most other conductors with comparable specifications. The pipe designed for use in fluid power systems is made from mild steel. It is manufactured as seamless.

Tubes are similar to pipes, but are considered to be semirigid. Tubing for use as a conductor in fluid power installations is made from thin steel. Tubing is lightweight, easy to install, has a good appearance, and develops few leaks during system operation, **Figure 2-11**.

Goodheart-Willcox Publisher

Figure 2-11. Tubing is used in both hydraulic and pneumatic systems for transfer of pressurized fluids. Here, both tubing and hoses are used to allow required machine member movement.

Hoses are flexible conductors made from layers of materials. They contain system fluid while withstanding system pressure and allowing easy movement of system components. Hose construction includes an inner tube to contain the fluid, a middle section of braided fabric or wire to withstand pressure, and an outer layer of material for protection from dirt and abrasion, **Figure 2-12**.

Goodheart-Willcox Publisher

Figure 2-10. A typical fluid power motor (red component). The construction of motors and pumps is similar. Several different designs are used in fluid power systems.

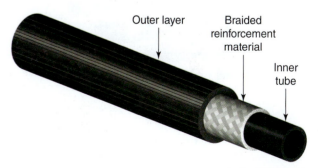

Outer layer

Braided reinforcement material

Inner tube

Goodheart-Willcox Publisher

Figure 2-12. Hydraulic hose is made using multiple layers of synthetic rubber, braided fabric, and wire to provide a flexible high-pressure conductor.

Fittings and adapters are needed to assemble conductors and other system components. *Fittings* are parts needed to assemble similar conductors, while *adapters* are required when connecting different types of conductors or when attaching conductors to system components.

2.3.4 Control Valves

Control valves are the components that make it possible to establish the direction of movement and the maximum force and speed of a system actuator. These characteristics are achieved by controlling fluid flow direction, pressure, and flow rate in the system. Many different designs of valves are available to achieve these characteristics. These designs range from simple, basic valves to extremely complex devices. However, each of these designs can be placed under a basic classification of directional control, pressure control, or flow control.

Directional control valves regulate the direction of actuator movement by creating flow paths to and from the actuator. A basic control valve consists of a valve body, internal elements that open and close fluid flow paths through the valve, an external device for shifting the internal elements, ports to connect the valve to the system, and sealing devices to prevent fluid leakage, **Figure 2-13**.

When a directional control valve is used in a fluid power system, it directs fluid to one side of the internal elements of an actuator, causing the actuator to move in one direction. Fluid from the other side of the actuator's internal elements flows through the valve and is returned to the system (or atmosphere). Shifting the valve creates a flow path to the opposite side of the internal elements of the actuator, causing the actuator to move in the opposite direction.

Pressure control valves regulate pressure in the system or a part of the system. These components are used to control maximum system pressure, limit pressure in a part of the system, or delay the movement of an actuator until a desired pressure is reached. A basic pressure control valve consists of a valve body, internal elements that control fluid flow through the valve, an external device to allow adjustment of the valve, ports to connect the valve to the system, and appropriate sealing devices.

In a hydraulic system, the basic pressure control device is a closed valve that does not open to allow fluid to flow through it until a desired pressure is reached, **Figure 2-14**. When the system reaches the desired pressure, the internal elements of the valve are forced open, allowing passage through the valve. Fluid that is not needed to maintain the desired system pressure then passes through the valve to a lower-pressure section of the system. The amount of fluid that passes through the valve varies according to the amount of fluid needed to maintain the desired system pressure. The smaller the amount of fluid needed to maintain the desired system pressure, the higher the flow through the valve to the low-pressure section of the system.

In a pneumatic system, the basic pressure control device is called a *pressure regulator*, **Figure 2-15**. Regulators are located at each workstation in a pneumatic system. A regulator is an open valve that allows air to move through without resistance until a desired system pressure is reached. Once the desired pressure is reached, the internal elements of the regulator begin to close, allowing only enough air to pass through to maintain the desired pressure.

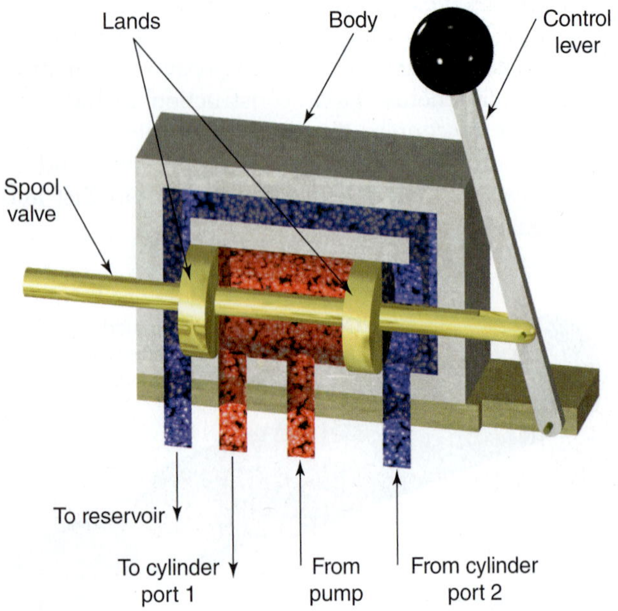

Lands   Body   Control lever

Spool valve

To reservoir

To cylinder port 1   From pump   From cylinder port 2

Goodheart-Willcox Publisher

Figure 2-13. Cutaway of a two-position directional control valve showing the flow passageways that allow cylinder extension and retraction.

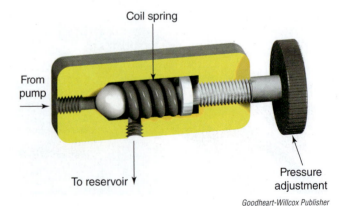

Coil spring

From pump

To reservoir

Pressure adjustment

Goodheart-Willcox Publisher

Figure 2-14. The internal construction of a hydraulic pressure control valve used to set the maximum operating pressure of a system.

A basic flow control valve consists of a valve body, internal elements to control flow through the valve, an external mechanism to allow adjustment of the valve, ports to connect the valve into the system, and appropriate seals to prevent fluid leakage, **Figure 2-16**.

The passageway in a basic flow control valve provides an open route for fluid flow through the valve. The volume of fluid flow through the valve is controlled by the size of the passageway opening. Adjusting the size of the passageway varies the flow volume, which controls the speed of system actuator(s). The larger the opening, the higher the volume of flow through the valve and the higher the speed of the actuator(s).

2.3.5 Fluid Conditioning Devices

Various devices are used to ensure the system fluid is free of contaminants and operating within an acceptable temperature range. Proper care of the fluid has a major effect on the performance and service life of components and the overall system. The type of conditioning devices required in a system is directly related to the complexity of the components and the environment in which the system operates.

Filters are the most common conditioning devices incorporated in fluid power systems, **Figure 2-17**. They remove foreign particles and other contaminates. Filters are located in various places of a system to safeguard specific components, as well as the overall system. Three typical locations are:

- In the intake line for the system.
- In high-pressure working lines.
- In the lines that return fluid from system actuators.

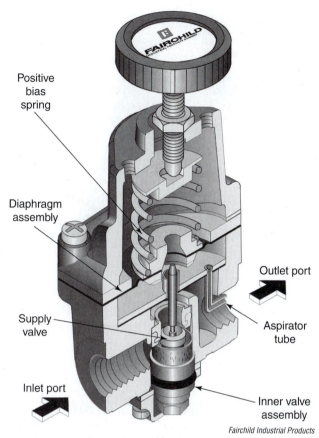

Fairchild Industrial Products

Figure 2-15. A pressure regulator is used to set the operating pressure of a pneumatic system workstation.

Flow control valves regulate fluid flow in the system or a part of the system. These valves control the volume of flow by varying the size of an orifice in the passageway through which the system fluid flows.

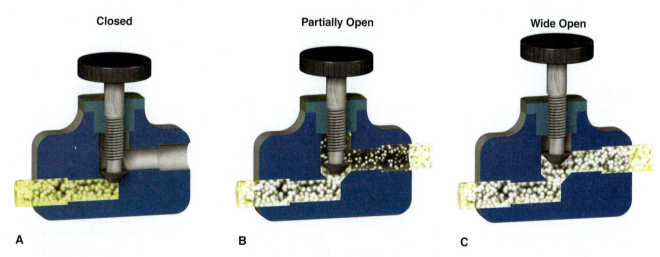

Goodheart-Willcox Publisher

Figure 2-16. Fluid flow through a flow control valve is adjusted by varying the size of an orifice in the valve. A—In the closed position, fluid does not flow. B—With the orifice partially open, restricted flow is allowed through the valve. C—When the orifice is fully open, flow through the valve is not restricted.

Figure 2-17. Filters, such as this compressor inlet filter, are used in several locations in both hydraulic and pneumatic systems to remove contaminates, which increases both the performance and service life of system components.

A basic filter consists of a filter housing, filter element, bypass valve, and ports to connect the filter into the system.

Many different filter designs are used in fluid power systems. In a typical basic filter, fluid enters the filter housing through the inlet port, is routed through the filter element, which removes contaminates, and is then directed back into the system through the outlet port. If the filter element becomes clogged with contaminants, the bypass valve opens and routes fluid around the filter element.

A pneumatic system includes two additional fluid conditioning devices: a separator and a lubricator. The *separator* removes water droplets from the compressed air. The filter and separator are often combined in one unit called a separator-filter. The lubricator adds a fine mist of oil to the air for lubrication of system components, **Figure 2-18.**

The hydraulic system reservoir also serves as an important component for fluid maintenance. A well-designed reservoir helps control temperature by dissipating heat, trapping contaminants, and allowing air trapped in the system oil to separate and escape.

The receiver also serves as a fluid maintenance device in the pneumatic system. The receiver allows water droplets and other contaminates to settle out of the compressed air. The device also helps control the temperature of the system by allowing heat to dissipate.

Figure 2-18. A lubricator in a pneumatic system adds a fine mist of oil to the compressed air before it is used at a system workstation.

2.4 Basic System Operation

Individual fluid power components must be assembled into a system to convert, distribute, and control the energy provided by the prime mover and produce the work desired from a machine. The basic systems described in this section include components from each of the five component groups and provide directional, pressure, and speed control of an actuator. Basic system examples are provided for both hydraulic and pneumatic systems to illustrate the structural characteristics of each system and the inherent differences between the systems.

2.4.1 Basic Hydraulic System

A basic hydraulic system is illustrated in **Figure 2-19**. The system includes an electric motor that serves as the prime mover. The prime mover produces the energy needed to operate the system and perform the work required of the actuator. The system contains basic components from each of the component groups to allow operation and control of a cylinder.

As the electric motor turns the pump, pressure differences are created within the system. As a result, oil moves from the reservoir into the intake line of the pump and through a system filter where impurities are removed. Continued rotation of the pump forces oil into the system lines (conductors), where it is distributed to other system components.

The first system component the oil encounters as it is forced through the system lines is the pressure

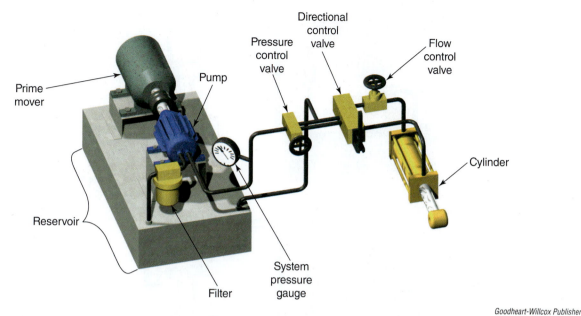

Goodheart-Willcox Publisher

Figure 2-19. The components of a basic hydraulic fluid power system that can provide directional, flow, and pressure control for a cylinder.

control valve, which is used to set system pressure. Next is the directional control valve. This valve directs the oil to the actuator (a cylinder in this example). Forcing oil into the actuator causes the actuator piston and rod to move (either extend or retract). Oil already in the actuator on the other side of the piston is moved out into system lines and returned to the reservoir through the directional control valve, **Figure 2-20A**. Shifting the directional control valve directs the oil flow to the other side of the actuator piston, which forces the piston and rod to move in the opposite direction, **Figure 2-20B**. Oil from the other side of the piston is returned to the reservoir. The extension and retraction of the actuator does the work that the system was designed to perform.

When the actuator is fully extended or retracted, or when it encounters a load heavier than it can move, there is no place for the oil that the pump continues to force into the system. This condition could cause pressure in the system to rise to a level high enough to damage system components. A pressure control valve located between the pump and the directional control valve protects the system from this potentially damaging high pressure. The pressure control valve is set at a pressure that is safe and appropriate for the system. The internal elements of the pressure control valve begin to open when system pressure approaches the valve setting. Oil that cannot be moved into the actuator passes through the pressure control valve and is returned to the reservoir. See **Figure 2-21**.

The speed at which an actuator moves depends on the rate of oil flow into the actuator. In a basic system, actuator extension speed is controlled by placing a flow control valve in the line between the directional control valve and the actuator. Closing the valve reduces the size of the orifice, which restricts oil flow though the valve. This restricted flow reduces the rate of actuator movement. Opening the valve enlarges the size of the orifice, which increases oil flow and the rate of actuator movement.

Restricting flow using a flow control valve affects the system in ways other than controlling actuator speed. The pump moves more oil into the system in front of the flow control valve than the lines and valves in that part of the system can accommodate. Attempting to force the oil into an inadequate space causes system pressure to rise. The pressure increase causes the pressure control valve to open, returning excess oil to the reservoir. The flow control valve continues to supply a set volume of oil to the actuator until the actuator is fully extended or the load on the actuator slows or stops movement.

During retraction of the actuator, the oil returning to the reservoir must pass through the flow control valve. Oil movement through the valve orifice is reversed but continues to be at a reduced flow rate, which slows the retraction speed of the actuator. A one-way check valve is often included in a flow control valve to allow free flow of oil around the restrictive orifice. This design allows the actuator to be retracted at maximum speed.

This system illustrates the structure and operation of a typical basic hydraulic system. The size, arrangement, and complexity of components varies from system to system, but the basic system functions remain the same.

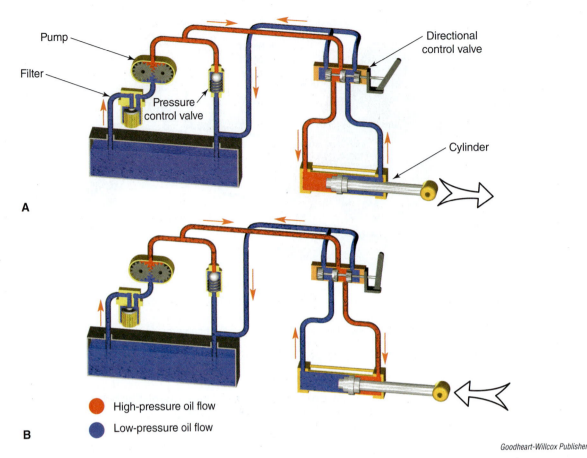

Figure 2-20. Oil flow through a basic hydraulic system. A—During cylinder extension. B—During cylinder retraction.

2.4.2 Basic Pneumatic System

A basic pneumatic system is illustrated in **Figure 2-22**. The system includes an electric motor as the prime mover, which provides the energy to compress the air used for system operation. The system also contains basic components from each of the other component groups to allow operation and control of a cylinder.

As the electric motor turns the compressor, pressure differences are created within the system. As a result, atmospheric air enters the compressor through an air-intake filter that removes airborne dirt. The compressor compresses the air and forces it into an air receiver. The air is held in the receiver in compressed form until it is distributed through lines to other system components. See **Figure 2-23**. A pressure switch controls the prime mover. The compressor functions only when additional air is needed to maintain the desired pressure in the receiver and the distribution lines leading to system workstations.

The first system component the compressed air encounters after it moves out of the receiver through system lines to a workstation is a separator filter. See **Figure 2-24**. This component removes droplets of condensed water and any remaining dirt particles. The air then moves to a regulator, which controls the operating

pressure at the workstation. The operating pressure maintained by the regulator is always less than the pressure at the receiver.

The compressed air next moves to the lubricator. This device adds a fine mist of oil to the air. The oil lubricates any components through which the air flows.

A directional control valve is the next component in the basic pneumatic circuit. This valve directs the air to the actuator, such as a piston in a cylinder, causing the actuator to move. Existing air from the other side of the piston is returned through a system line to the same valve, where it is exhausted to the atmosphere. Shifting the directional control valve directs airflow to the opposite side of the piston, which reverses the actuator. Air from the other side of the piston is exhausted to the atmosphere through the valve. See **Figure 2-25**.

When the actuator is fully extended or retracted, it simply stops moving. The actuator also stops moving when it encounters a load heavier than it can move. This is possible because of the compressibility characteristic of air and the overall design of the pneumatic system.

The receiver stores compressed air at a preset maximum system pressure established by a pressure-operated electrical switch, which controls the operation

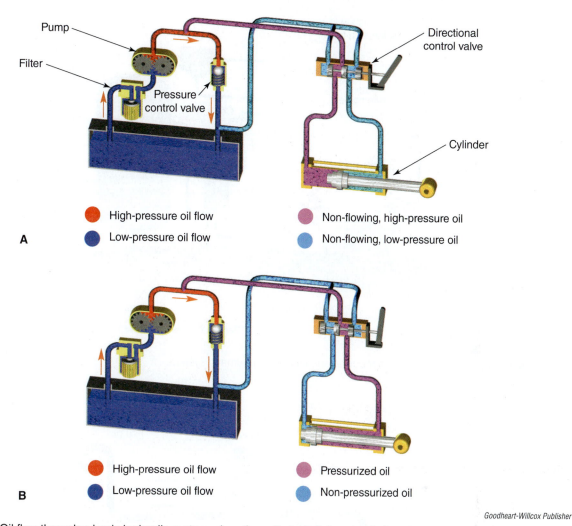

Figure 2-21. Oil flow through a basic hydraulic system when the cylinder is fully extended or retracted or when it encounters a load heavier than it can move. A—Fully extended or stalled on extension. B—Fully retracted or stalled on retraction.

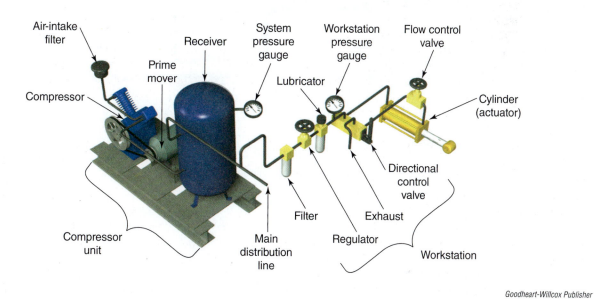

Goodheart-Willcox Publisher

Figure 2-22. The components of a basic pneumatic system that can provide directional, flow, and pressure control for a cylinder.

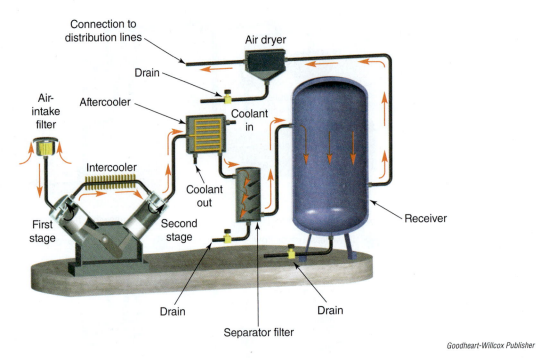

Connection to distribution lines

Air dryer

Drain

Aftercooler

Air-intake filter

Coolant in

Intercooler

Coolant out

First stage

Second stage

Receiver

Drain

Drain

Separator filter

Goodheart-Willcox Publisher

Figure 2-23. Airflow during compressor operation in a basic pneumatic system.

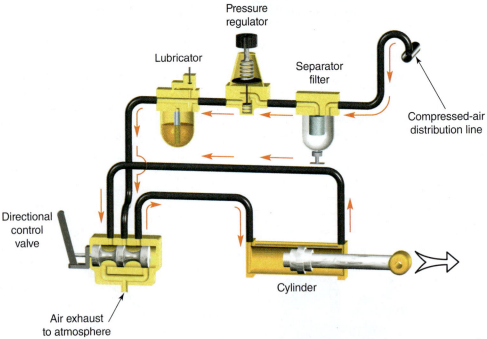

Pressure regulator

Lubricator

Separator filter

Compressed-air distribution line

Directional control valve

Cylinder

Air exhaust to atmosphere

Goodheart-Willcox Publisher

Figure 2-24. Airflow through a basic pneumatic system during cylinder extension.

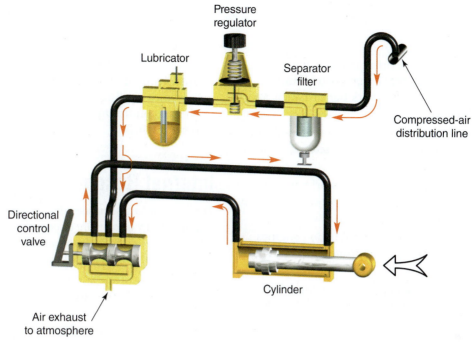

Figure 2-25. Airflow through a basic pneumatic system during cylinder retraction.

of the compressor. The regulator located at the workstation sets a working pressure at a level lower than the maximum system pressure. Air moves from the higher-pressure area of the receiver to the lower-pressure area of the workstation. When the actuator is overloaded or has reached the end of its travel, the pressure in the workstation area reaches the setting of the regulator. The regulator then blocks airflow from the receiver, protecting workstation components from excessive pressure.

Accurate control of actuator speed in a pneumatic system is difficult because of the compressibility of air. To provide the best control in a basic system, the flow control valve is placed in the return line between the actuator and the directional control valve. Adjusting the flow control valve changes the size of an orifice in the valve. The larger the orifice, the higher the airflow and the higher the actuator speed. Reducing the orifice size reduces actuator speed.

During actuator operation, air is routed through the flow control valve before it is returned to the directional control valve, where it is then exhausted to the atmosphere. This placement of the flow control creates pressure on both sides of the actuator piston, which produces a more uniform actuator speed. During actuator retraction, system airflow must pass through the flow control valve before entering the actuator. Flow through the valve orifice is reversed, but the size of the orifice still controls the airflow rate, slowing the retraction speed of the actuator. A one-way check valve allowing free flow of air around the restrictive orifice is often used to allow the actuator to be retracted at maximum speed.

This system illustrates the structure and operation of a typical pneumatic system. The size, complexity, and arrangement of the components varies from system to system, but the basic system functions remain the same.

Summary

- Fluid power systems perform five functions during operation including energy conversion, fluid distribution, fluid control, work performance, and fluid maintenance.
- Fluid power system components are grouped by the specific function or task performed within the system.
- The power unit of the system deals primarily with energy conversion and consists of a prime mover, a pump or compressor, and a reservoir or receiver.
- Electric motors and internal combustion engines are the most commonly used prime movers in fluid power systems.
- Pumps and compressors produce the fluid flow that transmits energy throughout the system. Fluid flow is created by internal pressure differences that cause fluid to move into these units and then forced out into the system.
- Although the primary function of both the hydraulic system reservoir and pneumatic system receiver is to store system fluid, they also contribute to system temperature control and fluid cleaning.
- System actuators consist of both cylinders and motors that perform the work of the system.
- Cylinders and motors perform work when pressurized fluid moves an internal part of the components from a high-pressure area toward a low-pressure area. Cylinders perform work involving linear motion; motors perform work involving rotary motion.
- Conductors consisting of pipes, tubes, and hoses distribute fluid throughout fluid power systems.
- Three groups of valves are used in fluid power systems to control fluid pressure, flow direction, and flow rate.
- Directional control valves vary the direction of movement of cylinders and motors by changing fluid flow paths to and from actuators.
- Pressure control valves control pressure in a fluid power system by restricting fluid flow into a part of the system or by allowing fluid to return to a low-pressure area after a desired pressure is reached.
- Flow control valves control fluid flow rate in a system by adjusting the size of an orifice in a passageway through which the fluid flows.
- Filters and other devices maintain system fluid by removing dirt, moisture, and excessive heat, as well as conditioning fluid to assure effective system performance.

Internet Resources

The following are some useful resources available on the Internet. Enter a company or organization name into a search engine to access its website. Explore the various areas of the sites to discover useful fluid power resources.

Eaton Corporation. Choose the "Hydraulics" product category to view resources available in the area of hydraulics.

How Stuff Works, Inc. Search the site for "hydraulics" to return a variety of articles on how hydraulic machines work. This site offers information on basic concepts, as well as short videos on machine operation.

Hydraulics and Pneumatics Magazine. Contains articles covering many aspects of both hydraulic and pneumatic components and systems.

Ingersoll-Rand Company. Search for "Compressed-Air Solutions" to review the impact of compressed-air equipment in various industries.

International Fluid Power Society. To access downloadable papers devoted to various fluid power topics, select the "Education and Training" tab, and then choose "Technical Papers". The materials available have been developed by a number of individuals and companies.

National Fluid Power Association. Presents online resources for self-paced learning. Select the "Education and Careers" tab, and then choose "Learning Resources". Search for the "Fundamentals of Pneumatics" program, which introduces basic concepts, terminology, applications, and automation processes used throughout the fluid power industry.

Chapter Review

Answer the following questions using information in this chapter.

1. *True or False?* Fluid power systems are made up of groups of components that perform the three tasks required of all hydraulic circuits.

2. *True or False?* The tasks of fluid power system components are all associated with the volume of fluid used in the system.

3. Which of the following is a basic function to both hydraulic and pneumatic systems?
   A. Energy conversion.
   B. Fluid distribution.
   C. Work performance.
   D. All of the above.

4. *True or False?* Linear (cylinder) or rotary (motor) are the two basic actuator types used in fluid power circuits.

5. The chief method used to distribute fluid through a fluid power circuit is a(n) _____.
   A. distributor
   B. pressure valve
   C. line
   D. actuator

6. *True or False?* Tubes are considered to be semirigid conductors that are lightweight, easy to install, and have a good appearance.

7. Shifting a _____ control valve creates a new fluid flow path in the system, which controls actuator movement.
   A. directional
   B. pressure
   C. flow
   D. conditioning

8. Control valves in a fluid power circuit may be designed to control the _____.
   A. direction of movement of an actuator
   B. force of an actuator
   C. speed of an actuator
   D. All of the above.

9. The maximum pressure in a basic pneumatic system is set by a(n) _____ that remains open until the desired system pressure is reached.

10. Flow control valves control the volume of flow by varying the size of a(n) _____ through which system fluid flows.
    A. orifice
    B. spring
    C. tube
    D. cylinder

11. Which of the following is a location a fluid power system filter would be found in a hydraulic system?
    A. Outlet port.
    B. Intake line to pump.
    C. Inside the pump.
    D. Inside pressure control valve.

12. During retraction of a cylinder, free flow around a flow control valve is provided by a one-way _____ valve.
    A. check
    B. pressure control
    C. exhaust
    D. bypass

13. In a pneumatic fluid power system, after the pressurized system air has completed its work, it is exhausted to the _____.
    A. cylinder
    B. pump
    C. atmosphere
    D. None of the above.

14. In a pneumatic fluid power system, a preset maximum system pressure is established by a pressure-operated electrical switch, which controls the operation of the _____.
    A. actuator
    B. compressor
    C. cylinder
    D. directional control valve

15. Accurate control of actuator speed in a pneumatic system is difficult because of the _____ of air.
    A. volume
    B. compressibility
    C. temperature
    D. velocity

16. During actuator operation, air is routed through the _____ before it is returned to the directional control valve.
    A. exhaust valve
    B. actuator
    C. pressure control valve
    D. flow control valve

Apply and Analyze

1. Examine the structure of fluid power systems in **Figure 2-2**.
   A. How do the basic structures of the two types of fluid power systems differ?
   B. What impact do these differences have on the design and operating characteristics of the two systems?

2. Fluid power systems transform and transmit energy to perform work. Most power transmission systems are not 100% efficient so that energy is lost or wasted during system operation. What are some sources of inefficiency (energy loss) in a hydraulic system? What are some sources of inefficiency (energy loss) in a pneumatic system?

3. Identify the components in a fluid power system that control the maximum load that can be moved and the speed at which it can be moved. Describe how each of these components determines system capacity.

4. **Figure 2-20** illustrates oil flow through a basic hydraulic system during cylinder movement. As the cylinder extends and retracts, oil flows unrestricted back to the reservoir. What effects would a flow control valve between the actuator and the reservoir have on system pressures and operation?

Research and Development

1. Identify a machine used in construction or manufacturing that uses hydraulic power as a significant part of its operation.
   A. Draw a schematic of the system, including as many components as you can.
   B. Describe these components and explain the role they play in the hydraulic system of the machine.
   C. Research a different model of this machine. Compare and contrast the basic design of the power system or one of the components used in both machines. Discuss how these variations contribute to the capacity of the machine to do work.

2. Identify a machine used in construction or manufacturing that uses pneumatic power as a significant part of its operation.
   A. Draw a schematic of the system, including as many components as you can.
   B. Describe these components and explain the role they play in the pneumatic system of the machine.
   C. Research a different model of this machine. Compare and contrast the basic design of the power system or one of the components used in both machines. Discuss how these variations contribute to the capacity of the machine to do work.

3. Automobile repair shops typically use both hydraulic and pneumatic equipment. Interview the operator of a local repair shop about their use of this equipment, the layout of the system, and how the choices may impact shop operation. Present your findings to the class.

Atlas Copco

Imaginative designs involving fluid power with mechanical and electrical components may produce robots that assume a human appearance.

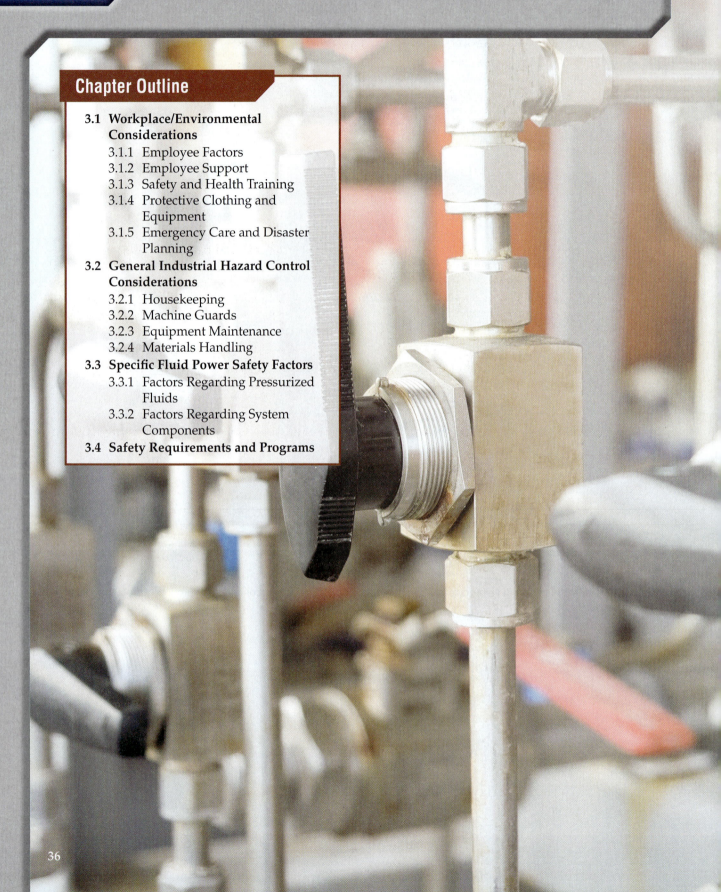

3 Safety and Health
Promoting Proper Practices

The use of equipment is an everyday activity in our industrialized society. Technology has produced a vast number of machines that range from simple hand tools to huge, complex machines. A large number of these devices have the potential to cause injuries to the operator or individuals working in the area. The number of workers injured on the job each year indicates that many people do not understand the environment in which they work or respect the equipment they operate or work near. Individuals working in the fluid power field will be exposed to a wide variety of work environments. They must be aware of safety and health issues related to those areas, as well as the specific issues dealing with fluid power. This chapter discusses items that influence general safety and specific issues that involve hydraulics and pneumatics.

Learning Objectives

After completing this chapter, you will be able to:

- Describe how the workplace environment influences the safety of individuals working in an industrial setting.
- Explain how awareness of safety issues in a situation influences the safety of an individual.
- Identify personal traits that may make an individual more prone to violate safety procedures.
- Identify general equipment considerations that influence safe working conditions in the fluid power industry.
- Describe safety rules that need to be followed when working with individual hydraulic and pneumatic components.
- Identify government regulations that influence safety regulations and practice in the fluid power industry.

Key Terms

air-line respirators
back injuries
cumulative injuries
dust mask
earmuffs
earplugs
emergency shower
equipment maintenance

eye protection device
eyewash station
first aid kit
hearing-protection device
housekeeping
lockout device
Occupational Safety and Health
    Administration (OSHA)

pressure safety valve
respirator face masks
safety helmets
safety shoes
self-contained breathing
    devices
training

3.1 Workplace/Environmental Considerations

Good equipment, shop layout, and working conditions are important elements in the prevention of both accidents and cumulative injuries. *Cumulative injuries* are those injuries that occur from long-term exposure to unsafe environmental conditions. Among the more important of these environmental work-area conditions are:

- General cleanliness.
- Quality of air.
- Level of light.
- Level of sound.

All of these conditions should meet or exceed state and federal standards, including those of the Occupational Safety and Health Administration (OSHA). The goals of any safety and health program should be to eliminate unreasonable risks and reduce accidents on the job.

Careful layout is important to avoiding accidents and injuries in any shop. Proper placement of equipment and adequate traffic lanes can reduce the number of traffic, falls, and collisions with equipment, **Figure 3-1**. The time spent planning a shop with effective machine placement and flow of workpieces reduces employee movement and helps eliminate injuries. In addition, the facility can be improved by providing adequate utility services. This will reduce the number of situations where improper or inadequate equipment connections are made because of distant or poorly placed service outlets.

General cleanliness is especially important in the hydraulic area because of oil leakage and spillage. Oil leaks at fittings and other hydraulic components should be promptly repaired. Any spilled oil should be completely wiped up to prevent slips and falls. The oil lost from attaching and detaching quick couplings also needs to be readily cleaned up to not only eliminate the hazards of slips and falls, but also to reduce the accumulation of dust and other debris in the pooled oil and eliminate the fire hazard of oil-soaked materials.

Yates Industries, Inc.

Figure 3-1. A shop that is carefully designed to include quality equipment, good lighting, sound control, and proper ventilation provides an environment that encourages safe working habits.

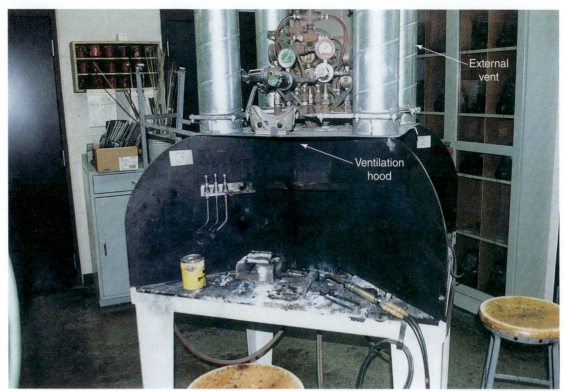

Goodheart-Willcox Publisher

Figure 3-2. Proper ventilation helps prevent accidents by providing a more comfortable temperature, but it can also reduce cumulative respiratory injuries by improving air quality.

Ventilation needs to be adequate at all times in all locations. Ventilation is critical in a work environment where fumes are a health issue, **Figure 3-2.** Air exchange and circulation are also important for the maintenance of suitable temperature and humidity levels. The effort spent on these two factors can improve production by increasing the comfort level of workers. This can also help reduce accidents by improving attention level and duration.

Noise can cause cumulative damage to the hearing of individuals who must work for extended periods in areas with high-decibel sounds. The extent of the sound should be tested and care taken to reduce the level or provide ear protection to meet local, state, and federal standards. Overall sound levels in a facility can be reduced by providing sound-absorbent materials on walls, ceilings, and partitions. Levels can also be controlled by constructing sound-reducing enclosures for devices that produce extreme noise levels. If the sound level cannot be reduced below the allowable maximum, ear protection must be provided and worn by individuals stationed in or frequenting the area, **Figure 3-3.**

Adequate illumination must be provided in all work areas. The required light intensity depends on the work being performed in the area. Applicable local, state, and federal standards should be obtained and closely followed. Regulations vary on light intensity,

Reprinted courtesy of Caterpillar, Inc.

Figure 3-3. Hearing protection must be worn when operating specialized equipment with high sound output. Hearing protection must also be provided in general work areas that have lower sound levels, but where those sounds may cause hearing problems due to cumulative effects.

but care should be taken to provide adequate general lighting in all areas and supplementary lights at work-stations for detailed work, such as machine or test bench locations. Higher light levels are also required in classrooms and training facilities.

Appropriate fire fighting equipment must be available in each work, training, or classroom area, **Figure 3-4**. Carbon dioxide, foam, or dry powder fire extinguishers need to be placed in the facility following the recommendations of local fire officials. Fire extinguishers have a class rating, which is the types of fire on which the extinguisher can be used. See **Figure 3-5**. A fire blanket, which can be used to put out a clothing fire, should be readily available. All flammable liquids, such as solvents and fuels, should be stored in metal containers in an area away from a heat source. These materials should be carefully handled, with any spill-age immediately wiped up. Rags soiled with oils and solvents should be removed from the work area at the end of the day and stored in fireproof containers. Each area of the facility should have fire-emergency proce-dures posted and all individuals working in an area should be familiar with those procedures.

All of these factors contribute to a safe working environment. Failure to provide the basic elements of efficient facility layout, cleanliness, lighting, noise

Goodheart-Willcox Publisher

Figure 3-4. Fire extinguishers of the appropriate type need to be located throughout a facility following regulations issued by the local fire department.

control, and fire prevention can considerably reduce individual safety. Care should be taken to provide an environment that meets at least the base requirements in each of these areas.

3.1.1 Employee Factors

Although often blamed, unsafe working conditions are *not* the chief cause of industrial accidents. Employees are really the chief cause. Statistics show that up to 85% of on-the-job accidents can be traced to unsafe acts by an individual. A wide variety of reasons are respon-sible for the acts that lead to these injuries. Four gen-eral areas of knowledge and activity that contribute to the safety record of employees are discussed in this section.

Knowledge

A primary reason for performing unsafely is a basic lack of knowledge about the job. It is the responsibility of the employer to provide employees with the infor-mation needed to safely perform the job. This informa-tion should include:

- Steps necessary to perform the job.
- Hazards associated with the job.
- Need for and use of personal protective equipment during the performance of the job.

The employer must provide training sessions so an employee has the opportunity to practice the procedures before actually attempting to do the job. However, employees must assume the responsibility of learning the procedures during the training ses-sions, as knowing the job is a critical part of job safety, **Figure 3-6**.

Awareness

A second safety factor is the lack of awareness of job hazards by employees. It is important that employees consider safety issues as they perform their jobs. The safety consequences of a task should be considered before performing the task. Awareness of what can go wrong increases the alertness of the individual and reduces the possibility of an accident.

Job hazards can be associated with the special-ized tasks of the job, but may be involved with very ordinary factors, such as equipment electrical cords or workplace housekeeping. Electrical cords are a simple example of a safety issue. Frayed cords or damaged connectors are dangerous. They contribute to the pos-sibility of electrical shock and fire. Also, allowing the workplace to become cluttered not only makes the area look unkempt, it increases the chances for falls caused by tripping on tools, cords, or raw materials.

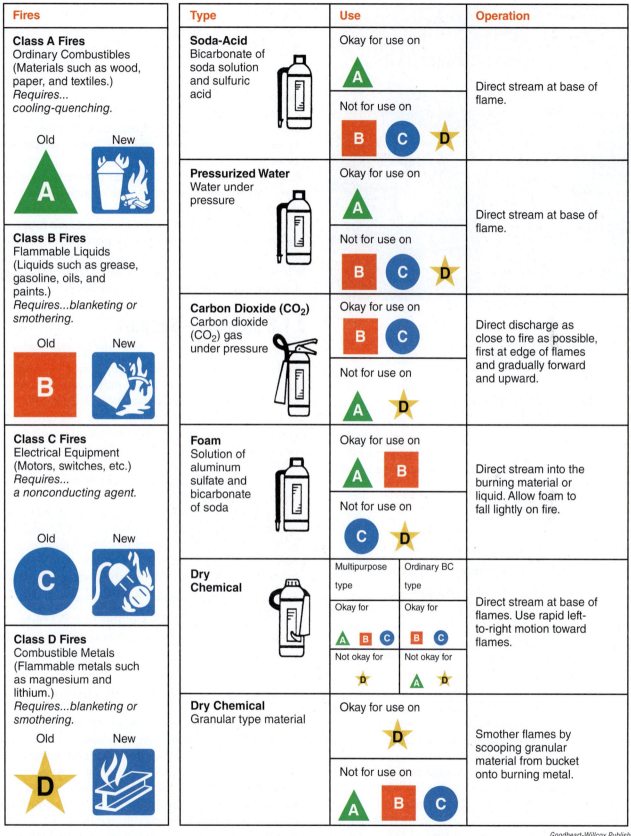

Fires	Type	Use		Operation
Class A Fires Ordinary Combustibles (Materials such as wood, paper, and textiles.) *Requires... cooling-quenching.* Old New	**Soda-Acid** Bicarbonate of soda solution and sulfuric acid	Okay for use on A		Direct stream at base of flame.
		Not for use on B C D		
Class B Fires Flammable Liquids (Liquids such as grease, gasoline, oils, and paints.) *Requires...blanketing or smothering.* Old New	**Pressurized Water** Water under pressure	Okay for use on A		Direct stream at base of flame.
		Not for use on B C D		
	Carbon Dioxide (CO$_2$) Carbon dioxide (CO$_2$) gas under pressure	Okay for use on B C		Direct discharge as close to fire as possible, first at edge of flames and gradually forward and upward.
Class C Fires Electrical Equipment (Motors, switches, etc.) *Requires... a nonconducting agent.* Old New		Not for use on A D		
	Foam Solution of aluminum sulfate and bicarbonate of soda	Okay for use on A B		Direct stream into the burning material or liquid. Allow foam to fall lightly on fire.
		Not for use on C D		
Class D Fires Combustible Metals (Flammable metals such as magnesium and lithium.) *Requires...blanketing or smothering.* Old New	**Dry Chemical**	Multipurpose type Okay for A B C Not okay for D	Ordinary BC type Okay for B C Not okay for A D	Direct stream at base of flames. Use rapid left-to-right motion toward flames.
	Dry Chemical Granular type material	Okay for use on D		Smother flames by scooping granular material from bucket onto burning metal.
		Not for use on A B C		

Figure 3-5. Fire extinguishers are rated for the type of fire on which they can be used. This chart shows the identification symbols that appear on the fire extinguisher and explains where each type may be used.

Eaton Fluid Power Training

Figure 3-6. Providing training to ensure employees understand their jobs will greatly improve the safety record of workers.

Personal characteristics

A third group of factors to consider is the personal characteristics of the individuals in a work group. A wide variety of factors influence the way individuals feel about their job, their fellow employees, and themselves:

- Lack of knowledge about the job.
- Lack of awareness of job hazards.
- Fear of operating a machine.
- Overconfidence in performing an operation or a job.
- Impatience with following procedures.

These characteristics vary considerably from individual to individual. For example, an inexperienced person may be fearful of a machine, while an experienced person may become overconfident of their skill. Both of these conditions can produce unsafe machine operation.

Concern for job security by both new and experienced workers can cause individuals to risk injury or health problems in order to retain a job. The need to be accepted by other workers is often a problem with new workers. They frequently find it difficult to follow safe methods if experienced workers are bypassing the procedures.

Impatience is another important element in many unsafe practices. This attitude is demonstrated by employees who feel production time is being wasted putting on protective equipment and setting up guards and safety devices on machines. However, working safely must be the first concern of both the employee and the employer.

Attention

A final group of safety factors is related to a person's attention. Those factors that divert a worker's attention cause the worker to be less aware of surroundings. Attention-diverting activities reduce concentration and increase the probability of accidents.

One of these diversions is thinking of personal problems while on the job. Conflicts with other workers can also be upsetting and may lead to unsafe job performance. Even if employees work well together, they may be dissatisfied with their job for various reasons. This may lead them to seek diversions that can lead to unsafe job performance.

Another factor that can lead to unsafe job functioning is simply the work location, **Figure 3-7**. A workstation located next to a heavily traveled aisle may result in inattention to detail and increased accidents because of socializing with coworkers who walk past.

3.1.2 Employee Support

Providing a work atmosphere in which issues of safety and health are understood and supported by employees is a good investment by an employer. The cost of establishing this type of atmosphere is expensive, but the alternative of not promoting such a program can be even more costly. Injuries and job-related health issues can result in a considerable increase in benefit and insurance fees, as well as the loss of valuable employee work time. In addition, the morale of sick or injured employees is often low.

A good safety and health program must start at the top management level. The involvement of management must be visible, continuous, and credible. Four areas that management can promote to help establish long-term employee support for an overall program are:

- Safety and health training programs.
- Protective clothing and equipment programs.
- Realistic emergency care and medical services.
- Established plan for disasters.

3.1.3 Safety and Health Training

Training can serve as an important influence and a substantial motivator in relation to workplace safety and health issues. Care must be taken to be certain the training is directed at a defined need of employees, rather than applied in an overall blanket approach used only to meet the requirements of an outside agency. As in other areas of employee development, general training should be used to solve problems dealing with lack of knowledge, skill, or motivation. A considerable amount of helpful and interesting safety and health-related content can be presented under these general classifications.

Goodheart-Willcox Publisher

Figure 3-7. Attention-diverting activities can be dangerous in many work situations. Accidents can be reduced if workers are constantly aware of activities in the immediate work area. These painters are working in an area with traffic. They must pay attention to not only their painting, but to the surrounding environment.

Training can be presented using a variety of formats, ranging from informal, one-on-one supervisor-employee contact to extensive, formal classroom training sessions. The style of the presentation is usually controlled by the complexity and extent of the topic. The first essential step in any training, however, is the need to define the objectives of the training.

Objectives should always be spelled out before content and methods are selected. These objectives should state how the training participants should perform on the job when the training is completed. For example, an objective might be stated as:

At the conclusion of this training, the participants will be able to select the correct type of fire extinguisher for use on specific classes of fires as defined by the National Fire Protection Association.

A wide variety of commercial training materials are available in both the safety and health fields. These materials provide a wide choice in the methods that may be used for instruction.

Safety and health-related training should be provided to all employees that could be adversely affected by the jobs they perform. Safety and health issues critical to the completion of these jobs should be considered as possible training topics. After appropriate topics have been selected, objectives for each of them must be developed and instructional content structured. The last step in the process should be the selection of the presentation method.

The selection of employees for safety and health-related training should be based on the needs of the group or individual. One obvious group is new employees, who need to be introduced to the general policies and procedures of the organization. In addition, new employees need to receive specific training related to their individual position. Experienced workers must receive training when new equipment or procedures are introduced or when they are transferred into a new work situation. See **Figure 3-8**. Fluid power specialists, because their work assignments cross into a number of different departments in an organization, need to receive broad-based training so they can safely function in their job.

3.1.4 Protective Clothing and Equipment

The need for protective clothing and equipment implies that hazards have not been totally eliminated in some occupations. The nature of some jobs is such that the probability of accidents and long-range health hazards are higher. Total elimination of risk in these jobs is difficult. Protective gear is, therefore, needed to ensure the safety of individuals performing these jobs. Fluid power specialists are exposed to most of the work environments in a factory. They need to be aware of what gear is needed to decrease the likelihood of an accident or a job-related health problem.

A

Eaton Fluid Power Training

B

Lab-Volt Systems, Inc.

Figure 3-8. A—Training programs need to be well planned and delivered to maintain employee interest. The group shown here is receiving hands-on experience at an Eaton/Vickers Training Center. B—Training participants should be provided materials with well-defined objectives so they are aware of what content will be presented.

Ear protection

Noise may be considered a health issue. Long-term exposure is typically the reason for hearing loss due to noise. However, a single, high-intensity incident may cause damage. Protecting employees from these noise hazards is a very common use for personal protective equipment.

A variety of *hearing-protection devices* are available. These devices reduce the decibel level of workplace noise reaching a worker's ears. Earplugs, earmuffs, and helmets are three devices that can be used for hearing protection.

Earplugs are the most popular and very commonly used by workers, **Figure 3-9**. Rubber, plastic, or foam plugs are readily available in most work situations. Earplugs are generally inexpensive and inconspicuous, which are important factors for many individuals that must wear noise control devices.

Earmuffs are larger, more expensive, and more conspicuous. However, they can be more effective than earplugs. Also, some workers feel earmuffs are more comfortable than earplugs.

Helmets are the most expensive form of hearing protection. They are usually used only in the most severe noise conditions or where a combination of a hard hat and sound protection is required.

Eye protection

Eye protection programs are widespread in industry. Many safety programs require that safety glasses be worn by all employees and visitors in any manufacturing area. *Eye protection devices* come in a wide variety of styles, ranging from plastic goggles to framed prescription eyeglasses, **Figure 3-10**. Face shields are also available for eye and face protection when working with dangerous liquids and chemicals.

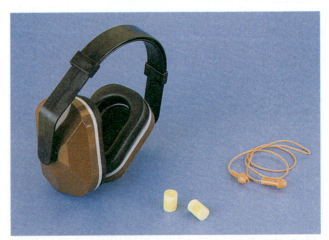

Goodheart-Willcox Publisher

Figure 3-9. Earplugs and earmuffs are two popular forms of noise protection for workers. They are available in a variety of materials and designs.

Goodheart-Willcox Publisher

Figure 3-10. A wide variety of safety devices are available to protect the eyes. They range from plastic safety goggles, to face shields, to prescription eyeglasses with safety lenses and side shields. Be certain the type of eye protection selected is appropriate for the work environment in which it is worn.

Care should be taken to be certain the lenses, frames, and side shields of eye protection meet ANSI standards for the work situation in which they are worn. Company policies on eye protection vary from required use in selected areas to general, plant-wide requirements. Whatever the detail of the policy, it is crucial that employees be aware of the critical nature of eye protection and that they follow established company policies.

Lung protection

Providing respiratory protection is a third important area for personal protection devices. The best way to deal with the air contaminants that make respiratory devices necessary is to change the process involved to eliminate the cause of the problem. An alternate solution is providing ventilation to reduce air contaminants to acceptable levels. However, if air contaminants cannot be eliminated or reduced, a variety of protection devices are available.

Respiratory devices can be classified as air-purifying or atmosphere-supplying units. The devices must deal with solid, liquid, and gaseous airborne hazards. The devices may be as simple as dust masks and face masks or as complicated as air-line respirators and self-contained breathing devices.

The most common respiratory device is a filtering facepiece respirator, or *dust mask*. These are made of a fibrous material and designed to cover the nose and mouth. See **Figure 3-11**. These masks are typically designed and approved only for filtering out particles such as dust, mist, and fumes. They are *not* appropriate for applications involving vapors or gases.

Respirator face masks are available in a variety of designs and materials. Some designs cover only the nose and mouth. See **Figure 3-12**. One design is a full-face mask that covers the nose and mouth and provides a shield for the eyes. The filters for these masks are usually cartridges or canisters that filter the air using adsorption, absorption, or chemical action.

Air-line respirators and *self-contained breathing devices* are used for exposure to high contaminants or where a breathable atmosphere is not available to a worker, **Figure 3-13**. The respirator supplies usable air to the worker through a hose attached to the face-piece of the device. The air for the system is supplied by either a compressor or compressed-air cylinder and metered to the user using several different designs. The self-contained device allows users to carry the apparatus with them, usually on their backs. Most current models of self-contained units discharge exhaled air to the atmosphere, although units that recycle the air are becoming more common. Atmospheric-supply units are generally used under only the most severe of working conditions or in emergency rescue situations.

Photo courtesy of 3M © 2007

Figure 3-11. The dust mask is a commonly used form of respiratory protection. Most of these masks are approved only for filtering particles such as dust, mist, and fumes.

Photo courtesy of 3M ©2007

Figure 3-12. Respirator face masks are designed for use in a variety of atmospheres. Here, one is being used during asbestos abatement.

Photo courtesy of 3M ©2007

Figure 3-13. Air-line respirators supply usable air to the worker through a hose attached to the facepiece of the device.

Head and foot protection

Safety helmets (hard hats) are used to protect the head from injuries caused by the impact of falling or moving objects. See **Figure 3-14**. They also provide limited protection from heat and electrical shock. Some incorporate sound and eye/face protection elements into their design. Helmets are commonly used throughout industry, with their use required in many factory areas. A fluid power specialist needs to be aware of safety policies requiring helmet use and observe those rules as they work throughout a complex.

Helmets are generally made from plastic or metal and use a strap suspension system to hold the helmet shell away from the head. Metal helmets are lighter in weight than plastic, but do not provide protection from electrical shock or corrosive liquids.

Safety shoes are another comon piece of protective equipment for the fluid power specialist. Specially designed shoes or shoes with built-in foot protection are needed in many industrial operations. See **Figure 3-15**. Wearing proper shoes can significantly reduce the number or severity of foot injuries. Manufacturing firms often provide proper safety footwear for their employees or assist in the purchase of such gear.

Steel-toed shoes provide protection from heavy weights. However, safety shoes are available in many styles and forms far beyond the traditional steel-toed work shoe. For example, metal-free footwear is available for use where there are severe electrical, fire, or explosion hazards. Wood-soled shoes are designed for working on wet floors or for jobs that require walking or standing on hot floors.

Metal safety guards worn over shoes to protect the toe and arch area of the foot are also available. These are worn in areas where heavy objects, such as metal castings and timbers, are routinely handled.

Atlas Copco

Figure 3-14. Safety helmets are primarily used to protect the head from impacts. A variety of designs are available to meet the special needs of particular industries.

Goodheart-Willcox Publisher

Figure 3-15. Steel-toed shoes are a common safety shoe found in manufacturing situations. Many other types of safety shoes are available.

Other protection

A variety of other personal protective clothing and equipment is manufactured for use in specialized situations. See **Figure 3-16**. These include complete garments, aprons, gloves, headgear, eye and face protection, and many more. The development of many new materials has allowed many of these items to provide substantially more protection than in the past. Fluid power specialists need to be aware of what clothing and equipment is available to provide a safer work situation and to diligently use the items recommended by their employer.

3.1.5 Emergency Care and Disaster Planning

Accidents where individuals are injured happen even in organizations that have good safety programs. Additionally, workers can suffer sudden and serious illnesses on the job and disasters, such as fires and storms, can occur. Rapid response is a primary consideration in these situations. Effective first aid and the time saved because disaster plans exist, are understood, and are followed can mean the difference between life and death.

The simplest of the tools needed is a *first aid kit*, **Figure 3-17**. These kits are commercially available or may be assembled following the advice of company medical staff. The kits need to be readily available and restocked on a regular basis. At least one person adequately trained in first aid is required in a facility if an on-site infirmary does not exist or a hospital is not close to the facility. *Emergency showers* and *eyewash stations* are required in facilities where exposure to corrosive material is a possibility, **Figure 3-18**.

Each supervisor and worker should know how to rapidly obtain medical assistance in routine first aid, emergency, and disaster situations. When serious injury or illness occurs, emergency medical assistance should be requested as soon as possible. First aid procedures should start immediately if the situation is life-threatening. The sick or injured worker should not be moved unless the location itself threatens further injury. Care of the sick or injured individual should be turned over to the emergency medical personnel as soon as they arrive on the

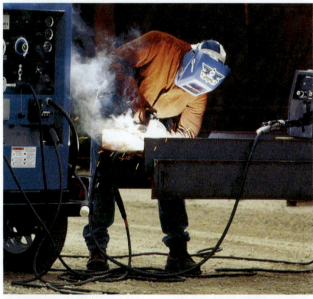

Miller Electric Mfg. Co.

Figure 3-16. A wide variety of specialized clothing is available to promote safety in various industries. Welders wear leather garments to protect themselves from hot slag and metal splatter encountered during the welding process.

Goodheart-Willcox Publisher

Figure 3-17. A first aid kit is a basic tool that should be part of any emergency care and disaster plan.

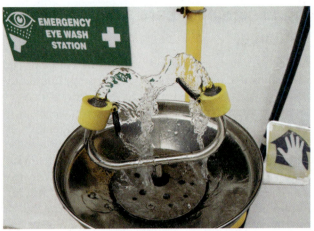

mark_vy/Shutterstock.com

Figure 3-18. Eyewash stations are required in areas where exposure to corrosive fluids is a possibility.

scene, **Figure 3-19**. Be prepared to provide information on any first aid administered to the victim.

The facilities, equipment, and procedures for first aid and medical treatment should be a part of the design and operating plan of all facilities. Also included in this is planning for the personnel required to care for employees in first aid, emergency, and disaster situations. The plan should indicate the position that has the primary responsibility for directing the operation of the team involved. Company personnel that work in several areas of a facility, such as the fluid power specialist, need to be familiar with the emergency plans for all the areas in which they work, **Figure 3-20**. This knowledge may prove to be extremely beneficial to both themselves and their fellow workers in any number of situations.

3.2 General Industrial Hazard Control Considerations

There are many different reasons for injuries in industrial settings. These include factors such as inattentive machine operators, inappropriate equipment or processes, defective or poorly designed equipment, and a multitude of other causes. Safety personnel generally feel that every effort should be made to promote safety by first removing as many hazards as possible in facilities and equipment through design before using employee safety promotion to reduce the remaining hazards. It must be kept in mind that there are situations in which hazards cannot be totally eliminated by design and the safe performance of workers must be relied on to eliminate accidents.

Because fluid power specialists work in such a variety of locations and situations in a factory, they should be aware of factors that can affect overall worker safety and morale. Problems in these areas can contribute to increased accidents and injuries in a facility:

- Housekeeping.
- Machine guards.
- Equipment maintenance.
- Materials handling.

Used with permission of CNH America LLC

Figure 3-19. The need for an on-site infirmary is dependent on the number of employees, the type of work carried on in the facility, and the proximity to a hospital.

Evacuation Plan

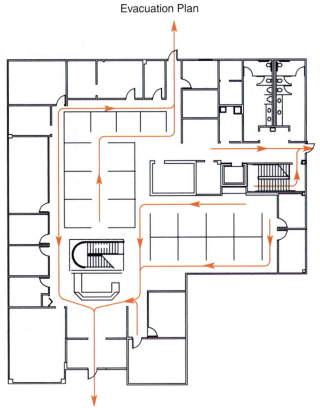

Goodheart-Willcox Publisher

Figure 3-20. The procedures to follow in case of a fire or a natural disaster should be readily available and understood by the employees. Often, evacuation routes like this one are posted throughout the facility.

Vladimir Melnik/Shutterstock.com

Figure 3-21. Keeping a facility clean, with aisles and walkways clear, promotes good general housekeeping and also helps prevent accidents.

Fluid power personnel can promote safety by being aware of these factors and providing information to the appropriate individuals so corrective steps may be taken to improve or eliminate the problem.

3.2.1 Housekeeping

General housekeeping can have a great influence on the number of accidents and injuries in manufacturing facilities. It should be ranked extremely high on the list of factors that can prevent accidents and increase productivity. *Housekeeping* not only includes keeping floors and work surfaces clean and free from dirt and debris, but also includes the proper storage of raw materials, partly finished products, and tools, **Figure 3-21**. Aisles and walkways need to be clean and clear of raw materials and products. Doing so prevents injuries from employees tripping over or running into these items or having the items fall on them. The idea that a busy and profitable manufacturing plant will be cluttered and dirty is *not* correct.

3.2.2 Machine Guards

All mechanical motion in a machine is, to some degree, hazardous. The *Occupational Safety and Health Administration (OSHA)* spells out when operators must be protected. If the hazard points in a machine cannot be eliminated, guards must be installed. These guards can take many forms, including:

- Simple, fixed enclosures that cover hazardous areas to prevent exposure to moving machine parts, **Figure 3-22**.
- Interlocking guards that shut off the power to the machine when they are opened.
- Automatic guards that pull or push an operator's hands, arms, or body from the hazardous area as the machine performs its work.
- Specialized machine designs that require two-handed operation or keep the operator away from hazardous areas by automatic placement, feeding, or ejecting of parts, **Figure 3-23**.

A multitude of devices must be considered for guarding, including items such as rotating shafts, keys and setscrews on shafts, gears, chain drives, belt drives, flywheels, and so on. Maintenance staff, including fluid power specialists, should be familiar with machine guarding, including lockout devices. *Lockout devices* prevent a machine from being accidentally turned on while maintenance work is underway. When machine service is complete, all guards must be replaced to the manufacturer's specifications.

Goodheart-Willcox Publisher

Figure 3-22. A multitude of fixed-enclosure guards are used to cover hazard points in machines in industry today. This guard covers the belt drive between the prime mover and compressor.

Used with permission of CNH America LLC

Figure 3-23. Several types of specialized guard designs are used on equipment to promote safe operation. One specialized design requires two-hand operation of the equipment. Both hands must be on control switches, which are placed well out of any danger, in order for the equipment to operate.

3.2.3 Equipment Maintenance

Proper *equipment maintenance* is a key factor in a good safety program. This means not only the service work done on large stationary pieces of equipment, but also maintenance of portable tools and hand tools. A systematic program of preventive maintenance should be set up to keep machines, tools, and equipment safe and operating at their maximum efficiency. Some equipment may need daily maintenance, while others can be serviced after a specified number of hours, days, or weeks.

The fluid power specialist should be heavily involved in developing and operating a preventive maintenance program. Equipment that could cause an accident if it fails during operation should be inspected and maintained on a closely controlled schedule. Parts should be carefully adjusted and replacement of key parts should be done on a definite schedule.

Care of portable tools is especially important, as many workers use some form of this equipment. Portable tools account for a high proportion of injuries, especially injuries to the hands. However, eye injuries and electric shock are also very commonly associated with operation of these tools. Extreme care should be taken while operating this type of equipment.

3.2.4 Materials Handling

The handling of materials in an industrial setting can be as hazardous as the manufacturing process itself. These hazards range from injuries caused by a worker applying improper lifting techniques to accidents involving mechanical material handling equipment, such as forklift trucks and cranes.

Back injuries represent the largest single group of injuries for which compensation is awarded. These injuries are often related to lifting. Care should be taken in all situations in which lifting must be done. It is possible to injure the back even with light loads if care is not taken to position the back and legs to obtain appropriate leverage. A conscious effort to position the legs and back comfortably while using the legs to do the lifting should reduce the number of back injuries. In addition, back supports are available to help reduce injuries, **Figure 3-24.** However, lifting injuries can occur even under the best of conditions.

The amount of materials physically lifted by humans in an industrial plant is usually tiny by comparison to the total weight involved in the manufacturing process. These materials are designated by tons, rather than in pounds and ounces common to the loads individuals lift. Hand- or electric-operated dollies, forklift trucks, conveyors, and cranes are commonly used to move the vast majority of materials in a factory. The selection of the system used to move the material is based on the weight and size of the material, type of operation being performed on the material, and configuration of the facility in which it must be moved. The equipment operator's skill, attitude, and awareness of hazards is critical to the safety performance of the equipment. Safety performance is also influenced by the awareness of workers to the factors involved in the operation of the systems. The more conscious the workers are of the operating characteristics of the systems, the more efficiently the systems can operate and the better the safety record.

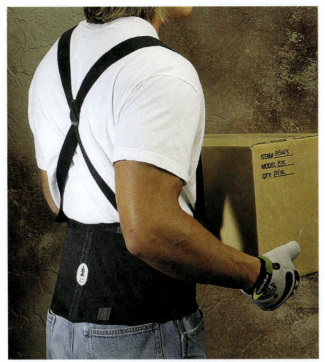

Ergodyne

Figure 3-24. Back injuries are common when work involves the handling and lifting of materials. Special care should be taken to correctly position the back and legs before performing the lift. Use the legs rather than the back for the actual lifting. Back supports such as the one shown here also help prevent back injuries.

3.3 Specific Fluid Power Safety Factors

Working safely as a fluid power specialist involves many factors. Several of these are general industrial safety items that have been mentioned earlier in this chapter. The importance of these safety issues cannot be overemphasized. A number of additional detailed factors relating to hydraulics and pneumatics need to be emphasized. These include specific hazards related to pressurized fluids and system components. The following sections discuss factors that are especially important to any individual working with fluid power.

3.3.1 Factors Regarding Pressurized Fluids

Individuals working with equipment that uses fluid power systems must constantly be aware of the fact that they are working with pressurized liquids and gases. Being aware of this fact and at all times treating the system as if it is pressurized and ready to function can eliminate many accidents, **Figure 3-25**.

Pressurized fluids must be treated with respect. Compressed gases in pneumatic systems and in accumulators in hydraulic systems can cause actuators to move even though the system appears to be turned off and out of service. Liquids compress very little, so

Yates Industries, Inc.

Figure 3-25. Operators of fluid power systems, from simple presses to large complex machines, will have fewer accidents if they treat the equipment as if it is pressurized at all times.

fluid compressibility is usually not critical in hydraulic systems without accumulators. However, in hydraulic systems using large-volume cylinders, movement is possible. This is especially true if small quantities of air have become entrained in the system oil.

Pneumatic systems, of course, use compressed air for their operation. The systems should always be approached as if they are pressurized and ready for operation. Shutting off the system and cycling it to eliminate compression should be a standard step before repair work begins.

Compressed air is commonly used for cleaning parts and equipment, **Figure 3-26**. Blowguns or air nozzles attached to air hoses can be used to remove dust, dirt, and liquid. Extreme care needs to be taken when using these devices to prevent eye and skin damage from flying liquid and dirt particles produced by the airstream. When using these devices, workers should avoid near or direct contact between the nozzle tip and the skin. It is possible for the air to penetrate the skin, causing severe medical problems. Pressurized air must be treated as a potentially dangerous tool. It is used for cleanup in a shop situation, not as a toy for any type of horseplay.

Unprotected, direct physical contact with pinhole leaks in hydraulic systems should also be avoided. The velocity of the liquid may cause penetration problems in cuts and abrasions on the hands. The pressure of the liquid is not a factor, as the pressure of the escaping liquid drops to atmospheric pressure as soon as it leaves the system.

Goodheart-Willcox Publisher

Figure 3-26. Using compressed air for cleaning parts is a common but very dangerous practice. Use adequate eye protection and treat the air nozzle with the respect of a tool that can easily produce injuries.

3.3.2 Factors Regarding System Components

Fluid power specialists and the operators of equipment involving fluid power are working with controlled forces. These forces can reduce the effort of the operator and produce tremendous gains in production. They can also produce situations in which the safety of individuals is compromised and equipment is damaged. Understanding equipment operation and knowing the basic safety guidelines for the system involved will greatly reduce the chances for accidents. A specific equipment training program is beyond the scope of this book and usually considered the responsibility of the employer. However, this section of the chapter discusses hazards associated with different types of fluid power components. Procedures and safety rules are suggested to reduce or eliminate problems.

Actuators

Actuators are generally considered the most dangerous of the fluid power components because they involve considerable linear or rotary motion. Cylinders and motors are the two most common actuators in both hydraulic and pneumatic systems. Various other actuator designs are used in special circuit applications. The safety hazards associated with actuators are usually related to high-speed movement or high-force output. Generally, pneumatic systems produce the highest speeds, while hydraulic applications yield the greatest forces.

The safe operation of actuators in an application begins with the proper selection of the component, its correct installation in the system, and its protection from heat, corrosion, abrasive dirt, and shock as it functions. Each actuator has manufacturer specifications that should be matched to circuit needs before installation. These specifications include factors such as pressure rating, cylinder bore and stroke, and motor torque output. Recommended mounting methods must be followed for the component as it is installed in the system.

Once the system is in operation, regular inspections should be performed during its service life. These inspections should include checking for items such as:

- Excessive system pressure.
- Excessive system operating temperature.
- Internal shock caused by the cylinder piston striking the cylinder end caps at the end of the stroke.
- External damage to the actuator caused by other machine members striking the component.

Pumps and compressors

Pumps and compressors are generally considered the second most dangerous group of components in fluid power systems. As with actuators, this designation is based on the fact that movement is involved in the operation of the units. Another factor that contributes to this designation is the coupling device used to connect the unit to the system prime mover. Coupling devices can range from simple, direct-drive shaft couplings or belt drives to complex, variable-speed couplings found on larger, stationary compressors. Often, coupling devices are exposed, allowing a machine operator or maintenance person to make direct contact with rotating parts, which can cause serious injury.

Proper sizing, alignment, and guarding of the couplings are important to both the efficient and safe operation of pumps and compressors. Correctly sized and installed couplings will provide many hours of safe power-unit operation. Guarding for the coupling may be a shield specifically designed for just the coupling or belt drive. A guard may also be a larger enclosure that covers or restricts access to the whole unit, including the pump or compressor and the prime mover. Whatever the guarding method, it is essential that the guard keeps hands and clothing away from the moving parts of the coupling or drive unit.

The prime mover and pump or compressor need to be properly sized for the application if the system is to operate both efficiently and safely. Units that are too large waste energy, while units that are too small may fail because they are operating at inappropriate pressures or temperatures.

Be certain a *pressure safety valve* is built into both hydraulic and pneumatic systems close to the pump or compressor. This valve is placed in the system to open if the normal maximum-system-pressure control device fails. If the normal pressure control method fails, system pressure could reach an unsafe level. The prime mover could then stall or the pressure could damage system components and possibly injure machine operators or other workers in the area.

Air receivers

An air receiver is the device in a pneumatic system that stores an adequate volume of compressed air and ensures a steady supply of pulse-free air to the system. This device needs to be carefully sized and located in a system to promote effective system operation. The receiver must be large enough to supply the air demanded by the system while reducing air pulsations that affect tool operation. It also must be appropriately located in the system to reduce pressure drops in the distribution line. Both of these situations contribute to a safer operating system, as they contribute to an air supply of uniform pressure and volume.

The air receiver also needs to be equipped with a valve that allows the release of condensed water vapor from the tank. These valves are available in both automatic and manual designs. However, under all circumstances, a procedure should be in place to routinely check the amount of condensed water vapor in the tank and drain it. The operating efficiency and service life of circuits and tools are affected by the condensed moisture circulated in the system. Both of these factors contribute to a safer operating system.

Air receivers are commonly used as the location of the system safety valve, an overall system pressure gauge, and a manually operated valve that can be used to relieve the pressure in the system. All three of these devices allow an operator to better control the system and provide a clean, dry, and pulse-free air supply that contributes to a safer pneumatic system operating environment.

Pressure, flow, and directional control

Pressure, flow, and directional control are three basic areas of fluid power. A distinct group of components supplies the control aspect for each of these areas. Terminology varies slightly between the hydraulic and pneumatic fields, but the concepts are very similar. Understanding these key concept areas can have a definite influence on how the operation of fluid power components and circuits is approached and handled. Understanding a system can result in the system being handled more safely.

Pressure control valves are used in both hydraulic and pneumatic systems to control the force output of actuators and to block their movement until a desired system pressure is achieved. Flow control valves are used to control the speed of actuators. Actuator directional control is achieved by changing the fluid flow routes through directional control valves so fluid flows to the desired section of the actuator to cause the unit to move in the desired direction. When initially setting up a circuit, it is safest to set pressure and flow control valves at their lowest setting and directional valves at their center or normal position.

Most pressure and flow control valves have few moving parts and seldom malfunction. However, they must be properly mounted and initially adjusted to the correct pressure and flow rate. It is considered good practice to install tamper-resistant devices on pressure and flow control valves once the machine has been set up. Operators should be advised not to change valve settings as the valves have been set at their safe operating levels. Reduced pressure or actuator speed indicates a problem somewhere else in the circuit. Changing the valve setting will not solve the problem and may produce a major safety issue.

Considerable care should be taken when locating and setting up directional control valves or other machine controls. These are general, safety-related rules to follow when locating and installing controls:

- Machine operators should be able to easily reach the controls without leaving their normal workstation.

- Controls should not interfere with the operator's normal work space or work surface.

- Controls should be located away from all moving machine components.

- Controls must be designed to prevent the accidental activation of the machine.

- The control mechanisms should be protected against accidental damage.

- Easy access for maintenance should be part of the control design.

Accumulators

An accumulator stores pressure in a hydraulic system through the use of weights, compressed springs, or compressed gases. This pressure can be maintained even after the system has shut down. Circuits that contain accumulators should be clearly marked to help prevent accidents from unexpected actuator movement resulting from the stored energy.

Circuits that contain accumulators should provide a means to automatically unload the accumulators when the system is shut down. Several different designs are available to provide this important safety feature. Fluid power specialists and maintenance staff should always be alert to the possibility of unexpected actuator movement when working on a system, especially if the circuit includes an accumulator. Always be certain the accumulator is vented before beginning work.

Portable air tools

Portable air (air-operated) tools are common in society today. These tools perform a full range of activities, including drilling, grinding, sanding, chipping, nut driving, stapling, and painting, **Figure 3-27.** They are found in a wide variety of applications, ranging from service industry situations to heavy manufacturing operations.

Because these tools are portable, tool manufacturers usually use compact and lightweight construction to make them more appealing to users. These design characteristics often promote poor safety practices by workers. Workers must learn to respect the speed and power of these small tools, keep their hands away from rotary or impact-producing parts, and wear safety glasses. From a safety standpoint, operators of portable air tools need to be especially alert for other workers in the area so those individuals are not injured

A   *Atlas Copco*

B   *Goodheart-Willcox Publisher*

C   *Goodheart-Willcox Publisher*

Figure 3-27. Portable, air-powered tools are common. A—Air ratchet and sockets. B—Air impact wrench and sockets. C—Suction-type paint gun.

by the tools or flying material being removed in the work process.

Portable air tools require a substantial volume of pressurized air for operation. The safety factor involved in the storage and movement of this air is a consideration when selecting pneumatic tools. Flexible hose is the typical delivery method between the air source and the tool. In typical manufacturing and service industries, a fixed air compressor station with overhead and wall-mounted lines provide air delivery. However, many work situations involving portable tools do not fit this format. For example, in a roofing operation, the stapler or nail driver may be supplied with air from a small, engine-powered compressor that can be easily carried onto a roof. Whatever the situation, it is important from a safety perspective to keep air supply hoses off the floor and surfaces of the workstation.

3.4 Safety Requirements and Programs

The Occupational Safety and Health Administration (OSHA) grew out of a federal legislative act passed in 1970. That act has been modified through the years, but still serves as the basis for safety regulation in the United States. The act outlines the minimum on-the-job requirement for factors such as:

- Fire protection.
- Emergency medical treatment.
- Personal protective equipment.
- Accident record keeping.

The regulations that have been developed since the act was implemented have extended into many other areas and now include factors such as the reporting of accident-related data, accident investigation, medical treatment, and many other factors.

In order for a company or other group to meet the requirements of OSHA, it is essential that some form of safety program exist in the organization. That program needs to be directed toward accident prevention at all levels of operation. The program should involve cooperative leadership that includes both employees and the employer, the promotion of safe and healthful working conditions, and encouragement of employees to follow safe working practices. All three of these elements play an important role in successful accident prevention.

OSHA offers online safety training and certification opportunities that can help working professionals learn to protect themselves while on the job. The 10-hour General Industry course emphasizes hazard identification, avoidance, control, and prevention. The course covers a wide range of topics.

Fluid power specialists, because of their involvement in many areas of a manufacturing plant, probably have more opportunities to influence safety and safety programs than most employees. They should make every effort to follow safe working procedures, use appropriate safety equipment, suggest changes that could improve safety throughout the plant, complete accident reports, and provide all other information requested concerning accidents and safety improvement. With this type of support, a safety program has a much better chance of reducing accidents and improving working conditions.

Summary

- Important elements in the prevention of both accidents and cumulative injuries are good equipment, appropriate shop layout, and satisfactory working conditions.

- Proper air exchange and circulation in a shop facility will protect the general health of all employees and can reduce accidents through improved attention levels of workers.

- Appropriate fire extinguishers and fire blankets must be provided in all shop, training, and classroom areas.

- Unsafe acts of employees cause up to 85% of on-the-job accidents. Lack of job knowledge, including an awareness of job hazards, is the primary reason that an employee performs a job unsafely.

- Safety and health-related training should be provided to all employees who could be adversely affected by the jobs they perform.

- Protective gear is used to ensure the safety and health of employees who perform jobs or work in areas where total elimination of risk is difficult.

- A systematic program of preventive maintenance promotes safety by keeping machines, tools, and equipment safe and operating at their maximum efficiency.

- Approach fluid power systems as if they are pressurized and ready for operation. Pressurized fluids can cause unexpected actuator movement even when the system is shut down.

- Actuators are generally considered the most dangerous of the fluid power components because of the linear and rotary motion they produce.

- Proper sizing and correct installation of components not only will improve the safety performance of a fluid power circuit, but also will increase system efficiency.

- When setting up a fluid power circuit, it is safest to initially set pressure and flow control valves at their lowest settings and directional controls in their normal or centered position.

- Circuits that contain accumulators should be designed to automatically unload system pressure during machine shut down to eliminate the danger of unexpected actuator movement caused by the release of fluid stored in a charged accumulator.

- OSHA regulates minimum requirements for industries in areas such as fire protection, emergency medical treatment, the use of personal protective equipment, and accident record keeping and reporting.

Internet Resources

The following are some useful resources available on the Internet. Enter a company or organization name into a search engine to access its website. Explore the various areas of the sites to discover useful fluid power resources.

Centers for Disease Control and Prevention. Presents various topics and resources on workplace safety and health, including chemical safety, eye safety, machine safety, and protective clothing. Search for "Workplace Safety and Health."

National Safety Council. A nonprofit, nongovernmental public service membership organization that provides safety and health information and training programs.

Occupational Safety and Health Administration (OSHA). OSHA provides extensive information on workplace health and safety, as well as training materials and other services. Search for hydraulics and pneumatics factors. OSHA is a part of the United States Department of Labor.

Wisconsin Department of Workforce Development. An example of state resources related to workplace safety. Search the site for "Hydraulics and Pneumatics Safety" to locate a variety of materials for this state.

Chapter Review

Answer the following questions using information in this chapter.

1. Which of the following workplace elements is important in the prevention of accidents and cumulative injuries?
   A. Good equipment.
   B. Shop layout.
   C. Working conditions.
   D. All of the above.

2. *True or False?* The amount of traffic, number of collisions with equipment, and number of falls can be reduced by proper equipment placement and adequate traffic lanes.

3. It is critical that spilled oil be quickly cleaned up in a shop area in order to reduce the _____.
   A. possibility of accidents caused by falls on slippery floors
   B. accumulation of dirt and debris in the pooled oil
   C. fire hazard of oil-soaked materials
   D. All of the above.

4. *True or False?* Unsafe working conditions are the chief cause of industrial accidents.

5. Which of the following is a factor that employees must know if they are to safely perform a job?
   A. Steps necessary to perform the job.
   B. Personal protective equipment required.
   C. Hazards associated with the job.
   D. All of the above.

6. *True or False?* It is the responsibility of the employee to learn and follow the safety procedures needed to perform a job safely.

7. The first step in any training is to define the _____ of the training.
   A. objective
   B. responsibilities
   C. cost
   D. All of the above.

8. The most popular hearing-protection devices are _____, which are both inexpensive and inconspicuous to wear.
   A. earmuffs
   B. helmets
   C. earplugs
   D. hard hats

9. Before requiring an employee to wear respiratory protection, an employer should make every effort to reduce _____ to a safe level in the work area.
   A. air contaminants
   B. temperatures
   C. humidity
   D. All of the above.

10. Which of the following is *not* a general factor contributing to accidents and injuries in an industrial setting?
   A. Inattentive machine operators.
   B. Employees properly trained on procedures.
   C. Inappropriate equipment or processes.
   D. Defective or poorly designed equipment.

11. When hazard points cannot be eliminated from the design of a machine, _____ must be installed to reduce or eliminate the danger. guards

12. When maintenance staff is working on equipment, _____ devices are commonly used so the machine cannot be accidentally turned on.
   A. sound-reducing
   B. lockout
   C. ventilating
   D. illuminating

13. To reduce the possibility of back injuries when lifting, position the legs and back comfortably, while using the _____ to do the lifting.

14. A major difference between hydraulic and pneumatic systems that is important to safety is the _____ characteristic of gases in pneumatic systems.
   A. inflammability
   B. penetrating
   C. cleansing
   D. compressibility

15. *True or False?* When working with a blowgun for cleanup, compressed air should be considered a potentially dangerous tool. The air can penetrate skin and cause severe medical problems.

16. A factor that is critical to the safe installation and operation of an actuator is _____.
   A. proper selection of the component
   B. correct installation in the system
   C. protection from heat, abrasive dirt, and shock as it functions
   D. All of the above.

17. Excessive system pressure in a fluid power system can be avoided by having a(n) _____ valve built into the system close to the location of the pump or compressor.

18. It is considered good practice to install _____ devices on pressure and flow control valves to prevent operators from resetting the system to unsafe pressure and flow levels.
   A. tamper-resistant
   B. self-correcting
   C. externally controlled
   D. remote electronic

19. Machine operators and maintenance workers must always be alert to the possibility of _____ in the system when an accumulator is included in the circuit.
   A. fluid
   B. pressure
   C. contaminants
   D. heat

20. A consideration when selecting pneumatic tools is the safety factor involved in the storage and movement of the _____ air. pressure

Apply and Analyze

1. You are setting up a basic hydraulic power system in an industry setting. The system will include cylinders and motors as actuators and will operate at high pressures and speeds. Describe the primary safety concerns you will take into account in your system design and operation.

2. Compare and contrast the safety issues associated with a pneumatic power system to those of a hydraulic system that performs a similar function.

Research and Development

1. Leaks in hydraulic power systems can be dangerous not only to the operator, but to the environment. Research common causes of hydraulic leaks and their solutions.

2. Visit an industrial site that uses fluid power in its operations. Tour the facility noting areas of operation or equipment that pose a risk to employees. Interview an operations manager or employee about the safety regulations and training that the site uses to minimize this risk. Based on your understanding of the operation, evaluate the strengths and weaknesses of the facility's safety program. Report your findings to the class.

3. Pneumatic powered hand tools are common today. What OSHA regulations and/or requirements address the risks associated with the use of these tools? Using the OSHA regulations/requirements as a guide, prepare a list of safety rules you might distribute to workers using these tools at a construction, mining, or other site.

4. Hydraulic fluid injection injuries are extremely dangerous. Research how these injuries happen, the health risks associated, and operating procedures that can reduce these incidents.

Joe Rossi, Saint Paul Pioneer Press

Casting can be dangerous and requires special safety equipment. Notice all of the different safety equipment used by these workers.

4

Basic Physical Principles
Applications to Fluid Power Systems

luid power is used in a wide variety of devices to aid in the performance of work. These devices range from very simple tools to complex machines. However, the operation of each device is based on fundamental scientific concepts. This chapter discusses these concepts in order for you to develop a general understanding of their application to fluid power systems and the tools and machines that are operated.

Learning Objectives

After completing this chapter, you will be able to:

- Identify and explain the design and operation of the six basic machines.
- Describe the factors that affect energy in fluid power systems.
- Explain how the potential power of a fluid power system is established.
- Describe the nature of heat and how it is measured in fluid power systems.
- Explain the nature of electricity and the basic methods used for electrical measurement.
- Identify and explain the operation of basic electrical circuits.
- Name and describe the characteristics associated with the fluids used in fluid power systems.

Key Terms

absolute pressure	Gay-Lussac's law	power
absolute zero	general gas law	pressure
alternating current (AC)	head	proton
atmosphere	heat	pulley
atom	horsepower	radiation
Boyle's law	ideal gas laws	resistance
Btu	inclined plane	screw
buoyancy	inertia	second law of thermodynamics
Charles' law	kinetic energy	sensible heat
circuit	latent heat	specific gravity
conduction	lever	specific heat
convection	magnetic pole	specific weight
current	magnetism	standard atmospheric
degrees	mass	pressure
direct current (DC)	mechanical advantage	temperature
electron	mechanical efficiency	thermodynamics
energy	molecule	torque
first law of thermodynamics	neutron	vacuum
flux	nucleus	velocity
force	Ohm's law	wedge
friction	Pascal's law	weight
fulcrum	potential	wheel and axle
gauge pressure	potential energy	work

4.1 Mechanics

Fluid power systems are used in a wide variety of equipment to do work. Many different designs are used in the construction of this equipment. These designs make use of the basic elements of mechanics, including the principles of:

- Simple machines.
- Energy transfer.
- Force.
- Power.

An understanding of these principles is required to fully comprehend the operation of hydraulic and pneumatic systems used in industry today.

4.2 Simple Machines

The number of different designs of complex machines used throughout the world today is unknown. It would be safe to estimate the number is in the millions. While there are many different designs of machines used, only six basic types exist, **Figure 4-1**. The basic machine types include the lever, inclined plane, wedge, screw, pulley, and wheel and axle. Each of these simple machines serves a specific purpose that cannot be duplicated by any of the others. The simple machines provide mechanical advantages that allow difficult tasks to be completed with a minimum of effort. They also provide for a gain in movement in exchange for a reduction in force.

4.2.1 Lever

The *lever* is constructed of a simple, rigid bar that pivots on a fixed axis called a *fulcrum*. It is one of the most commonly used machines. The parts of a lever include an effort arm, load arm, and fulcrum. The lever arm extends from the fulcrum to the point at which force is applied. The load arm extends from the fulcrum to the location of the load. The material from which a lever is constructed may vary considerably as it must only be solid and rigid enough to withstand the forces applied. See **Figure 4-2**.

Levers are divided into three classes determined by the placement of the fulcrum in relation to the effort and load arms. See **Figure 4-3**.

- First-class levers.
- Second-class levers.
- Third-class levers.

AGCO Corporation

Figure 4-1. Although fluid power equipment may be complex in its design, several simple machines such as the lever, screw, or wheel and axle are usually easily identifiable in its structure.

A first-class lever, **Figure 4-4**, is one in which the fulcrum is located between the effort and load arms. A common pair of pliers is a good example of this class of lever. A second-class lever has the load arm between the effort arm and the fulcrum. A common example of this lever is a nutcracker. A third-class lever is one in which the effort arm is placed between the load arm and the fulcrum. Salad tongs are a good illustration of a third-class lever.

4.2.2 Inclined Plane

The *inclined plane* is a slope used to increase the distance an object is moved while it is being raised or lowered by rolling or sliding. Loading ramps used to roll barrels up gradual inclines are excellent examples of incline planes. A ramp for handicapped individuals allowing ready access to different levels that normally require an elevator is another example of an inclined

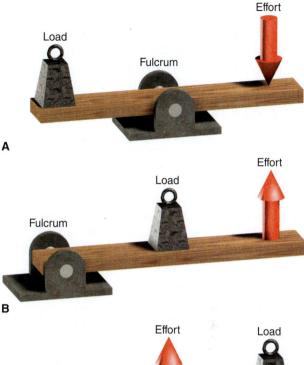

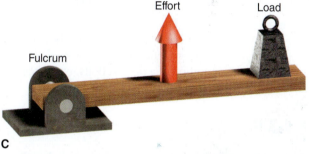

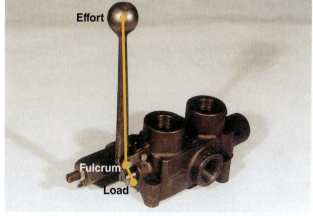

A

B

C

Goodheart-Willcox Publisher

Figure 4-3. Levers are grouped in three classes, depending on the location of the load and effort arms in relation to the fulcrum. A—First-class lever. B—Second-class lever. C—Third-class lever.

Goodheart-Willcox Publisher

Figure 4-2. Levers are often used in conjunction with fluid power cylinders to obtain the movement required by the machine designer.

Goodheart-Willcox Publisher

Figure 4-4. The application of simple machines in fluid power equipment may be as simple as the control handle on a manually operated directional control valve or as massive as the steering linkages on large earthmoving equipment. This directional control valve with manual actuator is an example of a first-class lever.

plane, **Figure 4-5A**. In each of these examples, the total effort to make the change in height is not changed by the incline plane. However, a mechanical advantage is obtained as the maximum effort needed at any given time is reduced due to the distance traveled on the incline.

4.2.3 Wedge

The *wedge* is made up of two inclined planes that share a common base. See **Figure 4-5B**. Wedges are generally considered small tools used against great resistance to separate objects. Splitting blocks of firewood using a wedge and sledgehammer is an example of an application of this simple machine. Other examples include axes, chisels, and other cutting devices, as well as nails that use a wedge shape to penetrate boards and structural timbers.

4.2.4 Screw

The screw is a form of the inclined plane. In a *screw*, the inclined plane is wrapped around a rod in a continuous spiral to form threads. The gradual slope of the threads provides a very high mechanical advantage. The principle of the screw is, therefore, often used when high force is required in the operation of a device. Bolts, wood screws, jar lids, and worm gears are examples of the screw.

4.2.5 Pulley

The *pulley* allows the direction of a force to be changed. When a single, fixed pulley is used, neither speed nor intensity of the force is changed. The distance over which the effort is applied is the same distance the load is moved. A single, movable pulley provides a mechanical advantage as the distance over which the effort is applied is twice the distance that the load moves. Multiple pulley sets are often used to provide an increased mechanical advantage when a high force is required to move or lift an object. Construction cranes make use of pulleys.

4.2.6 Wheel and Axle

The *wheel and axle* can be used to increase force or movement. The device is constructed of a wheel attached to an axle. This device is actually a form of a second- or third-class lever, depending on where the effort is applied. The fulcrum is the common center of the wheel and axle. The second-class lever principle applies when force to the wheel increases the force at the axle. The third-class lever principle applies when force is applied to the axle. The force at the surface of the wheel decreases, but the surface movement increases. The steering wheel of an automobile and the pulley drives on engine accessories are examples of using the wheel and axle to increase force. The drive wheels of a vehicle may be used to illustrate how movement is increased. Force applied to the axle turns the wheels, producing increased movement, but with a reduced force. Machines also use gears to apply the wheel and axle principle when increased force or speed is desired.

4.3 Basic Principles of Mechanics

Principles of mechanics are used extensively in machines that apply hydraulics and pneumatics in their operation. An understanding of these principles is essential to the successful conversion, transmission, and control of energy in fluid power systems. This section discusses the principles of energy, inertia, friction, and force.

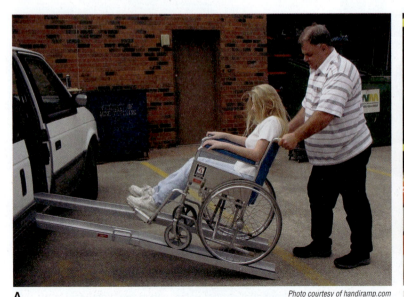

A — Photo courtesy of handiramp.com B — Goodheart-Willcox Publisher

Figure 4-5. A wedge consists of two inclined planes that share a common base. A—An inclined plane. B—A wedge.

4.3.1 Energy

Energy is defined as the capacity to do work. Energy is classified as potential energy or kinetic energy. The operation of fluid power systems involves both types.

Energy stored in a form that allows its release and use is *potential energy*. It is found in natural fuels such as crude oil and coal, **Figure 4-6**. Man-made devices, such as springs and hydraulic accumulators, also store potential energy. In these devices, the potential energy is created by work that has been performed on the equipment, **Figure 4-7**.

Energy in motion is *kinetic energy*. Its source is potential energy that has been released. Potential energy becomes kinetic energy as a tightly wound spring unwinds, a hydraulic accumulator discharges, or gasoline is burned in an engine.

Potential and kinetic energy exist in many forms. Fluid power systems are affected by a number of these forms, including electrical, heat, and mechanical. Other forms of potential and kinetic energy also exist, those related to chemicals, light, and sound. These forms may be included in specialized, accessory fluid power systems.

Energy cannot be created or destroyed, but it may be converted from one form to another. The wise use of energy is essential. Wasted energy cannot be salvaged and is lost as heat to the atmosphere. Careful planning of energy use is essential for effective fluid power system operation. Ineffective energy use increases the cost of system operation and is detrimental to the environment.

MTS Systems Corporation

Figure 4-7. A hydraulic accumulator is used to store potential energy. Energy accumulates in the device as it is pressurized, is stored under system operating pressure, and is then released in the form of kinetic energy during discharge of the unit.

A

Martin D. Vonka/Shutterstock.com

B

Deere & Company

Figure 4-6. The potential energy in consumable fuels is converted to several forms before it generates fluid flow and pressure in a hydraulic or pneumatic system. A—Coal is burned in electric-generating plants. B—Gasoline is consumed by internal combustion engines.

4.3.2 Inertia

Inertia is the tendency of a body to remain at rest or, if in motion, to maintain its speed and direction. An outside force must be applied to initiate the movement of an object or to change its movement. Designers must be aware of inertia so they can design equipment that provides appropriate power for start-up, braking, and directional control for effective and safe machine operation.

4.3.3 Friction

Friction is the resistance to the sliding movement of one or both of two surfaces in contact with each other. Friction is found, to some degree, in all machines. It converts kinetic energy to heat, which reduces the efficiency of the machine. Dry surfaces usually produce greater friction than wet surfaces. The careful lubrication of bearing surfaces reduces friction and increases the efficiency of equipment.

High levels of friction may be desired in specialized machine systems. A brake system is an example of a system that uses friction to perform work. Care must be taken in these systems to remove the generated heat to prevent damage to component parts.

4.3.4 Force

Force is defined as the effort that produces, changes, or stops the motion of a body. Force may be considered the effort needed to complete a task, but it may be applied without causing movement of an object. The force exerted to complete a task must overcome an opposing force that is produced by gravity, friction, or inertia, **Figure 4-8**. For example, lifting a weight involves an effort equal to the force of gravity on that weight. The start-up of a machine requires a force equal to the resistance of friction and the inertia of all the parts in the device.

4.4 Mechanical Measurements

Fluid power systems involve a number of basic principles of mechanics that include measurements and the units used to express those measurements. These principles include factors relating to:

- Force.
- Movement.
- Efficiency.

An understanding of these concepts and the measurements involved is essential to anyone who works with either hydraulics or pneumatics.

4.4.1 Pressure

Pressure is force exerted over a unit area. All materials apply pressure to the objects on which they rest. The weight, velocity, and temperature of materials effect the pressure they exert. Pressure is determined by dividing the force exerted by the area over which it is applied:

$$\text{Pressure} = \frac{\text{Force}}{\text{Area}}$$

A common unit of measure for pressure is pounds per square inch (psi). For example, a box weighing 1,000 pounds that rests on a $10'' \times 1''$ surface exerts a pressure expressed as 10 pounds per square inch, or 10 psi. See **Figure 4-9**.

Goodheart-Willcox Publisher

Figure 4-8. Cylinders are the devices commonly used to develop force in fluid power systems. The pressurized fluid in hydraulic and pneumatic systems acts on the face of the cylinder piston to generate the force needed to overcome a resistance and move a load.

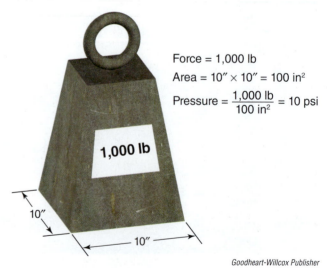

Force = 1,000 lb

Area = $10'' \times 10'' = 100 \text{ in}^2$

Pressure = $\dfrac{1,000 \text{ lb}}{100 \text{ in}^2} = 10$ psi

1,000 lb

10"

10"

Goodheart-Willcox Publisher

Figure 4-9. Pressure is defined as force divided by the area over which it is applied. It is expressed in units such as pounds per square inch (psi).

Pressure may also be stated using several other systems of measurement. Atmospheres and head are commonly used in the fluid power industry to indicate pressure. Both of these units of measurement indicate the pressure over an area created by the weight of a fluid.

4.4.2 Torque

Torque is defined as a turning or twisting force applied to a shaft, **Figure 4-10**. It is basically a rotary force that does not require motion. Torque is measured as the force applied to the radius of a circle centered on the shaft. The distance from the center of the shaft to the point where the turning force is applied is called the torque arm. Torque is calculated by multiplying the applied force by the length of the torque arm:

Torque = Force × Length of Torque Arm

The unit of measure for torque is the foot-pound or inch-pound. For example, a force of 10 pounds applied to a torque arm 1 foot in length produces a torque of 10 pound-feet or 120 pound-inches. Hydraulic motors apply torque as they perform work, **Figure 4-11**.

> **NOTE**
>
> Torque is commonly stated in foot-pounds or inch-pounds. However, you may also find it stated in pound-feet or pound-inches. There is no difference between the designations. You may also find torque expressed in the metric system as Newton-meters.

4.4.3 Work

Work is the application of force through a distance. Work is produced only when the force that is applied moves an object or device or causes a physical change in an object. It is measured by multiplying the force applied by the distance through which it is applied:

Work = Force × Distance

The unit of measure for work is the foot-pound. For example, a force of 100 pounds applied over a distance of 10 feet produces 1,000 foot-pounds of work.

4.4.4 Power

Power is the rate at which work is performed or energy is expended. Power output is calculated by dividing the amount of work completed by the time needed for its completion:

$$\text{Power} = \frac{\text{Force} \times \text{Distance}}{\text{Time}}$$

The unit of measure for power is foot-pounds per minute. For example, moving an object requiring a force of 100 pounds a distance of 10 feet in 1 minute produces 1,000 foot-pounds per minute of power.

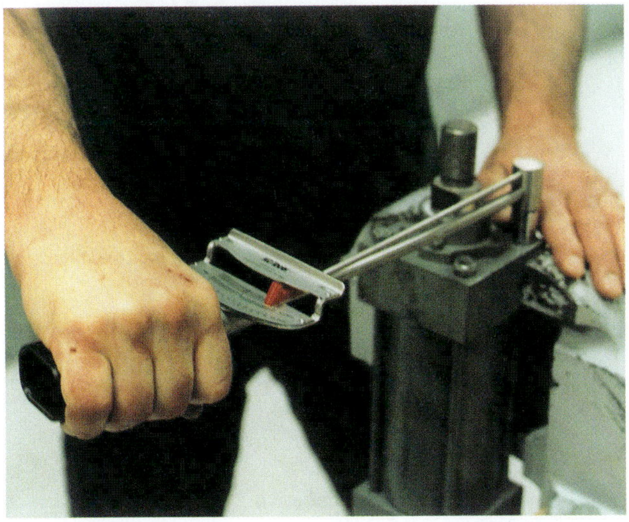

Goodheart-Willcox Publisher

Figure 4-10. A torque wrench allows bolts to be uniformly tightened by measuring the turning force applied. The length of the wrench serves as the torque arm and the scale on the wrench indicates the torque in pound-feet or pound-inches.

Goodheart-Willcox Publisher

Figure 4-11. Motors are the devices commonly used in fluid power systems to develop torque. Pressurized fluid acting on internal parts applies a turning force to the output shaft of the motor to rotate machine parts.

EXAMPLE 4-1

Work and Power

You must raise a 1,000 lb load a distance of 4'. How much work will be done?

$$\text{Work} = \text{Force} \times \text{Distance}$$

$$= 1{,}000 \text{ lb} \times 4'$$

$$= 4{,}000 \text{ ft-lb}$$

If the 1,000 lb load is raised 4' in 30 seconds, what is the power output in ft-lb per minute?

$$\text{Power} = \frac{\text{Force} \times \text{Distance}}{\text{Time}}$$

$$= \frac{1{,}000 \text{ lb} \times 4'}{0.5 \text{ min}}$$

$$= \frac{4{,}000 \text{ ft-lb}}{0.5 \text{ min}}$$

$$= 8{,}000 \text{ ft-lb/min}$$

4.4.5 Horsepower

Horsepower is a commonly used unit of work. The term is historically credited to James Watt. It was originally used to compare the work output of Watt's steam engine to horses, which were a primary source of power at the time. One horsepower is equal to 33,000 ft-lb of work per minute. The horsepower output of a device is calculated by dividing its power output in foot-pounds per minute by 33,000:

$$\text{Horsepower} = \frac{\text{Power}}{33{,}000}$$

The calculations for determining the horsepower requirements of prime movers and the output horsepower of fluid power systems are discussed later in this text.

4.4.6 Velocity

Velocity is defined as the distance traveled in a basic unit of time. It is calculated by dividing the distance traveled by the total units of time required for the travel:

$$\text{Velocity} = \frac{\text{Distance}}{\text{Time}}$$

The unit of measure for velocity commonly used in the fluid power field is inches per second, although inches per minute, feet per second, or feet per minute may also be used. For example, a cylinder that requires 10 seconds to extend 10" is operating at a velocity of one inch per second, or 1 in/sec.

4.4.7 Mechanical Advantage

Mechanical advantage is an expression of the relationship between the effort expended and the resistance that is overcome. Humans use machines to develop mechanical advantages in order to complete tasks that require forces greater than the direct application of their strength will produce. See **Figure 4-12**. Mechanical advantage may be calculated using either the forces or the distances involved:

$$\text{Mechanical Advantage} = \frac{\text{Output Force}}{\text{Input Force}}$$

$$\text{Mechanical Advantage} = \frac{\text{Effort Distance}}{\text{Resistance Distance}}$$

Both formulas provide the same result. Mechanical advantage is expressed as a numeric ratio without units of measure.

For example, a device is used to move a 500-pound weight a distance of one foot. An effort of 100 pounds applied over a five-foot distance is required to produce that movement. This device functions with a mechanical advantage of 5 to 1. Either the forces (weight and effort) or the distances involved can be used for calculation:

$$\frac{500 \text{ lb}}{100 \text{ lb}} = 5{:}1$$

$$\frac{5 \text{ ft}}{1 \text{ ft}} = 5{:}1$$

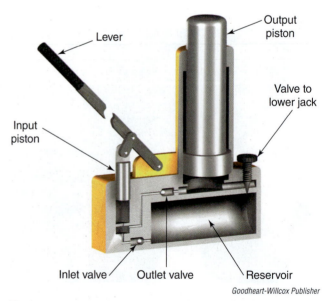

Lever

Output piston

Valve to lower jack

Input piston

Inlet valve   Outlet valve   Reservoir

Goodheart-Willcox Publisher

Figure 4-12. A hydraulic jack develops a high mechanical advantage through the use of both the area differences of the input and output pistons and the advantage gained in the first-class lever (the handle) operating the input piston.

4.4.8 Mechanical Efficiency

Mechanical efficiency is a comparison of the work input to the work output of a machine. Appropriate design and proper lubrication can increase the efficiency of a machine. Mechanical efficiency is calculated by dividing the work output of a machine by the work input:

$$\text{Mechanical Efficiency} = \frac{\text{Work Output}}{\text{Work Input}} \times 100$$

It is expressed as a percentage and is always less than 100% because of energy losses caused by friction.

4.5 Principles of Heat Transfer

The operation of a fluid power system involves the transfer and conversion of energy. Any movement of a solid, liquid, or gas produces heat, which is a form of kinetic energy. Excessive heat results in a waste of energy and a reduction in the efficiency of the system. Understanding and controlling heat in fluid power systems is desirable as heat is a form of energy not totally usable in the systems.

4.5.1 Thermodynamics

Thermodynamics is a science concerned with relationships between the properties of matter, especially those affected by temperature, and the conversion of energy from one form to another. Many systems for generating or transmitting power are influenced by the thermodynamic behavior of the fluids involved in their operation, **Figure 4-13**. Both hydraulic and pneumatic fluid power systems are affected by these principles.

The principle of the conservation of energy states:

> Energy can be transformed from one type to another, but can be neither created nor destroyed.

According to this principle, it is impossible to continuously extract energy from a source without replenishment. A good example of this principle is a battery, which must be recharged after it has been operating an electrical device. This basic principle is also known as the *first law of thermodynamics*.

EXAMPLE 4-2

Mechanical Measurements

A 1,000 lb load is lifted a distance of 4′ in 30 seconds at a constant rate. What is the velocity in units of inches per second?

Start by converting the distance in feet to inches.

$$\text{Distance (in)} = \frac{4 \text{ ft} \times 12 \text{ in}}{1 \text{ ft}}$$

$$= 48 \text{ in}$$

$$\text{Velocity} = \frac{\text{Distance (in)}}{\text{Time (sec)}}$$

$$\text{Velocity} = \frac{48 \text{ in}}{30 \text{ sec}}$$

$$= 1.6 \text{ in/sec}$$

If you are only able to apply a force of 250 lb to raise the 1,000 lb load, what mechanical advantage would you need to move the load?

$$\text{Mechanical Advantage} = \frac{\text{Output Force}}{\text{Input Force}}$$

$$= \frac{1,000 \text{ lb}}{250 \text{ lb}}$$

$$= 4:1$$

If you plan to raise the 1,000 lb load with the use of 4″ hydraulic piston, what pressure (in psi) must be developed in the fluid?

$$\text{Piston Area} = \pi \times \text{radius}^2$$

$$= \pi \times (2 \text{ in})^2$$

$$= \pi \times 4 \text{ in}^2$$

$$= 12.6 \text{ in}^2$$

$$\text{Pressure} = \frac{\text{Force}}{\text{Area}}$$

$$= \frac{1,000 \text{ lb}}{12.6 \text{ in}^2}$$

$$= 79.4 \text{ lb/in}^2$$

$$= 79.4 \text{ psi}$$

If the machine used to raise the 1,000 lb load is 80% efficient, how much work is actually done to lift the load 4′?

$$\text{Work Output} = \text{Force} \times \text{Distance}$$

$$= 1,000 \text{ lb} \times 4′$$

$$= 4,000 \text{ ft-lb}$$

$$\text{Work Input} = \frac{\text{Work Output} \times 100}{\text{Mechanical Efficiency}}$$

$$= \frac{4,000 \text{ ft-lb} \times 100}{80}$$

$$= 5,000 \text{ ft-lb}$$

Figure 4-13. All devices that involve energy transfer follow the laws of thermodynamics in their operation. Heating water for making coffee, as well as a heat exchanger in a complex fluid power system, must follow these basic laws.

Another principle critical to thermodynamics is less precise. It is known as the *second law of thermodynamics*. This law states:

Heat only flows between two bodies when one body has a higher temperature than the other, with heat moving from the warmer to the colder body.

This law is applied to the movement of heat throughout fluid power systems, but especially to heat exchangers used to maintain proper operating temperatures in the systems.

4.5.2 Heat

Heat is energy in transit due to a difference in temperature between two areas. It is kinetic energy. Heat is indicated by rapid molecular movement in a substance and is present in all materials that contain energy. Molecular movement may be slow and orderly, as in a substance at room temperature, or it may be forceful, as in a substance heated to a high temperature.

Fluid power systems generate considerable heat during operation. Heat is developed in these systems whenever a pressure drop that does not produce work occurs. Heat is generated by pressure drops caused by resistance to flow in pipes, hoses, and fittings; flow over relief valves and through flow control valves; and leakage in pumps and motors. See **Figure 4-14**.

Heat energy is measured in British thermal units (Btu) or calories. One *Btu* is the heat energy required to raise one pound of water one degree Fahrenheit at atmospheric pressure, **Figure 4-15**. One calorie is the heat energy required to raise one gram of water one

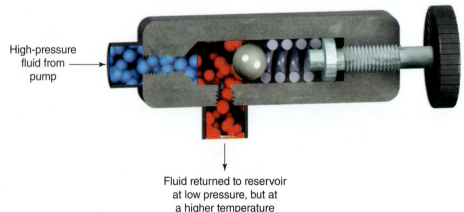

High-pressure fluid from pump →

↓
Fluid returned to reservoir at low pressure, but at a higher temperature

Figure 4-14. The relief valve in a hydraulic system generates considerable heat in the system. The energy in the fluid returned to the reservoir through the valve is not doing useful work and is converted to heat as it passes through the valve.

1 lb water at 70°F

+

=

1 Btu

1 lb water at 71°F

Figure 4-15. One Btu of heat is required to raise the temperature of one pound of water one degree Fahrenheit.

degree Celsius at atmospheric pressure. These values are used in calculating energy losses in systems and the sizing of heat exchangers to maintain proper system operating temperature.

The ability of a substance to absorb heat is known as *specific heat*. It is expressed as a number that compares the quantity of heat required to raise the temperature of one-unit weight of the substance one degree in temperature to the heat needed to raise an equal weight of water one degree Celsius.

4.5.3 Temperature

Temperature is the measure of hotness or coldness of a substance. Temperature is determined by the rate of molecular movement within the substance. The more rapid the molecule movement, the higher the temperature and the greater the kinetic energy.

Temperature is indicated by units called *degrees* and is measured using a thermometer or electronic device. Several degree scales are used to indicate temperature. The Fahrenheit scale is commonly used in the United States. The Celsius scale is used in the remainder of the world. The freezing point of pure water at atmospheric pressure is 32° on the Fahrenheit scale and the boiling point is 212°. The Celsius scale indicates the freezing point at 0° and the boiling point at 100°. See **Figure 4-16**. 1° on the Fahrenheit scale is equal to 5/9° on the Celsius scale.

Fahrenheit temperatures can be converted to the Celsius scale using the formula:

$$^{\circ}C = \frac{5}{9}\,(^{\circ}F - 32^{\circ})$$

Celsius temperatures may be converted to the Fahrenheit scale using the formula:

$$^{\circ}F = \frac{9}{5}\,^{\circ}C + 32^{\circ}$$

For example, converting the 212°F boiling point of water to degrees in Celsius is done as such:

$$^{\circ}C = \frac{5}{9}\,(212^{\circ}F - 32^{\circ})$$

$$^{\circ}C = 100^{\circ}C$$

Completing the calculations produces 100°C, which is the boiling point of water on the Celsius scale.

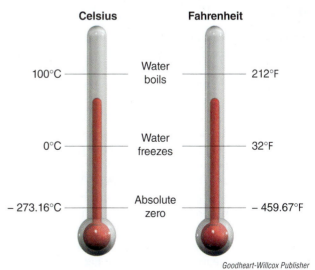

Celsius   **Fahrenheit**

100°C — Water boils — 212°F

0°C — Water freezes — 32°F

−273.16°C — Absolute zero — −459.67°F

Figure 4-16. Comparison of the Celsius and Fahrenheit temperature scales.

EXAMPLE 4-3

Specific Heat and Temperature Change

Heat is generated in fluid power systems whenever fluids move from high to low pressure without producing any mechanical work. If the fluid absorbs this heat, its temperature can rise, affecting system operation. Calculate the temperature increase expected in 1 lb of a petroleum-based fluid that has a specific heat of 0.5 Btu/(°F·lb) assuming that 10 Btus of heat are absorbed by the liquid. State the answer in both degrees Fahrenheit and Celsius.

$$\text{Temperature Change} = \frac{\text{Heat Added}}{\text{Specific Heat of the Fluid}}$$

$$= \frac{10 \text{ Btu}}{0.5 \text{ Btu/(°F·lb)}}$$

$$= 20°F$$

To convert this temperature change in °F to °C, you must use the conversion factor 1°C = 5/9°F.

$$°C = 5/9(20°F)$$

$$= 11°C$$

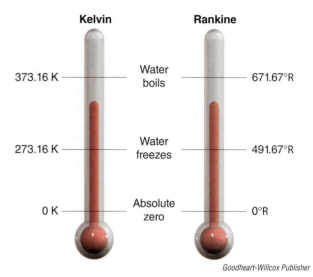

Goodheart-Willcox Publisher

Figure 4-17. Comparison of the Kelvin and Rankine absolute temperature scales.

Absolute temperature systems are also used in calculations in the scientific and engineering fields. These scales use absolute zero as their base. *Absolute zero* is the point where it is thought no molecular movement occurs in a substance, which indicates complete absence of heat. Two absolute temperature systems are used. See **Figure 4-17**. The Rankine system uses Fahrenheit degrees and the Kelvin system uses Celsius degrees. The freezing point of water on the Rankine scale is 491.67°R (459.67° + 32°) and 273.16 K on the Kelvin scale.

4.5.4 Forms of Heat

Two forms of heat are encountered in the operation of fluid power systems. The most common is sensible heat. It is also the form we are most aware of in our daily life. *Sensible heat* can be measured with a thermometer and is encountered when a room is too hot or too cold or when a substance feels cold, warm, or hot when it is touched, **Figure 4-18**. Changes in the application of heat energy in these situations results in changes in temperature, but does not usually change other factors.

A second form of heat is *latent heat*. This is heat that changes the state of a substance without changing its temperature. A common example of the application of latent heat is when water freezes or turns to steam. Both of these conditions are changes of state and involve latent heat without a change in temperature. See **Figure 4-19**. When the temperature of a substance reaches the point at which a change of state occurs, the temperature remains constant even though additional quantities of heat are being added or subtracted from the substance.

Goodheart-Willcox Publisher

Figure 4-18. Sensible heat can be measured with a thermometer and sensed by people as hot or cold.

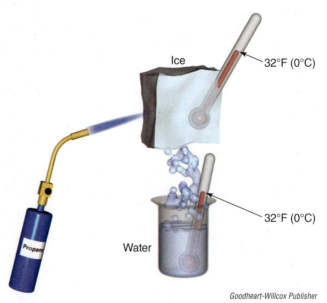

Goodheart-Willcox Publisher

Figure 4-19. Latent heat is the heat that changes the state of a substance without changing its temperature.

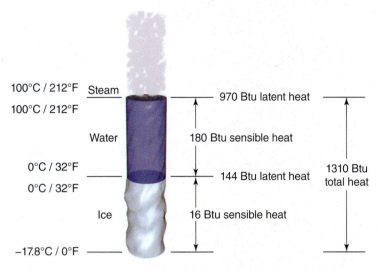

Figure 4-20. The amount of heat that must be added to or subtracted from water to change its state from ice (solid) to liquid to steam (gas).

For example, 970 Btus of heat must be *added* to one pound of water at 212°F to produce one pound of steam at 212°F. See **Figure 4-20**. The quantity of heat required to produce this type of change (liquid to vapor) is known as the latent heat of vaporization. The same amount of heat must be *removed* to change steam to water.

To produce one pound of ice at 32°F, 144 Btus of heat must be *removed* from 1 lb of water at 32°F. The hidden heat that produces this change (liquid to solid) is known as the latent heat of fusion. The same amount of heat must be *added* to convert ice to water.

4.5.5 Modes of Heat Transfer

There are three modes by which heat is transferred: conduction, convection, and radiation. Each mode provides a distinct means by which heat can flow from a warmer to a cooler substance. More than one mode may be used at a given time to produce a heat gain or loss. Fluid power systems employ all three transfer methods to maintain desired operating temperatures.

Conduction

The transfer of heat using molecular activity in solids is called *conduction*. A metal bar heated at one end will gradually increase in temperature at the other end as the bar transfers heat by conduction. See **Figure 4-21**. This transfer occurs as molecule movement on the heated end stimulates increased molecular activity along the bar. Molecules do not travel along the bar, but heat is gradually transferred from one portion of the substance to another as molecular activity increases in the metal and the bar warms up.

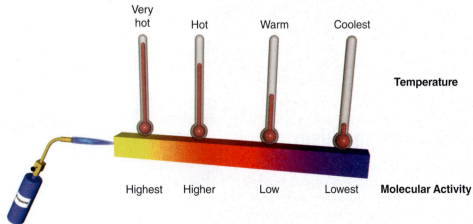

Figure 4-21. Heat transfer through solid objects is called conduction and involves increased molecular activity along the route of the heat movement.

Substances conduct heat at different rates. Metals such as copper are good conductors. Liquids and gases conduct heat to some degree, but are not considered conductors. Substances that do not readily transfer heat are considered heat insulators. Substances that contain microscopic chambers filled with a gas, such as foam plastics, are generally good heat insulators. Designers must be aware of the heat transfer characteristics of materials in order to produce component parts and systems that are efficient and provide an appropriate service life.

Convection

The transfer of heat using the physical movement of a gas or liquid to dispense the heat is called *convection*. The fluid is usually warmed by the heat source through direct contact using conduction. The warmed portion rises because it is less dense than the cooler portion. The cooler portion of the substance then moves to occupy the space vacated by the warmed material. This produces fluid movement or flow as the heating and replacement sequence continuously repeats. See **Figure 4-22**.

The principle of convection is used to aid in the cooling of fluid power systems. As the reservoir and other components in the system warm, they in turn warm the air they contact. The warm air rises and is replaced by cooler air, which in turn is warmed. This cycle is continued during the operation of the system and is a primary source of cooling in many basic systems.

Radiation

The transfer of heat using electromagnetic waves is called *radiation*. See **Figure 4-23**. The electromagnetic waves travel at the speed of light and can be transmitted even in a perfect vacuum, unlike conduction and convection that require a transfer medium. Radiant heat warms only the substance that absorbs the energy, not the material through which it travels.

All bodies emit radiant heat to some degree, but the emissions do not become significant unless temperatures are very high, such as in combustion chambers. The sun provides the most familiar example of radiant energy transfer, although electric radiant heating systems are available for home and commercial applications. Radiation plays only a small role in the operation of fluid power systems and is usually ignored when calculating heat transfer.

4.6 Electricity and Magnetism

Electricity is a form of energy that is used extensively in the operation of fluid power systems. It is used to power the motors that serve as the prime movers for many of the systems. It is also used to operate sensing devices and valves that control the systems. Finally, it is the power source for safety devices and temperature control equipment that ensure safe and efficient system operation.

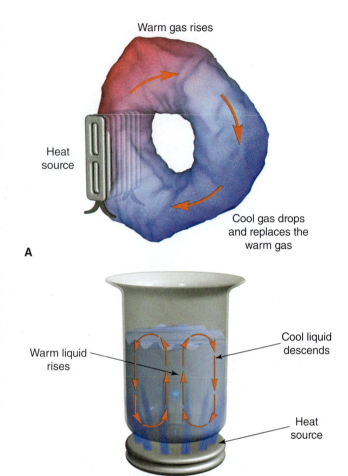

A

B

Goodheart-Willcox Publisher

Figure 4-22. Heat transfer by convection involves the physical movement of a liquid or gas. Movement occurs when the lighter, heated molecules rise and are replaced by cooler, heavier molecules. A—Convection in a gas. B—Convection in a liquid.

Goodheart-Willcox Publisher

Figure 4-23. Heat transfer by radiation involves electromagnetic waves that can transfer heat without the aid of a solid or liquid medium.

4.6.1 Nature of Electricity

An understanding of the nature of electricity requires an understanding of the basic structure of a substance. All substances are made up of *molecules*, which are compounds of the basic chemical elements. The structure and chemical elements that make up each molecule of a substance are identical. Changing the combination of the elements in a molecule produces an entirely different substance.

Molecules are composed of *atoms*, which are the smallest identifiable parts of an element. Atoms are composed of parts called electrons, neutrons, and protons. These atomic parts are the same in all elements. All protons are alike, all neutrons are alike, and all electrons are alike. Even though each element is structured from identical basic atomic parts, the characteristics vary because the number and arrangement of the parts are different for each.

Protons and neutrons form the *nucleus*, or center, of the atom. See **Figure 4-24**. The electrons follow orbital paths around the nucleus, much like the planets orbit the sun. Electrons, protons, and neutrons are very small even when compared to the size of an atom. Microscopic as these parts are, they play a major role in electrical theory.

Electrons, protons, and neutrons possess different electrical charges. *Protons* are positively charged particles. *Electrons* are particles with a negative charge. *Neutrons* are particles that have no charge. Charge refers to a characteristic of these particles that attracts or repels other particles. Like charges repel each other, while unlike charges attract each other. When positively and negatively charged particles are attracted to each other, they are "in balance" and neither attract nor repel other particles.

The atoms of substances are normally in electrical balance as each atom has an equal number of electrons and protons. The electrons continuously move, but the atoms are in electrical balance and stable because of the attraction of equal numbers of negatively charged electrons and positively charged protons. However, it is possible for electrons in the outer shell of the structure to modify their orbit and, in some cases, to actually become free. These free electrons are attracted to the outer shell of adjoining atoms. An atom is considered charged whenever it gains or loses electrons. An atom is considered negatively charged if it gains electrons and positively charged if it loses electrons.

Electrons move from a negatively charged substance to one that is positively charged. Electrical potential is the difference between the electron pressure levels of these substances. The level of electrical potential is established by the number of electrons in the substance, how easily the electrons in the outer shell can move, and the number of free electrons available. See **Figure 4-25**.

Electrical flow occurs when electrons begin to leave the outer shell of the atoms in a substance and move other free electrons with them through the substance. A substance that permits the flow of electrons is called an electrical conductor, while one that resists the movement of electrons is known as an electrical insulator.

Electrical devices use electron flow to do work. Electric motors that operate fluid power pumps, solenoids that shift directional control valves, and electronic sensing devices that automatically complete complex machine operations all function using this basic theory.

Copper

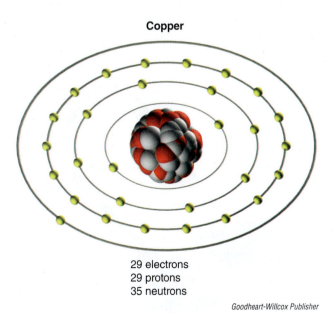

29 electrons
29 protons
35 neutrons

Goodheart-Willcox Publisher

Figure 4-25. The outer electron orbit of the copper atom contains only one electron, which is some distance from the nucleus. Copper is a good conductor because this electron can be easily forced out of its orbit as a free electron, which increases the conductivity of the substance.

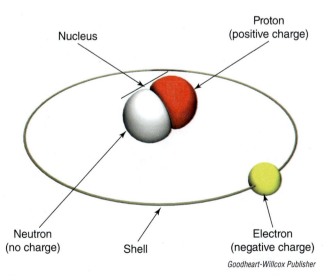

Nucleus

Proton (positive charge)

Neutron (no charge)

Shell

Electron (negative charge)

Goodheart-Willcox Publisher

Figure 4-24. The structure of an atom involves positively charged protons, negatively charged electrons, and neutrons that have no charge.

4.6.2 Nature of Magnetism

Applications in the fluid power field that use electricity also heavily depend on the principles of magnetism. The two phenomena interact in many situations. Magnetism is used to generate electricity. On the other hand, electricity is used to form the magnetic fields that operate electric motors, solenoids, and many other devices.

Magnetism is the characteristic of a substance that causes it to attract iron, **Figure 4-26**. Lodestone (magnetite), a magnetic iron oxide, is the only naturally occurring material that possesses polarity and has the ability to attract iron objects. It is possible, however, to magnetize certain other materials on either a temporary or permanent basis. Iron and iron alloys are most often used for this purpose as their permeability, or lack of resistance to magnetization, is high.

The ability of a magnet to attract another substance is concentrated in two areas called the *magnetic poles*. The poles are identified as the north, or north-seeking, pole and the south, or south-seeking, pole. Similar poles repel each other, while dissimilar poles attract each other. This is much like the negative and positive polarity factor associated with the electrons and protons of atoms.

Each magnetic pole is surrounded by a magnetic field. This is the area influenced by the attracting and repelling forces of the magnet. The field is made up of lines of force, or *flux*. In theory, the lines of force extend from the north pole to the south pole with movement occurring from north to south. The lines of force create a magnetic circuit similar to the movement of electrons when electrical potential exists between two points in a substance. See **Figure 4-27**.

Natural magnets are rare and usually quite weak. They are not considered satisfactory for commercial use. The materials are not suitable and the physical size required to produce adequate magnetic field strengths is not practical. However, it is possible to readily create suitable magnets in almost unlimited shapes and sizes. Artificial magnets made from either iron or metal alloys can be produced that have either temporary or permanent magnetic states.

Creating a magnet

A magnet can be created by exposing a hardened steel bar or other special metal alloy to a strong magnetic field. This may be done by simply rubbing the bar against an existing magnet or by exposure to a magnetic field produced by passing electrical current through a wire that has been coiled around the bar. When electricity is passed through a wire, a magnetic field is generated, **Figure 4-28**. In either case, the lines

Goodheart-Willcox Publisher

Figure 4-27. A magnet is surrounded by a magnetic field that forms paths between the magnetic poles.

Goodheart-Willcox Publisher

Figure 4-26. A permanent magnet has the ability to attract and hold iron substances. This attraction is concentrated in two areas of the magnet called poles.

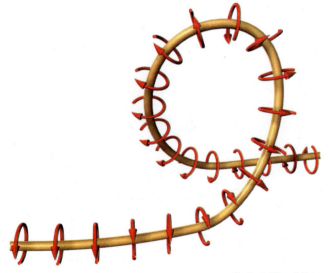

Goodheart-Willcox Publisher

Figure 4-28. A wire carrying electricity is surrounded by magnetic lines of force.

of force (flux) rearrange the atoms in the metal and magnetize the bar to produce a permanent magnet.

When wire is coiled around a metal bar, a solenoid is created, **Figure 4-29**. Devices using solenoids are commonly found in industry and the home. These solenoids use a soft iron bar or other material rather than hardened steel for a core. This produces an electromagnet that quickly becomes magnetized when electrical current flows through the coil, but loses magnetism almost instantly when the electricity stops flowing. The strength of magnetism in these devices can be easily varied by controlling the number of turns of wire in the coil and the amount of electricity that flows through the coil. The larger the number of turns in a coil and higher the electrical flow, the stronger the magnetism.

Producing electricity

Magnetism is also used to produce electricity. The principle involved is called electromagnetic induction and involves the cutting of magnetic lines of force by an electrical conductor. Electron flow is created in a conductor when it physically cuts (moves) through magnetic lines of force or when the lines cut (move) across the conductor. The amount of electricity produced by induction depends on the rate of movement of either the conductor or the magnetic field and the number of lines of force cut. Electrical output increases as both the speed at which the lines of force are cut and the number of lines increases.

4.6.3 Electrical Measurement

Working with electrical equipment and accessories in the fluid power area requires familiarity with the measurement of electricity, **Figure 4-30**. Although a wide variety of designs are used by the equipment manufacturers, operating theory and terminology are often similar. The basic electrical terms and the units of measurement are identical. These terms and measurements are related to the basic concepts of electrical current, potential, and resistance.

Electrical *current* can be defined as electron flow in a conductor. The unit of measure for current flow is the ampere (A). The amperage of the electrical current in a circuit is measured by an ammeter. Electrical *potential* can be defined as the pressure that causes electrons to move through a conductor. The unit of measure for electrical potential is the volt (V). The voltage of electricity flowing through a circuit is measured with a voltmeter. Electrical *resistance* can be defined as the opposition to electrical flow through a conductor. The resistance of a substance is determined by its basic structure, cross-sectional area, and physical temperature. The unit of measure for electrical resistance is the ohm (Ω). Electrical resistance is measured using an ohmmeter.

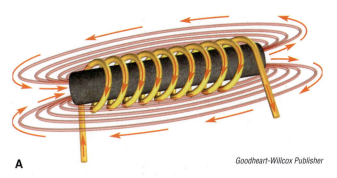

A   *Goodheart-Willcox Publisher*

B   *emel82/Shutterstock.com*

Figure 4-29. A—An electromagnet can be produced by coiling wire around a soft iron core. When electricity is flowing through the coiled wire, the metal concentrates the lines of magnetic force, producing a temporary magnet called a solenoid. B—Solenoids are used to operate many control devices in fluid power systems.

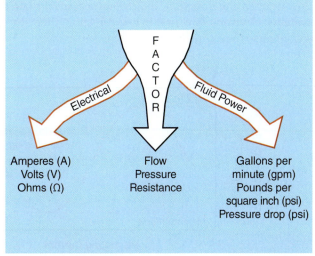

Goodheart-Willcox Publisher

Figure 4-30. Electrical amperage, voltage, and resistance may be compared to fluid flow, pressure, and resistance to flow in a hydraulic fluid power system.

Electrical current (amperage), potential (voltage), and resistance (ohms) are mathematically related. This relationship is stated in an equation referred to as *Ohm's law*. Ohm's law can be stated in any of the following forms, where E = potential, I = current, and R = resistance.

$$I = E \div R \qquad \text{Amperes} = \text{Volts} \div \text{Ohms}$$

$$E = I \times R \qquad \text{Volts} = \text{Amperes} \times \text{Ohms}$$

$$R = E \div I \qquad \text{Ohms} = \text{Volts} \div \text{Amperes}$$

These equations show that the current flow in a circuit is directly proportional to the voltage and inversely proportional to the resistance. This means that the current flowing in a circuit will increase when the voltage increases and decrease as the voltage decreases. In contrast, when the resistance in a circuit increases, the current flow decreases. Current flow increases when the resistance decreases.

The usefulness of electricity in fluid power systems relates to its ability to provide power and do work. The unit of measure for electrical power is the watt (W). One watt of work is produced by a current flow of one ampere at a pressure of one volt. The wattage used in an electrical circuit is measured by a wattmeter, but can be easily calculated by multiplying the amperage of the circuit by the voltage:

$$W = I \times E \qquad \text{Wattage} = \text{Amperes} \times \text{Volts}$$

4.6.4 Types of Electrical Current

Two types of electrical current flow are found in electrical systems. These are referred to as direct current and alternating current. The difference between the two types is the manner in which the electrons flow in a conductor. In *direct current (DC)* systems, electrons flow in only one direction. In *alternating current (AC)* systems, electrons flow first in one direction, then stop and reverse their direction.

Direct current maintains a fairly constant voltage. However, the voltage of alternating current fluctuates following the peaks and valleys of the electron flow. See **Figure 4-31**. These peaks and valleys are inherent in the start, stop, and reversing cycle. Alternating current electrical systems used in the United States commonly operate using 60 cycles per second, or 60 hertz (60 Hz).

Both direct and alternating currents are used in fluid power systems. Stationary fluid power equipment usually uses alternating current supplied by the local power company. The alternating current is converted to direct current by rectifying units to supply the DC needs of accessory systems. Internal combustion engines provide the power needed to drive mobile fluid power systems. These engines also drive alternators (providing AC) that, through a rectifier, maintain the charge of batteries, which in turn provide direct current for accessory system operation.

4.6.5 Basic Electrical Circuits

Circuits provide channels or routes for electrical current to follow as it performs work or completes a control function in a system. See **Figure 4-32**. A circuit must form a complete route that provides for the return of electrons to their source. The return route is called the ground. The flow of electrons is stopped if a break occurs at any point in the circuit.

Circuit components

A circuit is made up of four basic elements:

- Conductors.
- Control devices.
- Loading devices.
- Source of electrical energy.

Conductors can be any substance that permits the flow of electrons. In most circuits, the conductors are

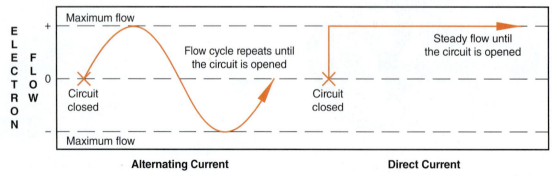

Figure 4-31. Direct current begins at zero and flows in one direction at a steady rate until it is stopped. Alternating current begins at zero, reaches a peak flow in one direction, decreases to zero, reverses its direction of flow, and reaches a peak in that direction before again returning to zero and repeating the cycle.

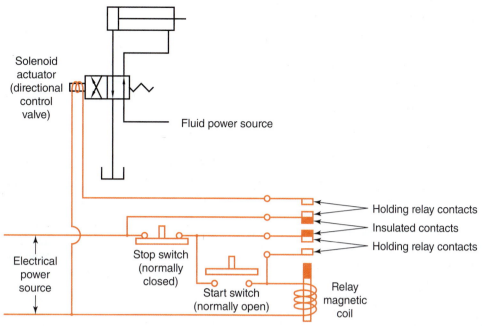

Figure 4-32. An electrical circuit is made up of a source of electrical energy, conductors to distribute the energy, a loading device to perform the desired work, and control devices to obtain the desired operating characteristics.

made from metals such as copper or aluminum. In some designs, they may be a special metal alloy that readily conducts electricity.

Control devices vary with the design of a circuit. These devices are used to start, increase, decrease, or stop the flow of electrons. They include such devices as switches, fuses and circuit breakers, and rheostats. Switches complete and break a circuit. Fuses and circuit breakers protect the circuit from overloading.

Rheostats allow the resistance in a circuit to be varied without breaking the circuit.

Loading devices are the elements that use electrical energy to perform work or convert the energy into another desired form. An electric motor or a valve-operating solenoid are loading devices common in the electrical circuits of fluid power systems. See **Figure 4-33**. Indicator lights or heating elements are other typical examples of electrical loads.

Figure 4-33. A number of electrical components can act as loading devices in electrical circuits found in fluid power systems. These include motors (A), solenoids (B), and indicator lights (C).

The source of electrical energy in fluid power systems is typically the local power company. This is the source of power for most stationary systems. Batteries and engine-driven alternators are typically the source of power for mobile applications. Both of these energy supplies are adequate to provide the electricity required to operate the electrical systems used with a fluid power system.

Circuit designs

The components of a circuit may be connected using either a series or a parallel configuration. A series circuit provides only one path for the current to follow as it moves through the circuit, **Figure 4-34**. If the circuit is made up of several loads, the current must flow through each, one after the other, in order to complete the circuit. The total resistance or load of this circuit design is high. The total load is equal to the sum of all of the loads. A problem with this design is the fact that failure of any one of the loading devices causes electron flow to stop throughout the circuit. Thus, the remaining loads no longer function. Think of a string of holiday lights where, if one bulb burns out, the remaining bulbs cease to shine.

A parallel circuit provides a structure where the current can follow individual routes to each load simultaneously, **Figure 4-35**. This design allows the current to be divided among the loads instead of being forced to pass through each load, as in the series circuit. A common primary conductor delivers current to the loads and returns it to the source after it has passed through the loads. The resistance in a parallel circuit is far less than in a series circuit with identical loads. The current flow through each of the multiple routes is affected only by the resistance in that route plus the resistance of the primary conductor.

In a series circuit, the voltage drops across each load, while the amperage remains the same throughout

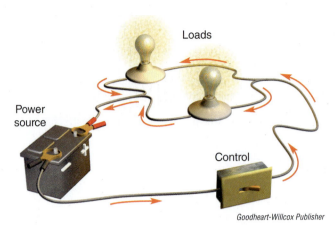

Goodheart-Willcox Publisher

Figure 4-35. A parallel circuit provides individual routes through an electrical circuit for multiple electrical loads. Each load has a complete and separate route from the power source, through the control devices, and back to the power source.

the circuit. Connecting the same loads in parallel causes the current to divide equally among the loads, while full circuit voltage is available to each load.

Series and parallel circuits are often combined in electrical equipment. For example, electrical switches are placed in series with the loads they control. Multiple loads in a single circuit are usually placed in parallel to obtain the most efficient performance. A number of series, parallel, and series-parallel circuits are typically found in the electrical equipment used in conjunction with fluid power systems.

4.7 Fluid Power Transmission

The primary topic of this text is the use of fluid power in society, especially in industrial operations. However, the design and operation of many consumer products also make use of the properties of liquids and gases. These fluids enable industrial designers to provide effective methods of power transmission to meet a variety of needs. The result is a range of applications from a simple hydraulic jack to complex applications in automated manufacturing lines.

4.7.1 General Characteristics of Fluids

A fluid is a substance that can flow and adjust its shape to fit the form of the container in which it is stored, **Figure 4-36**. A fluid may be either a liquid or a gas. Both of these states have a number of characteristics that are very similar, if not identical. Both liquids and gases possess weight, occupy space, and can be poured from a container. However, neither can possess a fixed shape. Both also have the ability, when pressurized, to exert that pressure on all the surfaces they contact.

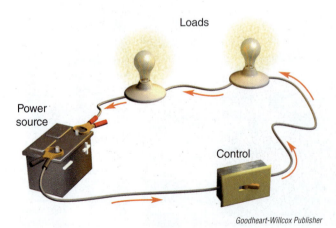

Goodheart-Willcox Publisher

Figure 4-34. A series circuit design provides a single path for the electrical current to follow through a circuit.

Goodheart-Willcox Publisher

Figure 4-36. A liquid assumes the shape of the container in which it is stored. Liquid has a stable volume and can be stored in open containers.

Important differences also exist between liquids and gases. Liquids are visible and often colored. Gases are frequently colorless and odorless, making them invisible and difficult to detect. The molecules of liquids are tightly held together providing them a stable volume. This makes it possible to store liquids in open containers. The liquid varies little in volume and stays in the container, except for the amount lost through evaporation. Gases, however, cannot be held in open containers. Their molecules will migrate to fill the container, spill over into the surrounding area, and be lost to the atmosphere. See **Figure 4-37**.

Another important difference between liquids and gases is compressibility. A liquid reacts much like a solid. It does not undergo a significant volume change when pressurized. Gases, on the other hand, compress as the pressure on them increases. See **Figure 4-38**. Both the incompressibility of liquids and the compressibility of gases are important characteristics in the operation of hydraulic and pneumatic systems.

4.7.2 Basic Properties of Materials

All materials possess a number of properties that serve to describe them and aid in their identification. These properties include a number of factors that primarily deal with appearance, rather than the structure of the material. However, another set of factors deals with basic properties that permit comparisons and allow the identification of materials suitable for specific situations.

Goodheart-Willcox Publisher

Figure 4-37. A gas assumes the shape of the container in which it is stored. However, it does not have a stable volume and will migrate to the atmosphere if the container is not covered.

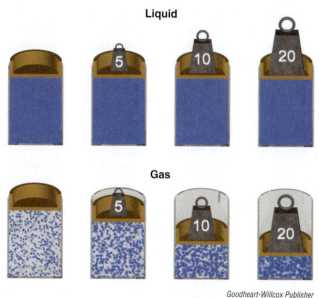

Goodheart-Willcox Publisher

Figure 4-38. A gas is compressible, while a liquid resists compression and is generally considered incompressible.

Mass

Mass is a fundamental concept in physics, mechanics, and other sciences. In the simplest of definitions, it is how much matter is in an object. The mass of an object is the same wherever the object is located. This applies to earthbound objects subjected to gravity or objects floating in space affected very little by gravity. Common units of mass that may be encountered are pounds, kilograms, and slugs.

Weight

Weight is one of the basic properties that allow comparisons. The *weight* of a device, structure, or body is the result of its mass being acted on by gravity. Weight is stated as a single figure, which represents the total gravitational pull. Common units of measure for weight in the fluid power field are pounds and kilograms.

Specific weight

The weight of a specific volume of a substance at a specific temperature and pressure is known as its *specific weight*. The need for the specific temperature and pressure reference is critical in calculations for gases, as they are greatly affected by these factors. However, liquids and solids are only slightly changed by pressure and temperature. The term incompressible fluid has evolved, as temperature and pressure are often ignored when working with fluids because of the slight effect.

$$\text{Specific Weight} = \frac{\text{Total Weight}}{\text{Total Volume}}$$

Specific weight is usually expressed as pounds per cubic foot. The maximum specific weight of water is 62.4 pounds per cubic foot. Similarly, dry air has a specific weight of .08 pounds per cubic foot at a temperature of 32°F and a standard atmospheric pressure of 14.7 psi.

Specific gravity

Specific gravity of solids and liquids is a comparison of the weight of a substance to the weight of an equal volume of water:

$$\text{Specific Gravity} = \frac{\text{Weight of Test Substance}}{\text{Weight of Equal Volume of Water}}$$

Water has a specific gravity of one and is used as the standard to measure other substances.

EXAMPLE 4-4

Specific Weight and Specific Gravity

If 2 ft³ of hydraulic oil weighs 107.2 lb, what is its specific weight and specific gravity?

$$\text{Specific Weight} = \frac{\text{Total Weight}}{\text{Total Volume}}$$

$$= \frac{107.2 \text{ lb}}{2 \text{ ft}^3}$$

$$= 53.6 \text{ lb/ft}^3$$

The specific gravity of the hydraulic oil can be found by using the weight of the oil and the weight of water. Water at 32°F has a specific weight of 62.4 lb/ft³ and a specific gravity of 1.

$$\text{Specific Gravity} = \frac{\text{Weight of Hydraulic Oil}}{\text{Weight of Equal Volume of Water}}$$

$$= \frac{53.6 \text{ lb}}{62.4 \text{ lb}}$$

$$= 0.86$$

The specific gravity of substances varies depending on their atomic and molecular structure. Solids tend to possess higher specific gravities, while liquids tend to be lower. The specific gravity of liquids can be easily tested using an instrument called a hydrometer, **Figure 4-39**.

The density of a substance changes with temperature. This change can influence the results obtained when using a hydrometer or other device to test specific gravity. Be aware of the temperature of the sample and the procedure for correcting the test instrument readout.

Specific gravity in operating compressed air and gas systems is often defined as the specific weight of the air or gas in the system compared to that of dry air at the same pressure and temperature. Be aware of this variation as you collect data from scientific data tables for system calculations.

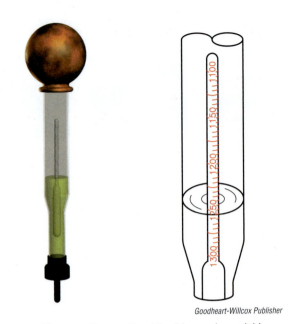

Figure 4-39. The specific gravity of liquids can be quickly determined by using a hydrometer.

Buoyancy

Buoyancy is another basic factor that affects the design and operation of many fluid power systems. *Buoyancy* is the force vertically exerted on a body by the fluid in which it is partially or wholly submerged. The force is equal to the weight of the fluid displaced by the body. A floating body displaces only enough fluid to equal its weight. In fluid power systems, float-operated valves that control fluid levels in tanks and reservoirs depend on the principle of buoyancy for their operation.

4.7.3 Pressure Measurement in Fluids

Several different scales are used to measure and designate the pressure of fluids, **Figure 4-40**. Remember, pressure is defined as force exerted over a unit area. All materials, including fluids, apply pressure to the objects on which they rest. The amount of pressure a fluid exerts depends on the specific weight of the fluid, when in an open container, or the force applied to it, when in a closed container.

Gauge pressure scale

The one fluid that affects all substances on earth is the atmosphere. The weight of the various gases that make up the atmosphere produces a pressure of approximately 14.7 pounds per square inch at sea level. This is considered *standard atmospheric pressure*. Pressure measurements that are made using standard atmospheric pressure as the base are referred to as *gauge pressure*. Standard atmospheric pressure registers 0 psi on the gauge pressure scale. Gauge pressures are identified as pounds per square inch gauge (psig). Most fluid power gauges are calibrated for gauge pressure.

Absolute pressure scale

Absolute pressure scale readings are required in many fluid power calculations. The *absolute pressure* scale uses a theoretical "zero pressure" as the base. Standard atmospheric pressure registers as 14.7 psi on this scale. For example, 100 pounds per square inch gauge equals 100 psi plus 14.7, or 114.7 pounds per square inch absolute (psia). Refer to **Figure 4-41**. The units used in the gauge and absolute systems are equal in size.

Vacuum scale

Pressure below standard atmospheric pressure is called a *vacuum*. Refer to **Figure 4-42**. Zero pressure, or the theoretical "perfect vacuum," exists in a space in which there is no matter in either solid, liquid, or gaseous form. This condition has never been achieved, although it is not difficult to produce a space that contains only minute quantities of matter. The concept of a vacuum is important to understand as lower-than-atmospheric pressure plays a large role in both hydraulic and pneumatic systems.

The intensity of a vacuum is usually measured by the height of a column of mercury it will support. Standard atmospheric pressure supports a mercury column 29.92" tall in a tube that is closed on the top end. The intensity of the vacuum is stated in inches of mercury, with 2.035 inches of mercury (Hg) equal to 1 psi.

Other scales

Pressure may also be stated using several other systems. Atmospheres and head are commonly used in the fluid power industry to indicate pressure. Both of these systems indicate the pressure created by the weight of a fluid. One *atmosphere* is the pressure created by the weight of the atmosphere at sea level. The average atmospheric pressure is considered to be 14.7 pounds per square inch. The pressures generated during the operation of fluid power equipment may be stated in atmospheres or parts of an atmosphere. For example, a pressure of one atmosphere is equal to 0 psig.

Head is stated as a unit of distance representing the height of a column of fluid. For example, a pressure may be stated as 25 feet head of water. In certain situations, head may be stated in inches of mercury. The concept of head can be better understood if pressure is examined in a container forming a one-foot cube. The container holds one cubic foot of liquid. If the liquid is water, the liquid weighs 62.4 pounds. The pressure generated by the water at the bottom of the cubic container can be stated as a one foot head of water. Dividing the weight of the water by the area of the bottom surface of the cube (62.4 pounds ÷ 144 square inches), you find that the water produces a pressure of .433 pounds per square inch. This is equivalent to one foot head of water.

4.7.4 Fluid Reactions to Pressure and Temperature

Fluids are able to effectively move energy from one location to another in hydraulic and pneumatic systems. The practical applications of this concept are based on several basic scientific principles, theorems, and laws. The following sections discuss several of these.

Dresser Instruments; Dresser, Inc.

Figure 4-40. A variety of pressure gauges are available to provide accurate readings of pressure under many different operating conditions.

Test Condition	Reading	
	psig	psia
100 psi gauge	100	114.7
100 psi absolute	85.3	100
Atmospheric pressure	0	14.7
Full vacuum	−14.7	0

Goodheart-Willcox Publisher

Figure 4-41. Comparison of the absolute and gauge pressure scales.

Test Condition	Reading		
	Inches Hg	psig	psia
Atmospheric pressure	0	0	14.7
	10.18	−5.0	9.7
	20.36	−10.0	4.7
Full vacuum	29.92	−14.7	0

Goodheart-Willcox Publisher

Figure 4-42. Comparison of the vacuum scale to the absolute and gauge pressure scales.

Pascal's law

Pascal's law reveals the general principle of how fluids move energy. This law states:

> Pressure applied to a confined, nonflowing fluid is transmitted, undiminished, to all points in the fluid with the pressure acting at right angles to all of the surfaces of the components in which it is confined.

Refer to **Figure 4-43**. The type of fluid, liquid or gas, and the shape of the components do not alter this law. The law is critical to the whole concept of energy transmission in fluid power systems.

Pascal's law is also important in terms of the amplification of force in systems. The law allows a small force applied to a small area in one part of a system to produce a proportionally larger force in another part of the system, **Figure 4-44**. For example, a force of 10 pounds applied to a piston with a surface area of one square inch produces a pressure of 10 pounds per square inch throughout the system.

$$\text{Pressure} = \frac{\text{Force}}{\text{Area}}$$
$$= \frac{10\ \text{lb}}{1\ \text{in}^2}$$
$$= 10\ \text{psi}$$

This allows a working cylinder with a piston area of 10 square inches to produce a force of 100 pounds.

$$\text{Force} = \text{Pressure} \times \text{Area}$$
$$= 10\ \text{psi} \times 10\ \text{in}^2$$
$$= 100\ \text{lb}$$

However, while the force increases, energy is not created. What is gained in force is lost in distance moved. The piston of the working cylinder moves only 1/10 of the distance of the piston receiving the input force. This ratio is established by the volume of the fluid displaced by the input piston and the volume needed to move the working cylinder piston.

Pascal's law is applied throughout fluid power systems. For example, based on this law, the small internal working area of a pump can be used to generate higher pressure and realistic power output. See **Figure 4-45**.

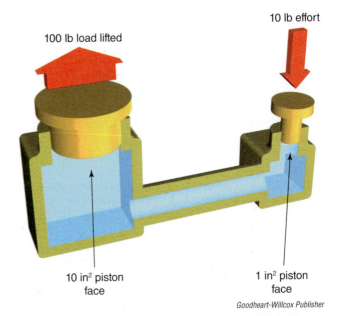

Goodheart-Willcox Publisher

Figure 4-44. A direct application of Pascal's law is the force amplification provided by the equal application of pressure to all surface areas contacted by pressurized fluid in a common chamber.

Goodheart-Willcox Publisher

Figure 4-43. Pascal's law is basic to the operation of fluid power systems. The law accounts for the ability of a confined, nonflowing fluid to transmit energy.

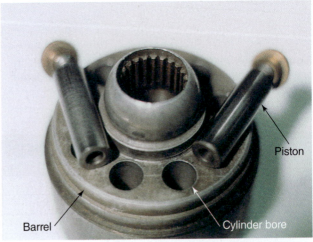

Goodheart-Willcox Publisher

Figure 4-45. Pascal's law allows the small internal working surfaces of hydraulic pumps and pneumatic compressors to develop maximum pressures with realistic power input.

EXAMPLE 4-5

Pascal's Law

A 5,000 lb load is lifted by a hydraulic piston with a diameter of 10″. What fluid pressure is required to lift this load?

$$\text{Piston Area} = \pi \times \text{radius}^2$$
$$= \pi \times (5 \text{ in})^2$$
$$= \pi \times 25 \text{ in}^2$$
$$= 78.5 \text{ in}^2$$

$$\text{Pressure} = \frac{\text{Force}}{\text{Area}}$$
$$= \frac{5,000 \text{ lb}}{78.5 \text{ in}^2}$$
$$= 63.7 \text{ lb/in}^2$$
$$= 63.7 \text{ psi}$$

If this load is lifted by applying a force to a piston with a diameter of 2″, what input force is required?

$$\text{Piston Area} = \pi \times \text{radius}^2$$
$$= \pi \times 1 \text{ in}^2$$
$$= 3.14 \text{ in}^2$$

$$\text{Force} = \text{Pressure} \times \text{Area}$$
$$= 63.7 \text{ psi} \times 3.14 \text{ in}^2$$
$$= 200 \text{ lb}$$

If the load must be lifted 2″, assuming 100% efficiency, how many total inches must the smaller piston move?

$$\text{Work Input} = \text{Work Output}$$

$$\text{Input Force} \times \text{Input Distance} = \text{Output Force} \times \text{Output Distance}$$

$$\text{Input Distance} = \frac{\text{Output Force} \times \text{Output Distance}}{\text{Input Force}}$$
$$= \frac{5,000 \text{ lb} \times 2 \text{ in}}{200 \text{ lb}}$$
$$= 50 \text{ in}$$

Bernoulli's theorem

Bernoulli's theorem establishes a relationship between fluid velocity, pressure, and elevation in a system. Details of the theorem have many implications for component and system designers. Equipment service personnel and operators, as well, need to be aware of the energy conversion concepts involved and how they can affect operating systems.

Bernoulli's theorem states:

In a stream of steadily flowing, frictionless fluid, the total energy contained in a given volume of the fluid is the same at every point in its path of flow.

This assumes that energy is neither created, destroyed, nor transferred out of the system. The energy exists in both kinetic and potential forms. The kinetic energy is related to fluid weight and velocity. The potential energy consists of fluid pressure caused by the pressurization of the fluid by a prime mover in the system or fluid elevation.

In a fluid power system, the principles of the theorem mean that each time the diameter of a line changes, the fluid flow velocity and pressure change to maintain an energy balance in the system. See **Figure 4-46**. A decrease in line or orifice size causes an increase in velocity (kinetic energy) and a decrease in pressure (potential energy). These situations occur in an actual operating system, but many times the changes are small and go unnoticed by a machine operator. If this information is needed, formulas and example calculations may be found using the materials discussed in the *Internet Resources* section at the end of this chapter.

A problem may occur in operating equipment because system fluid is not frictionless and resists being pumped through the system. This resistance to flow causes problems when an excessive amount of the potential energy is converted to heat, resulting in overheated fluid and component parts. In other situations,

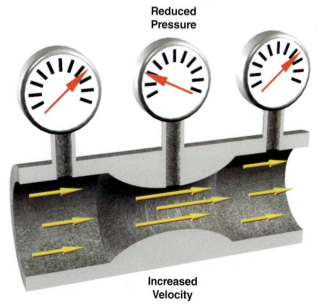

Reduced Pressure

Increased Velocity

Goodheart-Willcox Publisher

Figure 4-46. Bernoulli's theorem concerns the flow of fluid through a system. The velocity of the fluid is higher in restrictions to maintain flow rate, while the pressure is lower in those areas in order to maintain energy equilibrium.

increased fluid velocities through valves may cause them to malfunction.

Ideal gas laws

An ideal gas, also referred to as a perfect gas, is a theoretic gas with infinitely small molecules that exert no force on each other. The *ideal gas laws* are mathematical equations based on this theoretic gas that provide a good approximation of how a real gas reacts to pressure, temperature, and volume changes. All gas law calculations are completed using absolute pressure and temperature figures.

Several laws and equations are related to the gas laws. These laws bear the names of the individuals historically responsible for their development. The following sections describe the laws and equations commonly involved in fluid power system calculations. The laws apply to many situations in pneumatic systems and to hydraulic systems where gases are used in component operation, such as gas-filled accumulators. See **Figure 4-47**. The symbols in the equations represent these terms:

P_1 = Initial pressure

P_2 = Final pressure

V_1 = Initial volume

V_2 = Final volume

T_1 = Initial temperature

T_2 = Final pressure

Boyle's law deals with the changing of a gas from one pressure and volume condition to another while holding the temperature constant. The law states:

> When the temperature of a gas is held constant, the volume varies inversely with the absolute pressure of the gas.

This relationship is shown by the equation:

$$\frac{P_1}{P_2} = \frac{V_2}{V_1}$$

The equation can be rearranged in these forms:

$$V_2 = \frac{(P_1 \times V_1)}{P_2}$$

or

Figure 4-47. The principles established by the gas laws are used in sizing and determining the gas charge for hydraulic system accumulators.

$$P_2 = \frac{(P_1 \times V_1)}{V_2}$$

Charles' law deals with the changing of a gas from one temperature and volume condition to another while holding the pressure constant. The law states:

> When the pressure of a gas is held constant, the volume of the gas is directly proportional to the absolute temperature.

This relationship is shown in the equation:

$$\frac{V_1}{V_2} = \frac{T_1}{T_2}$$

The equation can be rearranged in these forms:

$$V_2 = \frac{(V_1 \times T_2)}{T_1}$$

or

$$T_2 = \frac{(V_2 \times T_1)}{V_1}$$

Gay-Lussac's law deals with the changing of a gas from one pressure and temperature condition to another while holding the volume constant. The law states:

> When the volume of a gas is held constant, the pressure exerted by the gas is directly proportional to the absolute temperature.

This relationship is shown in the equation:

$$\frac{P_1}{P_2} = \frac{T_1}{T_2}$$

The equation can be rearranged in these forms:

$$P_2 = \frac{(P_1 \times T_2)}{T_1}$$

or

$$T_2 = \frac{(P_2 \times T_1)}{P_1}$$

General gas law can be developed by combining the elements of Boyle's, Charles', and Gay-Lussac's laws. Combining these equations yields the equation:

$$\frac{P_1 \times V_1}{T_1} = \frac{P_2 \times V_2}{T_2}$$

Any of the terms can be solved through the use of algebraic substitution. For example, the final volume of a gas can be found through manipulating the equation to produce this formula:

$$V_2 = \frac{(T_2 \times P_1 \times V_1)}{(T_1 \times P_2)}$$

EXAMPLE 4-6

Gas Laws

An air cylinder has an initial temperature of 70°F with a gauge pressure of 2,600 psi. On a sunny day, the gas temperature may rise to 120°F. At that temperature, what pressure should the gauge read?

$$\frac{P_1 \times V_1}{T_1} = \frac{P_2 \times V_2}{T_2}$$

Because the air cylinder volume is constant, $V_1 = V_2$, the general gas equation can be reduced to Gay-Lussac's law by removing volume from the equation.

$$\frac{P_1}{P_2} = \frac{T_1}{T_2}$$

$$P_2 = \frac{(P_1 \times T_2)}{T_1}$$

To complete this calculation, pressure must be calculated in absolute pressure and temperature must be converted to the Rankine scale.

$$P_A = P_G + 14.7 \text{ psia}$$
$$= 2,614.7 \text{ psia}$$

$$\begin{aligned}
T_1 &= T_F + 459.67° & T_2 &= T_F + 459.67° \\
&= 70°F + 459.67° & &= 120°F + 459.67° \\
&= 529.67°R & &= 579.67°R
\end{aligned}$$

After the pressure and temperature have been converted to the proper units, Gay-Lussac's law can be used to find the pressure.

$$\frac{P_1}{P_2} = \frac{T_1}{T_2}$$

$$P_2 = \frac{(P_1 \times T_2)}{T_1}$$

$$= \frac{2,614.7 \text{ psia} \times 579.67°R}{529.67°R}$$

$$= 2,861.5 \text{ psia}$$

This measurement must be converted back to gauge pressure to determine the pressure reading on a gauge.

$$P_G = P_A - 14.7$$
$$= 2,861.5 - 14.7$$
$$= 2,846.8 \text{ psig}$$

Summary

- Only six basic types of machines exist even though millions of designs of complex machines exist in our society today. These basic machines include the lever, the inclined plane, the wedge, the screw, the pulley, and the wheel and axle.

- Energy is the capacity to do work and is classified as potential (stored) energy and kinetic (in motion) energy. Energy can be transformed from one type to another, but can be neither created nor destroyed.

- Force is the overall effort needed to produce, change, or stop the motion of a simple or complex device while pressure is force exerted over a unit area.

- Work is the application of force through a distance while power is the rate at which work is performed. Work is measured in foot-pounds and power is designated in foot-pounds per minute.

- Heat is kinetic energy measured in British thermal units (Btu) or calories. Heat is generated in a fluid power system whenever a pressure drop occurs that does not produce work.

- Heat may be moved using three modes of transfer: conduction, convection, and radiation.

- An atom is the smallest identifiable part of an element. It is composed of electrons, neutrons, and protons, which possess different electrical charges and play a major role in electrical theory.

- Electrical flow occurs when electrons leave the outer orbital paths of the atoms of a substance and move other free electrons through the substance with them.

- Electricity flowing in a conductor involves electrical current (amperage), potential (voltage), and resistance (ohms).

- Electrical circuits are made up of four basic elements: conductors, control devices, loading devices, and the electrical energy source.

- Liquids and gases have different compressibility characteristics. Liquids act more like solids, while gases undergo significant volume changes when pressurized.

- Fluids can be described using a number of different basic properties, including mass, weight, specific weight, and specific gravity.

- Pascal's law states that pressure applied to a confined nonflowing fluid is transmitted undiminished to all points in the fluid.

- Bernoulli's theorem, which involves the concept that the total energy in a given volume of flowing fluid is the same at every point in its path of flow, has implications for how changes in line diameter or orifice size affect fluid velocity and pressure.

- The ideal gas laws provide an approximation of how real gases react to pressure, temperature, and volume changes as they are pressurized and used in a system.

Internet Resources

The following are some useful resources available on the Internet. Enter a company or organization name into a search engine to access its website. Explore the various areas of the sites to discover useful fluid power resources.

Engineers Edge. Introduces properties and laws of thermodynamics. Provides a broad base of information on the basics of energy, work, and heat.

Wikipedia: The Free Encyclopedia—Ideal Gas Laws. Presents information on various gas laws.

Wikipedia: The Free Encyclopedia—Bernoulli's Principle. Provides information on Bernoulli's principle, including equations.

Chapter Review

Answer the following questions using information in this chapter.

1. Although millions of complex machines exist in the world today, they are constructed using the principles found in the _____ basic types of simple machines.
   A. 2
   B. 4
   C. 6
   D. 8

2. The parts of a lever include an effort arm, load arm, and _____.

3. *True or False?* A wheelbarrow is an example of a third-class lever.

For questions 4–7, match the terms with the description that best describes them.

A. Screw.   C. Wedge.

B. Lever.   D. Wheel.

4. Two inclined planes fastened together.

5. Spiral inclined plane.

6. Circular form of lever.

7. Bar that pivots on fixed axis.

8. Energy is defined as the _____.
   A. capacity to do work
   B. ability to produce heat
   C. unit of basic horsepower
   D. speed of rotary motion

9. *True or False?* Energy in motion is known as kinetic energy.

10. *True or False?* The tendency of a body to remain in motion once it has achieved operating speed is known as inertia.

11. Force exerted over a unit area is known as _____.
    A. power
    B. speed
    C. energy
    D. pressure

12. An example of torque is the _____ applied to the rotating shaft of a machine by a prime mover.
    A. power
    B. force
    C. pressure
    D. energy

13. The distance traveled in a standard basic unit of time, such as 3″ per second, is known as _____.
    A. velocity
    B. force
    C. pressure
    D. torque

14. The principle that heat always moves from a warmer to a colder body is known as the _____.
    A. first law of thermodynamics
    B. second law of thermodynamics
    C. third law of thermodynamics
    D. fourth law of thermodynamics

15. Which of the following is a way in which heat is generated in a fluid power system?
    A. Flow over relief valves.
    B. Flow through control valves.
    C. Internal leakage of pumps and motor.
    D. All of the above.

16. A ratio of the heat required to raise the temperature of equal weights of a substance and water the same amount is known as _____.
    A. absolute heat
    B. latent heat
    C. direct heat
    D. specific heat

17. The temperature scale that is used in scientific work, such as gas law calculations, is the _____ scale.
    A. absolute heat
    B. latent heat
    C. direct heat
    D. specific heat

18. Heat is transferred by _____.
    A. conduction
    B. convection
    C. radiation
    D. All of the above.

19. The three parts of the atom are the neutron, proton, and _____.

20. A substance that resists the movement of electrons is known as a(n) _____.
    A. electrical conductor
    B. electrical insulator
    C. electrical line
    D. magnetic field

21. Which of the following factors does *not* control the amount of electricity produced by electromagnetic induction?
    A. Rate of movement of the conductor.
    B. Rate of movement of the magnetic field.
    C. Rate of electrical current.
    D. Number of magnetic lines of force cut.

22. *True or False?* The unit of electrical power is the watt, which is equal to the work produced by a current flow of one ampere at a pressure of one volt.

23. *True or False?* The four elements of a basic electrical circuit are conductors, control devices, loading devices, and source of electrical energy.

24. The electrical _____ in a parallel circuit is always less than in a series circuit with equal loads.
    A. force
    B. wattage
    C. current
    D. resistance

25. The weight of a standard volume of a substance at a specific temperature and pressure is known as its _____.
    A. absolute weight
    B. specific weight
    C. working load
    D. capacity

26. The pressure produced by 10 feet head of water is equal to a gauge pressure reading of _____ psi.
    A. 2.33
    B. 3.33
    C. 4.33
    D. 5.33

27. The basic concept of _____ law is that pressure applied to a confined, nonflowing fluid is transmitted equally to all points in the fluid.
    A. Boyle's
    B. Bernoulli's
    C. Pascal's
    D. Charles'

28. According to Bernoulli's theorem, each time the diameter of a line in a fluid power system decreases, the velocity of the fluid in the line increases and the pressure _____.

29. Boyle's law assumes that the _____ of the gas is held constant between the two situations expressed in the problem.
    A. amount
    B. pressure
    C. velocity
    D. temperature

2. A hydraulic system is used to lift a 3,650 lb vehicle in an auto garage. Assume the vehicle sits on a piston of area 5.25 ft² and a force is applied to a piston of area 0.32 ft².
    A. What is the minimum force in foot-pounds that must be applied to lift the vehicle?
    B. Calculate the mechanical advantage achieved.
    C. If the vehicle must be lifted 4', how far must the smaller piston advance?
    D. If the maximum force that could be applied to the small piston is 150 foot-pounds, how would you modify the device to allow you to raise the vehicle?

3. Using the principles of Bernoulli's theorem, explain how a garden hose sprayer operates.

4. A 2" diameter pneumatic piston is used to lift a load of 200 lb.
    A. What gas pressure is required to accomplish this task?
    B. What volume of gas at this pressure is required to extend the piston 14"?
    C. Assume the piston is extended 14" while supporting this 200 lb load. If the temperature drops from 70°F to 60°F while the piston is extended, how far will the piston retract due to the contraction of the gas?
    D. If the temperature instead remains constant at 70°F, but the load increases to 220 lb, how far will the piston retract? Assume the piston is initially extended to 14".
    E. What are the implications of your calculations for the operation of pneumatic systems?

Apply and Analyze

1. Three different fluids have the following specific weights: Fluid A's specific weight is 50 lb/ft³; Fluid B's specific weight is 80 lb/ft³; Fluid C's specific weight is 10 lb/ft³.
    A. Calculate the specific gravity of each fluid.
    B. Calculate the pressure in pounds per square inch (psi) that each fluid would exert on the base of a cubic container that holds 1 ft³ of fluid. Assume that the container's base has dimensions of 12" by 12".
    C. Which of the three fluids would exert the greatest buoyancy force on a completely submerged object? Why?

Research and Development

1. Examine your residence for examples of simple machines. List these examples, identify the simple machine they represent, and describe the mechanical advantage gained by their use.

2. Determine the power output of yourself and your classmates by timing how long it takes each student to run up a flight of stairs. Prepare a report that details your experimental procedure, lists the variables you needed to measure, states your results, discusses any sources of experimental error, and describes how your power output in horsepower compares to common machines.

3. Design and conduct an experiment that illustrates the concept of latent heat. Share your results with the class.

Atlas Copco

The principles of levers are used in this machine. How many levers can you identify in the arm? Can you identify the fulcrum and load for each lever? Which classification is each lever?

5 Fluid Power Standards and Symbols

Language of the Industry

Chapter Outline

The use of fluid power systems is common in many industries in this country and throughout the world. This widespread use results in the involvement of many people in the design, operation, and maintenance of both components and systems. The need for guidelines so these individuals have common, basic rules to follow in their work has resulted in specialized standards for the fluid power industry. These standards have become a part of the documents available from national and international standardizing organizations. This chapter discusses the standardizing groups involved in the fluid power industry and the structure of standards, with a special emphasis on component symbols and circuit diagrams.

Learning Objectives

After completing this chapter, you will be able to:

- Describe the meaning of a standard and the importance of standardization in an industry.
- Identify the primary groups that provide standards for the fluid power industry.
- Explain the symbols used to designate components in fluid power circuit diagrams.
- Identify and explain the parts of circuit diagrams typically used with fluid power equipment.

Key Terms

American National Standards
   Institute (ANSI)
circuit diagram
control mechanism
cutaway symbol
electrical control

energy conversion device
feedback control
graphic symbol
International Organization for
   Standardization (ISO)
mechanical control

muscular control
pictorial symbol
pressure control
standard
symbol

5.1 Evolution of Standards

As the utilization of fluid power systems increased, so did the use of *standards* developed to guide the practices of the industry. Defined as anything that can be used as a basis for comparison, standards may be concerned with identification of size, shape, weight and strength, or any combination of these or other factors, **Figure 5-1**. For manufactured products, standards are usually a set of specifications that define a product or product part. However, standards also can relate to commonly used procedures in an industry or business, or to cultural concerns, such as social behavior, treatment of the environment, or health and safety issues for workers.

Standards are established by numerous public and/or private organizations. Examples of these organizations include government regulatory agencies, trade and business associations, scientific and professional societies, general membership associations, testing and certifying groups, or associations of different groups.

The fluid power industry is affected by general standards developed by a broad range of groups. In addition, specific fluid power industry standards are imposed by such groups as the National Fluid Power Association (NFPA), the Fluid Power Distributors Association (FPDA), the International Fluid Power Society (IFPS), and the Fluid Power Education Foundation (FPEF). Moreover, the development of standards is coordinated by several organizations, including the *American National Standards Institute (ANSI)* and the *International Organization for Standardization (ISO)*.

Standards may be based on custom, result from general consent of an affected group, or be based on legal or regulatory requirements. When developing a standard, organizations follow procedures that ensure input by individuals and organizations affected by the standard. The format that results can vary considerably in both length and technical complexity, but efforts are made to provide a final form that is easy to interpret and that allows for accurate and precise display of text, data, and drawings in one or more languages.

5.2 Fluid Power Symbols and Circuit Diagrams

Symbols and circuit diagrams are areas that have received a great deal of attention in relation to standardization in the fluid power field. They are also areas where considerable variation has been displayed over the years. Symbols and diagrams have been developed within companies to meet the needs at a specific time. Some of these symbols and diagrams have been adopted or adapted by other users. This section of the chapter examines several of the more common systems that have been in use for several years, as well as ISO graphic symbols and diagrams that are the latest versions of an internationally accepted system.

5.2.1 Types of Symbols

Symbols are used throughout the world in all areas and levels of the fluid power field. A knowledge of symbols is important as they serve as a special technical language. The *symbols* are used to communicate component design features, system construction, and system operation when constructing, operating, or troubleshooting circuits and systems. A knowledge of the common symbols is a valuable tool for any individual working in the hydraulic and pneumatic areas.

Tarzhanova/Shutterstock.com   *testing/Shutterstock.com*

Figure 5-1. Standards affect many aspects of daily life, from standard clothing sizes to safety standards in car manufacturing.

Four different types of symbols are encountered in fluid power documents:

- Graphic symbols.
- Pictorial symbols.
- Cutaway symbols.
- Combination symbols.

Graphic symbols

Graphic symbols are the most common. They are made up of a series of lines and outlines of standard graphic figures, such as squares, rectangles, triangles, and circles. Symbols of this type can be relatively simple to draw as they do not show the actual appearance of the component. They are suited to freehand sketching and both manual and computerized drafting procedures. The symbols for individual components can be easily connected by lines to represent a complex fluid power system. Circuit analysis is possible as internal operating characteristics of components can be shown in the graphic symbol. See **Figure 5-2**. Standardized forms of graphic symbols can be easily developed. These are especially suitable for international applications involving a variety of nations and languages. Graphic symbols are the most standardized form of fluid power symbols.

Pictorial symbols

Pictorial symbols are made up of line drawings of the exterior shape of a component and show its major details. See **Figure 5-3**. All external interconnecting lines are shown when these symbols are used to illustrate a fluid power system. Pictorial symbols provide a simplified method of showing the location of components in a circuit or system and the placement of all connecting piping. Circuit analysis is not easily accomplished when these symbols are used as only external features of the components are shown. Standardized forms of these symbols are limited as the external shape and placement of connections on components vary between manufacturers and product models.

Cutaway symbols

Cutaway symbols are really miniature section drawings of component parts, **Figure 5-4**. The symbols provide an excellent means of studying component and circuit design, examining system operation, and troubleshooting. Standard symbols do not exist for this classification of symbols. However, the following guidelines can assist a technical illustrator in preparing a drawing:

- Symbols should show all internal and external ports and clearly indicate flow paths through components.
- Actuator elements need to be shown to illustrate component operation.
- Component construction may be modified to clarify function or permit the shifting of movable parts to show flow paths during various phases of operation.
- The cutaway symbol should represent the relative size of the component compared to the other components shown in the circuit or system.

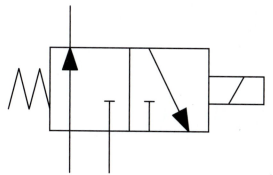

Goodheart-Willcox Publisher

Figure 5-2. The method that is most frequently used to represent components in fluid power systems involves graphic symbols. Standard graphic figures, such as lines, squares, triangles, and circles, are combined into a symbol to represent a component, such as the directional control valve shown in this symbol.

Single-Acting Cylinder

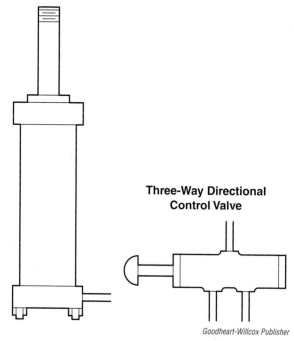

Three-Way Directional Control Valve

Goodheart-Willcox Publisher

Figure 5-3. Pictorial symbols are line drawings showing the exterior shape and characteristics of a component. The appearance of these symbols will vary as the construction of the components vary among manufacturers.

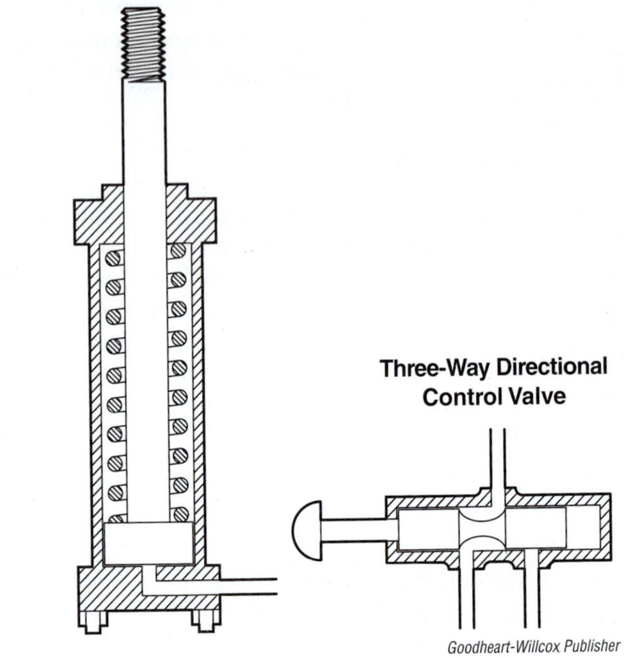

Single-Acting Cylinder

Three-Way Directional Control Valve

Goodheart-Willcox Publisher

Figure 5-4. Cutaway symbols are section drawings showing the basic internal parts of a component. The drawings show the internal construction of the components. These symbols are very helpful when studying the operation of a valve or system.

Combination symbols

Graphic, pictorial, or cutaway symbols may be used in combination in a circuit diagram. This may be done to highlight a portion of a component or system. No particular standard exists for the methods to use when developing combination symbols. The combination should produce a diagram that results in a clear understanding of the design or operation of the highlighted portion. An example is shown in **Figure 5-5**. In that figure, cutaway symbols are used for the directional control valve and pump, pictorial symbols are used for the two cylinders, and graphic symbols are used for the remainder of the circuit diagram.

5.2.2 Structure and Use of Basic Graphic Symbols

Graphic symbols are commonly encountered in fluid power applications. The symbols lend themselves to standardization. Several organizations publish versions of graphic symbols that are very similar. This section follows ISO 1219-2 international recommendations and shows the basic elements of that standard. This discussion does not present all of the possible combinations, but it shows how standard graphic figures can be

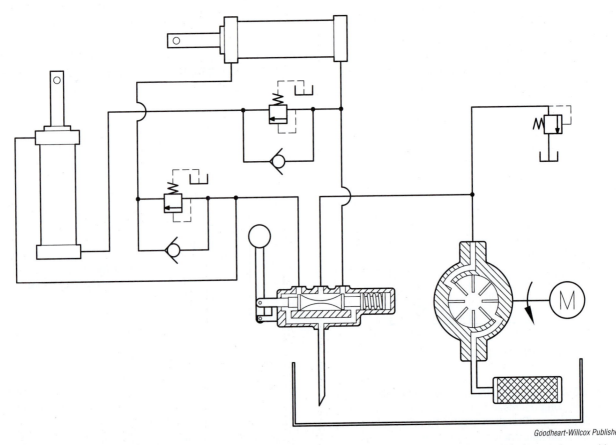

Goodheart-Willcox Publisher

Figure 5-5. A combination of symbol types may be used in a circuit diagram. For example, cutaway symbols may be used in combination with graphic and pictorial symbols to make component and system operation easier to understand.

combined to represent a hydraulic or pneumatic circuit. The standard should be consulted when complete or composite functional symbols are required.

General rules

Symbols represent component type, function, method of operation, and external connections. They are *not* intended to show the actual construction of a component. The symbols should generally be shown non-actuated in the de-energized (at rest or neutral) position.

Basic symbols and functional elements may be combined to represent complex functions. When two or more symbols are contained in one unit, they are usually drawn in an enclosure that consists of a thin chain line, which is a broken line with alternate short and long dashes.

External ports should be shown in the symbol, but they need not represent the actual location of the ports on the component. Ports are indicated by the junction of fluid lines with the basic symbol of the component or the enclosure symbol that surrounds complex symbols. Only the connections that are functional need to be shown in the symbol.

Unless specifically stated in the standard, symbols may be drawn in any orientation without affecting their meaning. However, increments of 90° are preferred by most users in the fluid power field.

Symbols are *not* used to indicate quantities such as pressure, flow rate, or the setting of a component. Letters, when used, are merely labels and do not describe parameters or values.

Basic symbols

A variety of basic figures serve as the components of graphic symbols. Lines, circles, squares, triangles, dots, and arrows are commonly found as parts of a typical symbol. These figures are combined to produce the desired symbol, which is combined with other symbols to produce a diagram of a fluid power circuit.

A variety of graphic lines are used to represent hydraulic and pneumatic lines in fluid power diagrams. Refer to **Figure 5-6**. Working and return lines use solid, continuous lines as a symbol. Electrical lines shown in a fluid power circuit are also represented by this type of solid line. Pilot control, drain, bleed, and filter lines use a dashed line as the symbol. Enclosures for units that involve two or more functions are formed using a chain line, which is a broken line with alternate short and long dashes. The final line symbol is used to identify mechanical connections, such as shafts, levers, and piston rods. These are represented by double, parallel solid lines.

Circles are used to represent a number of different components. The size of the circle and its location in the diagram can be used to help identify the type of unit the symbol represents. The largest-diameter circles designate energy conversion units such as pumps,

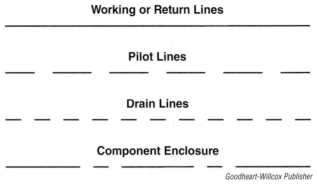

Working or Return Lines

Pilot Lines

Drain Lines

Component Enclosure

Goodheart-Willcox Publisher

Figure 5-6. The graphics used to show hydraulic and pneumatic lines in fluid power graphic symbol diagrams.

compressors, and motors. Medium-size circles, about 3/4 of the size of the conversion unit symbols, are used to show measuring units such as pressure gauges. Smaller circles are used to indicate a number of different components including nonreturn valves, rotary connections, and mechanical link rollers. Semicircles are used to designate motors or pumps with a limited angle of rotation.

A square resting on its side is used to indicate control components or a prime mover other than an electric motor. See **Figure 5-7**. Multiple positions of a control component, such as a multiple position directional control valve, are shown with multiple squares side-by-side to indicate the functioning of the valve in each of its positions.

A square resting on its corner represents conditioning apparatus. These include filters, separators, lubricators, and heat exchangers. An additional symbol is shown within the square to indicate specific information about the apparatus.

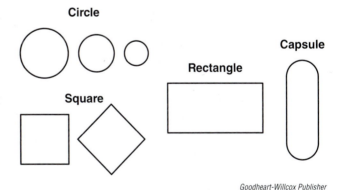

Circle

Square

Rectangle

Capsule

Goodheart-Willcox Publisher

Figure 5-7. A variety of basic shapes serve as the base for fluid power graphic symbols. Circles represent energy conversion units, such as pumps, compressors, and motors. Squares resting on their sides indicate control components or prime movers, while squares resting on their corners represent components involved in fluid conditioning. Rectangles are used for work devices and valves. Capsules represent a pressurized device, such as an accumulator.

Cylinders and valves have a rectangle as their basic symbol shape. No recommendations are included in the standard concerning the size of these symbols in relation to the graphic shapes representing other components. Rectangles are also used to represent the piston of cylinders, control mechanisms for valves, and cushioning devices for actuators.

Two additional basic graphic shapes are often used as component outlines. The first of these shapes is one-half of a rectangle that is open at the top with the height being one-half of the length. This shape serves as the symbol for a reservoir in a hydraulic system. The second form is a capsule. This looks like a rectangle with a semicircle on each end, as shown in **Figure 5-7**. The capsule may be placed in either a horizontal or vertical position, depending on the component it represents. The shape is used to designate a pressurized reservoir, air receiver, accumulator, or auxiliary gas bottle.

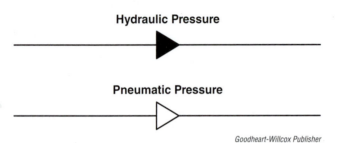

Goodheart-Willcox Publisher

Figure 5-8. The source of energy for a system is indicated by a small, equilateral triangle. An open triangle indicates a pneumatic system, while a solid triangle indicates a hydraulic system.

Functional elements

A variety of graphic elements are used to designate the operating medium, direction of movement, source of energy, and the way a part of a component is operated. These elements are added to the basic graphic shapes to provide more detail about the characteristics of a system component.

A small, equilateral triangle is used in conjunction with the previously discussed basic graphic shapes to show the direction of fluid energy movement through a component. This symbol is also used to show the type of fluid used in a system or component element. An open triangle indicates a pneumatic power source, while a solid triangle represents hydraulic operation. See **Figure 5-8**.

A variety of arrows are used as symbols to indicate factors such as fluid flow direction, equipment motion, heat flow, and the adjustability of pumps and other components. See **Figure 5-9**. Straight-line and sloping arrows indicate the direction of fluid flow, rectilinear equipment movement, and heat movement. Longer straight-line arrows are used to indicate variability or adjustability of equipment or component parts. These arrows are normally shown in a sloping position, rather than vertical or horizontal. Curved-line arrows are used to show shaft and equipment rotation.

Symbols for a number of additional, miscellaneous functions are shown in **Figure 5-10**. The closed path or port, spring, and restriction symbols are very commonly found in system diagrams. These symbols may be either attached to or enclosed in one of the basic graphic shapes, as shown in the figure.

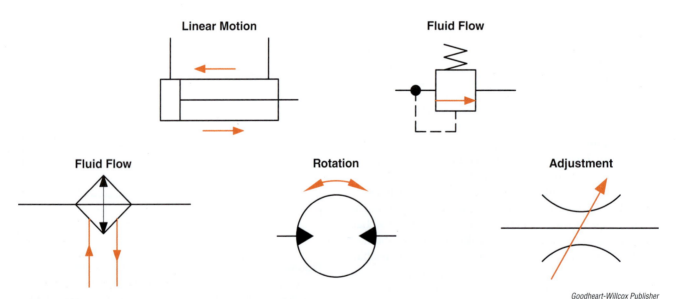

Goodheart-Willcox Publisher

Figure 5-9. Arrows are used to indicate several functions in fluid power graphic symbols. These include rectilinear motion, direction of fluid flow, direction of heat flow, direction of rotation, and the adjustability of components.

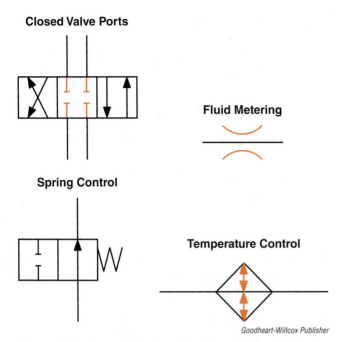

Figure 5-10. Symbols for a variety of functions are shown in this figure. The functions represented include closed ports to block fluid flow, restrictions for the metering of fluid, springs for control and activation of components, and temperature indication or control.

Flow lines and connections

Fluid power circuits are made up of a number of individual components. Standardized flow line and connection symbols provide a means to show how the components are interconnected to form the desired circuit.

The junction and crossing of lines is common in the structure of a fluid power circuit. The joining of lines in a circuit is indicated by a dot that covers the point where the lines connect. Care should be taken to avoid having this junction symbol indicate the connection of four lines at a single intersection point. The preferred form to follow is shown in **Figure 5-11** where two lines connecting to a third line are shifted slightly to eliminate the possibility of confusing them with lines that are simply crossing. Lines that cross each other but do not join are shown as crossing without the junction symbol dot. Hose, which is usually used to connect moving parts, is shown by a curved line connecting junction symbol dots at the end of fixed lines.

Symbols are not used to show the plumbing involved in the various lines of a fluid power system. Separate detailed drawings usually provide the information needed to assemble this aspect of the equipment. However, symbols are used for a number of connections that are important in the structure of a system.

Quick-disconnect couplings are connection devices represented by a graphic symbol. **Figure 5-12** shows a quick-disconnect coupling in both the connected and disconnected situations. The first part illustrates the connected state with an associated check valve that limits loss of fluid during the connection process. The second part shows the disconnected state. These devices are especially common in mobile applications where implements are attached and disconnected from tractors and other power sources. On mobile equipment, these connectors commonly include a mechanically opened, nonreturn check valve to allow easy connections with a minimum loss of fluid during attachment and detachment. Air bleeds and air exhaust ports are other connection devices that have symbols, as shown in **Figure 5-12**.

Rotary connections allow limited movement between sections of the plumbing in a circuit. These connections are used when a fluid power system is on equipment that swivels or turns in its operation. These devices are used in both hydraulic and pneumatic systems.

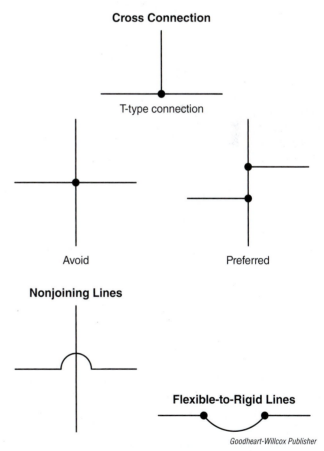

Figure 5-11. Junction points in lines are indicated by a dot that covers the points where the lines contact. Lines that are not connected simply cross each other. The correct method to use when showing two lines that cross without joining is with a semicircle. A hose connecting to two rigid lines is shown with an arc and two dots.

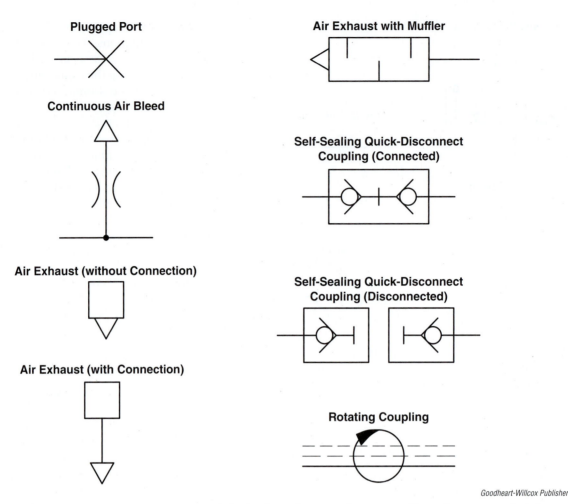

Plugged Port

Continuous Air Bleed

Air Exhaust (without Connection)

Air Exhaust (with Connection)

Air Exhaust with Muffler

Self-Sealing Quick-Disconnect Coupling (Connected)

Self-Sealing Quick-Disconnect Coupling (Disconnected)

Rotating Coupling

Goodheart-Willcox Publisher

Figure 5-12. A number of different connections are used in hydraulic and pneumatic systems. Shown here are symbols for typical types of connections.

Control mechanisms

A number of different *control mechanisms* are used to control the valves found in fluid power systems. Several basic symbols are used, separately or in combination, to indicate the control method. The valve control symbols are drawn across the end of the basic valve symbol. Some of the control symbol information is enclosed in standard basic control rectangle symbols, while other symbols are stand-alone figures.

Several basic symbols are used to show control-related elements of fluid power valves. See **Figure 5-13**. These symbols show a number of mechanical components and control methods. The mechanical components are the physical devices that shift the parts of a component to obtain the desired operation. The control method is the means used to move the mechanical controls in the valve.

The mechanical components of control mechanisms include rods, shafts, detents, latches, and over-center devices. Rods and shafts are common devices that allow a control device to be pushed, pulled, rotated, or moved to obtain the desired reaction. Detents, latches,

and over-center devices are used to ensure the control mechanism holds the position where it has been set.

Many different methods are used to control the operation of fluid power valves. However, the symbols for these devices are classified as either:

- Muscular.
- Mechanical.
- Electrical.
- Pressure.
- Feedback.

The symbols for these devices are shown in **Figure 5-14**. *Muscular controls* are those involving direct effort of a human operator, such as a hand-operated pushbutton or a foot pedal. *Mechanical controls* are operated by machine elements and include devices such as plungers and rollers. *Electrical controls* primarily involve solenoids and torque motors. *Pressure controls* can be either hydraulic, pneumatic, or a combination of both. They can involve either pressurizing the control, releasing pressure, or using opposing differential pressures. *Feedback controls* involve sensing

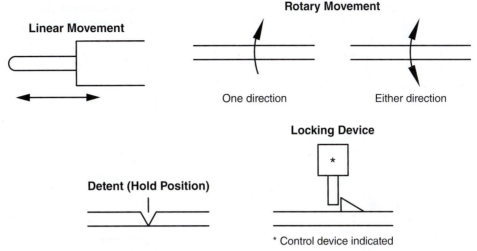

Goodheart-Willcox Publisher

Figure 5-13. Four common mechanical devices used to direct the movement of control devices in components are shown here. These include a rod to cause linear movement, shaft to cause rotary movement, detent to hold a position against a limited force, and latch to lock a device into a selected position.

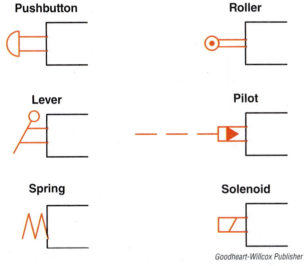

Goodheart-Willcox Publisher

Figure 5-14. A wide variety of control methods are used with fluid power valves. The six symbols shown here represent methods commonly found in hydraulic and pneumatic systems.

values internal and external to the valve, with the control device responding to these values and accordingly adjusting the operation of the valve.

Energy conversion

The *energy conversion devices* are hydraulic pumps, air compressors, motors, semirotary actuators, and cylinders. These devices either convert the energy of the prime mover into system pressure and flow or use the energy of the system fluid to do work. The symbols for these devices must be able to represent components from very basic pumps, motors, and cylinders

to complex units that are pressure compensated and provide fully variable displacement.

The basic graphic shape for a pump or motor is a circle. The basic triangle symbol is used to indicate a hydraulic or pneumatic application. A pump is designated when a corner of the triangle points to the circumference of the circle. A motor is indicated when a corner of the triangle points to the center of the circle and one side of the triangle is in contact with the circumference. The direction of rotation for both pumps and motors is indicated by a curved arrow shown over a shaft symbol protruding from the side of the circle. Examples of pump and motor symbols are shown in **Figure 5-15**.

The basic graphic shape for a cylinder is a rectangle. Double-acting cylinders are drawn with a closed rectangle and ports at either end. The piston is a small rectangle drawn across the short dimension of the cylinder rectangle, while the cylinder rod is a double line extending from the piston through one end of the basic cylinder rectangle. A single-acting cylinder is drawn with the rod end of the cylinder exhausting to atmosphere in a pneumatic system and to a drain in a hydraulic installation. There are simplified versions of the symbols as well. See **Figure 5-16**.

Energy storage

In a pneumatic system, the principle energy storage device is the air receiver. The symbol is a capsule drawn horizontally with lines extending from both ends. See **Figure 5-17**. No additional internal graphic figures are normally used to designate this component.

Accumulators serve as energy storage devices in specialized hydraulic circuits. The basic symbol for

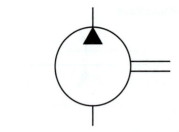

Hydraulic pump, one-flow direction, fixed displacement, one-rotation direction

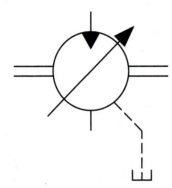

Hydraulic motor, one-flow direction, variable displacement, undefined control mechanism, external drain, two shaft ends

Goodheart-Willcox Publisher

Figure 5-15. Rotary energy conversion devices are represented by circles. Many variations of these units are found in circuits. The two symbols shown here represent two of the possible variations.

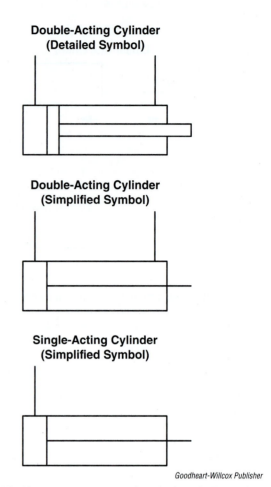

Double-Acting Cylinder (Detailed Symbol)

Double-Acting Cylinder (Simplified Symbol)

Single-Acting Cylinder (Simplified Symbol)

Goodheart-Willcox Publisher

Figure 5-16. Linear energy conversion devices (cylinders) are represented by rectangles. As shown here, the symbols can be detailed or simplified.

these components is also a capsule. However, in this situation, the shape is drawn vertically with a single line extending from one end of the unit. Internal graphic figures are used to indicate the energy-storage method used in the device.

Energy control and regulation

Directional, pressure, and flow control devices are grouped in the energy control and regulation category of graphic symbols. These symbols are made up of one or more boxes. When the symbol is made up of more than one box, the boxes are drawn with contiguous sides. These boxes can be either rectangles or squares, depending on the valve depicted. The number of boxes usually corresponds to the number of operating positions of the valve. Two parallel lines along the length of a valve symbol indicates that it has two or more distinct operating positions and a number of intermediate positions. See **Figure 5-18**.

External lines connected to the valve are usually attached to the box that represents the at-rest position. When visualizing the operation of a valve, each of the box configurations may be mentally shifted so the flow paths shown in the box align with the external lines.

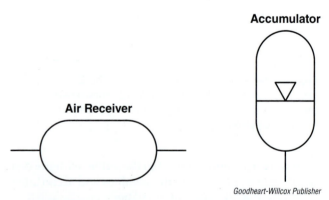

Air Receiver

Accumulator

Goodheart-Willcox Publisher

Figure 5-17. The left symbol is for an air receiver, an energy-storage device for a pneumatic system. The right symbol is for a gas-loaded accumulator, a common energy-storage unit in a hydraulic system. This symbol can be modified to represent a spring- or weight-loaded device by simply changing the internal gas charged triangle to show a spring or weight symbol.

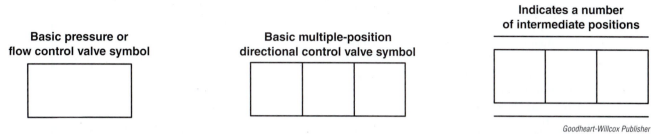

Figure 5-18. A control valve symbol may use one or more boxes to show various positions of operation. The left symbol is a single rectangle used to show a pressure or flow control valve. The middle symbol is three boxes with contiguous sides used to represent a three-position directional control valve. The parallel lines drawn along the length of the right symbol indicate that the valve has an infinite number of intermediate throttling positions, in addition to the positions indicated by the three boxes.

Multiple-position directional control valves typically include two, three, or more positions of operation. In a two-position valve, the symbol is drawn to show that the fluid moves through the actuator in one direction when in the first position and in the opposite direction when the actuator is in the second position. In valves with three or more positions, these additional positions provide additional modes of operation for the pump, actuators, or other parts of the system. Many different configurations are used for these additional positions. **Figure 5-19** shows the symbols for two typical designs for a three-position valve. The open-center design allows free fluid movement between all of the

external ports of the valve. The closed-center design has all of the ports blocked in the at-rest position.

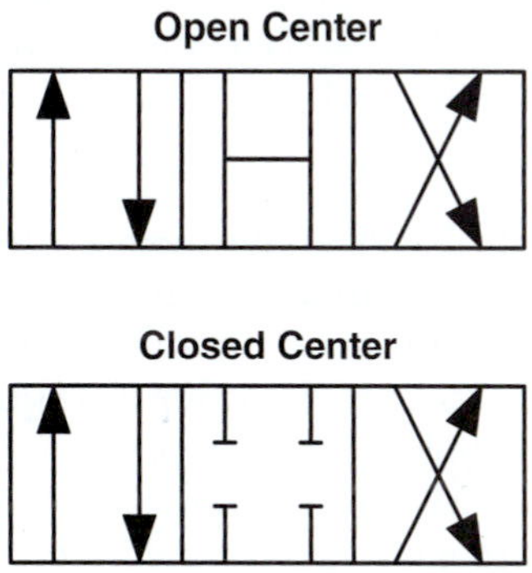

Many different configurations of directional control valves can be shown using the basic symbol graphics. Multiple boxes can be shown to illustrate valve operation in the various operating positions of the valve, while control mechanism symbols can show how the valve is shifted. **Figure 5-20** shows symbols for several standard, multiple-position directional control valve types.

Other types of common directional control valves are nonreturn valves, shuttle valves, and quick exhaust valves. The nonreturn valve is also called a one-way or check valve. The simplest form of a one-way valve opens if the inlet pressure is higher than the outlet pressure. Several other versions of nonreturn valves exist in spring-loaded and pilot-operated forms. **Figure 5-21** shows symbols for several of these components. A detailed symbol shows more of the structure and operational principles of these components than a simplified version. Generally, simplified symbols are adequate for most purposes.

Pressure control valves are devices used to control or limit pressure. The basic graphic shape used as the symbol for these valves is a square. The common valves that are shown using these symbols are relief, sequence, unloading, and pressure-reducing valves. The usual operating position for these valves in a functioning circuit is described as either normally closed or normally open. This simply means that a normally

Figure 5-19. A variety of center positions are available for multiple-position directional control valves. The symbol for the valve should represent the correct center position.

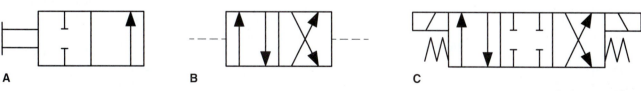

Figure 5-20. Three typical directional control valve symbols are shown here. A—A simple, two-position shut-off valve with a muscular control device. B—A two-position, four-way valve that is pilot-operated. C—A simplified symbol for a solenoid-operated, three-position, spring-centered, four-way valve with a closed center.

closed valve is usually closed, blocking flow through the unit, and normally open valves are usually open allowing flow. During operation, the normally closed valve opens providing its design function, while the normally open valve partially or fully closes to perform its function. In the symbol for a normally closed valve, the arrow that shows flow through the valve is not aligned with the external ports of the valve. In the symbol for a normally open valve, the arrow is aligned with the ports to indicate flow through the valve. See **Figure 5-22**.

Control valve symbols also indicate other features of the valves, such as springs, orifices, pilot lines, and drains. Both simplified and detailed versions of these symbols are used. The detailed symbols can become complex. Complex symbols often require an enclosure to define the parts included in the unit, as shown in **Figure 5-23** for a two-stage pressure control valve. Simplified symbols for several types of basic pressure control valves are shown in **Figure 5-24**.

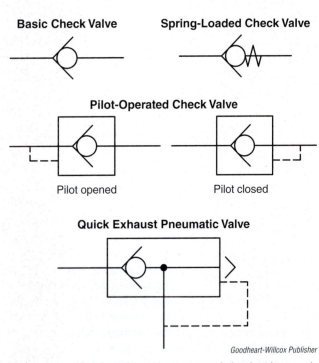

Goodheart-Willcox Publisher

Figure 5-21. Symbols for various types of check valves and a quick exhaust valve.

Goodheart-Willcox Publisher

Figure 5-22. Symbols used for normally open and normally closed valves.

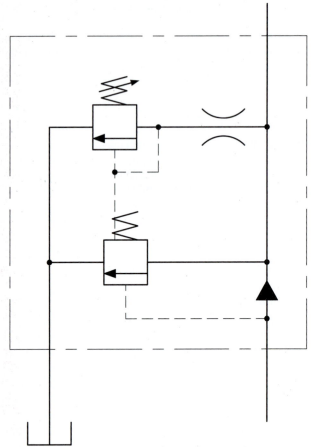

Goodheart-Willcox Publisher

Figure 5-23. A detailed symbol for a two-stage pressure relief valve. The component enclosure is used to indicate that all of the functions depicted by the various symbols are located within a single unit.

Flow control valves are devices used to control fluid flow in a circuit that, in turn, controls the speed of actuators. The preferred basic symbol for these valves includes the features of the valve in a square or rectangular box. The box is often dropped in simplified versions. **Figure 5-25** shows a number of variations that may be used to show the capability of a flow control valve. The curved lines represent the orifice (restriction) that controls fluid flow. The arrow passing through the orifice at a 45° angle indicates that the orifice is adjustable. The dimension shown in **Figure 5-25** indicates the physical diameter of the orifice. All of the symbols shown are acceptable. The level of information needed and company preference are the final determining factors in which symbol to use.

The orifices in flow control valves normally restrict flow in both directions through a valve. Often, it is desirable to have controlled flow in one direction and free flow in the opposite direction. **Figure 5-26** shows the symbol for a typical one-way restrictor valve. This valve combines flow control elements and a check valve to allow nonrestricted flow when the fluid flow reverses.

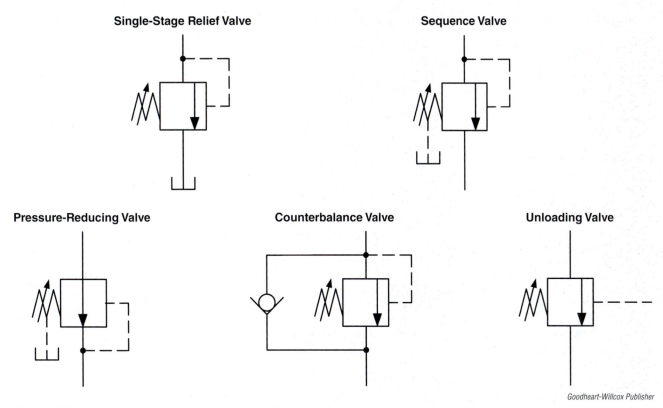

Figure 5-24. Simplified symbols are shown in this illustration for several common pressure control valves.

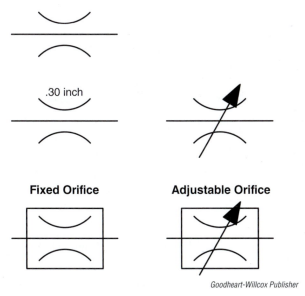

Figure 5-25. This illustration shows several versions of symbols that may be used for flow control valves. An adjustable valve is shown with a slanted arrow passing through the orifice. The size of a fixed orifice can also be indicated with a dimension.

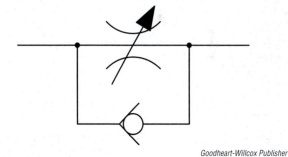

Goodheart-Willcox Publisher

Figure 5-26. This illustration shows the symbol for a flow control valve that allows control of flow in one direction and free flow (through a check valve) in the opposite direction.

Pressure and temperature changes in a fluid power system affect the flow rate through the system. Flow control valves use compensation devices in an attempt to maintain accurate flow. Pressure compensation is shown in a simplified symbol by placing an arrow across the flow line of the symbol. Temperature compensation is shown by placing a thermometer symbol across the symbol flow line. **Figure 5-27** shows simplified symbols for pressure-compensated and temperature-compensated, adjustable flow control valves. If a single valve has both of these features, they can be combined into a single symbol, as shown in the figure. This is an indication of how easy a symbol may be structured to match the features of the component.

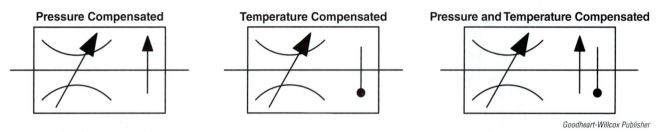

Pressure Compensated   **Temperature Compensated**   **Pressure and Temperature Compensated**

Figure 5-27. The left symbol is a simplified symbol for a flow control valve that is pressure compensated to help maintain a uniform flow rate through the valve. Pressure compensation is indicated by the arrow perpendicular to the flow line through the symbol. The middle symbol is for a temperature-compensated valve. The temperature-compensation feature is shown in the symbol by the temperature indicator symbol positioned across the flow line. The right symbol is for a valve that is both pressure and temperature compensated.

Fluid storage and conditioning

A number of different devices are used to condition and store both liquids and gases for fluid power systems. Simple symbols indicate these storage devices, which keep a ready supply of hydraulic fluid and compressed air. Somewhat more complex symbols show the components used for fluid conditioning, which involves removal of dust and dirt, maintenance of proper temperatures, and separation of moisture (from pneumatic system air).

Hydraulic reservoirs are shown as an open rectangle with the cut made along the long dimension of the figure. The open top indicates that the reservoir is not a sealed unit. Pickup, return, and drain lines touch the bottom of the open figure to indicate that they are below the surface of the hydraulic fluid. See **Figure 5-28**.

A number of reservoir symbols are shown in a hydraulic circuit, even though only a single reservoir is used in the system. This situation exists as return and drain lines from complete component symbols in a circuit may each terminate locally with a small reservoir symbol. This practice simplifies the diagram by eliminating a number of lines that otherwise would need to be shown returning to the main system reservoir.

A sealed or pressurized hydraulic reservoir is designated by a horizontally oriented capsule symbol. System line connections to this symbol are perpendicular to the long sides of the graphic. Below-fluid-level connections to the unit are indicated by lines that pass through the figure and contact the line representing the bottom side of the component.

A number of different fluid conditioners are used in hydraulic and pneumatic systems. These components include filters, separators, air dryers, lubricators, and heat exchangers. The basic graphic for these units is a square shown resting on one corner. The preferred orientation of the inlet and outlet lines is horizontal for all of these components. Symbols of separators or assemblies with separators are drawn *only* in the horizontal position. Symbols representing a number of the common conditioning components are shown in **Figure 5-29**.

A simplified symbol is often used for a group of common conditioning components used in many pneumatic systems. This group usually consists of a filter, separator, pressure regulator, pressure gauge, and lubricator. **Figure 5-30** shows both the detailed and simplified symbols for this common pneumatic combination unit.

Supplementary equipment

The supplementary equipment category of symbols primarily contains measuring instruments and specialized electrical switches. The basic symbol for

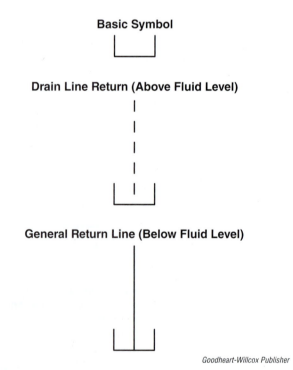

Basic Symbol

Drain Line Return (Above Fluid Level)

General Return Line (Below Fluid Level)

Figure 5-28. The hydraulic reservoir is drawn as an open rectangle, as shown in the top illustration. The middle illustration shows the reservoir with a drain line return that is above the fluid level. The bottom illustration shows the reservoir with a general return line from the system that is below the fluid level.

Figure 5-29. These symbols represent common fluid-conditioning devices found in hydraulic and pneumatic systems.

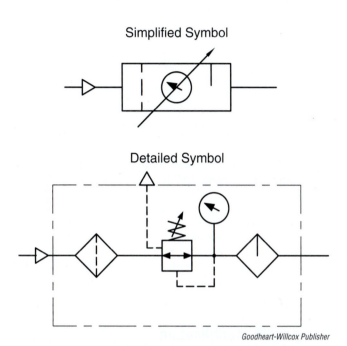

Goodheart-Willcox Publisher

Figure 5-30. The top symbol is a simplified symbol for a common grouping of pneumatic system components used for the conditioning of air at the workstation. This grouping often includes a filter, separator, pressure regulator, pressure gauge, and lubricator. The bottom symbol is the detailed symbol for this functional group of components. Notice the complexity of the symbol.

measuring instruments is a circle, while a square is used for switches. The symbols in this category are used to indicate instruments that measure fluid pressure, flow rate, temperature, and level. Symbols also exist for a tachometer (used to measure the rotational speed) and torque meter (used to establish the torque output of motors and other rotational devices). **Figure 5-31** shows the symbols for common measuring instruments.

Two types of switches that often appear in fluid power circuits are the pressure switch and limit switch. Both are used in circuits to provide control of system operation by sensing system pressure and actuator movement. The basic symbols for these components are shown in **Figure 5-32**.

5.2.3 Circuit Diagrams

Circuit diagrams give very complete information on a fluid power system. This information is useful for system assembly, operation, and testing. Circuit diagrams include schematics to illustrate the structure of a system, showing each component and interconnecting line. Diagrams also contain information on conductor sizes and a list of components. Data are provided on component sizes, flow capacity, pressure settings, and speed. In addition, information is provided on the

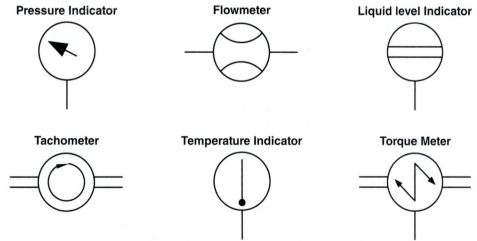

Goodheart-Willcox Publisher

Figure 5-31. A wide variety of measuring instruments are used in fluid power systems. The six symbols shown here represent instruments that may be found in system diagrams.

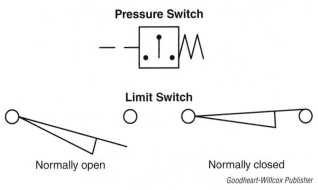

Pressure Switch

Limit Switch

Normally open   Normally closed

Goodheart-Willcox Publisher

Figure 5-32. The electrical components represented by these symbols are often found in fluid power systems.

sequence of operation for the circuit. The ISO standard 1219-2 indicates that these items should be included in circuit diagrams:

- Schematic diagram.
- List of components.
- Sequence of operation.
- Conductor size.
- Component sizes.
- Flow capacity.
- Pressure settings.
- Component speed.

Other standards use a slightly different format, but include very similar content.

General rules

The schematics used in circuit diagrams should be clear. They must include all components and connections so it is possible to follow circuit movement and commands throughout the entire sequence of the system's operating cycle. The schematics need not take into account the exact physical arrangement of the equipment, but must include adequate drawings and related details to form a comprehensive document detailing system construction and operation. See **Figure 5-33**.

The lines or connections between the different components of the system should be drawn with a minimum of crossing points. The position of codes and indices should not overlap the space reserved for equipment parts and lines. In complex systems, the schematic should be divided into sections based on related control functions. When multiple pages are required, line connections between sheets should be carefully identified.

The symbols for the component parts in the fluid power system should be logically arranged from the bottom to the top and from left to right on the schematic. This places the energy source on the lower-left side of the drawing and the actuators on the upper-right side.

ISO 1219-2:2012(E/F)

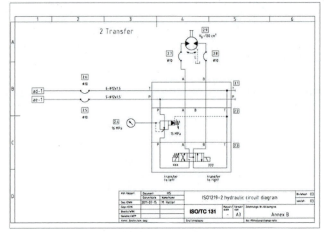

Figure 5-33. This is a hydraulic system circuit diagram based on the ISO standard 1219-2 for symbol and diagram formats.

The control components of the system should be in sequential order, moving upward from the power source and right toward the actuators. The symbols should show a powered-up system in the at-rest or neutral position.

Component identification

Identification should be provided for all equipment component parts in the circuit diagram. A code number should be located next to the symbols on the schematic. The ISO standard 1219-2 recommends a four-element code. See **Figure 5-34.** The first position in the code is reserved for a number to identify the installation in which the component is located. The second position is reserved for identification of the type of media used

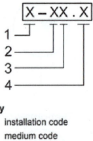

Key

1   installation code
2   medium code
3   circuit number
4   component number

Figure 5-34. This is the identification code of components and hose assemblies. This identification code should be used in diagrams and any related documentation.

in a circuit. The component is normally identified by a letter using a code such as:

- H for hydraulics.
- P for pneumatics.
- C for cooling.
- K for cooling lubricant.
- L for lubrication.
- G for gas engineering.

The third position in the code should refer to the number of the circuit in the installation. The normal procedure is to start circuit numbering with 0 and proceed with numbers in sequential order. The fourth code position is a different number for each component in the circuit. For example, four limit switches in a circuit are consecutively numbered 1 through 4. See **Figure 5-35**.

In addition to the coding system for components, several additional items of information should be provided about system piping. Lines should be labeled to indicate their function, such as P for pressure supply lines in both hydraulic and pneumatic systems. In hydraulic systems, tank return (reservoir) lines should be labeled T and leakage drain lines indicated with an L. Connections for the piping of circuit subassemblies should also be identified on the diagrams.

Technical information

A wide variety of information can be included in a circuit diagram. ISO standards suggest a complete list of technical items that can be included. The following items should be considered basic. They provide valuable information about a circuit or system.

- Prime mover rated power, direction of rotation, and speed.
- Flow and pressure ratings of the pump or compressor in the system.
- Fluid capacity recommendations for the hydraulic reservoir.
- Type and rating of hydraulic fluid used in the system.
- Level of filtration required in the system.
- Capacity and maximum pressure rating of the pneumatic receiver.
- Nominal sizes of pipes, tubing, and hoses used in the system.
- Details of the actuators, including type, size, pressure rating, and function.
- Pressure control valve information, including function and pressure settings.
- Flow control valve information, including type and flow rate.
- Directional control valve information, including activation method and center position configuration.
- Accumulator volume, precharged pressure, and working pressure range.

Additional support information also can be included to assist in circuit installation, operation, and maintenance. For example, a component list, sequence of operation, function chart, and suggested circuit or system arrangement may be included.

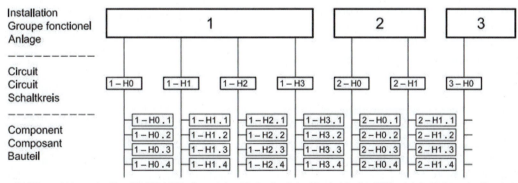

Figure 5-35. This chart shows the relationship among the individual parts of the identification code of components.

Summary

- Standards are considered to be anything that can be used as a basis for comparison.
- General standards are established by numerous public and/or private organizations. Specific fluid power industry standards are imposed by such groups as the National Fluid Power Association (NFPA), the Fluid Power Distributors Association (FPDA), the International Fluid Power Society (IFPS), and the Fluid Power Education Foundation (FPEF).
- Domestic and international standards coordinating organizations exist to assist the many groups involved in standards development. The American National Standards Institute (ANSI) is the primary domestic coordinating group, while the International Organization for Standardization (ISO) is the international group.
- Symbols are used for designating components in fluid power circuits in all parts of the world. The types of symbols most often encountered in the fluid power field are the graphic, pictorial, cutaway, and combination.
- Pictorial symbols are made up of line drawings of the exterior shapes of fluid power components while cutaway symbols are miniature section drawings of the components. Though useful for training, these symbols are not easily standardized.
- Graphic symbols are made up of basic figures including lines, circles, squares, triangles, dots, and arrows. These symbols are intended to represent the type, functions, operation, and external connections of a fluid power component, but not show the actual construction of the unit.
- Circuit diagrams provide a variety of information about fluid power systems for use during system assembly, operation, and testing. These documents include schematic diagrams of the fluid power circuits, component lists, and a sequence of operation of the system.
- All system components in a circuit diagram should be identified with a code number to allow easy identification. ISO standards provide a four element coding system to provide a logical, easily applied identification system.

Internet Resources

The following are some useful resources available on the Internet. Enter a company or organization name into a search engine to access its website. Explore the various areas of the sites to discover useful fluid power resources.

American National Standards Institute (ANSI). Provides information about ANSI, including an overview of the organization, structure and management, membership, and standards activities.

Compressed Air and Gas Institute (CAGI). Provides information about CAGI, which is dedicated to promoting improved performance of air and gas compressors and related equipment.

Engineers Edge. Shows ISO hydraulic schematic symbols organized in several categories.

International Fluid Power Society (IFPS). Provides information for individuals interested in education, training, and certification in the fluid power workforce.

International Organization for Standardization (ISO). Provides links to the general structure of the organization, standards development, current news and media, frequently asked questions, and products.

National Fluid Power Association. Provides an overview of the association, including structure of the organization, details of the industries served, membership, sponsored events, and involvement in fluid power education.

Norgren Graphic Symbol Library. Provides an extensive library of ISO pneumatic symbols.

Chapter Review

Answer the following questions using information in this chapter.

1. The primary coordinating work with standardizing groups in the United States is done by the _____.
   A. National Fluid Power Society (NFPA)
   B. International Organization for Standardization (ISO)
   C. International Fluid Power Society (IFPS)
   D. American National Standards Institute (ANSI)

2. Standards are based on _____.
   A. custom
   B. general consent of an affected group
   C. regulatory requirements
   D. All of the above.

3. *True or False?* When developing a standard, organizations follow procedures that assure input by individuals and organizations affected by the standard.

4. *True or False?* Fluid power symbols are used only to communicate system construction.

5. What are the symbols called that are really miniature section drawings of component parts?
   A. Graphic.
   B. Pictorial.
   C. Cutaway.
   D. Combination.

6. Graphic symbols are *not* intended to show the
   _____.
   A. function of the component
   B. location of the component
   C. appearance of the component
   D. type of fluid

7. The symbol for a component involving two or
   more functions is surrounded by a chain line that
   represents a(n) _____.
   A. enclosure
   B. working line
   C. drain line
   D. mechanical connection

8. The preferred orientation of graphic symbols is in
   increments of _____.
   A. 15°
   B. 45°
   C. 60°
   D. 90°

9. The symbol for a control component or a prime
   mover appears in a graphic diagram as a _____
   resting on one side.
   A. rectangle
   B. square
   C. triangle
   D. parallelogram

10. *True or False?* Direction of fluid flow is indicated
    by a small equilateral triangle.

11. The symbols for the devices used to operate
    directional control valves are drawn on the end of
    the basic valve symbol. These devices are referred
    to as _____.
    A. lever handles
    B. control mechanisms
    C. energy converters
    D. shifting units

12. Describe the basic symbol for a hydraulic pump.
    A. Components surrounded by a chain line.
    B. Circle containing a solid triangle.
    C. Small, solid equilateral triangle.
    D. Small, open equilateral triangle.

13. Directional, pressure, and flow control device
    symbols are made up of one or more _____.
    A. boxes
    B. circles
    C. triangles
    D. dots

14. *True or False?* In a simplified symbol for a flow control
    valve, an arrow that passes through the orifice in the
    symbol indicates the valve is not adjustable.

15. In a hydraulic system drawing, multiple _____
    symbols are used to terminate fluid return lines to
    reduce the number of lines returning to the main
    reservoir.
    A. valve
    B. enclosure
    C. control component
    D. reservoir

Apply and Analyze

1. Several different organizations are involved in
   determining standards within the fluid power
   industry. What would be the impact on the
   industry if these organizations did *not* coordinate
   their efforts in standardization? Describe at least
   three different impacts.

2. Draw a graphic representation of the circuit
   shown in **Figure 2-19**. Are there other graphical
   symbols that could be used to represent some of
   the components? If so, identify the components
   and draw a second symbol that could represent
   these components.

3. Redraw a circuit shown in **Figure 5-5**, replacing
   the cutaway and pictorial symbols with graphical
   symbols. Label all of the components. What
   information contained in the cutaway and pictorial
   symbols is not displayed by the graphical symbols?

4. Describe a situation in which you would choose to
   use pictorial or cutaway symbols in a circuit drawing
   rather than graphical symbols. Explain your choice.

Research and Development

1. Find a hydraulic or pneumatic circuit diagram on
   the Internet that includes at least 10 components.
   Identify and label the components. Redraw the
   circuit using pictorial and/or cutaway symbols
   for at least five different components.

2. Research one of the organizations involved in
   developing or coordinating fluid power industry
   standards. Prepare a report that describes the
   organization's mission, priorities, and members.

3. Standards are developed with input from
   individuals and organizations affected by the
   standard. Investigate how American National
   Standards Institute (ANSI) or International
   Standards Organization (ISO) develop fluid
   power industry standards. Prepare a flowchart
   or other diagram that illustrates the process and
   includes the various participants in the process.

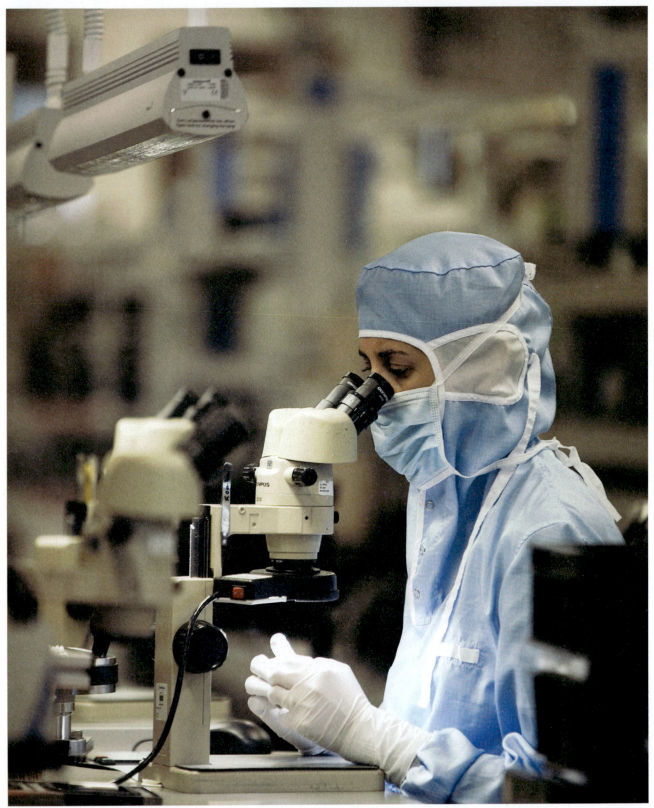

Atlas Copco

As fluid power standards are revised to reflect new materials, manufacturing techniques, and filtration standards, extra care must be taken to assure quality production.

6

Hydraulic Fluid
The Energy Transmitting Medium

In hydraulic systems, energy is transmitted through a liquid, rather than through shafts, gears, cables, and belts, as in a mechanical power transmission system, or the wires and complex control components used in electrical systems. Although hydraulic systems may sound less complicated, they are still complex devices requiring consideration of many elements. One of the fundamental elements that must be considered is the liquid to be used in the system. This chapter deals with the many factors relating to liquids that must be considered when selecting a hydraulic fluid for a particular application.

Learning Objectives

After completing this chapter, you will be able to:

- Describe the various functions a hydraulic fluid performs in a fluid power system.
- Identify and explain the general properties of a liquid that would make it suitable as a hydraulic fluid.
- Name and describe the general categories of materials that are commonly used as hydraulic fluids.
- Explain the terms used to describe the basic characteristics of hydraulic fluids.
- Explain procedures to follow for the selection and performance monitoring of hydraulic fluids.
- Describe appropriate procedures for handling, storage, and disposal of hydraulic fluid.

Key Terms

absolute viscosity
additives
anti-wear agents
API gravity
biodegradable fluids
capillary viscometer
catalyst
emulsion
extreme-pressure agent
film
fire point
flash point

Four-Ball Method
friction
high-water-content fluids
  (HWCF)
inverted emulsion
kinematic viscosity
lubrication
lubricity
oil-in-water emulsion
oxidation
oxidation inhibitor
phosphate ester

polyglycol
pour point
rust inhibitor
Saybolt viscometer
sludge
spontaneous ignition
synthetic fluids
Timken method
viscosity
viscosity grades
viscosity index number

6.1 Functions of a Hydraulic Fluid

The primary function of the liquid in a hydraulic system is to transmit the energy to do the work the system is designed to complete. The fluid is the medium by which the energy is transmitted from the pump to the cylinders and motors that do the actual work. To do the best possible job, the fluid needs to easily flow and quickly respond to pressure and directional variations. The fluid in a system should be considered just as important to the system as any of the hardware components, **Figure 6-1.** In addition to transmitting energy, hydraulic fluids act in a number of other roles to ensure the effective operation of a hydraulic system:

- Lubricating component parts.
- Helping maintain proper system temperature.
- Sealing clearances between internal parts.
- Helping prevent rust and corrosion of parts.
- Cleaning the internal parts of the system.

A key function is the lubrication of the internal parts of the various components. The amount of movement of internal parts differs between components. Some valves involve only slight movement for pressure or flow control, while pumps and actuators can involve both high-speed rotation and sliding movement. The hydraulic fluid must provide lubrication to reduce friction and minimize wear of parts in each of these situations.

Assisting in the maintenance of proper system temperature is another basic function of the fluid. Heat generated by friction, pressure drops across valves, or external heat sources can raise the temperature of system components to a level where system efficiency is lost or damage occurs to parts. The fluid must be capable of readily absorbing the heat and transferring it to other sections of the system where it can be dispersed. Heat can be transferred from the fluid to the atmosphere or other medium, such as the water in a heat exchanger.

Sealing clearances between the internal parts of components is another role a hydraulic fluid must perform. Many valve designs depend on close manufacturing tolerances to control internal fluid leakage. An example is directional control valves, where fluid-film strength and small clearances, rather than separate seals, control leakage between the spools and the valve body. See **Figure 6-2.**

Another important function of a hydraulic fluid is to protect the internal surfaces of component parts from rust and corrosion. Rust and corrosion may occur when water enters the system through condensation or directly during machine cleaning. Other contaminants can also enter the system during machine operation. Combinations of these materials often result in the formation of chemicals that can cause rapid deterioration of component parts. Fluids must be formulated to protect systems against these elements.

A hydraulic fluid must also function as a basic system cleaner. Fluid movement through a hydraulic system constantly flushes the system. Insoluble contaminants that enter the system, materials that are present from system assembly, or other materials that have formed because of chemical changes slowly move through a system. A quality fluid allows rapid settling and separation of these contaminants once they have reached the relative calm fluid found in the system reservoir.

Goodheart-Willcox Publisher

Figure 6-1. All hydraulic systems use some type of fluid to transmit power within system circuits and perform a variety of additional functions. This fluid is as important as any of the mechanical components of the system.

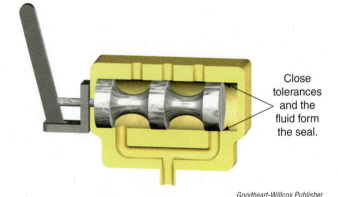

Close tolerances and the fluid form the seal.

Goodheart-Willcox Publisher

Figure 6-2. The hydraulic fluid acts as a seal when close machining tolerances, such as in this spool design, are used to control internal leakage.

6.2 Performance Characteristics of a Hydraulic Fluid

The numerous functions required of a hydraulic fluid mean it must be carefully selected if it is to perform satisfactorily in a system. A variety of characteristics must be considered when choosing a fluid for a particular system, including:

- Lubricating power.
- Resistance to flow.
- Viscosity stability.
- Low-temperature-operating ability.
- Resistance to oxidation.
- Reaction to condensation and other water.
- Resistance to foaming.
- Resistance to fire.

6.2.1 Lubricating Power

Lubrication deals with the reduction of friction between the surfaces of two bodies. *Friction* is the resistance to movement between two surfaces in contact. It is evident when you attempt to slide one body of material that is resting directly on the surface of a second body. Friction occurs in this situation because of the roughness of the surfaces in contact. This roughness exists even in highly polished surfaces. The amount of friction varies, depending on the force pushing the

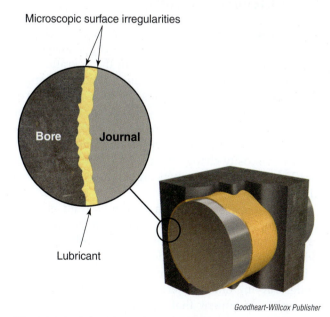

Microscopic surface irregularities

Bore   Journal

Lubricant

Goodheart-Willcox Publisher

Figure 6-3. A fluid fills the irregularities on the surfaces of bearings and provides a film that separates the surfaces. The molecules of the fluid slide over each other providing smooth operation.

surfaces together and the surface roughness. A smooth surface or light load produces less friction than a rough surface or heavy load.

The resistance to movement between two surfaces is greatly reduced by placing a thin layer of liquid between them. This layer of liquid, called a *film*, separates the surfaces of the materials. The film is made up of molecules that are strongly attracted to each other, but can freely move past each other. The molecules in the film fill the irregularities in the surfaces of a bearing, for example, and form a layer to separate the parts of the bearing. See **Figure 6-3**. When the liquid film is in place, the two surfaces of a bearing are separated from each other. The molecules of the liquid slide over each other and allow the parts of the bearing to move with little or no actual contact between their surfaces. The resistance to movement of the bearing is then determined by the resistance of the molecular movement within the film, rather than the rough surfaces of the bearing material.

The ability of a liquid to form a strong film between two bearing surfaces and adhere to the surface of the material is known as *lubricity*. This characteristic is very important in a hydraulic fluid, as the fluid must have the ability to minimize the wear of component parts under a variety of conditions. Wear in pumps, cylinders, motors, and control valves causes leakage, pressure loss, and reduced accuracy of control. Hydraulic fluids of all types must have the ability to adequately lubricate these basic system components in order to minimize friction and provide smooth, efficient component operation with a minimum of wear.

6.2.2 Resistance to Flow

The internal resistance to flow of the molecules of a liquid is known as *viscosity*. A fluid with high viscosity resists flow, while one with low viscosity flows more easily under the same conditions. Viscosity is one of the more important properties of the fluid used in hydraulic equipment, **Figure 6-4**. A fluid with the proper viscosity forms a strong film to provide an internal seal in components and reduce wear by protecting parts from direct contact. However, it must not be excessively viscous, requiring excessive energy output to move the parts against the resistance of the fluid.

Changing the temperature or pressure of a hydraulic fluid also changes its viscosity. Warming a fluid causes it to flow more easily, while cooling it causes the fluid to become more viscous. Increasing the operating pressure of a system causes the viscosity of the system fluid to increase. Care must be taken to select a fluid based on the expected normal operating temperature and pressure of the system, **Figure 6-5**.

Goodheart-Willcox Publisher

Figure 6-4. Using the proper viscosity fluid in a hydraulic system is important, as excessively low viscosity will not provide adequate lubrication and too high of a viscosity will require excessive energy output to operate. The oil on the right is more viscous than the oil on the left.

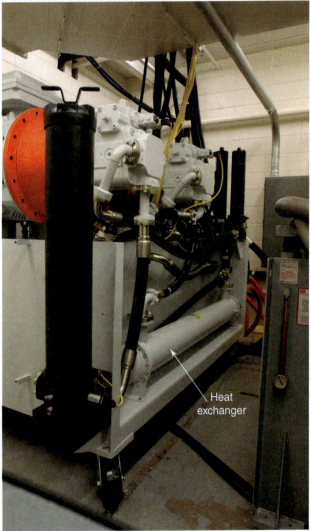

Heat exchanger

MTS Systems Corporation

Figure 6-5. The viscosity of a fluid changes with its temperature. It is important to monitor the operating temperature of a system. If correct temperature levels cannot be maintained, a heat exchanger may be required in the system.

If a fluid has a viscosity that is too low for system operating conditions, a number of problems can result:

- Increased wear.
- Increased internal leakage.
- Decreased pump efficiency and control accuracy.
- Increased system temperature.

There can be an increase in the overall wear in the system, as the fluid will not provide a sufficiently strong film to protect parts from friction. There can also be an increase in internal leakage in components because the fluid cannot provide an adequate seal of the clearances between close-fitting, moving parts. This leakage will cause less efficient operation of pumps and motors, less accurate control and slower response time for all actuators, and an increase in system operating temperature.

Selecting a fluid with a viscosity that is too high for system operating conditions may also cause a number of problems. A majority of these problems are caused by the higher internal friction of a high-viscosity fluid. Increased friction may result in increased power consumption, higher pressure drops across components, overall sluggish system operation, and an increase in system operating temperature. The higher viscosity fluid can contribute to pump cavitation problems because of increased resistance to flow through the inlet line of the pump. Separation of air and dirt particles from fluid returned to the reservoir can also be inhibited by an excessively high viscosity.

6.2.3 Viscosity Stability

It would be very desirable if the viscosity of a hydraulic fluid was fixed and did not vary during the changing conditions encountered throughout the operation of a fluid power system. However, fluid viscosity changes as system temperature and pressure vary. A measure of the relative change in viscosity that can be expected for a given change in temperature is provided by a viscosity index number. A fluid with a high index number has less viscosity change than a fluid with a low number. Viscosity index numbers are discussed later in this chapter.

The viscosity index of a hydraulic fluid is especially important when the operating temperature range is great. Special attention to viscosity index is wise when selecting fluid for mobile equipment operating with minimum system temperature control and under varying ambient temperatures, **Figure 6-6**. For example, equipment that operates in both a warm manufacturing area and a cold warehouse must have fluid with the proper viscosity index. The importance of the viscosity index becomes less in a system with carefully designed temperature controls and a relative constant ambient temperature.

Atlas Copco

Figure 6-6. A fluid that maintains a uniform viscosity is especially important in hydraulic systems that experience a wide range of ambient temperatures. Viscosity stability is indicated by a viscosity index number, with a higher number indicating greater stability.

6.2.4 Low-Temperature-Operating Ability

When hydraulic systems operate in a cold environment, the fluid must be able to flow under atmospheric pressure to the intake side of the pump. The characteristic that indicates the ability of a liquid to flow under these conditions is called *pour point*. This fluid property must be carefully considered when extremely cold system-start-up temperatures are a possibility. It is an especially important consideration for equipment stored outside or in unheated facilities in cold climates. Generally, the pour point of the fluid used in a system should be 20°F below the lowest-expected ambient temperature.

6.2.5 Resistance to Oxidation

Oxidation is a chemical reaction that increases the oxygen content of a compound. Many hydraulic fluids are susceptible to oxidation. System operating conditions that promote oxidation and can lead to general deterioration of hydraulic fluids include:

- High temperatures.
- Entrainment of air in the fluid.
- Reactions caused by contact with different metals in the system.
- Contaminants in the system, such as water and dirt.

All of these conditions tend to reduce the chemical stability of system fluid and shorten its service life.

High-operating temperatures in any area of a system increase the oxidation rate of the fluid in the system. The fluid temperature in the reservoir of an operating hydraulic system should ideally be 110–140°F, **Figure 6-7**. In many systems, varying load conditions cause the temperature to sporadically exceed this level.

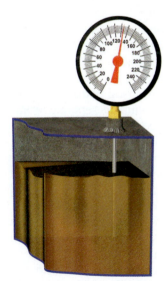

Goodheart-Willcox Publisher

Figure 6-7. The oxidation rate of hydraulic fluids increases as the temperature of the fluid increases. Most systems recommend a reservoir fluid temperature range of 110–140°F.

In some systems developed for special applications, design limitations may require higher general operating temperatures. Higher temperatures may also occur in localized areas of a system where design features do not allow the fluid to freely circulate to dispense heat.

Hydraulic fluid normally contains small quantities of air that are dissolved as the fluid passes through the system. The higher the system pressure, the greater the quantity of air that can be dissolved. When the fluid moves to a low-pressure portion of the system, the dissolved air forms bubbles that then move with the fluid through the system. Air bubbles can also be mixed into the fluid by improperly designed return lines in the reservoir. A reservoir with inadequate capacity or leaks in the pump intake line or any other line or component that operates below atmospheric pressure will also increase oxidation. These *entrained* air bubbles increase the surface area of the contact between the fluid and the air and tend to increase the rate of fluid oxidation.

A number of metallic materials commonly used in the construction of hydraulic systems act as a ***catalyst***, promoting the oxidation of the system fluid. Copper is generally recognized as the greatest promoter of chemical reaction in systems. Iron, aluminum, and some other materials act to a lesser degree as oxidation promoters. Care must be taken to minimize the amount of these materials in system construction.

6.2.6 Reaction to Condensation and other Water

Hydraulic systems are usually vented to the atmosphere at the reservoir. When this is the case, the system fluid and atmosphere have direct contact. As a result, moisture in the air can condense and mix with the hydraulic fluid. Water can also enter a system during equipment cleanup or exposure to weather in exterior installations. Regardless of the type of fluid used in the system, this water can cause operational and service problems.

In a system using petroleum-based fluid, part of this moisture remains separate from the hydraulic fluid and can be drained from the bottom of the reservoir. However, the water and hydraulic fluid can form an *emulsion*, which is a mixture of fluids that usually cannot be mixed. This is difficult to remove and can promote corrosion, decrease heat dispersion, and generally shorten the service life of the fluid. In situations where exposure to water is possible, selecting a fluid that resists emulsification is important to ensure proper system operation.

In systems using water-based fluids, additional water modifies the fluid. The fluid characteristics can be changed to the point where system performance, component service life, and fluid service life are negatively affected. Viscosity and lubricating properties are

primary among these changes, although many of the other basic properties are reduced as the water content increases in the fluid.

6.2.7 Resistance to Foaming

Fluid foaming and air entrainment can cause a variety of problems in a hydraulic system. Air can be drawn into a system through leaks in any line that operates below atmospheric pressure. Air can also enter at a pump inlet line that is partially above fluid level because of low system fluid. Return lines that end above fluid level because of low fluid levels can also cause air to be churned into the fluid.

In addition to increased oxidation, the foam that forms can, in extreme cases, actually overflow through the vent of the reservoir. The resulting fluid loss creates a messy work environment. Additionally, a serious system problem results when the foam is drawn into the pump and forced out into other system components. The air in the foam/fluid mixture is compressible and can cause slow, spongy, and erratic system operation. Foam combined with other system contaminants, such as dirt and water, can also form undesirable residue, known as *sludge*, that reduces the performance and service life of both the fluid and the system components.

6.2.8 Resistance to Fire

The danger of fire exists, to some extent, in any situation where hydraulic fluid power is used to operate equipment. These potential fire situations range from typical pieces of equipment on the floor of a manufacturing plant or assembly operation to coal mining equipment and aircraft installations. The level of fire resistance required of a fluid varies considerably by application. In many situations, petroleum-based fluids can provide the required safety level, while in other situations water-based or synthetic liquids must be used.

Three terms are used to indicate a fluid's susceptibility to heat and fire. The first of these is ***flash point***. This is the lowest temperature at which a liquid gives off sufficient vapors to ignite when an open flame is applied under specific test conditions. The second term is ***fire point***. This refers to the temperature at which a substance is vaporizing rapidly enough to ignite and continuously burn if an open flame or spark is applied. Flash and fire points are discussed in more detail later in the chapter. The third term is ***spontaneous ignition***. This is the temperature at which a fluid will ignite and burn without the application of an external flame or spark. One or more of these terms is often used to designate the required fire resistance of a fluid for a particular operation.

6.3 Commonly Used Hydraulic Fluids

Many liquids can be used in hydraulic equipment to transmit power, thus performing the primary function of a hydraulic fluid. However, many liquids do not have essential properties that allow them to provide functions key to the effective operation of a typical hydraulic system. For example, water is readily available and inexpensive for many installations. However, water alone is a poor lubricant, promotes rust, freezes, and rapidly evaporates at temperatures within the operating range of many systems. Petroleum-based liquids are very suitable for hydraulic fluids, as they typically possess many characteristics that are desirable for efficient system operation. However, under certain operating conditions, fire resistance greater than that of petroleum-based fluids is essential. Fire-resistant fluids may be water based or synthetic.

6.3.1 Petroleum-Based Fluids

The most common hydraulic fluid consists of a blend of petroleum oils and additives that produces a desired set of properties, **Figure 6-8**. This blend produces a combination of properties that allow the fluid to perform satisfactorily in an operating hydraulic system. A quality petroleum oil blended for use as a hydraulic fluid should:

- Readily transmit pressure and energy.
- Provide lubrication for the bearing surfaces in system component parts.
- Seal close-fitting parts against leakage.
- Act as a cooling agent for the system.
- Remove dirt and wear particles from internal parts.
- Protect component parts from rust and other corrosion.

In addition, petroleum-based fluids can also withstand the effects of oxidation, resist foaming, and readily separate from entrained air, dirt, and water particles once the fluid returns to the system reservoir.

6.3.2 Biodegradable Fluids

Biodegradable fluids have been developed to meet the need for environmentally safe fluids for use in hydraulic systems in mining applications, offshore platforms, national parks, wildlife refuges, ski resorts, and farming operations, **Figure 6-9**. Small spills of these fluids

Grayling Recreation Authority, Hanson Hills Recreation Area

Figure 6-9. Biodegradable fluids have been developed for use in situations where environmentally safe fluids are required. Parks, nature trails, and resorts are typical examples of these situations.

Element	Function
Refined mineral oil	Primary carrier that provides the base of the fluid
Viscosity-index improver	Reduces viscosity changes due to temperature variation
Anti-wear additive	Reduces bearing surface wear under light to normal loads
Extreme-pressure additive	Protects bearing surfaces from severe-loading conditions
Corrosion inhibitor	Prevents corrosion of components from oxygen and water
Oxidation inhibitor	Prevents oxidation of hydraulic fluid
Pour-point improver	Lowers the pour point of the hydraulic fluid
Defoamer	Prevents formation of foam in the fluid
Demulsifier	Allows rapid separation of oil and water
Dispersant	Reduces deposit buildup by holding contributing elements in suspension

Goodheart-Willcox Publisher

Figure 6-8. Elements of a typical petroleum-based hydraulic fluid.

will not harm soil or waterways. The fluids are broken down by organisms that normally exist in nature. However, larger spills should be treated the same as for a regular petroleum-based product. Used oil should also be disposed of according to federal, state, and local regulatory codes.

Biodegradable fluids are vegetable- or synthetic ester-based oils that provide good service in moderate to severe conditions. Many of the fluids have a tendency to darken during normal service. Therefore, viscosity change is a better indicator of oxidation changes than fluid color. Care needs to be taken not to contaminate these fluids with mineral-based oils as this will adversely affect the biodegradable characteristics of the products.

6.3.3 Soluble-Oil Emulsions

Some fluid power systems that involve huge component parts, such as major forging and extrusion presses, use water as the primary hydraulic fluid, **Figure 6-10**. However, water by itself cannot provide sufficient lubrication, sealing, and rust prevention characteristics to ensure effective system operation and a reasonable service life for the component parts. The water is, therefore, mixed with 1–5% soluble oil to form an *oil-in-water emulsion*. This combination can function well with specialized pumps at pressures in the range of 2,000 to 3,000 psi. The high water content of this fluid requires that it not be subjected to freezing temperatures. Also, the system operating temperature should not exceed 140°F. Higher operating temperatures produce an excessive rate of evaporation, requiring considerable maintenance time.

6.3.4 High-Water-Content Fluids

High-water-content fluids are primarily water to which 2–5% soluble chemicals have been added to form a solution. These chemical additives provide the lubrication, rust prevention, and wear resistance needed to protect the various parts of system components. Because of the high water content and the water-to-additive ratio, the fluids are commonly known as *high-water-content fluids (HWCF)*, or 95/5 fluids.

These fluids can function in specially designed reciprocating pumps at pressures as high as 10,000 psi. The limited anti-wear characteristics of most of these fluids restricts the operating pressure of systems using typical piston and vane pumps to 1,000 psi. They are not recommended for use in systems that use gear pumps. The high water content of these fluids requires that freezing temperatures be avoided and the maximum system operating temperature be limited to 140°F to avoid excessive water loss by evaporation.

6.3.5 Fire-Resistant Fluids

Fluid power systems operate under diverse conditions with many situations involving high ambient temperatures and, in some cases, actual exposure to open flames. To reduce the possibility of fires in those applications, a fire-resistant fluid is typically used in the system. The fluids are not generally fireproof. However, they will not burn without sustained exposure to an ignition source.

Certain properties of fire-resistant fluids differ considerably from petroleum-based fluids. This results in the need to reduce system operational specifications, such as pressures and operating speeds. The expected service life of components is also reduced in many cases. Lower lubricity; chemical reactions with seals, gaskets, and nonferrous metals; and environmental toxicity are several of the problems that may be encountered when using fire-resistant fluids.

Fire-resistant fluids include water-oil emulsions, water-glycol fluids, and synthetic fluids, **Figure 6-11**. The following sections include information on fluid formulation, properties, and areas of application for these types of fluids.

Brand X

Figure 6-10. Large forming presses often use soluble-oil emulsions as a hydraulic fluid. These fluids are susceptible to both evaporation and freezing because of their high water content.

Fluid Type	Composition	Comments
Water Content	Oil-in-water (5% oil, 95% water)	• Extremely high fire resistance • Requires typical hydraulic fluid additives • Has strengths and weaknesses of plain water
	Water-oil (40% water, 60% oil)	• High fire resistance • Requires typical hydraulic fluid additives • Has inherent instability and maintenance needs
	Water-glycol (35–50% water, remainder glycol)	• Most popular of the fire-resistant fluids • Can be used with most equipment designed for oil • Must be closely monitored for evaporation
Synthetic	Phosphate esters, chlorinated hydrocarbons, or various combinations	• Contain no water or petroleum oil • Some available brands are biodegradable • Cost is a selection factor

Goodheart-Willcox Publisher

Figure 6-11. Characteristics of common fire-resistant hydraulic fluids.

Water-in-oil emulsions

Water-in-oil emulsions contain approximately 40% water in an oil base to provide the desired fire resistance. The oil surrounds the finely divided water droplets to form what is often called an *inverted emulsion* because the water is suspended in oil instead of the oil in water. These fluids should not be confused with the soluble oil and high-water-content fluids, which have a much higher water content and the oil is suspended in the water.

The initial viscosity of water-in-oil emulsions is usually higher than the viscosity recommendations for most fluid power systems. However, the viscosity of the fluid tends to decrease during system service to the point where it is approximately the same as a comparable petroleum product.

Pumps need to be specially selected for systems that use water-in-oil emulsions. Also, system operating temperatures should be held below 140°F and water evaporation must be carefully monitored. A laboratory test is required to determine the amount of water to add to the system in order to maintain correct fluid viscosity. This test is essential, as the viscosity of an inverted emulsion is higher than the viscosity of either of its primary fluids.

Water-glycol fluids

Water-glycol fluids are made of 35–50% water with the remainder being some type of polyglycol. *Polyglycol* has a chemical structure similar to automotive antifreeze. The glycol and water mix to form a uniform fluid rather than a suspension of one fluid within the other, as is the case with an emulsion. The viscosity of water-glycol fluids is comparable to petroleum-based fluids. In addition, they have better stability, are less expensive, and provide better fire protection than inverted emulsions.

The water-glycol fluids adversely affect some types of seal material and paint. Also, system pressure and load durations need to be reduced to ensure adequate service life of pumps using these fluids. The maximum recommended operating temperature for systems using this type of fluid is 140°F to prevent excessive water evaporation.

Synthetic fluids

Synthetics have been used in a variety of fluid power applications, including engine oils and hydraulic fluids. The term *synthetic* has been defined in many ways. For purposes of this text, *synthetic fluids* are considered to be refined products containing additives that produce high levels of certain factors desired by end users. Synthetic fluids may contain highly refined petroleum as a carrier for the formulated elements.

Phosphate ester is the most common synthetic fire-resistant fluid manufactured for use in hydraulic systems, although other synthetic fire-resistant fluids are produced. These other fluids include chlorinated hydrocarbons, blends of various phosphate-ester-based fluids, and other materials such as mineral oil. All of the synthetic products have very high flash points and provide excellent fire resistance. They function well as hydraulic fluids and provide viscosities similar to petroleum-based products, satisfactory performance at high system pressures, and good component lubrication.

Synthetic fire-resistant fluids are very expensive in relation to other fluid types. Several other factors must also be considered before installing these fluids in a system. Special seal and gasket materials are required in all system components, as conventional sealing materials are not compatible with the chemical makeup of the synthetic fluids. The fluids also tend to dissolve most paints. Environmental factors must also be considered as the fluids are often toxic and may require special handling for disposal.

6.4 Hydraulic Fluid Additives

A number of *additives* are used to increase the chemical stability and overall performance of hydraulic system fluids. Typically, these materials are used to enhance most of the performance characteristics of the fluids. A wide variety of chemicals are used in these additives to improve the performance and extend the life of system components, as well as increase the service life of the fluid, **Figure 6-12**.

Extreme-pressure agents are used in fluids to help prevent metal-to-metal contact between the surfaces of the moving parts of components. These additives react with the metal surfaces to form a protective coating on the part. The agents are generally organic compounds containing one or more of elements such as sulfur or phosphorus.

Another group of additives that also prevent metal-to-metal wear are the *anti-wear agents*. These additives function at lower operating pressures and temperatures than the extreme-pressure agents. These materials are also typically formulated from organic compounds, including materials such as phosphates and ammonia-based salts.

Viscosity-index-improver additives are blended into hydraulic fluids to reduce the rate of viscosity change caused by system temperature variations. These additives allow the fluid to have the flow characteristics of a low-viscosity fluid when cold and a higher-viscosity fluid at operating temperature. A number of organic chemicals are used to provide this quality.

Low-temperature performance of fluids is increased by the use of pour-point depressants. These additives allow a hydraulic fluid to flow more freely at low temperatures, permitting cold-weather operation of equipment. Most of these additives use chemicals that prevent the formation of crystals in fluids. The crystals reduce the fluid's ability to freely flow. The ability to flow is especially important in pump intake lines. If a fluid becomes so thick it cannot be drawn into the system, noisy system operation and pump damage will result.

Oxidation inhibitors are used to reduce the tendency of hydraulic fluids to oxidize. These materials are formulated to interfere with the complex chemical reactions that continually occur in an operating hydraulic system. The rate at which these reactions occur is influenced by system operating temperature, the amount of contact the fluid has with air, and the catalytic effects of the various metals used in the system. The service life of hydraulic fluids can be extended by controlling these factors, as well as selecting a fluid that contains appropriate oxidation inhibitors.

Demulsifier additives improve the surface tension characteristics of petroleum-based hydraulic fluids. The higher-than-normal surface tension they create allows water that has entered the system to readily separate from the fluid and collect in the bottom of the reservoir. Water may enter the system through internal system condensation or from external sources. Rapid separation of this water reduces the possibility of an emulsion forming in the system, which can lead to sludge formation.

Surface tension modification is also the principle involved in anti-foaming agents. These agents *reduce* the surface tension of a fluid so any bubbles that form will quickly break down before foam is created in the reservoir. Preventing foam formation helps minimize fluid oxidation by reducing the air-to-fluid contact caused by the large surface area of the bubbles. It also eliminates spongy system operation that results if foam is pumped into the system. Also eliminated are the safety issues created when excessive foaming causes fluid overflow through the reservoir vent. This overflow can result in slippery floors, which are a hazard.

A group of additives is used to protect the metal parts of system components. Iron and steel parts are protected by various *rust inhibitors* that either neutralize acids or form a film on metal surfaces to protect them from damage. Nonferrous metal component parts are protected by corrosion inhibitors that chemically react with the metals to create protective surface films. Some hydraulic fluid producers also describe these film-producing additives as metal deactivators, as they decrease the catalytic effect of the metal on the oxidation rate of the fluid.

6.5 Hydraulic Fluid Specifications

Many different hydraulic fluids are used in the multitude of fluid power systems operating in the world today. This chapter has discussed a number of typical fluids and their performance characteristics. Comparing these fluids and selecting one that best fits the needs of a system can become a major task. Selection of a fluid should be based on recommendations from the

Vane tip and
cam ring contact

Goodheart-Willcox Publisher

Figure 6-12. Extreme-pressure and anti-wear agents help protect bearings and parts, such as pistons and vanes in pumps, from high loads and surface speeds that can cause rapid wear.

system or component manufacturer. Of course, these recommendations must be examined in relation to the operating environment of the system.

The companies formulating and marketing hydraulic fluids provide information about their products, including details regarding the performance characteristics of the fluids. The final selection of a fluid requires matching the requirements of the system, as described by the manufacturer, and the characteristics of a fluid, as established by the fluid producer. The match involves a recognized group of terms, test procedures, and units of measure. This section discusses the meaning of the indicated tests and the information commonly provided on the specification sheets of system/component manufacturers and hydraulic fluid producers.

6.5.1 Viscosity

Viscosity is a major factor in describing the performance characteristics of a hydraulic fluid. Its relative importance is indicated by the fact it is commonly listed as the first, or at least one of the first, items on a fluid specification/data sheet or bulletin. Several different systems of rating viscosity may be encountered when working in the hydraulic area. Each system provides consistent information within the rating system, although the specific meaning of the ratings varies among the systems.

Absolute viscosity

The *absolute viscosity* rating is also called dynamic viscosity. This rating is often used in engineering and scientific references. The rating probably will not be found on typical hydraulic fluid data sheets, but it is a term with which you need to be familiar. Absolute viscosity is defined as:

> The force per unit of area required to move one molecular layer of a substance over another layer of the material.

The usual description of this concept involves the force needed to move a flat plate over a fixed, parallel, flat surface at a constant velocity when the space between them is filled with the specified liquid. The unit of absolute viscosity is the poise, which has the dimensions grams per centimeter per second. The centipoise is 1/100 of a poise and is the unit most commonly used for the rating.

The concept of absolute viscosity involves a fluid film located between the moving and fixed surfaces described above and shown in **Figure 6-13**. This fluid film should be considered a series of molecular fluid layers separating the surfaces. The fluid is adhered to both surfaces, with the fluid velocity zero at the fixed surface. The fluid velocity at the moving surface is equal to the speed of that surface. The movement of the fluid layers between the two surfaces varies on a straight

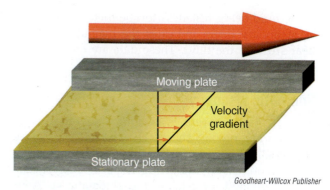

Goodheart-Willcox Publisher

Figure 6-13. The concept of viscosity involves the force necessary to shear molecular layers of fluid that form a film between two surfaces.

line between zero at the fixed plate and the maximum speed at the moving plate. As the moving plate slides along the fixed surface, the various molecular layers of fluid *shear*, allowing the plates to move without touching each other. Absolute viscosity is the force per unit area necessary to shear these layers of fluid.

There are a number of methods that may be used to measure absolute viscosity. Some of these use fundamental principles of fluid mechanics that are beyond the scope of this book. Others use mechanical devices to produce values that can be used to compare different fluids. One of these devices is a commercially available rotating drum viscometer. **Figure 6-14** illustrates the concepts involved in the design and operation of one of these test instruments. The device involves an inner stationary drum and a rotating outer drum separated

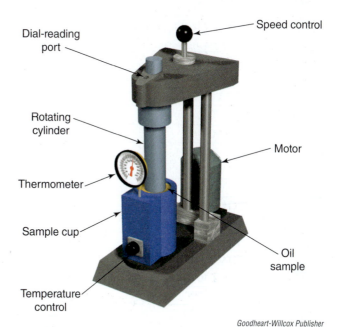

Goodheart-Willcox Publisher

Figure 6-14. The rotating drum viscometer uses a torque meter to measure the shear exhibited by a fluid, which is then used to calculate the material's absolute viscosity.

by the fluid being tested. A torque meter attached to the stationary drum indicates the magnitude of the shear stress in the fluid. These data are then used to calculate the viscosity.

Kinematic viscosity

Kinematic viscosity ratings are commonly reported on fluid information data sheets. The *kinematic viscosity* rating method is the most precise indicator of viscosity found on most hydraulic fluid specification sheets. The test method for establishing kinematic viscosity consists of measuring the time required for a fixed amount of fluid to flow under gravity through a calibrated, glass *capillary viscometer* while under a fixed pressure and temperature, **Figure 6-15**. The test is normally conducted at 40°C and 100°C. The final viscosity rating is calculated by multiplying the flow time by a calibration constant established for each individual viscometer. The unit of measure used is the stoke (St) or centistoke (cSt). ASTM standard D445 outlines the procedure commonly used in the United States for conducting a kinematic viscosity test.

ISO viscosity grades

The need for a wide range of liquid lubricant viscosities becomes more critical as new materials, designs, and applications develop for industrial equipment.

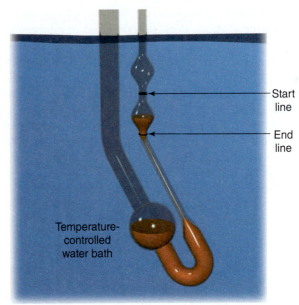

Goodheart-Willcox Publisher

Figure 6-15. The kinematic viscosity of a fluid may be determined by the use of a calibrated, glass capillary viscometer.

Internationalization of trade also requires a system accepted by business and industry around the world. These criteria have been met by ISO standard 3448, which includes twenty *viscosity grades*, **Figure 6-16**.

ISO Viscosity Grade	Nominal Viscosity at 40°C	Minimum Viscosity at 40°C	Maximum Viscosity at 40°C
ISO VG 2	2 cSt	1.8 cSt	2.2 cSt
ISO VG 3	3 cSt	2.7 cSt	3.3 cSt
ISO VG 5	5 cSt	4.5 cSt	5.5 cSt
ISO VG 7	7 cSt	6.3 cSt	7.7 cSt
ISO VG 10	10 cSt	9 cSt	11 cSt
ISO VG 15	15 cSt	13.5 cSt	16.5 cSt
ISO VG 22	22 cSt	19.8 cSt	24.2 cSt
ISO VG 32	32 cSt	28.8 cSt	35.2 cSt
ISO VG 46	46 cSt	41.4 cSt	50.6 cSt
ISO VG 68	68 cSt	61.2 cSt	74.8 cSt
ISO VG 100	100 cSt	90 cSt	110 cSt
ISO VG 150	150 cSt	135 cSt	165 cSt
ISO VG 220	220 cSt	198 cSt	242 cSt
ISO VG 320	320 cSt	288 cSt	352 cSt
ISO VG 460	460 cSt	414 cSt	506 cSt
ISO VG 680	680 cSt	612 cSt	748 cSt
ISO VG 1000	1000 cSt	900 cSt	1100 cSt
ISO VG 1500	1500 cSt	1350 cSt	1650 cSt
ISO VG 2200	2200 cSt	1980 cSt	2420 cSt
ISO VG 3200	3200 cSt	2880 cSt	3520 cSt

Typicall Hydraulic Fluid

Goodheart-Willcox Publisher

Figure 6-16. ISO viscosity grades indicating nominal, minimum, and maximum kinematic viscosity for each of the twenty classifications used in the system.

This system has been gaining acceptance in world markets for lubricants, hydraulic fluids, electrical oils, and other applications. The standard defines each of the grades in kinematic viscosity at 40°C. These ISO grades are defined in this country by ASTM standard D2422. Each of the grades includes the prefix ISO VG followed by a nominal viscosity number. These grade numbers range from a low of 2 to a high of 3200. The range of viscosity within each grade can vary plus or minus 10%. For example, an ISO VG 10 fluid could range from 9 to 11 cSt. Hydraulic fluid viscosity typically ranges from 22 to 150 on this viscosity grading scale.

SAE viscosity grades

The Society of Automotive Engineers (SAE) grading system is extensively used in the automotive field to rate engine oils, gear lubricants, and axle lubricants. Through the years, the system has also been used to rate hydraulic fluids. Although the system is not used as much in the fluid power field as it once was, the ratings are still found on some fluids.

This system defines grades ranging from a low of 0 to a high of 60, **Figure 6-17**. Each of the grades designates a defined kinematic viscosity (in cSt) at 100°C. The grades are defined by SAE standard J300.

The system also uses a W suffix. This indicates the ability of that fluid to perform at specified cold temperatures. These cold temperatures range from –5°C to –35°C.

Fluids graded by this system may be designated by two numbers, such as 5W-20. This grade indicates the material can perform as a 5W-grade fluid under cold conditions and a regular 20-grade fluid under normal working temperatures. Fluids with two grade numbers are called multiviscosity fluids.

Saybolt universal viscosity

The Saybolt universal viscosity system has seen widespread use in the fluid power field. The *Saybolt viscometer* is an apparatus containing a cup with a calibrated orifice in the bottom. See **Figure 6-18**. Ratings are established by heating the test fluid in the cup to a standard temperature of 100°F or 212°F and then measuring the number of seconds needed to fill a 60 mL sample flask with fluid flowing through the calibrated orifice. The resulting time measurement, called Saybolt Seconds Universal (SSU), serves as a means to compare fluids, but does not relate to the standard definition of viscosity.

Until recently, the Saybolt test procedure was listed as ASTM standard D88. However, that standard is no longer being supported by ASTM. Refiners and manufacturers are being encouraged to use more accurate means of measuring viscosity, such as those covered in ASTM standards D445 and D446. These standards outline test procedures and equipment for determining kinematic viscosity. ASTM also provides standard D2161 that lists the formula and tables needed to convert SSU to cSt. Even though the Saybolt test standard is no longer supported by a standardizing group, the procedure continues to be used in the profession. Most hydraulic fluid specification sheets list SSU information. Some data sheets carry the listing as Saybolt Universal Seconds (SUS).

Minimum Viscosity at 100°C	SAE Viscosity Grade
0W	3.8 cSt
5W	3.8 cSt
10W	4.1 cSt
15W	5.6 cSt
20W	5.6 cSt
25W	9.3 cSt
30	9.3 cSt
40	12.5 cSt
50	16.3 cSt
60	21.9 cSt

Goodheart-Willcox Publisher

Figure 6-17. Ten SAE viscosity grades that may be used with hydraulic oils in some applications.

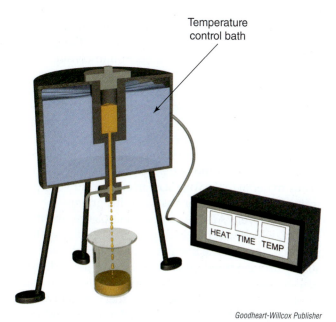

Goodheart-Willcox Publisher

Figure 6-18. Saybolt viscometer measures the time required for 60 mL of fluid to flow through a calibrated orifice at a specified temperature.

6.5.2 Viscosity Index

A viscosity index number appears as part of the specifications for most hydraulic fluids. The *viscosity index number* is widely accepted as an indicator of kinematic viscosity variation between 40°C and 100°C. The number is used as a measure of the rate of viscosity change as the temperature of the fluid changes. When comparing several fluids, the viscosity of the material with the highest index number changes less than those with lower index numbers. The higher the index number, the lower the rate of viscosity change. It must be emphasized that viscosity index is only an *indicator* of viscosity change due to temperature change. It is not a direct measure of viscosity or any other quality of a fluid.

ASTM standard 2270 is usually used in the United States to determine the viscosity index number for hydraulic fluids. The standard outlines two procedures for calculating the index number. Procedure A is for fluids with a viscosity index number up to and including 100. Procedure B is for fluids with an index number greater than 100. The kinematic viscosity of the fluid for which the index number is being determined must be known at 40°C and 100°C. Both procedures also require viscosity information about a reference fluid that is used in the calculations. The index number for the fluid can then be calculated using the procedures and formulas outlined in the standards document.

6.5.3 Specific and API Gravity

Both specific gravity and *API gravity* compare the weight of a given volume of a substance to the weight of an equal volume of distilled water at specified temperatures. Neither of the gravity measurement systems produce numbers that are indicators of the quality of a fluid, unless a number of other factors are taken into consideration. However, they are important for determining weights when systems are moving large volumes of fluid or when large quantities of a fluid are being purchased.

The process used to determine the specification and the unit of measurement varies between the two methods, even though the underlying concept is similar. Specific gravity is the ratio of the weight of a given volume of any material to the weight of an equal volume of pure water at 4°C. Pure water has a specific gravity value of 1.00. However, API (American Petroleum Institute) gravity, which was designed primarily for use with oil products, uses a scale that rates pure water at a value of 10.00. Also, the standard temperature for the API test is 60°F. Because of this, the test is often reported with a suffix of "60/60 degrees F," indicating that both substances in the test were at the recommended test temperature.

The scale used for reporting specific gravity provides a ratio between the weight of the test material and that of distilled water. Substances lighter than the water have a decimal figure less than 1.00, while those that are heavier than water have a figure greater than 1.00. In contrast to this, when using the API gravity test procedure, liquids that are lighter than water have values greater than 10.00, while those heavier than water have values less than 10.00. On either scale, the higher the number, the lighter the weight of the substance. See **Figure 6-19**.

The test for determining specific gravity or API gravity uses a hydrometer to measure the gravity of the fluid being evaluated. A hydrometer, thermometer, and clear-glass-hydrometer cylinder are all the equipment that is needed for these tests. Care needs to be taken to ensure the fluid is at the correct test temperature, no air bubbles have adhered to the surface of the hydrometer bulb, and the scale on the float is correctly read in relation to the meniscus (curved, upper surface) of the test fluid. ASTM standards D1298 and D287 provide detailed procedures, data tables, and conversion formulas for these tests.

6.5.4 Pour Point

The pour point indicates the lowest temperature at which a test sample of a hydraulic fluid moves when a specific cooling procedure is followed. This temperature serves as the baseline figure to establish the lowest temperature at which the fluid can practically be operated for cold-weather installations.

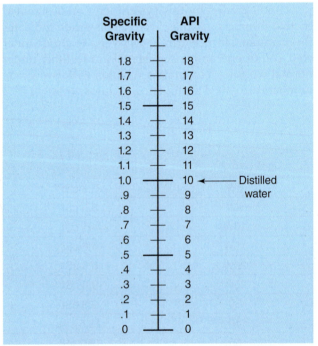

Goodheart-Willcox Publisher

Figure 6-19. Specific gravity and API gravity scales vary, with distilled water measuring 1.0 on the specific gravity scale and 10.0 on the API scale. The API gravity scale is designed for use with crude petroleum, liquid petroleum products, and mixtures of petroleum and non-petroleum products.

Goodheart-Willcox Publisher

Figure 6-20. Pour point is established when a cooled fluid will not flow when a test container is tilted using a specific procedure.

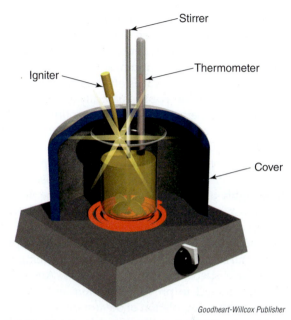

Goodheart-Willcox Publisher

Figure 6-21. The flash and fire points of a fluid are tested by slowly heating a fluid while periodically passing a small open flame over its surface to determine when sufficient vapors are being given off to ignite (flash point) and sustain combustion (fire point).

The test requires use of a sample test jar, a cooling bath, and thermometers for observation of the temperature of the sample and coolant, **Figure 6-20.** The hydraulic fluid sample is placed in the test jar and cooled in 3°C increments until the sample does not flow when tilted. The pour point is reported as 3°C above this temperature. ASTM standard D97 provides detailed procedures for this test.

6.5.5 Flash and Fire Points

Both the flash point and the fire point relate to the fire resistance of a fluid. Flash point indicates the fluid temperature needed to form a flammable mixture with air. Fire point establishes the fluid temperature required to support combustion. Flash point information is useful when shipping and safety regulations define flammable and combustible materials. Flash point can also be used to identify the presence of volatile and flammable substances in materials that are normally considered either nonflammable or having a high fire resistance.

Testing for flash and fire points requires a standardized test cup, heater, test flame applicator, and thermometer, **Figure 6-21.** The fluid sample is placed in the test cup and heated. As the sample approaches the expected flash point temperature, a test flame is applied to the surface of the fluid at 2°C increments. The flash point of the fluid is the temperature on the thermometer when a flash appears at any point on the surface of the fluid. Fire point is established by continuing to increase the temperature of the fluid until it ignites and burns for at least five seconds after the test flame is applied to the surface of the sample. ASTM standard D92 provides details of the procedures and test apparatus for these tests.

6.5.6 Rust Prevention

A specific test is available to indicate the ability of a fluid to aid in preventing the rusting of ferrous (iron based) parts. Rusting may occur should water become mixed with the hydraulic oil. The procedure is also used to test heavier-than-water hydraulic fluids. The test is used for monitoring in-service fluids as well as being a specification of new fluids.

The apparatus involved in this test includes a heated oil bath to maintain the temperature of the test sample, beaker to contain the sample, powered stirring device, thermometer, and steel test rod. A designated amount of test fluid is placed in the beaker. The thermometer, stirring device, and steel test rod are mounted in the beaker. A specified amount of distilled water is then added to the test fluid. The test is run at the designated temperature for 24 hours or longer with constant stirring. At the end of the test period, the steel test rod is removed and examined for rusting. If no rust is evident, the fluid passed the test. The test procedures allow for reporting of the degree of rusting, if that is desired. ASTM standard D-665 provides details of the procedures and test apparatus for this test.

6.5.7 Corrosion and Oxidation

The petroleum-refining process removes most of the sulfur compounds normally found in crude oils. However, traces of these materials remain and can cause problems with metal parts of hydraulic system components. This is especially true of any part made of copper. Standard tests have, therefore, been developed to test the effect of oils on copper and other metals, such as aluminum, bronze, magnesium, titanium, and steel.

ASTM standard D130 is often used to test the effect of the oil on copper. The test basically consists of placing a polished copper strip in a given quantity of the test oil and maintaining a specified temperature for a designated test period. The strip is then removed and compared in appearance to photographs of standard samples.

Another ASTM standard, D4636, is used to more thoroughly test hydraulic oils to determine their resistance to oxidation and any tendency to corrode various metals. This test uses washer-shaped samples of the test metals. A glass tube containing the oil sample and the metal samples is heated in a constant-temperature bath for a specified number of hours. Air is passed through the oil to provide agitation and a source of oxygen. After completion of the test period, the oil and the metal samples are examined to determine change in weight, any corrosion that appears on the metals, change in oil viscosity, weight loss of the oil due to evaporation, and measurable sludge caused by reactions between the oil and metal samples. Test results are reported as classifications matching descriptions found in the tables included in the standards document.

6.5.8 Demulsibility Characteristics

The ability of petroleum-based and synthetic hydraulic fluids to separate from water, or demulsify, is an important operating characteristic. ASTM standard D1401 provides a test method for determining and reporting this water-separation characteristic. The test is used to report specifications for both new fluids and for monitoring of oils already in service.

The procedure involves stirring a measured amount of distilled water into a test sample of hydraulic fluid at a specified temperature. The stirring duration is five minutes at a specified rate. The test sample is then set aside and examined at five-minute intervals until the test fluid and water are separated or until the test time expires. The results are reported in five-minute increments for clearing or as the total test time and volume of emulsified fluid remaining after the expired time.

6.5.9 Foaming Characteristics

A fluid that foams excessively can be a serious problem in hydraulic systems involving high-volume pumping rates. ASTM standard D892 provides a means for empirically rating the foaming characteristics of hydraulic fluids. Air at a measured flow rate is blown through a specified volume of fluid that is maintained at a constant test temperature. After five minutes, the air is shut off and the volume of foam measured.

The fluid is then allowed to settle for 10 minutes before the volume of foam is again measured. This procedure is repeated at two additional specified test temperatures. The test results are simply reported as passed or as the volume of foam present in each of the test sequences.

6.5.10 Lubricity and Wear-Prevention Characteristics

Several ASTM standards can be used to indicate the lubrication and wear-prevention characteristics of fluids. Some of these standards use specially designed laboratory equipment to conduct extreme-pressure tests to determine the capacity of a fluid to provide lubrication. Others use typical industrial hydraulic components to load a standard-production hydraulic pump with measured pump wear used to indirectly indicate the lubricating qualities of the fluid.

ASTM standard D2782 provides a test method with laboratory equipment for measuring the extreme-pressure properties of lubricating fluids. The procedure uses an extreme-pressure-testing device and is referred to as the *Timken method*. The test device contains a steel test cup that is rotated against a steel test block while being flooded with the test fluid. The test cup is pressed against the test block with slowly increasing pressure until scoring occurs. Additional test blocks are used to repeat the procedure to establish a precise maximum load the fluid film will withstand without breaking down. The results of the test are reported in pounds for both the maximum safe load and the minimum load that causes scoring.

Another standard for determining the wear-preventive characteristics of hydraulic fluids in situations relating to sliding steel-on-steel applications is ASTM standard D2266. This test procedure, referred to as a *Four-Ball Method*, involves rotating a steel ball under load against three stationary steel balls covered with the fluid under test. See **Figure 6-22**. Following specific test load and temperature conditions, the rotating ball is forced against the stationary test balls and turned at 1,200 rpm for one hour. The stationary balls are then removed from the tester and the diameter of the wear scars is carefully measured under a microscope. The results of the test are reported as the arithmetic average of the scar diameters. The test results can be used to establish relative wear-preventing properties for fluids. Correlations have not been established between the test results and expected service in the field.

Weight loss of internal pump parts may be used to establish the wear characteristics of hydraulic fluids. ASTM standard D2882 provides the procedures for such a test using a constant-volume, high-pressure vane pump. This standard makes use of a pump manufactured by Vickers. However, similar procedures are used with pumps made by other companies.

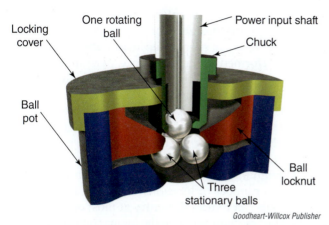

Figure 6-22. Four-ball laboratory test equipment may be used to establish the lubricating qualities of a fluid. The test equipment rotates one ball against three stationary balls that are covered with the test fluid. The degree of scaring on the balls indicates the lubricity of the fluid.

The test circuit required for this procedure includes the pump, system relief valve, reservoir, filter, heat exchanger, and pressure, flow, and temperature sensors. See **Figure 6-23**. A new pump cartridge is used for each test, with the cam ring and pump vanes accurately weighed before the test begins. The test fluid is circulated through the pump for 100 hours at a pressure of 2,000 psi. The system heat is controlled to provide a constant fluid temperature at the pump inlet. The results of the test are reported as the weight loss of the cam ring and vanes. The test report also includes observations of any unusual wear, scuffing, seal deterioration, or deposits.

The D2882 standard is not currently supported by ASTM. A replacement standard may be made available when a new pump model has been selected for use in the test circuit.

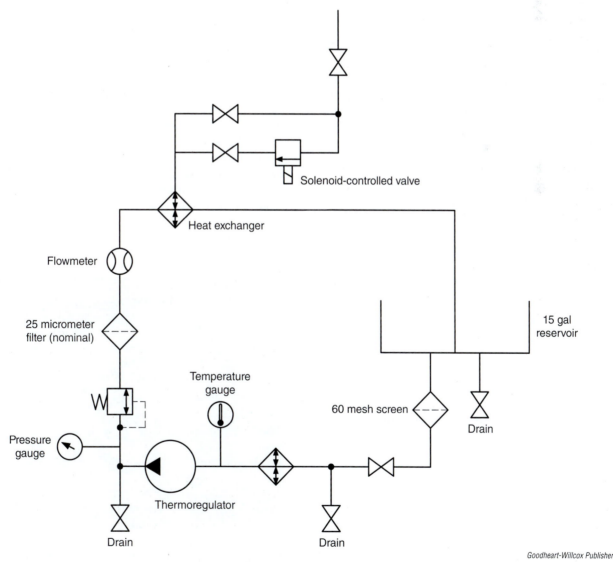

Figure 6-23. The wear characteristics of hydraulic fluids can be established by using a pump test circuit made up of standard fluid power components operating under carefully controlled temperature, load, and time conditions.

6.6 Handling and Maintaining Hydraulic Fluids

Hydraulic fluids of all types are expensive. They can also be costly to maintain, if care is not exercised in their handling and upkeep. Proper care not only reduces costs of new replacement fluids, but reduces the time maintenance personnel spends in flushing systems and changing fluids. In many cases, proper fluid care also extends the service life of the total fluid power system, further reducing costs. Awareness of a number of simple rules and other factors can reduce the costs related to the fluid used in a system and help extend the service life of equipment.

6.6.1 Storage and Handling of Fluids

Storing new, unused hydraulic fluids is an important consideration in any organization that makes use of fluid power systems. Improperly stored fluid can become contaminated. Contaminated fluid not only results in the loss of the fluid, but the loss associated with increased maintenance time if contaminated fluid is installed in a system. Considerable time can be lost due to system flushing that may be required or the downtime of equipment idled waiting for the arrival of a new supply of fluid. In general:

- Store drums in a cool, clean, dry place.
- Place drums on their sides to reduce chances of contamination.
- Carefully clean drum tops before removing bungs.
- Use clean fluid-transfer equipment.

Drums used for fluid storage should be stored indoors in a cool, clean, dry place. Storing the drums on their sides reduces the accumulation of dirt around the bung, **Figure 6-24**. It also eliminates any seepage of water or other liquids into the barrels from spills in the storage area. In all situations, to reduce contamination, the barrel tops and bungs should be carefully cleaned before the bung is opened. General rules of cleanliness should be followed when transferring fluid from the barrel to the system reservoir. The use of clean containers, transfer pumps, and hoses is a must to ensure no contamination occurs while changing or replenishing fluid.

6.6.2 System Cleanliness

The cleanliness of the fluid in a fluid power system involves factors related to:

- System design.
- System construction.
- Housekeeping practices of a facility.

Contaminants can enter a system through many routes. A poorly designed or constructed system can allow dirt to enter even when reasonable housekeeping is performed by maintenance personnel. A new fluid power system that has not been thoroughly cleaned after construction can also contain many contaminants. Cleaning solvents, core sand, metal chips, paint flakes, and rust scale are just a few of the foreign materials that can appear in the fluid of a newly constructed system if not properly cleaned.

Foreign materials may enter a fluid power system through a number of locations. Attention to construction details can reduce dirt entry at most of these locations. The reservoir is one of the major components in a system where special attention must be directed if fluids are expected to have a reasonable service life. For example, the cover of the reservoir should fit well with a gasket sealing the space between the cover and the tank. Gaskets are also needed to seal the space around the pump inlet, system return, and drain lines where they pass through the cover.

The reservoir filler hole should be constructed with both a screen and a tight-fitting cover. The reservoir also needs to be fitted with a breather including an air filter that can be easily cleaned. See **Figure 6-25**.

Cylinder piston rods are also an entry point for contaminants that can shorten the service life of hydraulic fluids and system components. Contaminants can enter the system during cylinder retraction. These contaminants vary with the type of installation, but can include dust, soil, water, cutting oil, metal chips, and a variety of other materials. Equipment should be designed to protect piston rods from excessive contact with these materials. In situations where cylinders are constantly exposed to quantities of foreign materials,

Figure 6-24. Fluid storage drums should be kept in a cool, clean, dry place. Tops and bungs should be carefully cleaned before opening the drums.

Goodheart-Willcox Publisher

Figure 6-25. Many contaminants that are troublesome in a hydraulic system enter through the reservoir. Be certain that all reservoir fittings and openings are sealed or filtered to prevent the entrance of water, dirt, cutting oils, and other foreign materials.

piston rod wiper rings can be used to prevent entry of the materials. These devices wipe contaminants from the rod as it is retracted. Boots that cover the rods can also be used. See **Figure 6-26**.

The hydraulic systems of equipment should be regularly inspected to ascertain the condition of system fluid, as well as the overall condition of mechanical components. The frequency of these inspections depends on the type of equipment and the intensity of operation. A sample of fluid should be taken from the upper portion of the fluid in the reservoir after the system has reached operating temperature. The appearance of the sample should be visually checked with that of an unused sample of the fluid. If, by comparison, the used fluid is cloudy, muddy, or an inappropriate color or contains suspended particles, a closer examination of the fluid is justified. If the reason for the contamination cannot be identified by a closer visual examination of the fluid and system, an analysis by a qualified laboratory is required and legitimately justified.

A & A Manufacturing Co., Inc., Grotite

Figure 6-26. Dirt and fluids adhering to the rod of a cylinder can enter a hydraulic system when the cylinder retracts. Piston rod wiper rings or boots (shown here) can reduce this contamination.

6.6.3 Fluid Operating Temperature

Operating temperatures have considerable influence on the service life of hydraulic fluids and often reflect the efficiency of systems. The heat in the fluid of an operating fluid power system comes from the energy lost by forcing oil to flow through the system. Heat is generated when energy is used to move the fluid against a resistance within the system without completing work outside of the system. If a system is not properly designed, this heat may increase fluid temperatures above recommended levels. Higher-than-recommended temperatures cause fluid oxidation and general fluid deterioration. Another factor directly related to temperature is evaporation. This must be considered when fluids containing water are used in a system.

Most system designers indicate the ideal fluid temperature in the reservoir is between 110–140°F. This operating temperature may be slightly higher if a quality fluid is used that contains additional oxidation inhibitor additives. However, reservoir temperature does not tell the whole story regarding fluid temperatures in a system. Hot spots in system control or actuator components can raise the temperature of system fluid over the recommended level in localized areas. Fluid temperatures in these areas may be at a level that causes rapid fluid deterioration.

If the system reservoir fluid temperature is above the recommended level, several factors need to be examined:

- Ambient temperature in the area where the system is operating.
- Clearance for air circulation around the reservoir.
- General cleanliness of the unit.

The system may need to be shielded from external heat sources. It is also possible that the reservoir needs to be moved away from equipment to allow unrestricted air contact with all surfaces of the tank. Any accumulation of dust and dirt should be removed from the exterior surfaces of the unit. If these factors do not appear to be the source of the excessive temperature, several reservoir design elements should be examined:

- Reservoir fluid capacity.
- Relationship of the pump inlet line to the system return line.
- Physical shape of the reservoir.

A reservoir that is too small recirculates the fluid into the system too quickly. This does not allow adequate heat dispersal through the reservoir walls. As a result, the desired fluid temperature cannot be maintained. A general rule for reservoir sizing is a fluid capacity of two to three times the gallons per minute flow rate of the system pump.

The pump inlet and system return lines should be positioned as far apart as possible or separated by a baffle. This design element forces fluid returning from the operating system to traverse through the reservoir. Doing so facilitates cooling and the separation of dirt and other foreign materials.

The physical shape of the reservoir needs to be considered. See **Figure 6-27**. Much of the heat transfer that keeps system fluid within the proper operating temperature range occurs through the reservoir sidewalls.

A shallow, broad-based tank provides a proportionally smaller sidewall area than a deeper, narrow-based unit. However, a tank that is too narrow and deep restricts the separation of entrained air and encourages fluid foaming.

Additional factors that can contribute to high fluid temperatures relate to improper system circuit design

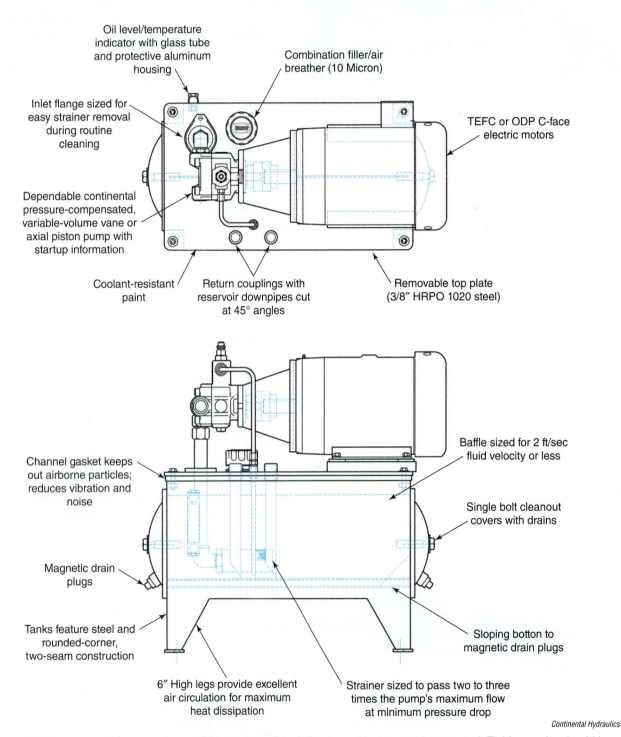

Continental Hydraulics

Figure 6-27. The reservoir is a very important component in relation to system temperature control. Fluid capacity should be two to three times the pump output (in gallons per minute) and have adequate sidewall surface area for cooling.

or operation. A larger-than-necessary pump, higher-than-required relief valve setting, smaller-than-necessary lines or components, and slower-than-necessary sequencing of the circuit operation are all factors that produce unnecessary heat in a system. A careful analysis of these factors plus an awareness of ambient temperature conditions and system reservoir design should lead to identification of most excessive fluid temperature problems, **Figure 6-28**.

6.6.4 Maintaining Fluid Cleanliness and Operating Temperature

During system operation, fluids will become contaminated to some degree regardless of the efforts to keep them clean. It is standard practice to include filtration devices in most systems to promote the cleanliness of both the internal parts of system components and the fluid used in the system. Specialized heat-exchange devices are also used in situations where fluid temperatures cannot be held within the recommended operating range. The design, construction, and operation of both of these types of devices and their related circuits are discussed in Chapter 12, *Conditioning System Fluid*.

Goodheart-Willcox Publisher

Figure 6-28. Excessive heat may be generated if component selection and pressure settings are not appropriate for a system. For example, a pump with a capacity larger than required or a relief valve setting that is too high can cause system overheating.

Summary

- The primary function of liquid in a hydraulic system is to transmit energy to do the work of the system.
- When selecting a fluid for a specific hydraulic system, a variety of characteristics must be considered: lubricating power, viscosity, viscosity stability, ability to operate in extreme temperatures, oxidation resistance, ability to separate from water and dirt easily, resistance to foaming, and fire resistance.
- Viscosity is the internal resistance to flow of a liquid produced by the molecules of the substance. A liquid with a high viscosity resists flow while one with low viscosity flows easily. Viscosity impacts the ability of the liquid to reduce friction and to provide sealing between component parts.
- Temperature and pressure affect the viscosity of a liquid. A fluid that is warm will flow easier than when it is cold. The rate of change of viscosity in relation to its change of temperature is known as viscosity index.
- Pour point is the ability of a fluid to flow when cold. The pour point of a hydraulic fluid should be 20°F below the coldest expected ambient system operating temperature.
- The oxidation rate of a hydraulic fluid is affected by temperature, air, contact with metals used in the construction of a system, and contaminants that enter a system.
- The fluid temperature in the reservoir of an operating hydraulic system should ideally be 110–140°F for a longer fluid service life.
- Selecting a fluid that can resist emulsification and can easily separate from unwanted water that might enter the system is important to proper system operation.
- The most common fluid used in hydraulic systems consists of blends of petroleum oils and additives that produce the desired set of operating properties.
- Water is not used alone as a hydraulic fluid because it is a poor lubricant, promotes rust, and freezes and evaporates rapidly at temperatures within the operating range of many typical hydraulic systems.
- A number of chemicals are used as additives in hydraulic fluids to increase their stability and overall performance.
- Drums used for fluid storage should be stored on their side in a cool, clean, dry place. General rules of cleanliness should be followed when transferring fluid from the barrel to the system reservoir.

Internet Resources

The following are some useful resources available on the Internet. Enter a company or organization name into a search engine to access its website. Explore the various areas of the sites to discover useful fluid power resources.

IHS Engineering 360. Select the "Reference Library," and then search for "hydraulic fluids" to obtain a listing of articles.

ThomasNet News. Search for "hydraulic fluids." This returns several articles from a variety of hydraulic fluid producers.

United States Department of Agriculture—News & Events. From the News & Events page, search "biodegradable hydraulic fluid." This search returns several articles on the development of commercial-grade, biodegradable hydraulic fluid for use in heavy equipment.

Wikipedia: The Free Encyclopedia—Hydraulic Fluid. Provides a basic review of the purpose of hydraulic fluids, and the various types and trade names used in industry.

Chapter Review

Answer the following questions using information in this chapter.

1. The primary function of the liquid in a hydraulic system is to _____.
   A. clean system parts
   B. remove heat from the system
   C. transmit energy to do the work the system is designed to complete
   D. prevent rust and corrosion of system components

2. *True or False?* The amount of friction is constant, even with varying forces pushing the surfaces together.

3. *True or False?* Lubricity is the ability of a liquid to form a strong film between two bearing surfaces.

4. If a fluid with too low of a viscosity is used, the hydraulic system may experience _____.
   A. increased internal leakage
   B. decreased pump efficiency and control accuracy
   C. increased system temperature
   D. All of the above.

5. *True or False?* A hydraulic system fluid with a viscosity that is too low can increase the operating temperature of the system.

6. The amount that the viscosity of a hydraulic fluid will change as the temperature of the fluid changes is indicated by the _____.
   A. condensation number
   B. viscosity index
   C. flash point
   D. absolute index

7. A condition that promotes rapid deterioration of a hydraulic fluid is _____.
   A. high system operating temperatures
   B. dirt and water contaminants
   C. air entrainment
   D. All of the above.

8. *True or False?* Air bubbles entrained in a hydraulic fluid increase fluid oxidation because of the increased surface contact between the fluid and the internal surface area of the components.

9. *True or False?* In hydraulic systems using water-based fluids, additional water will *not* cause operational and service problems.

10. *True or False?* The lowest temperature at which a liquid gives off sufficient vapors to ignite when a flame is applied is called the ignition point.

11. Water makes a poor hydraulic fluid in a fluid power system because it _____.
    A. is a poor lubricant
    B. promotes rust
    C. rapidly evaporates at temperatures within the operating range of many systems
    D. All of the above.

12. *True or False?* Hydraulic fluids that are formulated by adding 2–5% soluble chemicals to water are called high-water-content fluids (HWCF).

13. *True or False?* A water-in-oil type of fire-resistant fluid is often called an inverted emulsion because the water is suspended in the oil.

14. Typically, what is the initial viscosity of a water-in-oil emulsion compared to that recommended by system manufacturers?
    A. Lower.
    B. Both are the same.
    C. Higher.
    D. Zero.

15. *True or False?* The most common of the synthetic fluids is phosphate ester.

16. The hydraulic fluid additives that are used to prevent metal-to-metal contact in system components are called anti-wear or _____ agents.
    A. reduced wear
    B. extreme-pressure
    C. pour point
    D. oxidation-controlling

17. The hydraulic fluid additive that reduces the rate of change of fluid viscosity is known as a(n) _____ additive.
    A. viscosity-index-improver
    B. pour point
    C. viscosity controller
    D. fire point reducer

18. Improving surface tension characteristics increases the rate at which water separates from hydraulic oils and collects in the _____.
    A. actuators
    B. conductors
    C. filter elements
    D. reservoir

19. *True or False?* Rust and corrosion inhibitors in hydraulic oils protect the metals in system parts by neutralizing acids that may develop as the system operates.

20. *True or False?* Pour-point depressants are used as additives in hydraulic oils to provide better performances at high-system operating temperatures.

21. The most precise viscosity rating that is usually found in specification sheets for hydraulic equipment is _____ viscosity.
    A. absolute
    B. Saybolt universal
    C. kinematic
    D. SAE

22. *True or False?* The viscosity index number appears as part of the specifications of most hydraulic fluids.

23. Corrosion tests used with petroleum-based hydraulic fluids determine the effect on metal parts of components from _____ compounds that have remained in the fluid after refining.

24. Which of the following is a general method used to establish the lubrication and wear-prevention qualities of hydraulic fluids?
    A. Capillary viscometer.
    B. Extreme-pressure test.
    C. Saybolt universal viscosity system.
    D. All of the above.

25. When storing and handling replacement hydraulic fluid drums, place drums on their sides to reduce chances of _____.

26. When cylinders are exposed to extreme conditions, contamination of system fluid can be reduced by using _____ to prevent the entry of dirt from the surfaces of the piston rods.

27. Fluid samples taken from hydraulic systems for routine testing purposes should be collected only from the _____ of the fluid in the reservoir and only when the system is at operating temperature.

28. _____ in the fluid of a hydraulic system is caused by forcing fluid through the system without completing work outside of the system.

29. _____-than-recommended temperatures cause fluid oxidation and general fluid deterioration.

30. The cleanliness of the fluid in most hydraulic systems is ensured by the addition of _____.
    A. temperature control devices
    B. adequate storage drums
    C. filtration devices
    D. transfer pumps

Apply and Analyze

1. You are working with a hydraulic system that operates at relatively high temperatures.
    A. How does the high-operating temperature affect your hydraulic fluid?
    B. How might these effects on the hydraulic fluid impact circuit components and operation?
    C. Describe the characteristics of the hydraulic fluid you would choose to use in this system.
    D. What additives would you use to increase system performance? Explain your answer.
    E. Describe any aspects of the system or component design that could be used to lessen the impact of a high-operating temperature.

2. Contaminants in a hydraulic circuit can lead to poor performance and failure of equipment.
    A. Make a list of the top 6 preventative measures you would take to reduce the impact of contaminants in your system.
    B. Describe how the priorities you listed might change, depending on the environment your system is operating in or the function the system performs.

Research and Development

1. Select 3–5 common household liquids, such as dish detergent, cooking oil, shampoo, or lotion. Design and execute some simple tests to determine the relative viscosity, demulsifying ability, and foaming properties of the liquids. Write a report that includes an explanation of your experiment design, an analysis of your results and any possible errors, and a brief discussion of the suitability of these liquids for use in a hydraulic power system based on your test results.

2. Research a synthetic fluid used in hydraulic systems. Determine the composition, characteristics, applications, benefits, and limitations of this fluid. Report your findings to the class.

3. Heavy equipment operators working in environmentally sensitive areas may need to use an eco-friendly, biodegradable fluid to comply with government regulations. Prepare a presentation you might give to a company needing to make this change in hydraulic fluid. Be sure to address the options they have, the equipment considerations they may need to make to allow the use of this different fluid, and the costs and benefits associated with the switch.

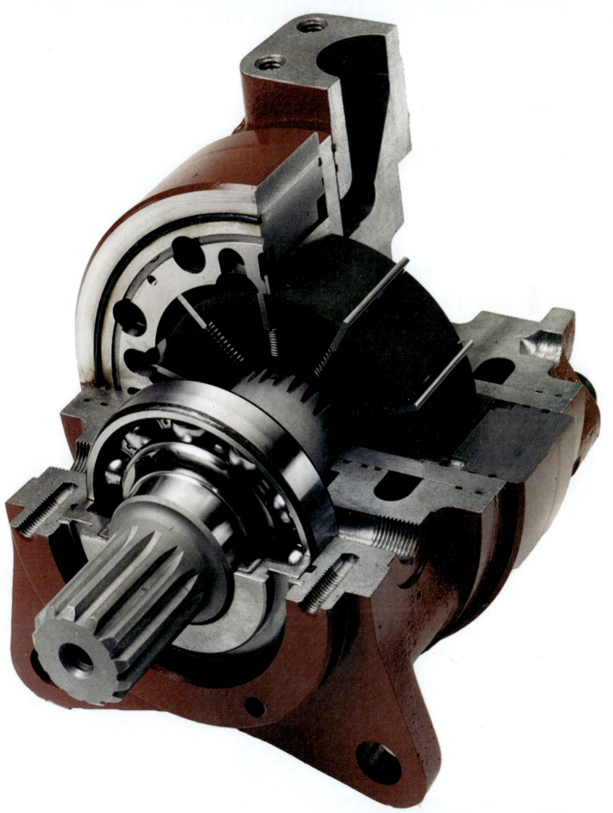

This is a vane-type hydraulic motor that develops high torque. Hydraulic fluid acts on the vanes, which causes the output shaft to turn. There are many important factors related to hydraulic fluid that must be considered when designing or servicing a hydraulic system.

7

Source of Hydraulic Power
Power Units and Pumps

The power unit is the source of power for the remainder of the hydraulic system. The devices that comprise this unit are fundamental to the operation of the system. These include the prime mover that supplies the energy to operate the pump, which creates fluid flow in the system. In addition, the unit usually includes components to store system fluid, control maximum system pressure, and clean the fluid. This chapter introduces each of these components and then discusses pump designs in detail.

Learning Objectives

After completing this chapter, you will be able to:

- Describe the function of a hydraulic power unit and identify its primary components.
- Explain the purpose of a pump in a hydraulic system.
- Explain the operation of a basic hydraulic pump.
- Compare the operating characteristics of positive-displacement and nonpositive-displacement hydraulic pumps.
- Compare the operating characteristics of rotary and reciprocating hydraulic pumps.
- Compare the operating characteristics of fixed- and variable-delivery hydraulic pumps.
- Explain the principles involved in the operation of a pressure-compensated hydraulic pump.
- Describe general construction for each of the various hydraulic pump designs.
- Interpret performance data supplied by a pump manufacturer.
- Explain cavitation and its effect on pump performance and service life.

Key Terms

axial-piston pump
balanced-vane pump
bent-axis design
cavitation
centrifugal pump
coupler
crescent design
cylinder barrel
displacement
dual pump
entrained air
external-gear pump
fixed-delivery pump
fluid filter
fluid flow weight
gear pump
general hydraulic horsepower
gerotor design
helical gear
herringbone gear
impeller

inline design
internal-gear pump
jet pump
lobe pump
mechanical efficiency
nonpositive-displacement
   pump
overall efficiency
pintle
piston pump
piston shoe
positive-displacement pump
power unit
pressure balancing
pressure control valve
prime mover
prime mover horsepower
propeller pump
pump
radial-piston pump
reciprocating pump

reciprocating-plunger pump
relief valve
revolving-cylinder design
rotary pump
safety valve
screw pump
shoe plate
spur gear
stationary-cylinder design
strainer
swash plate
unbalanced-vane pump
valve plate
vane pump
vapor pressure
variable-delivery pump
venturi
volumetric efficiency
wear plate

7.1 Power Unit

The design and appearance of the power unit of a hydraulic system varies considerably, depending on the equipment in which the system is operating, **Figure 7-1**. However, even though the appearance is different, the *power unit* must provide the energy for operation of the system, move the fluid through the system, and provide a safe maximum limit on system pressure while assisting in maintaining proper system operating temperature and fluid cleanliness.

The hydraulic power unit is often supplied by manufacturers as a package. The package includes all of the desired components and required plumbing. Also, system controls are often included. Standard units are available or components can be ordered to meet specific requirements. Each of the components of the power unit are discussed in detail in various sections of this book.

7.1.1 Prime Mover

In order to serve as the overall source of energy for a hydraulic system, the power unit must contain a prime mover. The *prime mover* converts energy into mechanical movement. This mechanical movement then acts as the carrier of the energy through the system. In most systems today, the prime mover is an electric motor or internal combustion engine. See **Figure 7-2**. These sources may be individual units dedicated to operating the system or they can be power take-off arrangements where the motor or engine operates several different units. An example of a power take-off setup is found in mobile farm or earth-moving equipment. The engine supplies power to move the equipment from one location to another, but also operates auxiliary systems, such as leveling devices, digging buckets, plows, or other attachments.

7.1.2 Coupler

In most hydraulic power units, the prime mover is mechanically connected directly to the input shaft of the system pump using a *coupler*. See **Figure 7-3**. One of several different designs of coupler can be used to maintain proper alignment between the two units. Care must be exercised to ensure the coupler is adequate in relation to rated speed, load, and allowable misalignment. Belt, chain, and gear drives are

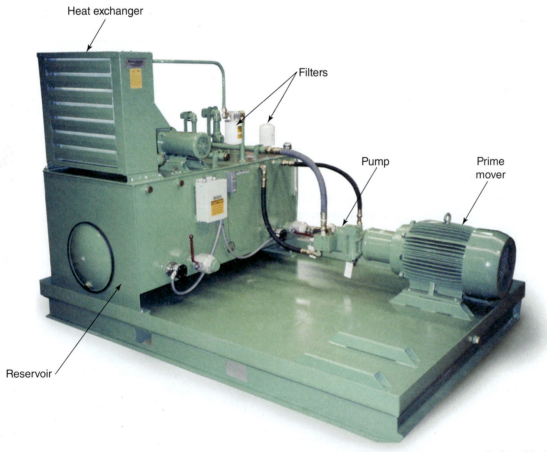

Kim Hotstart Manufacturing Company

Figure 7-1. The source of power for the hydraulic system is the power unit, which is made up of a pump driven by a prime mover and includes a reservoir, filters, driveshaft coupler, and basic pressure control valves.

also used to connect some prime movers and pumps. These types of connections are generally less preferred than direct drives because they are less efficient, have a lower safety factor, and place higher stress on the equipment. However, these devices may be required in situations where changing to a higher or lower speed is required.

Prime
mover

A

Continental Hydraulics

DEERE

800C

B

Prime
movers

Deere & Co.

Figure 7-2. The prime mover converts energy into mechanical movement. A—Manufacturing installations generally use electric motors. B—Mobile equipment generally use internal combustion engines.

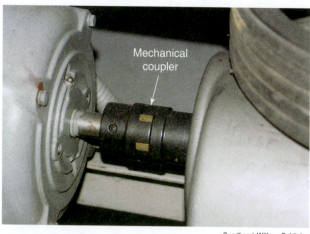

Mechanical
coupler

Goodheart-Willcox Publisher

Figure 7-3. The coupler transfers power between the prime mover and the hydraulic system pump. Careful selection is important to provide quiet and trouble-free system operation.

7.1.3 Pump

The *pump* is the device that causes the fluid to move through the system, carrying the energy to the point where it is needed. A wide variety of pump designs are available for use on power units. The design selected for use depends on the required system flow rate, system operating pressure, and a variety of other requirements.

7.1.4 Reservoir

Another major component of a power unit is the reservoir, which serves multiple purposes in a power unit. First, it is a tank that holds system fluid. The pump draws fluid from the tank and circulates it through the various components to operate the system. Fluid is then returned to the reservoir to await continued system use. Second, while the fluid is being held in the reservoir, it cools while dirt and other unwanted materials settle out. Third, the top of the reservoir often serves as a mounting location for the prime mover and the pump.

7.1.5 Pressure Control Valve

The power unit normally includes a *pressure control valve* to establish a maximum operating pressure for the system. See **Figure 7-4.** The valve may function as the *relief valve*, establishing the operating pressure for the system, or it may be set at a slightly higher pressure to act as a *safety valve.* If the normal system relief valve fails, the safety valve will open, preventing damage to the system from excessive pressure. In both situations, the valve serves to protect system components from damage and the prime mover from being overloaded by high pressures.

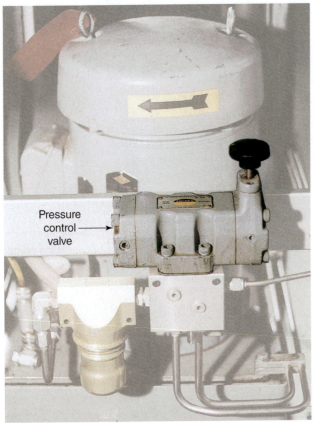

Pressure control valve

Goodheart-Willcox Publisher

Figure 7-4. A basic pressure control valve in the form of a relief valve or a safety valve is a part of most power unit packages. These valves are included to establish the maximum operating pressure of a system and/or protect from system overload.

7.1.6 Filters and Strainers

Fluid filters are often included as a part of the power unit of a hydraulic system. The importance of keeping system fluid free of contamination cannot be overemphasized. A *strainer* is often attached to the end of the intake line that is submerged in the reservoir fluid. This device is constructed of fine-wire screen and designed to remove larger dirt particles. Additional filtering devices are used in the pump intake or system return line to remove fine particles that have entered the fluid, **Figure 7-5**.

7.2 Basic Pump

A wide variety of pump designs are used to convert the mechanical power output of the prime mover into fluid flow. The appearance and structure of these pumps may be different, but the operating principles that cause the fluid to move through the unit are very similar. The basic operating principles require that the pump have areas within its structure where less-than-atmospheric pressure, as well as above-atmospheric pressure, can be produced and maintained.

7.2.1 Structure of a Basic Pump

A hand-operated piston pump can be used to illustrate the basic concepts involved in pump operation. This pump is very similar to a simple tire pump used to inflate bicycle tires. The pump consists of a pumping chamber made up of a cylinder, closed at one end, in which a movable piston operates. Two one-way valves are located in the closed end of the chamber. One of these serves as the pump inlet valve and is attached to a line connected to a fluid reservoir. The second valve serves as the outlet valve and is connected to the hydraulic system. A handle is attached to the piston to allow an operator to easily slide the piston back and forth in the chamber, **Figure 7-6**.

7.2.2 Operation of a Basic Pump

The operation of a pump involves intake and pressurization phases. The discharge pressure is based on several factors. The next sections discuss these phases and the factors related to discharge pressure.

Intake phase

The inlet phase of basic pump operation is illustrated in **Figure 7-7**. As the piston is pulled toward the open end of the pumping chamber, the volume increases between the piston and the end of the chamber containing the valves. As the volume increases, pressure in the closed chamber drops below atmospheric pressure. This reduced pressure acts on the outlet valve, allowing it to be closed by atmospheric pressure if the pump is just beginning operation or by higher system pressure if the pump is attached to an operating system. The reduced pressure in the chamber also allows the inlet valve to be forced open by system fluid moved

Goodheart-Willcox Publisher

Figure 7-5. Protecting the system from dirt and other contaminants in the hydraulic fluid begins in the power unit. A strainer is often used on the intake end of the pump inlet to remove larger particles.

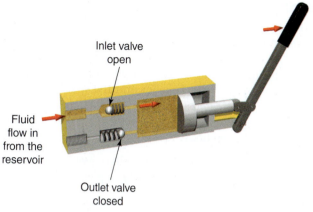

Figure 7-7. During the basic pump intake phase, the piston moves to enlarge the pumping chamber volume. This allows fluid to move from the reservoir, through the inlet, to the chamber.

through the pump inlet line by atmospheric pressure pushing against the surface of the fluid in the reservoir. Fluid flow caused by this pressure difference fills the pump chamber.

Pressurization phase

The pressurization phase of basic pump operation is illustrated in **Figure 7-8.** When the piston is pushed back toward the closed end of the pumping chamber, the volume decreases between the piston and the chamber end. Force applied to the piston produces a higher-than-atmospheric pressure in the fluid trapped in the pump chamber. The pressure causes the inlet valve to close, blocking the return of fluid to the reservoir, and the outlet valve to open, allowing fluid to flow through it into the system.

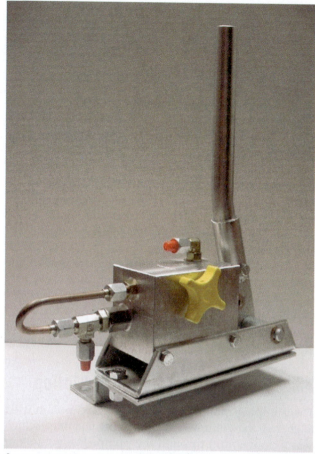

Figure 7-6. The basic hydraulic pump can be compared to a simple, hand- or foot-operated pump. A piston moves within a cylinder to provide a variable-volume chamber. An inlet valve connects the chamber to a reservoir and an outlet valve connects it to the system. A—Hand-operated pumps. B—Foot- and hand-operated pumps.

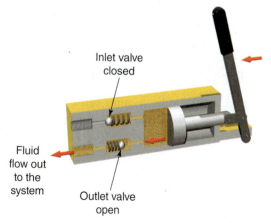

Figure 7-8. During the basic pump pressurization phase, the piston moves to decrease the volume of the pumping chamber, forcing the fluid to move through the discharge valve to the system.

Discharge characteristics

The maximum pressure developed in the system is determined by the resistance to the flow of fluid in the system and the force that can be applied to the piston. The output flow rate of the pump is determined by the size of the pumping chamber and the speed at which the cycle is repeated (the speed of the prime mover). The volume of the pumping chamber is often measured in cubic inches (in³).

$$\text{Pump Flow Rate (in}^3/\text{min)} = \text{Displacement Volume (in}^3) \\ \times \text{Cycles per Minute (rpm)}$$

The pump flow rate, or output, is typically reported in gallons per minute (gpm). Given that 1 gallon is equal to 231 in³, the following equation can be used for conversion.

$$\frac{\text{Pump Flow Rate}}{\text{(gpm)}} = \frac{\text{Pump Flow Rate (in}^3/\text{min)} \times 1 \text{ gallon}}{231 \text{ in}^3}$$

EXAMPLE 7-1

Discharge Characteristics

Determine the output flow rate in gallons per minute (gpm) of a pump that has a pumping chamber volume of 8 in³ and that is operating at a speed of 1,750 cycles per minute.

$$\begin{aligned}
\frac{\text{Pump Flow Rate}}{\text{(in}^3/\text{min)}} &= \frac{\text{Displacement Volume (in}^3) \times}{\text{Cycles per Minute (rpm)}} \\
&= 8 \text{ in}^3 \times 1{,}750 \text{ rpm} \\
&= 14{,}000 \text{ in}^3/\text{min}
\end{aligned}$$

$$\begin{aligned}
\frac{\text{Pump Flow Rate}}{\text{(gpm)}} &= \frac{\text{Pump Flow Rate (in}^3/\text{min)} \times 1 \text{ gallon}}{231 \text{ in}^3} \\
&= \frac{14{,}000 \text{ in}^3/\text{min} \times 1 \text{ gallon}}{231 \text{ in}^3} \\
&= 60.6 \text{ gpm}
\end{aligned}$$

7.3 Basic Pump Classifications

All pumps used in hydraulic systems can be classified using three methods. The various categories involve the design, structure, and operation of pumps relating to:

- Displacement.
- Pumping motion.
- Fluid delivery characteristics.

7.3.1 Displacement

Displacement is the volume of fluid displaced per unit of time. The displacement classification relates to the physical volume of the chamber, or output, of a pump per revolution under varying system operating conditions. The factor is dependent on the design characteristics of the pumping chamber of the pump and its ability to deliver fluid. Pumps are classified as either positive or nonpositive displacement.

Positive-displacement pumps produce a constant output for each revolution of operation over a wide range of operating pressures. These pumps have very close tolerances between moving and stationary parts. The operating efficiency of these pumps causes some variation in output as system pressure changes. However, output remains relatively constant. Most hydraulic system pumps are positive displacement, **Figure 7-9**.

Nonpositive-displacement pumps have larger clearances between moving and stationary parts. Because of this design factor, considerable slippage can occur in the pumping chamber, resulting in pump flow variations as system operating pressures change. Pumps classified in this category are usually used in low-pressure applications, such as fluid transfer, rather than for actuator operation, **Figure 7-10**.

7.3.2 Fundamental Pumping Motion

The pumping motion relates to the basic motion used to change the volume of the pumping chamber, which acts to move fluid through the pump. The volume of the chamber increases during the intake phase of operation and decreases during the pressurization phase. Pumps are classified as either rotary or reciprocating.

There are several different rotary pump designs. *Rotary pumps* use rotating parts in the pumping chamber to produce the increase and decrease in volume needed to move fluid through the pump, **Figure 7-11**.

Goodheart-Willcox Publisher

Figure 7-9. The component parts of the pumping chamber of positive-displacement pumps are machined with close tolerances, allowing uniform flow output per pump revolution.

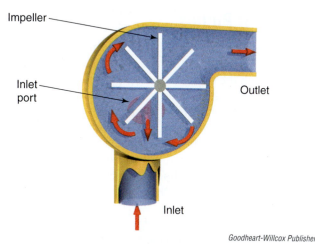

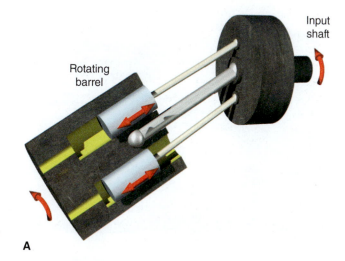

Figure 7-10. The pumping chamber parts of a nonpositive-displacement pump have considerable clearance, allowing variation in pump flow output as system load conditions change.

A

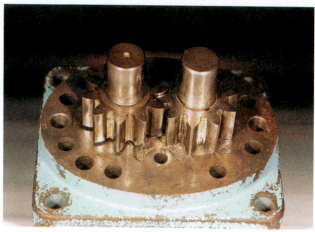

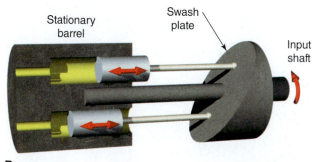

Goodheart-Willcox Publisher

B

Figure 7-12. The pumping action of a reciprocating pump is based on the back-and-forth movement of a piston or other component in a pumping chamber. Several designs are used to convert the rotary motion of the pump input shaft into reciprocating movement. A—Bent axis. B—Swash plate.

Figure 7-11. The pumping action of a rotary pump is based on the continuous rotating movement of pumping-chamber components, such as these gears.

7.3.3 Fluid Delivery Characteristics

Fluid delivery characteristics are associated with the fluid output of a pump. This is determined by the size of the pumping chamber of the unit. In some pumps, the volume of the chamber is fixed, **Figure 7-13A**. The fluid output of these *fixed-delivery pumps* is constant for each pump revolution. In other designs, the volume of the pumping chamber can be varied, **Figure 7-13B**. *Variable-delivery pumps* can produce fluid outputs ranging from zero, or no flow per revolution, to the maximum allowed by the size of the unit. Variable-delivery pumps include a mechanism that allows an operator to vary the internal pump geometry, thus changing the size of the pumping chamber.

When the construction of a pump allows, the manufacturer may produce both a fixed and a variable model. However, the inherent construction characteristics of some pump designs may not allow for a pumping chamber that can be varied. In these situations, the pump is available only as a fixed, or constant-output, unit. Specific information on the design and operation of this feature of hydraulic pumps is discussed later in this chapter.

Rotary motion may also be used to drive component parts that develop centrifugal force. This is another method that can be used to move fluid through a pump.

Reciprocating pumps use the up-and-down linear movement of a piston in a cylinder or similar structure to move fluid through the pump. As the piston moves downward, the volume of the chamber above the piston enlarges. This creates the low-pressure area needed to bring fluid into the pump from the reservoir. The upward movement of the piston reduces the volume of the chamber. This pressurizes the fluid and moves it out of the pump under pressure. A number of hydraulic pump manufacturers use variations of this concept to produce a wide variety of pump designs. See **Figure 7-12**.

Figure 7-13. Pumps are available in fixed- and variable-delivery designs. Some designs can be produced only as fixed, while others are available as either. A—Gear pumps are fixed delivery only. B—Vane and piston pumps are available as either fixed or variable delivery.

7.4 Pump Design, Operation, and Application

A variety of different styles of hydraulic pumps are available to system designers for use in a system. The design that is selected for a particular hydraulic system is the result of a careful analysis of that system. Operating pressure, flow requirements, circuit cycle rate, expected length of service, environmental conditions, and cost are all important considerations when selecting a pump for use in a system. This section examines the basic design and features of typical positive- and nonpositive-displacement hydraulic pumps.

7.4.1 Gear Pumps

Gear pumps are positive-displacement, rotary units. They use the rotary motion of gears to directly move fluid through the unit. These pumps consist of a housing with inlet and discharge chambers and a pair of meshed gears that serve as the pumping mechanism. One gear of the pumping mechanism is turned by the prime mover of the hydraulic system. The second gear serves as the driven gear. The pumping action of the mechanism results from the unmeshing and meshing of the gear teeth as the gears rotate. See **Figure 7-14**.

As the gears unmesh in the inlet chamber of the pump, the volume of the chamber increases, thus creating a low-pressure area. The low-pressure condition causes hydraulic fluid to be forced into the chamber by atmospheric pressure. Hydraulic fluid is then carried from the inlet chamber to the discharge chamber in the space formed between the gear teeth and the closely fitted pump housing. Fluid is forced out of the pump outlet when the volume of the discharge chamber is reduced as the gear teeth mesh.

Several different basic designs of gear pumps are used in hydraulic equipment. Although the design and construction may vary, the operation of each design closely follows the above explanation. The details of gear pump construction result in two classifications: external-gear and internal-gear designs.

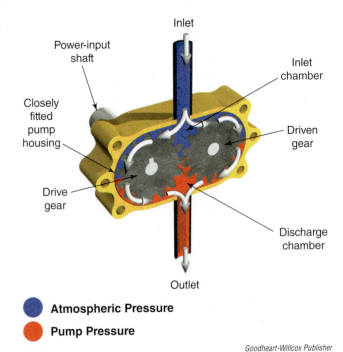

- ● **Atmospheric Pressure**
- ● **Pump Pressure**

Figure 7-14. As the gears of the pumping unit separate in the inlet chamber, a low-pressure area is formed, allowing fluid to enter. The fluid is carried to the discharge chamber and forced out of the pump as the gear teeth mesh.

External-gear design

External-gear pumps have teeth on the outside of the rotating shafts. See **Figure 7-15**. These pumps are the most popular in the hydraulic field. They are generally low cost when compared to other designs of similar displacement and operating characteristics.

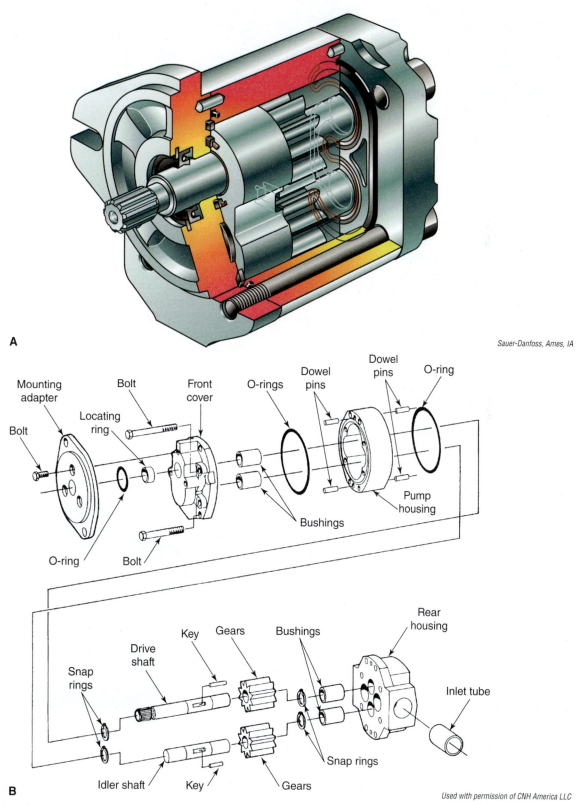

A

Sauer-Danfoss, Ames, IA

Used with permission of CNH America LLC

Figure 7-15. A typical gear pump is a relatively simple mechanism involving a minimum of moving parts and points of wear. A—Cutaway view of a typical gear pump. B—Exploded view of a typical gear pump.

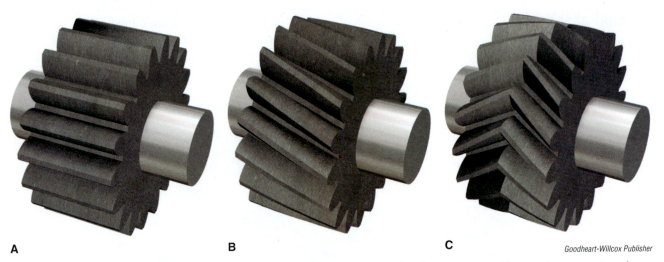

A   **B**   **C**   *Goodheart-Willcox Publisher*

Figure 7-16. The spur, helical, and herringbone gear designs are used in hydraulic gear pumps. The most common type is the spur gear because of simplicity of design and cost of manufacture. A—Spur. B—Helical. C—Herringbone.

External-gear designs provide a good life expectancy, have a reasonable operating efficiency, and stand up well when subjected to typical fluid contamination. External-gear pumps are available for pressures up to 3,000 psi and flow capacities up to 150 gpm.

Spur, helical, or herringbone gear designs are used as the pumping mechanism in external-gear pumps. See **Figure 7-16**. The gear teeth are of an involute design, meaning that the teeth are rolled or curled outward in a spiral pattern. This ensures that rolling contact is made between the gear surfaces to provide a positive seal and minimal tooth wear. The *spur gear* design is the easiest to produce. Therefore, it is the most common and least expensive of the three gears. It is also the noisiest of the designs. The *helical gear* design is like a spur gear that twists along its axis. It allows for more power than a spur gear. However, a disadvantage of the helical gear design is the need to provide thrust bearing surfaces between the sides of the gear and the pump case. This is required because of the inherent side-thrust characteristics of the helical gears caused by the slanted design. The *herringbone gear* design provides the quietest operation.

Internal-gear design

Internal-gear pumps use a gear arrangement in which one of the gears has external teeth and the other has internal teeth. See **Figure 7-17**. The gear with the external teeth is powered by the system prime mover and turns the gear with the internal teeth. These pumps use special gear designs that appear very different from the involute pattern. The external gear in these pumps usually has one or two teeth less than the gear with the internal teeth. This design results in a very low difference in relative speed between the gears, which produces low gear wear. Two variations of this design are commonly found in hydraulic systems: gerotor and crescent.

The most common of the two internal-gear pump design variations is the *gerotor design*. The pumping chamber of this design consists of two rotating gears. The inner gear contains external teeth and is mounted within the outer gear, which contains internal-gear teeth. As the inner gear rotates, it forces the outer gear to turn within the pump housing. There is one less tooth on the inner gear than on the outer gear. These pumps generally function in the same pressure range as external-gear designs and operate at a lower noise level.

The gear tooth form on a gerotor is such that one point on each tooth of the external gear is always in contact with the surface of the internal gear. This contact

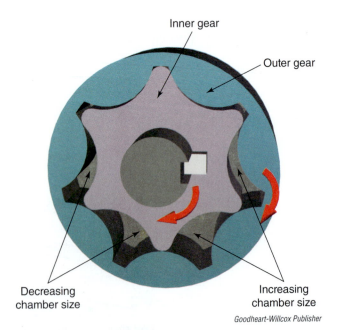

Goodheart-Willcox Publisher

Figure 7-17. The gerotor internal-gear pump uses a specialized internal gear design to obtain a higher flow output than other gear pump designs of comparable physical size.

forms a number of sealed chambers that vary in size during each pump revolution. Each chamber enlarges during the first part of its rotational cycle to form a low-pressure area. This low pressure allows fluid to flow into the chamber. As the rotation continues, a maximum volume is reached in each chamber, followed by a decrease in volume. The decrease in volume forces the fluid out into the hydraulic system. See **Figure 7-18**.

The *crescent design* also uses an inner gear with external teeth and an outer gear containing internal teeth. However, the pumping chamber also has a fixed, crescent-shaped element, **Figure 7-19**. The inner gear is directly attached to the pump drive shaft. The external teeth of the inner gear mesh with the internal teeth of the outer gear. The smaller diameter of the inner element allows it to be located off-center within the outer gear. The fixed crescent is located in the space between the two gears that results from the off-center location. Minimum clearances between the pump housing, crescent, and inner and outer gears provide the internal pump seal. These pumps operate at lower noise levels than external-gear designs. However, the operating pressure range of the design is generally considerably lower with typical operating ratings of 150–500 psi.

As the pump rotates, the external teeth of the inner gear are withdrawn from the internal teeth of the outer gear as they pass through the intake section of the unit. This causes the volume of the intake section to increase, resulting in a low-pressure area. This low pressure allows fluid to flow into the chamber, filling the spaces between the gear teeth. As the pump continues to rotate, the fluid is trapped between the gear teeth, fixed crescent, and pump housing and carried into the discharge section. In this section, the teeth of the inner and outer gears again mesh, reducing the chamber volume. This results in the formation of a high-pressure area, causing the fluid to move out into the hydraulic system.

7.4.2 Vane Pumps

Vane pumps are positive-displacement units that use the rotary motion of vanes to produce a pumping action, **Figure 7-20**. They are commonly used in industrial situations with pressure ratings ranging from 1,000–2,200 psi and delivery rates from 1.5–75 gpm. The internal configuration of some designs requires a minimum operating speed of 1,200 rpm. The typical design consists of:

- Housing.
- Cam ring.
- Slotted rotor.
- Vanes.
- Inlet and discharge ports.

The cam ring contains a pumping chamber. The slotted rotor contains the movable vanes. The rotor and vanes form the pumping mechanism. The pumping mechanism

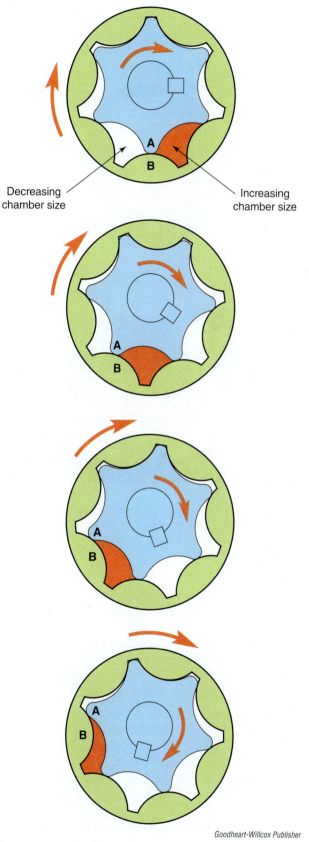

Decreasing chamber size

Increasing chamber size

Goodheart-Willcox Publisher

Figure 7-18. The one tooth difference between the drive gear and driven gear elements of the gerotor mechanism provides minimum surface speed differences and helps to reduce component wear.

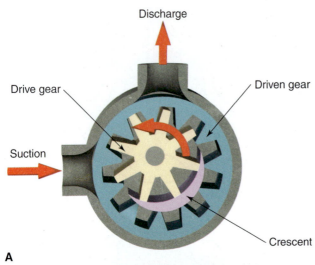

B

Goodheart-Willcox Publisher

Figure 7-19. The crescent of the crescent design separates the inlet and discharge chambers to provide a seal between the driven gear and the internal gear element of the pumping mechanism.

is directly attached to the input shaft of the pump, which is turned by the system prime mover. The mechanism is offset in the cam ring. This results in a number of variable-size chambers between the vanes, rotor, and ring. Pumping action occurs as the size of the chambers increase and decrease as the rotor is turned.

Vane pumps are manufactured following two varying designs: unbalanced or balanced. These classifications are based on the load placed on the bearings that support the rotating pumping mechanism.

Unbalanced-vane pump design

An *unbalanced-vane pump* is a pump with a circular pumping chamber located in a cam ring fitted into the pump housing. See **Figure 7-21**. The width of the cam ring matches the width of the slotted rotor and vanes. The diameter of the rotor is smaller than the diameter of the pumping-chamber opening in the ring. The pumping capability of the design is created by placing the slotted rotor and vanes in an offset position in the pumping chamber. As the pump turns, the vanes are thrown against the inner surface of the ring by centrifugal force. As a result, small, individually sealed chambers are formed between each of the rotor slots. The size of these chambers varies as the unit rotates because of the offset location of the rotor. As a chamber enlarges, a low-pressure area is formed, causing fluid to flow into the chamber through the pump inlet. As the size of a chamber decreases, the fluid is forced out of the pump to the system through the pump discharge.

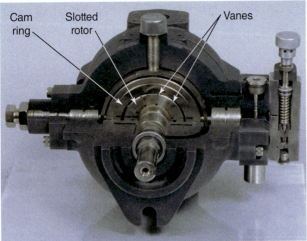

Courtesy of Eaton Fluid Power Training

Figure 7-20. Vane pumps are commonly used in both stationary and mobile applications. The pumping mechanism in these pumps is made up of chambers formed by sliding vanes.

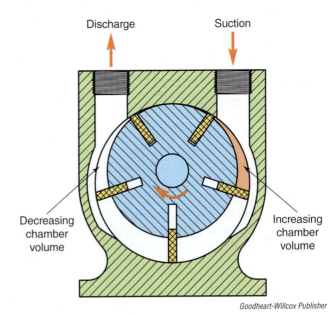

Goodheart-Willcox Publisher

Figure 7-21. As the rotor turns in an unbalanced-vane pump, the chambers formed by the vanes are enlarging as they pass the inlet chamber. They are decreasing in volume as they pass the discharge port. These variations cause the pressure differences that move fluid through the pump.

The unbalanced aspect of the design is the result of forces acting on the bearings of the pump during operation. When no load is placed on the pump, the pressure variation between each of the small individual pumping chambers is relatively low. As such, only a light, radial load is placed on the pump bearings. As system operating pressure increases, the difference in pressure between the chambers on the inlet side and the discharge side increases. The pressure difference becomes unbalanced, resulting in a heavy radial load on the pump shaft and bearings.

Balanced-vane pump design

A *balanced-vane pump* has a pumping mechanism that includes a cam ring with an elliptical opening slightly wider than it is tall. The ring is fitted into a pump housing that has two inlet areas located 180° apart and

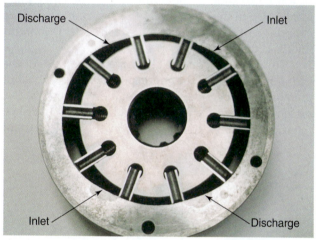

Figure 7-22. A balanced-vane pump uses an elliptical cam ring that allows two inlet chambers and two discharge chambers opposite of each other. This balances the pressure placed on the pump shaft and reduces bearing load.

two discharge areas also located 180° from each other. See **Figure 7-22**. The diameter of the slotted rotor and vanes of the pumping mechanism is slightly less than the narrow width of the elliptical opening.

The rotor/vane mechanism is placed in the center of the opening. This produces two pumping chambers, one on either side of the rotor. The inlet ports and the discharge ports of the two chambers are connected through passageways in the pump housing. The general operation of the pump is the same as the unbalanced design, except two pumping chambers are involved.

The balanced aspect of the design is based on the fact that equal forces are acting on the pump shaft and bearings. With the two inlet and two discharge points 180° apart, the radial loads are equalized, greatly reducing the stress on the pump components.

7.4.3 Piston Pumps

Piston pumps are positive-displacement units that use the reciprocating movement of pistons in a pumping mechanism to move system fluid. These pumps are the oldest of the pump designs, having been used since the early 1900s or before, **Figure 7-23**. The various designs currently in use provide excellent control of clearances and the application of modern materials. The pumps can offer volumetric efficiencies in the high 90% range, operating pressures of 10,000 psi or higher, and high operating speeds that in specialized situations can exceed 10,000 rpm.

A typical piston pump consists of a housing that supports a pumping mechanism and a motion-converting mechanism. The pumping mechanism alternately causes fluid to move into the unit and then to be forced out to the system. The motion-converting mechanism converts rotary movement to the reciprocating motion needed to power the pump mechanism.

The pumping mechanism typically consists of a metal casting in which several blind-end cylinders

Figure 7-23. A wide variety of piston pumps is available for hydraulic systems. A full range of both pressure and flow ratings provides a variety of options for the hydraulic system designer.

are bored. Each cylinder is equipped with a close-fitted piston that moves back and forth (reciprocates) in the bore. The blind ends of the cylinders are fitted with one-way inlet and discharge valves or ports to control the flow of fluid through the mechanism. As the pump rotates, the motion-converting mechanism causes the pistons to be alternately withdrawn from the cylinder bores. As a piston is withdrawn, the volume of the space above the piston increases, resulting in a low-pressure area. Fluid flows into this low-pressure area from the system through the inlet valve or port. As the pump continues to turn, the piston is eventually forced back into the cylinder, reducing the volume of the space above the piston. Reducing the volume in the cylinder forces fluid through the discharge valve or port and out into the system.

Classification of piston pumps is based on the relationship between the axis of the input-power shaft and the plane on which the reciprocating motion of the pistons occurs. Three classifications are generally used to describe these pumps: axial, radial, and reciprocating plunger.

Axial-piston pumps

Axial-piston pumps are commonly found in the hydraulic field. They are available in both full flow and pressure ranges. Flow capacities can be over 100 gpm and pressure ratings can be up to 5,000 psi.

The reciprocating motion of the pistons in axial pumps occurs parallel or at a slight angle with the axis of the input shaft of the pump. The motion in an *inline design* is parallel to the input shaft. The motion of a *bent-axis design* is at a slight angle to the input shaft.

The inline design is the most popular piston pump. It consists of a housing containing an input power shaft, swash plate, cylinder barrel, pistons, piston shoes, shoe plate and shoe-plate springs, and valve plate. Refer to **Figure 7-24**. The *cylinder barrel* is attached to the input shaft, with both components rotating on the same axis. The barrel has several precisely machined bores fitted with pistons. The *piston shoes* are attached to the ends of the pistons that rest on the swash plate. These serve as the bearing surfaces between the pistons and swash plate. The shoes are held in contact with the swash plate by the *shoe plate* and springs. The *swash plate* is mounted in the pump housing and provides a fixed angular surface (commonly 15°) on which the piston shoes slide as the pump turns. The *valve plate* is mounted in the pump housing toward the end of the cylinder barrel opposite of the swash plate. It contains inlet and discharge ports or valves.

During pump operation in the inline design, the prime mover turns the input power shaft. The shaft then turns the cylinder barrel and the attached pistons, piston shoes, and shoe plate and springs. The shoe plate and springs hold the piston shoes in constant

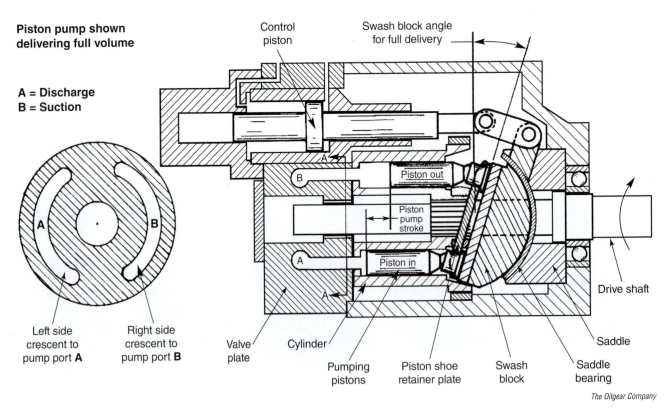

The Oilgear Company

Figure 7-24. The inline axial-piston pump provides a pump suitable for many applications. The axis of the power-input shaft and the centerline of the pistons are parallel, providing a very compact unit.

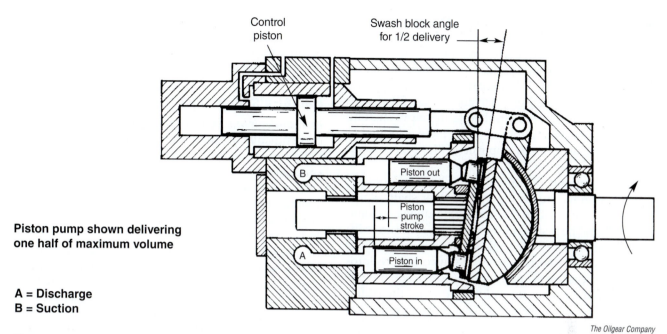

Control piston

Swash block angle for 1/2 delivery

B

Piston out

Piston pump stroke

A

Piston in

Piston pump shown delivering one half of maximum volume

A = Discharge
B = Suction

The Oilgear Company

Figure 7-25. The angle of the swash plate in relation to the centerline of the power input shaft controls the length of the piston stroke in an inline axial-piston pump.

contact with the fixed swash plate. The angular position of the swash plate causes the pistons to reciprocate in the bores of the cylinder barrel as the barrel rotates.

As a piston is withdrawn from the bore, the volume above the piston increases. This causes a reduction in pressure. The reduced pressure allows system fluid to flow into the pump from the reservoir. As the angle of the swash plate forces the piston back into the bore, the volume above the piston becomes smaller. This causes the fluid to flow out of the pump into the system. The valve plate passageways are positioned such that the inlet line is open to the area above the piston when bore volume is increasing. The area above the piston is open to the discharge line as the volume is decreasing. See **Figure 7-25**.

One of the critical design factors in this pump design is the piston shoes that operate against the surface of the swash plate. Contact pressure between these surfaces is high during high-pressure system operation, which can cause rapid wear. It is, therefore, critical that adequate lubrication is available at the point of contact. This is accomplished in many designs by a passageway drilled through the piston that delivers a small quantity of oil to the area to ensure lubrication, **Figure 7-26**. Proper operation and service, including adequate fluid filtration and recommended pump operating speeds, will ensure maximum service from this style of unit.

The construction details of the inline style of axial-piston pump may vary between manufacturers. One of the more common major among from the design described above is one in which the cylinder barrel remains stationary and the swash plate rotates. This design requires variations in valve plate design and

piston shoes, even though most of the basic operating concepts are very similar.

The bent-axis design consists of a pump housing holding a power input shaft and a cylinder barrel. The housing is constructed so the axes of the input shaft

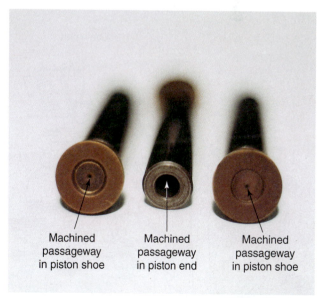

Machined passageway in piston shoe

Machined passageway in piston end

Machined passageway in piston shoe

Goodheart-Willcox Publisher

Figure 7-26. A critical lubrication region in an inline piston pump is the piston shoe/swash plate contact area. One solution is to route high-pressure oil from the pumping chamber to the area via the machined passageways shown in these pistons and shoes.

and cylinder barrel form an angle. This angle is often 30°, although other angles are used by some manufacturers. The cylinder barrel and input shaft are connected with a universal joint, which allows the barrel to be driven by the input shaft. The cylinder barrel contains several precisely machined bores fitted with pistons. The pistons are attached to the input shaft by individual connecting rods. A valve plate is located in the housing on the end opposite of the power input shaft. Refer to **Figure 7-27**.

During pump operation in the bent-axis design, the prime mover turns the input power shaft, which turns the cylinder barrel through the universal joint, **Figure 7-28**. The pistons and connecting rods turn with the combined units. The angle between the input shaft and the cylinder barrel causes the pistons to reciprocate in the bores as the unit rotates. As a piston is withdrawn from its bore, the volume between the piston and the valve plate increases. This causes a reduction in pressure, which allows system fluid to flow into the pump. As the piston moves back into the bore, the volume decreases. This causes the fluid to flow out of the pump to the system. The valve plate passageways are positioned such that the inlet line is open to the area above the piston when bore volume is increasing. The area above the piston is open to the discharge line as the volume is decreasing.

Many manufacturers consider the bent-axis, axial-piston pump more rugged than the inline type. In general, the units are capable of larger flow rates, greater

pressures, and higher operating speeds. The design also functions better as a self-priming unit. This is because the connecting rod drive is more positive than in some of the inline pump designs.

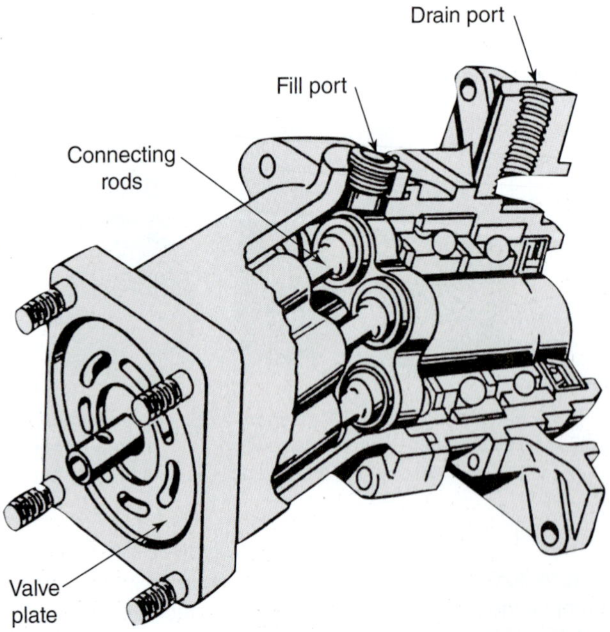

Radial-piston pumps

In *radial-piston pumps*, the motion of the pistons is perpendicular to the pump input shaft. The pistons radiate out from the axis of that shaft. Two basic design variations are also used with this classification. In the *stationary-cylinder design*, a cam on the input shaft operates the pistons, which are located in a fixed cylinder block. In the *revolving-cylinder design*, the piston and cylinder assembly rotates as a unit.

Radial pumps usually have the highest continuous pressure capability of any of the hydraulic pumps. Standard models of these pumps are available that will perform continuously in the 10,000 psi range. They are also highly efficient, making them desirable in circuits where very accurate control of component speed is required.

Piston pumps using the stationary-cylinder design consist of a housing that contains precision cylinders, pistons, piston springs, inlet and outlet check valves, and a drive shaft containing an eccentric cam. Typically, three to seven cylinders are symmetrically located around the drive shaft. See **Figure 7-29**. The piston springs hold one end of the pistons in constant contact with the cam. Rotating the drive shaft and attached cam causes a reciprocating movement of the pistons in the cylinders. Each time piston movement causes the volume of a cylinder to increase, low pressure occurs in the cylinder. This low pressure allows fluid to move into the cylinder from the system reservoir through the inlet check valve. When cylinder volume is decreasing, the fluid is forced out of the cylinder through the outlet check valve. Each cylinder in the pump functions the

Drain port

Fill port

Connecting rods

Valve plate

Goodheart-Willcox Publisher

Figure 7-27. The cylinder block and pistons of the bent-axis piston pump are turned by a universal joint. This allows an angle to exist between the centerlines of the block and the power-input shaft.

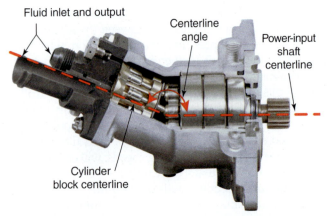

Fluid inlet and output

Centerline angle

Power-input shaft centerline

Cylinder block centerline

Courtesy of Eaton Fluid Power Training

Figure 7-28. In a bent-axis piston pump, the angle between the centerline of the power-input shaft and the cylinder block determines the length of the piston stroke.

same, with the combined flows producing the output of the pump.

The revolving-cylinder design of a radial-piston pump involves a pump housing in which a fixed, circular reaction ring is fitted. Piston movement is controlled by the reaction ring. This ring ensures the pistons are withdrawn from the cylinder bores as the cylinder block is rotated. Located off a center in the reaction ring is a fixed *pintle*, which serves as the bearing for the rotating cylinder block. The pintle also contains inlet and outlet ports. These ports control fluid flow into and out of the cylinders located in the cylinder block. Pistons are fitted into the cylinder bores with the ends contacting the reaction ring via some type of bearing surface. The system prime mover turns the pump through an input shaft attached to the cylinder block. Refer to **Figure 7-30**.

As the cylinder block rotates, the pistons remain in constant contact with the reaction ring. Since the reaction ring is off center, the pistons reciprocate in the cylinder bores. As a piston is moving out, the cylinder volume toward the pintle enlarges. This causes a low-pressure area to develop. At this point, the cylinder is aligned with the inlet port of the pintle and fluid from the system reservoir flows into the cylinder. As the cylinder block continues to rotate, the piston is forced back into the cylinder block. The volume of the cylinder is reduced and fluid is forced out into the system. Each cylinder functions in the same way, with the output of each cylinder contributing to the total output of the pump. Refer to **Figure 7-31**.

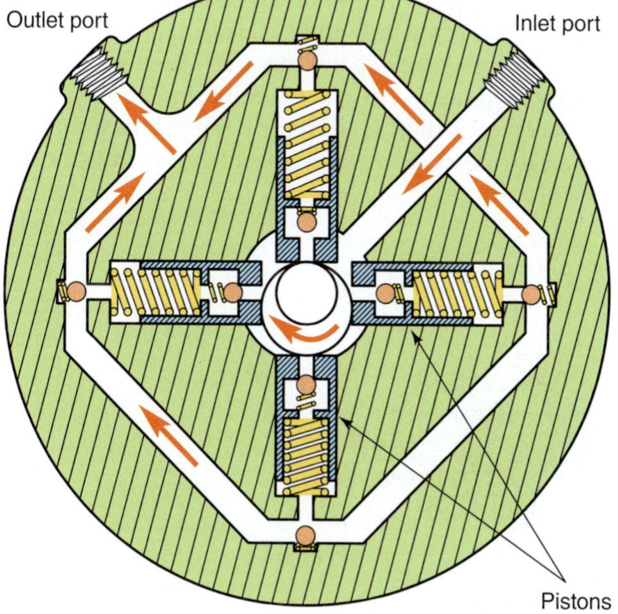

Reciprocating-plunger pumps

Reciprocating-plunger pumps have a design in which piston motion is perpendicular to the input shaft with the pistons extending along the axis of the input shaft of the pump. Unlike most of the other designs of piston pump, the cylinder bores of reciprocating-plunger

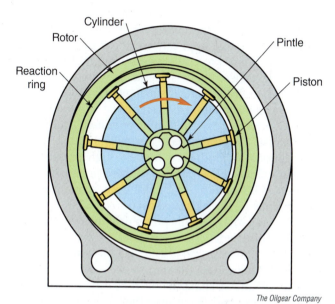

The Oilgear Company

Figure 7-30. In a revolving-cylinder radial-piston pump, fluid intake occurs when centrifugal force throws the pistons out against the reaction ring. Fluid is discharged to the system as the pump rotates and the reaction ring forces the pistons toward the center of the cylinder block.

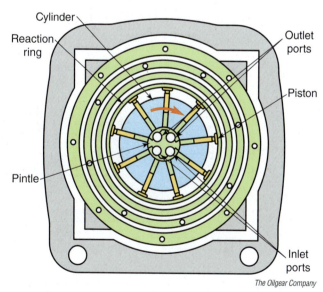

The Oilgear Company

Figure 7-31. The pintle of the revolving-cylinder radial-piston pump serves both as a bearing surface for the turning cylinder block and as the location of the inlet and discharge ports.

Goodheart-Willcox Publisher

Figure 7-29. In the stationary-cylinder design of radial-piston pumps, the cylinders are equally spaced around a cam located on the power-input shaft. As the input shaft turns, the cam operates the spring-loaded pistons, forcing fluid into the system.

pumps are located in a fixed block. Each of the cylinders is fitted with a piston that reciprocates in the bore.

One of the common methods used to produce the reciprocating motion is to attach each piston to a crankshaft via connecting rods. This design closely resembles the structure of an internal combustion engine. Other variations of the design use cams or a swash plate, rather than a crankshaft, to move the pistons during fluid discharge and springs to power the intake stroke. Each cylinder is fitted with inlet and discharge valves to control fluid movement through the pump.

The application and structure of reciprocating-plunger pumps varies considerably. During the early years of hydraulic applications, these pump designs were widely used to power centralized hydraulic systems in factories. Those systems delivered high flow rates and could operate at substantial pressure levels. The central systems were designed to supply fluid to several pieces of hydraulically operated equipment. In contrast, the pumps have also been used in special applications to produce extremely high pressures at very low flows. Flow outputs of these units vary from as little as 0.1 gpm to over 1,000 gpm, while pressure ratings may be as low as 200 psi or over 20,000 psi.

One advantage of this pump design is the fact that the major elements of the pump, other than the cylinder and piston areas, do not contact the material being pumped. This allows the unit to use a separate lubrication system. Often, specialized bearing designs and lubricants are used. Separating the lubricating function from the pumping function allows the pump to be used to move fluids that are not good lubricants or materials that cannot be contaminated. A second advantage to these designs is the positive seal provided by cylinder valves, which results in high operating efficiencies.

The large, centralized system where this type of pump was once prominent is now considered an outmoded design. However, this type of pump continues to be used in hydraulic system applications. It is still used in high-pressure, low-flow systems. In addition, this type of pump has found applications in many fields, providing high-pressure fluids for agricultural fertilizer injection, crop spraying, high-pressure cleaning equipment, waterjet material cutting, and numerous other tasks.

7.4.4 Screw Pumps

The *screw pump* uses a rotary design with intermeshing screws to form pumping chambers, **Figure 7-32**. These chambers linearly move fluid through the pump. The design provides continuous, positive displacement of the fluid to deliver a nonpulsating flow at all speeds. A benefit of the design is very quiet operation. This is due to little metal-to-metal contact of moving parts and the nonpulsating nature of the fluid flow.

Screw pump designs typically involve pumping elements consisting of one, two, or three rotating screws. The pumping elements operate within bores in the pump housing. Very close tolerances are maintained between the pumping element screws and the bore surface.

The single-screw design uses a spiral-shaped drive screw that rotates eccentrically (off center) within a specially shaped bore machined into a cylinder sleeve. The sleeve is fitted into the pump housing. As the screw rotates, fluid is trapped and carried through the pump. This action is much like how an auger moves dry materials, such as grain from a combine into a truck.

The two-screw design uses a primary and secondary shaft fitted with screw threads. The primary shaft is directly powered by the system prime mover, while the secondary shaft is driven by the primary through the timing gears. The shafts have integral screw threads

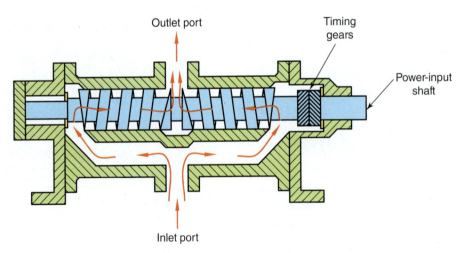

Goodheart-Willcox Publisher

Figure 7-32. The screw pump uses intermeshing screws to create pumping chambers that deliver a nonpulsating flow at all operating speeds.

and are positioned so the threads are closely meshed, but maintain minimum clearance. As the screws turn, a volume of fluid is trapped in the open spaces between the threads and pushed along the axis of the pumping element to the discharge of the pump.

The three-screw design has a pumping element containing a powered primary screw shaft that turns two secondary shafts. The screw threads on both of the secondary shafts mesh with the threads on the primary shaft. As in the other designs, fluid is moved through the pump when it becomes trapped in the spaces between the rotating screw threads.

Screw pumps are produced with flow outputs ranging from approximately 2 gpm to over 3,000 gpm. Pressure capabilities can be as high as 4,000 psi. The large-capacity pumps are typically used in applications where high-volume output is required at a relatively low pressure. An example of this situation is a large press that uses a high-volume screw pump for rapid closing and opening time and a low-volume piston pump to operate at high pressure for the actual pressing operation. High initial cost is a factor that eliminates screw pumps from consideration for many industrial and mobile applications.

7.4.5 Lobe Pumps

The *lobe pump* is a close relative of the external gear pump. It is a positive-displacement unit that uses a rotating pumping element to move fluid through the pump. The design has a pump housing and generally two three-lobe, gear-shaped units to form the pumping element. See **Figure 7-33**. One of the three-lobe units is directly connected to the system prime mover. The second is connected to the first by standard, external gears that synchronize lobe rotation. This allows for very close clearances to be maintained without lobe contact.

The geometry of the design allows the pumping-element lobes to maintain one close contact area between both elements at all times. In addition, the tips of the lobes are also in close contact with the pump housing. These factors allow a seal to be constantly maintained between the inlet and discharge of the pump. As the lobes rotate, the chamber connected to the inlet port is constantly enlarging, while the discharge chamber is constantly decreasing in volume. This causes a low-pressure area in the inlet chamber, which allows fluid to enter, while other fluid is being carried into the discharge chamber and creating discharge flow.

These pumps tend to have greater flow output than a gear pump of comparable physical size. This is due to the additional pumping-chamber volume inherent in the lobe design. The pumps also tend to operate quieter than gear pumps because of the clearance

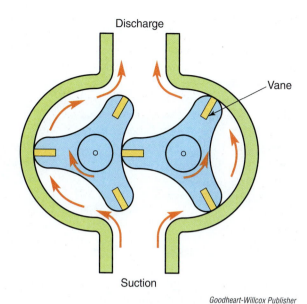

Figure 7-33. The lobe pump has a large output flow compared to its physical size. However, the smaller number of larger pumping chambers produces a more pulsating output flow. This is undesirable in many applications.

maintained between the lobes. Possible disadvantages of this design are a lower pressure range and a pulsating flow caused by the lobe configuration.

7.4.6 Centrifugal Pumps

Centrifugal pumps are nonpositive-displacement units that use the rotating motion of a pumping element to directly generate fluid movement. The design is not normally used in a hydraulic system as the fluid source for operating the system. The units are typically used to transfer fluids from one point in a system to another when the only resistance expected is the flow resistance and weight of the fluid. In some industrial situations, large centrifugal pumps transfer hundreds of gallons of fluid per minute. However, the pressure involved is usually below 200 psi and results only from the resistance to flow in the transfer lines. Outside of hydraulic system transfer pumps, common consumer applications of this design are the coolant pump on an automobile engine and the pump used in household washing machines to extract water from the tub, **Figure 7-34**.

A typical centrifugal pump consists of a pump housing containing an impeller mounted on the end of an input shaft. The *impeller* is disk-shaped, with one side containing curved blades radiating out from the center, and moves fluid through the pump. See **Figure 7-35**. The inlet of the pump is located in the housing near the center of the impeller. Fluid enters the inlet and floods the pump housing around the impeller. The discharge of the pump is located on the outer

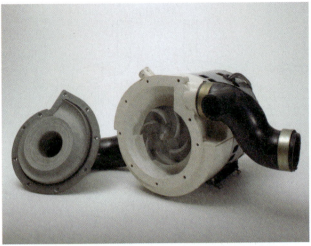

Figure 7-34. A centrifugal pump is often used as a fluid-transfer pump in hydraulic systems. The design is also commonly used in consumer applications, such as washing machines, dishwashers, and cooling system pumps in automotive engines.

circumference of the housing. The clearance between the impeller and the housing is generally large in these designs. This allows slippage when resistance to flow is met during pump operation.

As the pump input shaft and impeller are turned by the prime mover, the fluid filling the space between the impeller blades rotates with the impeller. Centrifugal force acting on the trapped fluid causes it to move outward toward the pump housing and the discharge. This fluid movement causes two things to occur. First, the centrifugal force causes an increase in

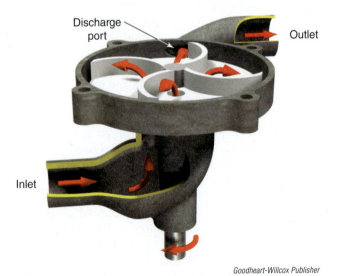

Figure 7-35. A centrifugal pump moves fluid using centrifugal force. Considerable clearance exists between the impeller and the pump housing, allowing slippage of fluid as resistance increases in the outlet line.

pressure at the housing, which moves fluid out of the pump discharge. Second, the movement of fluid away from the inlet causes a low-pressure area to develop, which allows new fluid to enter the pump through the inlet. The movement of fluid out of the pump while new fluid is brought in continues as long as excessive resistance to fluid flow does not develop in the system. Changing pump speed changes the amount of centrifugal force developed in the pump. This allows the flow through the pump to be varied.

When the fluid leaving the pump encounters resistance in the system, the pressure of the fluid in the pump increases. The maximum pressure that can be generated is established by the centrifugal force developed from the rotating impeller. When this pressure level is reached, the fluid in the spaces between the impeller blades continues to be carried as the impeller rotates. However, no fluid leaves or enters the pump housing. When system resistance decreases, pump output is simply reestablished. The substantial internal clearances between the impeller and the housing plus the centrifugal force used to move the fluid out of the pump allow this type of operation.

Centrifugal pumps are not self priming because of the clearance involved between the impeller and the housing. The pump impeller housing must be flooded with fluid at startup to ensure proper operation. Flooding is typically accomplished by submerging the pump in the fluid to be moved or by having the fluid in a reservoir located higher than the pump.

7.4.7 Other Pump Designs

Two additional nonpositive-displacement pump designs may be encountered in some fluid power systems: propeller and jet pumps. Both of these units serve more as accessory pumps than as the primary power source for the system. The units are used to transport fluid from one container to another. They are also used to evacuate system reservoirs or sumps that cannot be easily drained.

Propeller pumps

A *propeller pump* consists of a propeller-shaped pumping element located in a close-fitting, tube-shaped housing. See **Figure 7-36**. As the prime mover turns the pumping element, a low-pressure area is created on the inlet side of the pump. At the same time, a higher-pressure area is created on the discharge side. The low inlet pressure causes fluid to flow into the pump, while the higher pressure at the discharge moves fluid out into the system. The units are capable of moving large volumes of fluid at relatively low pressures. An important feature of this design is the clearance between the pumping element and housing. This provides

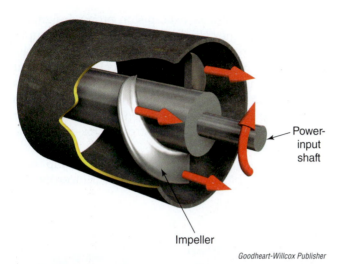

Power-input shaft

Impeller

Goodheart-Willcox Publisher

Figure 7-36. A propeller pump is a nonpositive-displacement unit that is primarily used as a fluid-transfer unit, rather than as a system power source.

sufficient slippage to allow the unit to continue rotating even when resistance at the output of the pump stops flow.

Jet pumps

A *jet pump* uses a variation of Bernoulli's principle to move fluid. The design involves forcing fluid through a pumping element that consists of a housing containing a nozzle, venturi, inlet port, and discharge port. See **Figure 7-37**. A *venturi* is a narrow area that restricts fluid flow. Fluid flow, often created by the regular system pump, is forced through the nozzle and venturi area in the pumping element. This fluid movement creates a low-pressure area while causing an increase in fluid velocity. The combination of the low pressure and increased fluid velocity causes additional fluid to be brought into the pumping element through the inlet port, mixed with the fluid entering

through the nozzle, and then forced out of the discharge port. The pumping element can be remotely located from the primary pump that supplies the fluid for operation. This pump design is used only in specialized situations in large equipment. However, jet pumps are commonly used in shallow water wells for home, farm, and industrial use.

7.5 Additional Design Features of Pumps

The discussions in the previous sections deal with basic pump designs and features. The descriptions included only those elements required for a fixed rate of pump output. However, many additional features are used by pump manufacturers to provide operating characteristics desired by hydraulic system users. This section examines four additional features that are commonly used in hydraulic pumps:

- Variable-flow delivery.
- Pressure compensation.
- Pressure balancing.
- Dual pumps.

7.5.1 Variable-Flow Delivery

Some pumps can be designed to provide the operator an easy means of changing the size of the pumping chamber. This allows variable-flow delivery from a single pump. Not all basic pump designs can accept this design modification. For example, the gear and balanced-vane pump designs cannot be produced as variable-flow units. This is due to the configuration of their pumping elements. Unbalanced-vane, axial-piston, and radial-piston pumps commonly include variable-flow output features.

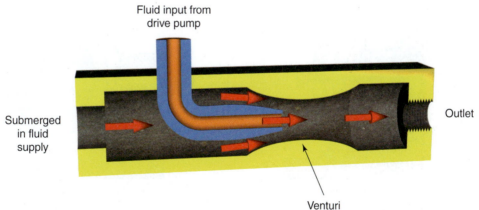

Fluid input from drive pump

Submerged in fluid supply

Outlet

Venturi

Goodheart-Willcox Publisher

Figure 7-37. The jet pump uses Bernoulli's principle to move fluid within a system. A nozzle discharging fluid through a venturi produces a low-pressure area. This causes additional fluid movement.

Unbalanced-vane pumps

The variable-flow feature is added to an unbalanced-vane pump design by incorporating a movable cam ring. The ring can slide back and forth within the pump housing to vary the size of the pumping chamber. See **Figure 7-38**. When the rotor shaft and ring are concentric (sharing the same center), the space between the rotor and ring is equal at all points. Under these conditions, no flow is produced when the pump is turning. This is because the volume does not vary in the pumping chambers between the vanes. When the ring is moved to the maximum offset position (eccentric, or not sharing the same center), a maximum variation in the chamber size is created. Turning the pump under this condition produces maximum flow. Moving the ring between the two positions, therefore, produces a full range of flows, from zero to the maximum pump output.

Piston pumps

Variable-flow delivery can be designed into both axial- and radial-piston pumps. To vary the pump output, these designs use mechanisms that change the length of the piston stroke in the pumping chamber. Positioning an internal pump mechanism to produce no stroke results in zero fluid flow from the unit. A position that creates the maximum stroke yields full fluid flow.

The inline, axial-piston pump design commonly uses an adjustable swash plate to vary the length of the piston stroke, **Figure 7-39**. Tilting the plate until it is perpendicular to the axis of the cylinder barrel results in a no-stroke situation. This yields no fluid output from the pump. Tilting the plate to the maximum angle produces the longest stroke. Therefore, this position produces the highest output flow.

With the bent-axis, axial-piston pump design, the angle between the cylinder block and the power input shaft is varied to change piston stroke, **Figure 7-40**. When the centers of the cylinder block and the input shaft are concentric, a no-stroke condition exists. This produces no pump flow. Tilting the cylinder block to the greatest angle allowed between the barrel and shaft results in the longest piston stroke. This creates the highest flow output.

The rotating-cylinder, radial-piston pump may also be a variable delivery. The feature is incorporated

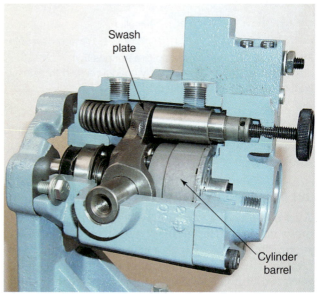

Continental Hydraulics

Figure 7-39. An inline-piston pump fitted with an adjustable swash plate can produce variable flow. If the swash plate is perpendicular to the cylinder barrel, there is no piston stroke. Varying the angle produces varying strokes and flow.

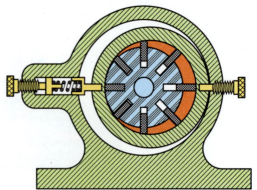

Goodheart-Willcox Publisher

Figure 7-38. The unbalanced-vane pump can be a variable-delivery pump if a cam ring that moves in relation to the center of the rotor is incorporated into the design. The greater the eccentricity of the parts, the greater the delivery of the pump.

Zero Fluid Output

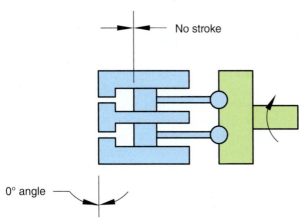

No stroke

0° angle

Slight Fluid Output

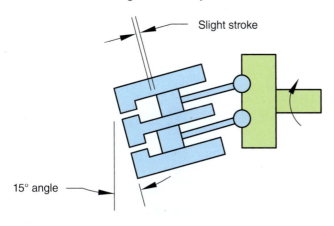

Slight stroke

15° angle

Higher Fluid Output

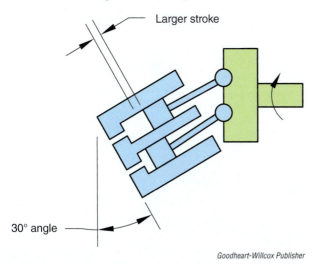

Larger stroke

30° angle

Goodheart-Willcox Publisher

Figure 7-40. A variable-displacement version of a bent-axis piston pump is produced by providing a device that can vary the angle of the cylinder block in relation to the power-input shaft. The greater the angle, the longer the piston stroke and the higher the pump output.

into this pump design with the use of a movable reaction ring, **Figure 7-41.** When the ring and fixed pintle are concentric, the cylinder block can rotate without the pistons moving. Moving the ring off center of the pintle results in piston movement and fluid flow from the pump. Like the other variable-displacement pumps, the flow output can vary between zero at the no-stroke position to maximum flow when the reaction ring is offset to produce the longest piston stroke.

7.5.2 Pressure Compensation

The pressure-compensation feature of a pump allows the unit to sense system pressure and vary the volume of the pumping chamber. This allows the pump to provide only sufficient fluid to maintain a desired system operating pressure. When system actuator resistance is low and flow demand high, the volume of the pumping chamber is the largest and the pump produces maximum flow. As the load on the actuator increases, the higher resistance to fluid flow causes system pressure to climb. In a pressure-compensated pump, this increased pressure causes the geometry of the pumping chamber to change. As a result, only the amount of flow needed to maintain a preset system pressure is produced. When the load on the system stalls the actuator, the volume of the pumping element approaches zero with only sufficient flow produced to compensate for system leakage. See **Figure 7-42.** Pressure

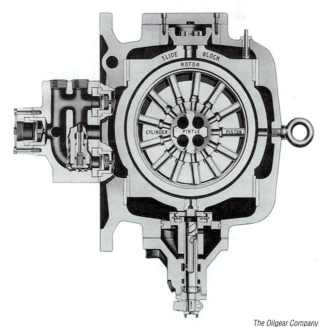

The Oilgear Company

Figure 7-41. The rotating-cylinder version of the radial-piston pump can be fitted with a mechanism to adjust the reaction ring to produce a variable-displacement pump. When the reaction ring (rotor) and cylinder block are concentric, no flow is produced. The greater the eccentricity, the greater the piston stroke and the higher the flow.

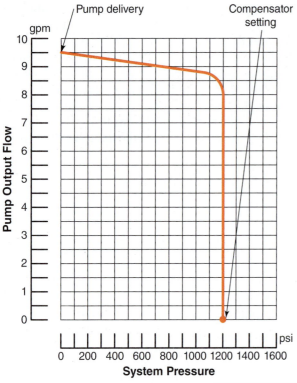

Goodheart-Willcox Publisher

Figure 7-42. The output flow of a pressure-compensated pump depends on the pressure in the system. When system pressure is below the compensator setting, the pump delivers maximum flow. As the pressure approaches the compensator setting, the pump output drops to only the flow needed to maintain system pressure.

compensation is available only in unbalanced-vane and piston pumps. This is because a pump must be capable of variable-flow output to allow the compensating mechanism to operate.

In a basic, unbalanced-vane pump, the compensating mechanism involves a spring that forces the pumping chamber ring to the maximum offset. See **Figure 7-43.** The pump produces maximum flow at this setting. As resistance to fluid flow begins to increase, reflecting an increased system load, pressure on the pumping chamber ring gradually shifts the ring until it is centered on the rotor and vanes. When the ring is centered, flow output of the pump is near zero with system pressure maintained at the preset level.

In an axial-piston pump design, the compensating mechanism senses system pressure and alters the angle of an internal mechanism, **Figure 7-44.** In an inline pump, the angle of the swash plate is changed. In a bent-axis pump, the angle between the input shaft and the cylinder bore barrel is changed. These changes vary the stroke of the pump, which changes the pump displacement. Compensation is provided in the rotating-cylinder, radial-piston pump by using the compensating mechanism to move the reaction ring.

Maximum flow output is produced when the ring is at the greatest offset. No flow is produced when the center points of the cylinder block and ring are concentric.

A major advantage of the pressure-compensated feature is the savings that result from reduced energy consumption. Basically, the energy consumed by the hydraulic system prime mover is determined by system load. System load is established by pressure and flow demands. The pressure-compensation feature reduces the amount of time that the prime mover must spend producing maximum flow against maximum system pressure. As soon as the system actuator encounters a load condition that stops its movement, the compensator reduces flow output, which reduces energy consumption and operating cost. See **Figure 7-45.**

The compensating feature of these pumps serves to control maximum pressure in a system. However, compensated pumps should *not* be considered the safety device for limiting maximum system pressure. A relief valve *must* be included to protect the system.

Continental Hydraulics

Figure 7-43. The compensator in a pressure-compensated vane pump senses system pressure and automatically positions the cam ring to produce the fluid output required to maintain a desired pressure level.

Continental Hydraulics

Figure 7-44. Pressure compensation is achieved in variable-displacement piston pumps by automatically setting the stroke length of the pump. The compensator senses system pressure and mechanically changes the angle of the swash plate or cylinder barrel, or the eccentricity of the reaction ring, to produce the volume of fluid required to maintain pressure.

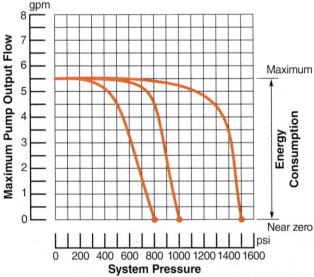

Goodheart-Willcox Publisher

Figure 7-45. The power required to drive a hydraulic system pump depends on the flow rate and pressure of the fluid. With a pressure-compensated pump, the flow rate and, therefore, power requirements are greatly reduced when the required pressure is being maintained. The curves shown here represent three system pressures for the same pump.

7.5.3 Pressure Balancing

The pressure developed in a hydraulic system pump may range from several hundred to several thousand pounds per square inch. The pressure is used in system actuators to generate force, which is one of the concepts basic to the operation of fluid power systems. While the pressure is needed to develop actuator force, it can cause problems in the operation of units such as pumps. Examples of these less-than-desirable situations are:

- Side loading of pump bearings.
- Leakage of fluids on the end surfaces of gears and vane rotors.
- Wear from the heavy loads placed on piston pump components, such as the piston shoes and swash plates.
- Leakage between the vane tips and cam ring of vane pumps.

Pressure balancing is a design technique used to reduce or balance the pressure differences and overcome a number of problems. The following describes several design approaches used to overcome typical pressure-related problems.

Side loading of the main shaft bearings is inherent in the design of gear, vane, and radial-piston pumps. This loading is caused by the fact that 180° of shaft rotation is devoted to the inlet of fluid at less-than-atmospheric pressure, while the other 180° of the rotation is concerned with the discharge of the fluid at a higher pressure. As a result, the pressure difference may range from near 0 psi to the maximum system pressure. This difference in pressure generates unequal forces that place considerable side load on the bearings. The balanced-vane pump design solves this problem by the use of two crescent-shaped pumping chambers that allow two inlet and two discharge ports. See **Figure 7-46.** The inlets are located 180° from each other, as are the discharge ports. This places an equal load on the shaft bearing under all system pressure conditions. Other pump designs compensate for uneven loading through the use of heavier bearings to withstand the increased load. In some cases, pressure lubricating of the shaft bearings is used to compensate for side loading.

The end surfaces of gears in gear pumps and the rotor and vanes in vane pumps cause special problems for the pump manufacturer. These surfaces are often referred to as thrust or wear surfaces. The clearance between these surfaces and the pump housing must be small to reduce internal leakage. At the same time, the metal of the two surfaces must be compatible to provide a long service life. A common method used to promote both a good seal and long wear is including a *wear plate* on one or both ends of the gear or rotor. See **Figure 7-47.** This plate is fitted into the pump housing and held against the end surfaces of the gear or rotor by light spring pressure. In addition, a passageway from the discharge section of the pump is connected to the space between the wear plate and the pump housing. During pump startup, the spring pressure ensures a good seal between the components. As actuator load causes the system pressure to increase, the wear plates are more tightly forced against the end surfaces of the

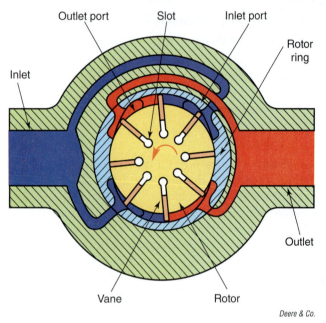

Figure 7-46. A balanced-vane pump has two inlet and two discharge areas that are opposite of each other. The pressures in the areas counteract each other, greatly reducing bearing loads.

gear or rotor. Careful design ensures that the force pushing the wear plates against the end surfaces does not become excessive, which could cause rapid wear and system power loss.

Heavy loads are placed on a number of pump components during operation. In several situations, pressure generated within the system can be used to balance forces and reduce the load on these components. An example is found in piston pumps. Piston shoes, slippers, or ball joints provide critical bearing surfaces connecting the pistons and the pump drive shaft. These bearing surfaces are heavily loaded when the pump is operating at maximum system pressure. Manufacturers incorporate features in the pump that deliver pressurized fluid from each associated piston chamber to these surfaces. That pressurized fluid is used at each bearing surface to counteract the load, as well as provide lubrication. Refer to **Figure 7-26**.

In other situations, system pressure can be used to ensure a positive seal between critical pump components. Vane pumps are usually designed to apply system pressure to the rotor end of the vanes to ensure that a positive seal is maintained between the cam ring and vane. The slot in the rotor in which the vane slides contains a circular opening at the base. This section of the slot is exposed to pumping-chamber pressure through channels machined into the end surface plates of the pump. When the rotor is in the discharge phase of pump operation, system pressure forces the vanes out against the cam ring, assuring a positive pumping-chamber seal. When the rotor is in the intake phase, the slot is exposed to lower pressure. This allows the vane to more easily move back into the rotor. See **Figure 7-48**.

A   B

Figure 7-47. A—Light spring tension is often used to hold wear plates in contact with the end surfaces of the gears in gear pumps. B—Pressures exerted on the surfaces of the wear plates by the inlet and discharge areas are balanced through lines leading to chambers behind the plates.

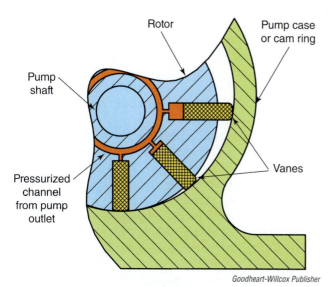

Figure 7-48. The ends of the vanes inserted into the rotor are subject to pressure from the inlet and discharge chambers in many vane pump designs. This pressure and centrifugal force work together to hold the vanes against the cam ring during pump operation.

7.5.4 Dual-Pump Designs

A *dual pump* is a unit made up of two pumps housed in a single case and driven by a single prime mover, **Figure 7-49**. The units are usually constructed using

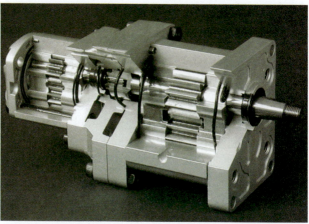

Sauer-Danfoss, Ames, IA

Figure 7-49. Dual pumps combine two pump units in a single case, usually for ease in system fabrication. In most of these units, one pump has high flow/low pressure ratings and the second has low flow/high pressure ratings.

a basic gear- or vane-pump design. Models are available as basic, two-pump units or with integral valving. Those with built-in valving are typically fitted with relief and unloading valves. The pumps are produced as a single unit to meet the requirement of applications needing two fluid flow rates for operation.

A typical example of the use of dual pumps is a high-low circuit. See **Figure 7-50**. A high-flow output

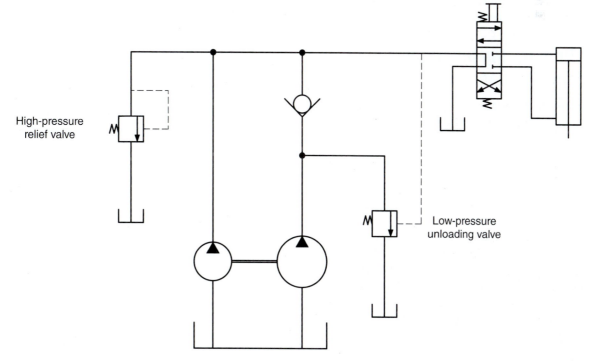

Goodheart-Willcox Publisher

Figure 7-50. A typical application for a dual pump is a high-low circuit, where the combined flow of the high- and low-volume pumps extend a large cylinder. Once the cylinder is extended, the low-volume/high-pressure pump is used to create the high system pressure required to finish the operation.

is required to extend a large cylinder a considerable distance under low pressure. After the initial movement, a much higher pressure is required to complete the work. In this circuit, one of the pumps in the dual-pump design has a large flow capacity, while the second has a much lower flow capacity. The flow from both pumps is combined to provide the initial high input to quickly extend the cylinder. At the point where the operation requires high pressure, the high-volume pump is unloaded to the reservoir and the low-volume pump continues to operate the system as resistance increases.

7.6 Pump and Power Unit Design and Operating Considerations

The effectiveness of the hydraulic system depends on the careful selection and assembly of components that make up the system. This is especially true of the pump, prime mover, and plumbing of the power unit. Each component needs to be adequately sized to ensure that the expected tasks can be performed, but not oversized to the point where initial component price and continuing operating costs are higher than necessary. Several concepts need to be understood to properly select components and then assemble, operate, and troubleshoot the power unit. These concepts include reading pump specifications, calculating system horsepower requirements, and designing pump inlet lines to eliminate pump cavitation.

7.6.1 Interpreting Information Sheets of Pump Manufacturers

Most pump manufacturers supply considerable information about the pumps they produce. Typically, this information is broken down into four categories:

- General specifications.
- Performance data.
- Installation drawings.
- Application information.

The amount of information and the presentation method varies among manufacturers. Often, this varies even among the models produced by the same company.

General specifications include operating information such as displacement per revolution, flow at rated speed and pressure, maximum rated speed, continuous rated operating pressure, and allowable inlet line vacuum. Specification tables may also list physical information, such as pump weight and type of

mounting flange. Specifications are usually presented in table form. See **Figure 7-51**. Often, a table is available that presents information about several variations of the same pump. This provides an easy means of comparing models during initial pump selection or when identifying a pump during service.

Graphs are typically used to present pump performance data. Overall pump efficiency and volumetric efficiency are two performance data that are commonly charted on all pump designs. Pressure-compensated designs usually plot power requirements to show performance during standby operation. See **Figure 7-52**. Typically, two or more operating pressures are shown for each factor. Considerable variations will be found in the form used by companies to illustrate what they consider to be critical performance information about the operation of their pumps.

Installation drawings provide detailed information about the shape and size of a pump. See **Figure 7-53**. This information is very useful to system designers for placement and mounting of the pump. It is also used for inlet and discharge line routing. Often, these drawings include dimensions for the various input shaft configurations available for the pump.

The application information section of data sheets lists general information about pump installation and operation. Information typically included is specific directions on the positioning of the pump, location of oil drain lines, recommended filtration level for the system, and recommendations for the type and viscosity of hydraulic fluid. Cold weather startup is another factor that is commonly discussed in application information.

7.6.2 Prime Mover Power Requirements

The size of the prime mover required for the power unit needs to be established when developing the specifications for a hydraulic system. The size is primarily determined by the peak power needed during the operating cycle of the components in the system. This approach is necessary because:

- System pressure changes as actuator loads fluctuate.
- Pump efficiency varies as system pressure changes.
- Varying system operating temperatures result in a range of hydraulic fluid viscosities that change fluid flow resistance.

Each of these changes results in a somewhat different power requirement. The following sections discuss the source of the basic horsepower formula, the involvement of pump efficiency in the calculation, and a general horsepower formula for the power unit.

PVR50 SERIES
VANE PUMPS
VARIABLE DISPLACEMENT PRESSURE COMPENSATED

**4-Bolt
Flange Mounted**

TYPICAL PERFORMANCE SPECIFICATIONS

			PUMP SIZE			
			32A15	42A15	50B15	70B15
VOLUMETRIC DISPLACEMENT*	cu. in./rev.		7.7	9.9	7.7	9.9
	ml/rev.		126	162	126	162
PUMP DELIVERY AT 1750 RPM*	91.5 psi	gpm	39	182	223	284
	6.3 bar	lpm	148	26.5	34	41
	rated	gpm	32	42	50	70
	pressure	lpm	121	159	189	265
COMPENSATED PRESSURE RANGES	Max.	psi	1500	1500	1500	1500
		bar	103	103	103	103
	Rated	psi	1500	1500	1500	1500
		bar	103	103	103	103
	Min.	psi	350	400	350	400
		bar	24	27.6	24	27.6
OPERATING SPEEDS**	Min. rpm		800			
	Rated rpm		1200	1200	1800	1800
	Max. rpm		2200	1500	2200	1800
POWER INPUT AT RATED FLOW & PRESSURE (1750 rpm)	hp		36	42	50	60
	kW		27	31	37	45
MAXIMUM POWER INPUT TO DRIVE SHAFT	Max.	hp	100			
		KW	75			
SUCTION PRESSURE	Max.	in./Hg	5		3	
		bar	-0.17		-0.10	
	Min. Specific Grav. < 1	psi	20	10	20	10
		bar	1.4	.07	1.4	0.7
	Min.	in./Hg	5		3	
	Specific Grav. > 1	bar	-0.17		-0.10	
CASE DRAIN FLUID VELOCITY	Max.	ft./sec.	5			
		m/sec.	1.5			
NOMINAL FLOW AT DEADHEAD PRESSURE	Max.	gpm	3			
	Pressure	mlpm	11			
	Min.	gpm	2.5			
	Pressure	mlpm	9.5			
MAXIMUM CASE PRESSURE		psi	10			
		bar	0.7			
WEIGHT		lbs.	119			
		kg	54			

NOTES:
*   Volumetric displacement is measured displacement at 91.5 psi (6.3 bar) and rated rpm per ANSI specification. Volumetric displacement varies with both pressure and rpm. Flow rates at any rpm other than the rated rpm may be approximated as follows:

$Q_2 = Q_1$ (N-142)/1667 where Q_1 = Flow (gpm) at rated rpm at 91.5 psi (6.3 bar).
Q_2 = Flow (gpm) at N rpm.
N = rpm at which Q_2 is to be determined.

**   When operating above 1500 psi (103 bar), it is recommended that a direct-acting differential relief valve be used at the pump to relieve pressure spikes and surges.
Maximum rpm at full displacement - 1900 rpm. For higher rpms up to 2000 rpm, pump displacement must be reduced to limit flow to 60 gpm (227 lpm) maximum.

OVERALL EFFICIENCY
At Maximum Displacement, Fluid Viscosity 130 SUS

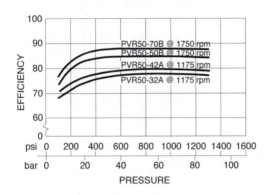

TYPICAL SOUND LEVEL

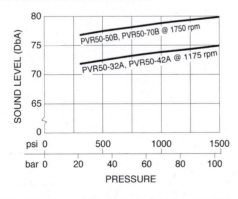

PRESSURE and VOLUME ADJUSTMENT SENSITIVITY

		PUMP SIZE	32A15	42A15	50B15	70B15
PRESSURE ADJUSTMENT	Press Change/Turn	psi (bar)	115 (8.0)		135 (9.4)	
	Max.Torque	ft./lbs.(kg/m)	26.5 (13.7)			
VOLUME ADJUSTMENT	Flow Change/Turn	gpm (lpm)	14 (53.0)		22 (83.0)	
	Min. Flow Adjust.	gpm (lpm)	6.0 (22.7)	8.0 (30.3)	9.5 (36.0)	12.5 (47.0)
	Max. Torque	ft./lbs. (kg/m)	28 (3.9)	16 (2.2)	28 (3.9)	16 (2.2)

Continental Hydraulics

Figure 7-51. A general specification sheet for a hydraulic pump lists a wide variety of information about a specific pump model. Typical information on these sheets includes displacement per revolution, recommended operating speed, and allowable inlet line vacuum.

PVR50 SERIES
VANE PUMPS
VARIABLE DISPLACEMENT PRESSURE COMPENSATED

NOTE: Typical performance curves are based on ISO VG46 oil at 120° F. (49° C.). Above 400 SUS (84 CS), add 2% hp/100 SUS.

NOTE: Deadhead horsepower is read from curves at 0 gpm flow and pressure compensator setting psi.

32A15 (at 1175 rpm)

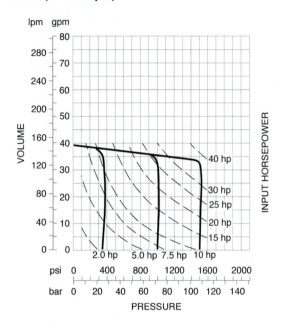

42A15 (at 1175 rpm)

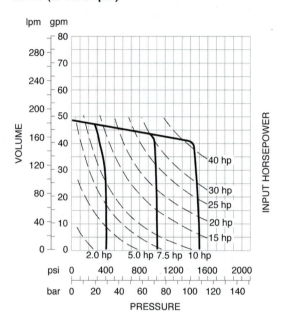

50B15 (at 1750 rpm)

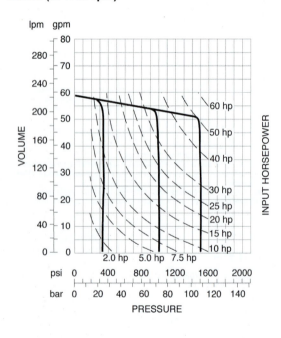

50B3L (at 1750 rpm)

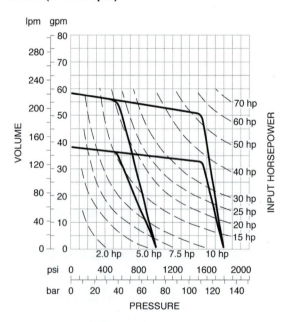

Continental Hydraulics

Figure 7-52. A performance data sheet for a hydraulic pump provides information about factors such as pump efficiency. This information is presented in chart and graph form, as well as in written descriptions.

PVR50 SERIES
VANE PUMPS
VARIABLE DISPLACEMENT PRESSURE COMPENSATED

® **CONTINENTAL HYDRAULICS**

PUMP DIMENSIONS
Right Hand Rotation (CW) - Code RF

Dimensions shown in: Inches
(millimeters)

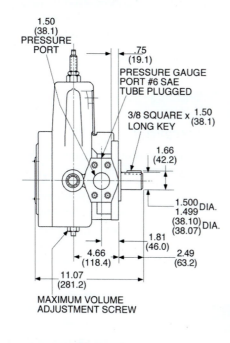

1.50
(38.1)
PRESSURE
PORT

.75
(19.1)

PRESSURE GAUGE
PORT #6 SAE
TUBE PLUGGED

3/8 SQUARE x 1.50
LONG KEY (38.1)

1.66
(42.2)

1.500 DIA.
1.499
(38.10) DIA.
(38.07)

1.81
(46.0)

4.66
(118.4)

2.49
(63.2)

11.07
(281.2)

MAXIMUM VOLUME
ADJUSTMENT SCREW

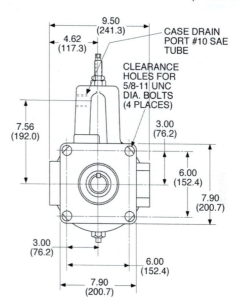

9.50
(241.3)

CASE DRAIN
PORT #10 SAE
TUBE

4.62
(117.3)

CLEARANCE
HOLES FOR
5/8-11 UNC
DIA. BOLTS
(4 PLACES)

7.56
(192.0)

3.00
(76.2)

6.00
(152.4)

7.90
(200.7)

3.00
(76.2)

6.00
(152.4)

7.90
(200.7)

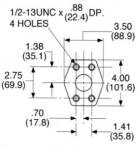

1/2-13UNC x .88 DP.
(22.4)
4 HOLES

3.50
(88.9)

1.38
(35.1)

2.75
(69.9)

4.00
(101.6)

.70
(17.8)

1.41
(35.8)

Pressure Port
1-1/2 Inch SAE 4-Bolt Connection Pad

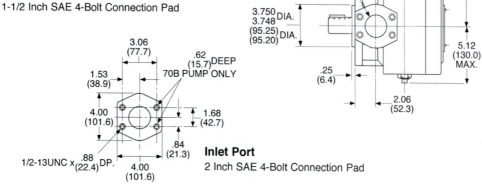

3.06
(77.7)

1.53
(38.9)

.62
(15.7) DEEP
70B PUMP ONLY

4.00
(101.6)

1.68
(42.7)

.84
(21.3)

1/2-13UNC x .88 DP.
(22.4)

4.00
(101.6)

Inlet Port
2 Inch SAE 4-Bolt Connection Pad

PRESSURE
ADJUSTMENT
SCREW

4.66
(118.4)

2.00 DIA.
(50.8)
SUCTION
PORT

11.91
(302.5)
MAX.

3.750 DIA.
3.748
(95.25) DIA.
(95.20)

.25
(6.4)

2.06
(52.3)

5.12
(130.0)
MAX.

Figure 7-53. Information concerning the physical size of a pump is listed on the installation drawings. This includes details on shaft sizes, keyways, port locations, and sizes, as well as all exterior dimensions.

General hydraulic horsepower formula

Horsepower is defined as 33,000 foot-pounds of work per minute. Moving an object weighing 33,000 pounds one foot in one minute or one weighing 550 pounds one foot in one second requires one horsepower of effort. In a fluid power system, a pressurized fluid is moved instead of an object. The weight of the moved fluid functions the same as the weight of the object. The pressure of the fluid converted to fluid head serves as the equivalent to distance moved. The time factor is the same in either example. This is a basic formula calculating *general hydraulic horsepower*:

$$\text{General Hydraulic Horsepower} = \frac{\text{Fluid Flow Weight (lb/min)} \times \text{Pressure (feet of head)}}{33{,}000 \text{ ft-lb/min}}$$

Fluid flow weight is determined for the basic formula as:

$$\text{Fluid Flow Weight} = \frac{\text{Pump Flow Rate } (F_R) \times \text{Weight of Water } (W_W) \times \text{Specific Gravity of Fluid } (S_G)}{\text{gallons/ft}^3 \, (G_F)}$$

The pressure in feet of head is determined for the basic formula as:

$$\text{Pressure (feet of head)} = \frac{\text{System Pressure } (S_P) \times \text{Square Inches in a Square Foot } (S_I)}{\text{Weight of Water } (W_W) \times \text{Specific Gravity of Fluid } (S_G)}$$

These symbols represent the various elements in the fluid flow weight and pressure in feet of head portions of the formula:

F_R = Pump flow rate (gpm)

W_W = 62.4 lb/ft^3

S_G = Specific gravity of the hydraulic fluid

G_F = 7.48 gal/ft^3

S_P = System pressure in lb/in^2

S_I = 144 in^2/ft^2 (used as a conversion factor)

Substituting these elements into the basic formula produces this formula:

$$\text{General Hydraulic Horsepower} = \frac{\dfrac{F_R \times 62.4 \text{ lb/ft}^3 \times S_G}{7.48 \text{ gal/ft}^3} \times \dfrac{S_P \times 144 \text{ in}^2/\text{ft}^2}{62.4 \text{ lb/ft}^3 \times S_G}}{33{,}000 \text{ ft-lb/min}}$$

Reducing the various elements of the equation produces this working formula:

$$\text{General Hydraulic Horsepower (hp)} = \frac{\text{Pump Flow Rate (gpm)} \times \text{System Pressure (psi)}}{1{,}714}$$

The formula can also be expressed using a decimal instead of the 1,714 denominator. The fraction 1/1,714 can be converted to 0.000583 and used in this format:

General Hydraulic Horsepower (hp) = gpm × psi × 0.000583

EXAMPLE 7-2

Hydraulic Horsepower

How much hydraulic horsepower would a pump produce if its output is 10 gpm at 1,500 psi?

$$\text{Hydraulic Horsepower} = \text{Pump Flow Rate (gpm)} \times \text{System Pressure (psi)} \times 0.000583$$
$$= 10 \text{ gpm} \times 1{,}500 \text{ psi} \times 0.000583$$
$$= 8.75 \text{ hp}$$

NOTE

The horsepower formulas presented here provide the theoretical horsepower required to move the fluid in the system. They do not take into consideration losses that occur in the system.

Power unit efficiency

The horsepower actually required of the prime mover will be greater than that calculated with the formulas presented in the previous section. This is because energy is lost in the operation of both the pump and prime mover. These losses must be taken into account as power units are designed and assembled.

This leakage is caused by basic pump design factors, manufacturing tolerances, and wear caused by operation over time. These losses expressed as a *volumetric efficiency* factor. Volumetric efficiency is a percentage of actual fluid flow out of the pump compared to the theoretical flow out of the pump. This number is less than 100% to show the amount of fluid lost through leakage. Volumetric efficiency is calculated using the following formula:

$$\text{Volumetric Efficiency } (V_E) = \frac{\text{Actual Pump Flow Rate}}{\text{Theoretical Pump Flow Rate}} \times 100$$

A second energy loss also occurs in the pump and is caused by internal friction. This friction is caused by factors such as shaft bearings; other close-contact surfaces; the viscosity of the hydraulic fluid, which causes resistance to flow; and fluid turbulence. These losses, identified as *mechanical efficiency,* are expressed as a percentage less than 100% and calculated with this formula:

$$\text{Mechanical Efficiency (M}_\text{E}) = \frac{\text{Theoretical Horsepower}}{\text{Actual Horsepower}} \times 100$$

Measuring the actual horsepower needed to operate the pump requires the use of a dynamometer to measure the actual power required to operate the unit.

The final efficiency calculation is for *overall efficiency.* This calculation considers *all* energy losses in the pump. This formula is used to determine the overall efficiency rating:

$$\text{Overall Efficiency (O}_\text{E}) = \frac{\text{Volumetric Efficiency} \times \text{Mechanical Efficiency}}{100}$$

Using pump efficiency information

Efficiency ratings contain a number of variables that often are unclear to the user of hydraulic systems. Data for each of these terms may be found for most hydraulic pumps, although some manufacturers supply only part of the information or use variations of the terminology. Becoming familiar with the terms and the information a manufacturer supplies is critical to the success of a fluid power technician.

The information supplied by volumetric efficiency tables or graphs allows the fluid power technician to calculate the actual output flow of pumps at different speeds and pressures. For example, the overall efficiency rating of a pump can be used in conjunction with the general hydraulic horsepower formula to establish the size of the prime mover needed for a hydraulic power unit.

This example shows how the general hydraulic horsepower formula can be used with the overall efficiency rating to determine the *prime mover horsepower* needed to operate the power unit of a hydraulic system:

$$\text{Required Prime Mover Input Horsepower (Rhp)} = \frac{\text{gpm} \times \text{psi}}{1,714 \times \text{O}_\text{E}}$$

or

$$= \frac{\text{gpm} \times \text{psi} \times 0.000538}{\text{O}_\text{E}}$$

EXAMPLE 7-3

Power Unit Efficiency

A gear pump has a displacement volume of 2.75 in³. If the actual pump flow at 1,800 rpm and 2,000 psi is 19.8 gpm, what is the pump's volumetric efficiency?

$$\text{Theoretical Pump Flow Rate (in}^3\text{/min)} = \frac{\text{Displacement Volume (in}^3\text{)}}{\times \text{Cycles per Minute (rpm)}}$$

$$= 2.75 \text{ in}^3 \times 1,800 \text{ rpm}$$

$$= 4,950 \text{ in}^3\text{/min}$$

$$\text{Theoretical Pump Flow Rate (gpm)} = \frac{4,950 \text{ in}^3\text{/min} \times 1 \text{ gallon}}{231 \text{ in}^3}$$

$$= 21.4 \text{ gpm}$$

$$\text{Volumetric Efficiency (V}_\text{E}) = \frac{\text{Actual Pump Flow Rate}}{\text{Theoretical Pump Flow Rate}}$$

$$= \frac{19.8 \text{ gpm}}{21.4 \text{ gpm}}$$

$$= 0.925$$

$$= 92.5\%$$

If the pump's mechanical efficiency is 85%, what is the overall efficiency of the pump?

$$\text{Overall Efficiency (O}_\text{E}) = \frac{\text{Volumetric Efficiency} \times}{\text{Mechanical Efficiency}}$$

$$= 0.925 \times 0.85$$

$$= 0.786$$

$$= 78.6\%$$

What is the actual required prime mover input in horsepower to operate this system?

$$\text{Actual Hydraulic Horsepower (hp)} = \frac{\begin{array}{c}\text{Theoretical Pump Flow Rate (gpm)} \times \\ \text{System Pressure (psi)}\end{array}}{1,714 \times \text{O}_\text{E}}$$

$$= \frac{21.4 \text{ gpm} \times 2,000 \text{ psi}}{1,714 \times 0.786}$$

$$= 31.8 \text{ hp}$$

7.6.3 Effects of Inlet-Line Design on Pump Operation

All of the fluid used in the operation of a hydraulic system enters the system through the inlet line of the pump. As the pump rotates, a low-pressure area is formed within its pumping chamber. Atmospheric pressure acting on the surface of the fluid in the reservoir then pushes fluid into the pump through the inlet line.

The relatively low pressure differential and the lower-than-atmospheric pressure (vacuum) that exists creates special problems for hydraulic system operation. System designers must be aware of the relatively low pressure differential that is available to move the fluid from the reservoir to the pump.

The inlet line must be carefully sized to eliminate excessive fluid velocities and pressure drop. Inlet lines should be no smaller than the diameter of the pump inlet with a fluid velocity calculated at no more than 4 feet per second.

The height of the pump above the level of the reservoir fluid must also be limited. A general rule is that the pump inlet should not be more than 3 feet higher than the fluid level of the reservoir.

Fluid viscosity must be carefully monitored so excessive pressure drops are not created. Strainers and inlet-line filters must be carefully selected to minimize pressure drop in the line.

Care must be taken with each of the points discussed above. Otherwise, it is possible that the pressure in the inlet line will drop to the point where adequate fluid cannot enter the pumping chamber of the pump.

When the pressure in the pump inlet section drops too far below atmospheric pressure, a principle called liquid vapor pressure becomes a factor. *Vapor pressure* is the pressure at which a liquid begins to form a vapor. See **Figure 7-54**. The point at which this occurs is controlled by the liquid type, temperature, and pressure. When the vapor pressure is reached in the inlet of a hydraulic pump, a condition known as *cavitation* begins to appear. This is identified by bubbles of vapor that form in the hydraulic fluid in the pump inlet. These bubbles produce three conditions that affect pump operation:

- The bubbles reduce the ability of the fluid to lubricate, causing increased part wear.
- When the bubbles encounter the high-pressure section of the pump, they suddenly collapse and generate tons of force that can produce severe damage to the metal surfaces of the pump chamber.
- The collapse of bubbles also produces a high noise level in the pump.

Continued operation of a pump that is cavitating can severely damage or destroy the pump, **Figure 7-55**. Metal particles released from damaged pump parts may also cause excessive wear in other system components.

Another factor somewhat similar to cavitation is *entrained air*. This condition results from air that enters the inlet line, is drawn into the pump, and remains suspended in the fluid. The air may enter the line through leaks in fittings, lines, or accessories, such as filters or due to a low reservoir level. A pump operating under this condition produces high noise levels similar to cavitation. Damage from this condition can be similar to that produced by cavitation. This condition can be differentiated from cavitation by determining inlet line pressure. Cavitation produces a high, negative inlet-line pressure (vacuum), while a slightly negative (closer to 0 psi) pressure may indicate a condition caused by entrained air.

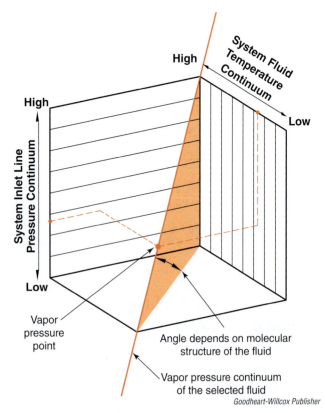

Goodheart-Willcox Publisher

Figure 7-54. The point at which the vapor pressure of a liquid is reached and gas bubbles begin to form depends on the molecular structure of the material, its temperature, and the external pressure applied (such as that encountered in the inlet portion of a hydraulic pump).

Courtesy of Eaton Fluid Power Training

Figure 7-55. Serious physical damage to pump components can occur if the reason for cavitation is not identified and eliminated from a hydraulic system.

Atlas Copco

The portable pump is used to circulate water through this system, rather than functioning as a means of pressurizing hydraulic fluid.

Summary

- A basic system power unit consists of a prime mover to power the system, a pump to move the fluid, a reservoir to store fluid, a relief valve to control maximum system pressure, a filter to clean the fluid, and various plumbing to transport fluid between components.

- All pumps operate using similar principles: an enlarging chamber in the pump allows fluid to enter, while reducing the chamber volume moves fluid out into the system. Inlet and discharge valves or ports control fluid movement through the pump.

- The maximum pressure developed in a hydraulic system is dependent on the resistance to fluid flow in the system and the force that the prime mover can exert.

- The output flow rate of a hydraulic pump is determined by the volume of the pumping chamber and the operating speed of the unit.

- Hydraulic pumps can be classified using three basic aspects: displacement, pumping motion, and fluid delivery characteristics.

- Positive-displacement pumps produce a constant output per revolution. Larger clearances in nonpositive-displacement pumps allow greater slippage and, therefore, greater flow variations with changing load.

- Hydraulic pumps use either rotary or reciprocating motion to change the volume of internal chambers and move fluid through the pump.

- Hydraulic pumps have either fixed or variable delivery. Fixed-delivery pumps have pumping chambers that have a volume that is constant, so their output is the same with each cycle. Variable-delivery pumps have chambers whose geometry may be changed to allow varying flow from the pump at a fixed speed.

- System operating pressure, flow rate, cycle rate, expected length of service, environmental conditions, and cost are factors that must be considered when selecting a pump.

- Gear pumps are positive-displacement, fixed-delivery rotary units that rely on the meshing and unmeshing of gears to move fluid through the pump. Both external- and internal-gear configurations are used in these pump designs.

- Vane pumps are positive-displacement, fixed- or variable-delivery rotary units that consist of a slotted rotor fitted with movable vanes that rotates within a cam ring in the pump housing. The off-center position of the rotor within the ring creates pumping chambers that vary in volume as the pump rotates.

- Piston pumps are positive-displacement, fixed- or variable-delivery units that use the reciprocating movement of pistons in a pumping mechanism to move system fluid. Piston pumps are classified as axial-piston, radial-piston, or reciprocating-plunger type.

- Screw pumps involve pumping elements that consist of one, two, or three rotating screws. As the screws rotate, fluid is trapped and carried along to the discharge of the pump.

- Lobe pumps use two three-lobed gear-shaped elements to pump fluid. These pumps generate larger output flow than a gear pump of comparable physical size, but operate at lower pressure and have pulsating output flow because of pumping-chamber geometry.

- Centrifugal pumps are nonpositive-displacement units that use centrifugal force generated by a rotating impeller to move fluid.

- Manufacturer data sheets provide general specifications, performance data, installation drawings, and application information needed for selecting, installing, and operating a hydraulic pump.

- Pump operation can be affected by cavitation if the system is not configured to allow adequate fluid to enter the pump. Cavitation occurs when fluid pressure at the inlet drops too low, causing bubbles to form in the fluid. Collapse of the bubbles in the high-pressure section of the pump causes noisy operation and releases a shock wave that damages pump parts.

Internet Resources

The following are some useful resources available on the Internet. Enter a company or organization name into a search engine to access its website. Explore the various areas of the sites to discover useful fluid power resources.

Machinery Lubrication Magazine. Provides a variety of articles and illustrations concerning cavitation.

Continental Hydraulics—Vane Pumps. Provides information for a series of vane pumps. Provides typical information available from manufacturers. Links are provided to product documentation, including specifications, control options, and accessories. Information provided for a series of different vane pumps.

Engineers Edge. Search for "screw-type pumps." Provides information on the design and operation of two- and three-screw pump designs.

Hydraulics & Pneumatics Magazine. Under "Fluid Power Basics," select "Pumps" to obtain information on various hydraulic pump designs with illustrations and explanations of operation.

Wikipedia: The Free Encyclopedia—Gear Pump. Provides detailed explanation of the design and operation of gear pumps. Additional sites listed provide considerable detail about various gear pump designs.

Wikipedia: The Free Encyclopedia—Cavitation. Provides information and examples related to cavitation.

Chapter Review

Answer the following questions using information in this chapter.

1. *True or False?* The device that is used to move fluid through a hydraulic system is called the actuator.

2. In order to serve as the overall source of energy for a hydraulic system, a power unit must contain a _____.

3. *True or False?* A safety valve found in some power units is set at a pressure slightly higher than the normal system operating pressure for protection if the relief valve fails.

4. *True or False?* During pump operation, the pressure in the pumping chamber is below atmospheric pressure only during the intake phase.

5. *True or False?* During pump operation, the pressurization phase increases the volume in the pumping chamber.

6. The maximum pressure developed in the system is determined by the resistance to the flow of fluid in the system and the force that can be applied to the _____.

7. All pumps used in hydraulic systems can be classified according to _____.
   A. displacement
   B. pumping motion
   C. fluid delivery characteristics
   D. All of the above.

8. A _____ pump produces a constant output for each revolution of operation over a wide range of operating pressures.
   A. positive-displacement
   B. nonpositive-displacement
   C. rotary
   D. All of the above.

9. The pump that includes a mechanism allowing an operator to easily vary the size of the pumping chamber is classified as a(n) _____-delivery design.
   A. variable
   B. external
   C. internal
   D. fixed

10. The most popular hydraulic pump uses _____ gears as the primary elements of the pumping mechanism.
    A. external
    B. internal
    C. variable
    D. inline

11. Two variations of this design are commonly found in hydraulic systems: _____ and crescent designs.

12. The _____-vane pump design achieves a pumping action by placing the slotted rotor and vanes in an offset position in the pumping chamber.
    A. crescent
    B. unbalanced
    C. sliding
    D. balanced

13. The _____-vane pump design achieves a pumping action by an elliptical opening in the cam ring, which allows two inlet and two discharge ports per pump revolution.
    A. crescent
    B. unbalanced
    C. sliding
    D. balanced

14. The pump design that uses reciprocating motion of a piston in a cylinder to move fluid through the unit is a _____ pump.
    A. gear
    B. gerotor
    C. vane
    D. piston

15. Some styles of piston pumps can function in systems requiring operating pressures as high as _____ psi or higher in some cases.
    A. 1,000
    B. 3,000
    C. 5,000
    D. 10,000

16. *True or False?* The cylinder barrel and the power input shaft of the inline-axial piston pump are parallel.

17. Components called _____ serve as the bearing surfaces between the pistons and swash plate of the inline axial-piston pump.
    A. piston shoes
    B. saddle bearings
    C. crescent plates
    D. control pistons

18. In the bent-axis piston pump design, the cylinder barrel is turned by power delivered through a(n) _____ attached to the input shaft of the pump.
    A. swash block
    B. universal joint
    C. control piston
    D. saddle

19. _____ pumps are usually considered to have the highest continuous pressure capability.
    A. Bent-axis piston
    B. Reciprocating-plunger
    C. Radial-piston
    D. Axial-piston

20. The _____ pump design provides continuous, positive displacement of the fluid to deliver a nonpulsating flow at all speeds, with very quiet operation.
    A. screw
    B. lobe
    C. bent-axis
    D. gear

21. In a centrifugal pump, the primary element of the pumping mechanism is called a(n) _____.
    A. venturi
    B. impeller
    C. cam
    D. propeller

22. *True or False?* Jet pumps are commonly used in shallow water wells for home, farm, and industrial use.

23. *True or False?* Variable flow can be achieved in an unbalanced-vane pump by incorporating a movable cam ring.

24. *True or False?* The pressure-compensating feature of a hydraulic pump can be used as the safety device for limiting maximum system pressure.

25. Pressure balancing is inherent in the balanced-vane pump design due to the _____-shaped pumping chamber that allows two inlets and two outlet ports.

26. *True or False?* Dual-pump designs incorporate two pumps in a single case and sometimes also include a relief and unloading valve.

27. Volumetric efficiency is typically found in the _____ section of hydraulic pump information sheets.
    A. performance data
    B. overall efficiency
    C. horsepower performance
    D. installation drawing

28. The general hydraulic horsepower formula can be simplified to pump output (gpm) multiplied by system pressure (psi) divided by a constant of _____.
    A. 62.4
    B. 550
    C. 1,714
    D. 33,000

29. *True or False?* A general rule is that the pump inlet should be more than 3 feet higher than the fluid level of the reservoir.

30. *True or False?* When bubbles of vapor appear in the outlet line of a hydraulic pump, it is an indication of pump cavitation.

Application and Analysis

1. A fixed displacement gear pump has a displacement of 0.114 in^3. If the pump's actual output at different operating conditions is as listed in the chart below, calculate the theoretical output of the pump and its efficiency at the different speeds and pressures.
   A. How does pump efficiency change as the speed of operation increases? Describe one or two reasons for the observed change.
   B. How does pump efficiency change as the system pressure increases? Describe one or two reasons for the observed change.
   C. What changes might occur in pump output and efficiency if the viscosity of the hydraulic fluid or the temperature of the operation were varied? Explain your answer.
   D. How does the required input from the prime mover change with operating speed and pressure?

2. A hydraulic system must move a load linearly in both forward and reverse directions at the same speed. Cylinder operation is intermittent. What type of pump would you choose to use in this system to maximize system energy efficiency? Why?

3. You are called to troubleshoot a pump that is very noisy during operation. What are some possible causes of this noise? What steps might you take to determine which of these is responsible?

Research and Development

1. Hydrostatic transmission systems can operate with either fixed- or variable-displacement pumps, though most use variable-displacement pumps. Investigate the pump used in a specific piece of equipment with a hydrostatic transmission. Prepare a report that describes pump design, operation, and capacity. Discuss the impact of pump choice on equipment operation.

2. Using a pump manufacturer's product listings found online, compare three positive-displacement pumps that have the same basic design, but different capacities of output and maximum operating pressures. Determine what aspects of pump design control the capacities of the three pumps, and for what application each pump might be best suited. Be prepared to explain your answer to your classmates.

3. Investigate the historical development of the piston pump. Write a report that includes any design changes and their impact on pump applications and the effect of new materials or machining techniques on pump capacity and operation. Include a description of the people involved in the development process if possible.

4. As an equipment designer, you are looking for a pump to use in a hydraulic system. The pump must be able to produce at least 15 gpm, speeds of 2,000–3,000 rpm, and system pressures of 2,000–2,500 psi. Find specific models of three different pump designs, such as vane, piston, or screw, that could be used in your circuit. Compare the three pumps to determine what features, such as size, efficiency, and cost, could be used to make your final decision. Prepare a presentation that shows pump options and justifies your final decision of pump design.

Speed (rpm)	Theoretical Output (gpm)	Actual Output (gpm)			Pump Efficiency (%)		
		100 psi	1,000 psi	2,500 psi	100 psi	1,000 psi	2,500 psi
1,200		0.58	0.48	0.32			
1,800		0.87	0.77	0.61			
3,600		1.73	1.65	1.52			

Goodheart-Willcox Publisher

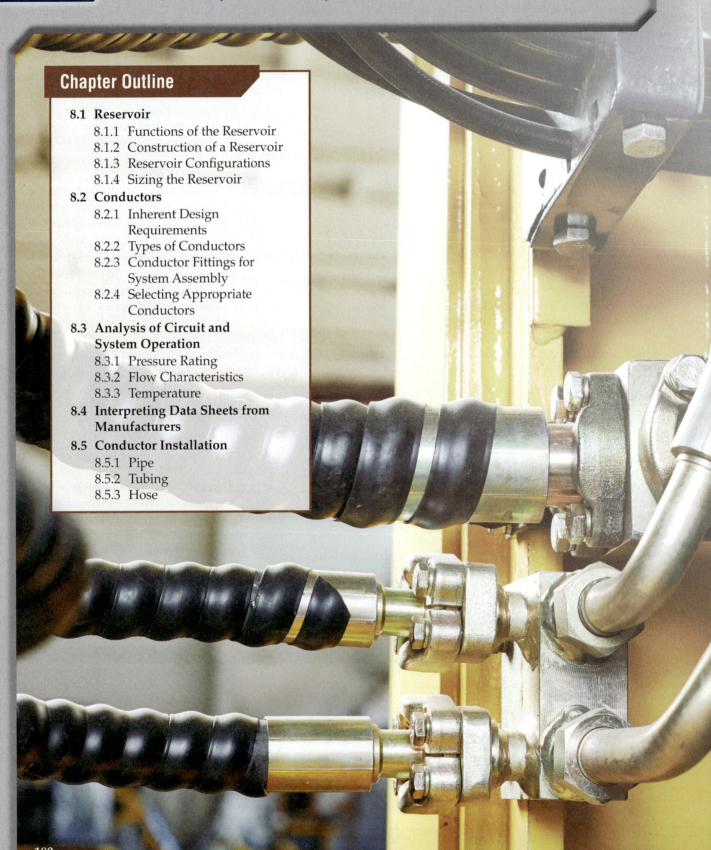

8

Fluid Storage and Distribution

Reservoirs, Conductors, and Connectors

The fluid in a hydraulic system serves several important functions, including energy transmission and component lubrication, cooling, and cleaning. These functions require that fluid be circulated through the system to remove excess heat, flush dirt from internal areas, and lubricate bearing surfaces while moving actuators to do the work for which the system was designed. These functions are facilitated by hardware such as the reservoir and fluid conductors. The reservoir acts as a storehouse for fluid while aiding in system cleaning and cooling. Conductors and associated connectors allow the movement of fluid between system components to transmit energy and also aid with cooling and cleaning. This chapter discusses each of these components and a number of design variations used by system component manufacturers.

Learning Objectives

After completing this chapter, you will be able to:

- Describe the function of the hydraulic system reservoir and identify the primary design features involved in its construction.
- Explain the factors that must be considered when establishing the size of a reservoir for use in a system.
- Describe the factors that are basic to the design and structure of fluid conductors used in hydraulic systems.

- Compare and contrast the various types of pipe, tubing, and hose used in hydraulic systems and explain the ratings for each conductor.
- Identify and explain the factors that must be considered when selecting a conductor for use in a hydraulic system.
- Explain the information typically found on data sheets provided by the manufacturers of hydraulic system pipe, tubing, and hose.

Key Terms

adapter fitting
baffle
burst pressure
compression fitting
conductor
drain return line
Dryseal pipe thread
ferrule
fittings

flared fittings
fluid turbulence
hose
hose-end fitting
inlet line
laminar flow
manifold
nominal sizing
packed-slip joint

pipe
quick-disconnect couplings
reservoir
return line
schedule number
shock pressure
spiral clearance
tubing
vortex

8.1 Reservoir

The *reservoir* is the component that holds the system fluid not currently in use in the system pump, control components, actuators, and lines. This reserve fluid supply is necessary to make up fluid lost in leakage during system operation. It is also needed to compensate for the varying volume between extension and retraction positions of cylinders. The availability of this reserve fluid supply is basic to the operation of any hydraulic system, even though the reserve quantities required will vary depending on circuit design and operation.

8.1.1 Functions of the Reservoir

The primary role of the reservoir is to provide storage for the hydraulic fluid required for system operation. However, a well-designed and well-constructed reservoir provides a number of important additional functions, as listed below. See **Figure 8-1**.

- Removal of heat.
- Separation of solid particles.
- Release of trapped air.
- Separation of water.

In even the least complex designs, the reservoir dissipates system heat to the atmosphere through radiation. This radiation primarily occurs through the reservoir walls. The amount of heat transferred may be increased by incorporating simple devices, such as cooling fins placed on reservoir walls. Baffles in the interior of the reservoir may also be used to increase cooling by directing warm fluid returning from the

system to the exterior walls of the tank to promote rapid heat transfer.

Fluid movement is slow in the reservoir when compared to the fluid velocity through system components and lines. This reduction in the rate of movement allows some of the heavier particles being carried by the fluid to settle to the bottom of the reservoir. This process helps clean the fluid, but it should *not* be considered a reliable fluid filtering method. Larger, heavier metal or dirt particles will settle out, but many of the smaller particles will not. This is especially true when the fluid has a high viscosity. Regular filters properly placed in the system provide much more reliable fluid cleaning.

Most reservoirs are vented to the atmosphere. Venting exposes the surface of the fluid in the reservoir to normal atmospheric pressure. This is important to the operation of a hydraulic system as it allows air that has become trapped in the fluid to escape. Typically, hydraulic fluid contains small quantities of air when it is being used in a system. The air may actually be dissolved in the fluid without a visible indication of its presence. This type of situation normally does not cause operational problems. However, when the fluid begins to appear cloudy, it is an indication that large quantities of small air bubbles have become entrained in the fluid. This can cause foaming, spongy system operation, and increased operating temperatures. A reservoir that is adequately sized and containing proper baffles will allow entrained air to form larger bubbles. These larger bubbles rise to the surface of the liquid and escape to the atmosphere, thus reducing or eliminating the problems associated with entrained air.

Water that may have become mixed with petroleum-based hydraulic fluids also has an opportunity to separate as the mixture passes through the reservoir. Water is heavier than the hydraulic fluid. This allows water to settle to the bottom of the reservoir. It can then be drained off, if a sufficient quantity exists. Separation of the two liquids is relatively slow, so it is important to have a reservoir of adequate size to allow the fluid sufficient time to separate as it moves through the tank. If the water and oil form an emulsion, this separation process is not possible.

8.1.2 Construction of a Reservoir

The construction details of a reservoir may vary considerably, depending on the application of the hydraulic system. In most manufacturing situations, the reservoir is a tank located in close proximity to the system pump. It is easily recognized as a tank. In other situations, the cavities in structural members of a machine may serve as the reservoir. However, each design must provide fluid storage and help condition system fluid by assisting in the removal of heat and contaminants.

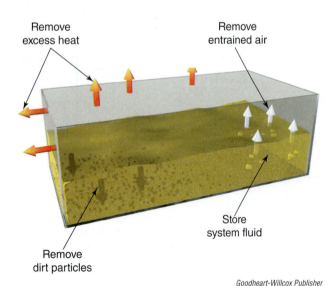

Remove excess heat

Remove entrained air

Remove dirt particles

Store system fluid

Goodheart-Willcox Publisher

Figure 8-1. Besides storage, the reservoir helps cool and clean the hydraulic fluid. This helps to prolong the life of both the system fluid and components.

The typical hydraulic system reservoir is a rectangular, covered tank made from welded or formed steel plate. The tank is often fitted with short legs to permit circulation of air for cooling of the bottom surface. Also, the bottom of the tank is often dished and includes a drain hole to allow easy drainage. The ends, sides, or top of the tank are often fitted with ports with removable covers to permit easy cleaning. The tank is also fitted with a vented breather that allows the tank to take in and release air as the fluid level changes during system operation. In many systems, the breather is incorporated into the filler cap.

An especially important part of the construction of a reservoir is the baffle. A *baffle* separates the pump inlet and system return lines. This baffle is often nothing more than a vertically positioned, flat plate that divides the tank lengthwise. See **Figure 8-2**. The baffle prevents the warm hydraulic oil from moving directly from the return line to the pump inlet. Forcing the oil to travel around the baffle provides longer contact with the sides of the tank. This allows better cooling as well as a longer time for separation of air and the settling of dirt and water. The reservoir is fitted with three lines connected to the system. These are the pump inlet line, system return line, and a drain return line.

The *inlet line* provides fluid to the pump. The inlet line should be large enough to provide fluid at a velocity and pressure drop lower than that suggested by the pump manufacturer. The hole where the inlet line enters the reservoir should be fitted with a *packed-slip joint*

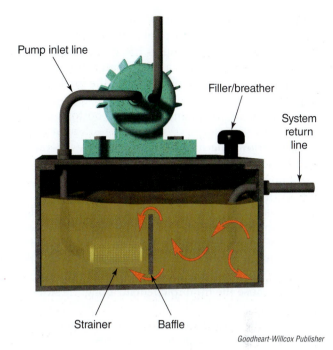

Goodheart-Willcox Publisher

Figure 8-2. Baffles are used inside reservoirs to direct the flow of hydraulic fluid. They control fluid movement in the tank to promote fluid cooling and cleaning.

to prevent dirt and water from entering the tank through the opening while absorbing normal pump and power unit vibration. See **Figure 8-3**. The pickup end of the inlet line should be positioned well below the normal

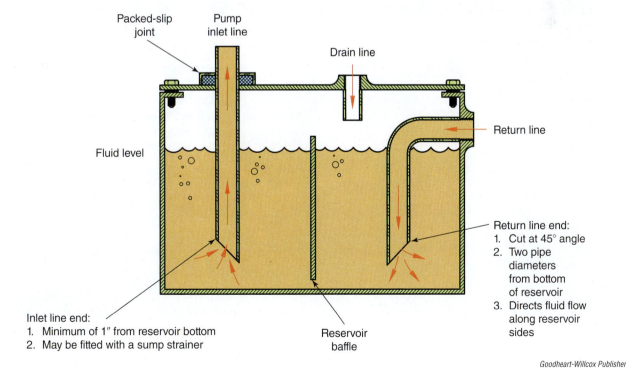

Goodheart-Willcox Publisher

Figure 8-3. Pump inlet, return, and drain return lines must be carefully positioned in a reservoir. The distance from the ends of these lines to the bottom of the reservoir will affect system performance.

fluid level in the tank to prevent the formation of a vortex. A *vortex* is a whirlpool that can form around the inlet line if the pickup is too close to the fluid surface. In extreme cases, a vortex can actually allow air to be drawn into the pump. Also, the inlet should be placed at least 1″ from the bottom to minimize recirculation of sediment from the bottom of the tank.

The *return line* returns fluid from the operating circuit back to the reservoir. The return line should be large enough in diameter so fluid velocity is not excessive during the maximum-expected return flow. It must always be positioned on the opposite side of the baffle from the inlet line. A common practice is to cut the end of the return line at a 45° angle. The return line should end approximately two pipe diameter off the bottom of the reservoir. The sloped end should be positioned to cause the fluid to circulate along the sides of the tank.

The *drain return line* returns fluid drained from externally drained components to the reservoir. It must provide drainage with a minimum of back pressure. This is necessary as drainage from components needs to escape as easily as possible to allow effective operation and lubrication.

The top surface of the reservoir is often used as the mounting base for the system power unit, **Figure 8-4**.

The pump is placed close to the fluid supply, minimizing inlet line pressure drop. Locating the power unit on the reservoir requires the addition of a metal support plate and other fittings to facilitate attachment. In larger-capacity units, the weight of the pump and prime mover become important factors in determining the strength of the structural members of the reservoir and the mounting plate.

When placing a reservoir in use for the first time, a great deal of care should be taken to be certain the inside of the tank is clean. Any machining chips and welding slag obviously should be removed as well as any cutting oils. The inside of the tank should be finished with a material that will withstand the chemical characteristics of the fluid used in the system. It must also be able to withstand the chemical activity promoted by contaminants that enter the system during equipment operation or cleaning.

8.1.3 Reservoir Configurations

A rectangular tank with the pump located above the fluid level is probably the most commonly used form. This configuration is found in both stationary and mobile hydraulic-powered equipment. The requirements regarding reservoir construction can easily be met using this form. However, other system

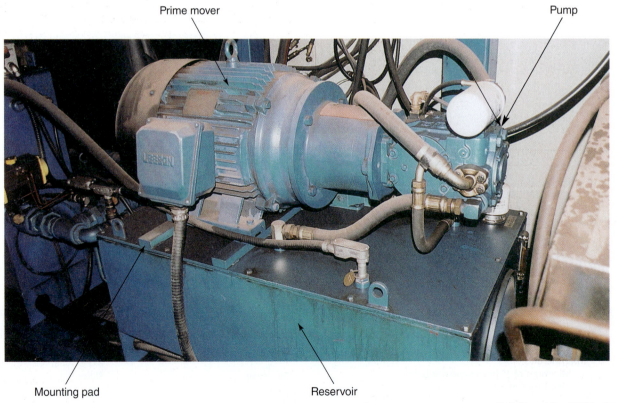

Used with permission of CNH America LLC

Figure 8-4. The reservoir is often used as a mount for the system prime mover, pump, and other system components. The tank must be made from adequately strong material to support these units.

characteristics, such as limited physical space or the need for positive pump inlet pressure, may require reservoir variations.

L-shaped or overhead reservoirs may be used if a positive pressure (greater than atmospheric) is needed on the pump inlet line. Both of the designs place the pump below the level of the oil in the tank. See **Figure 8-5**. This difference in height may be only a few inches, but it allows gravity to assist in moving fluid into the pump. Of course, these systems must have inlet line shutoff valves to minimize fluid loss from the reservoir when the pump must be serviced.

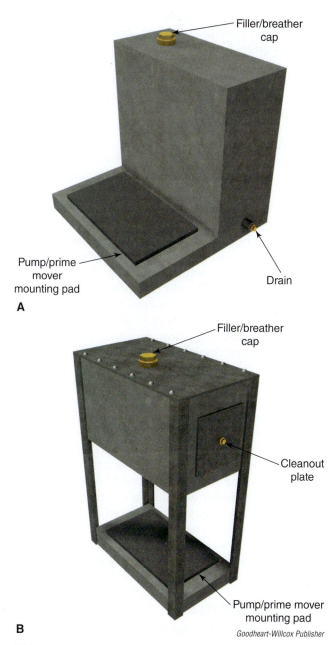

Figure 8-5. An L-shaped or overhead reservoir design is often used when the pump inlet line needs a slight positive pressure for more efficient operation. A—L-shaped reservoir. B—Overhead reservoir.

Other shapes and forms may also be used for hydraulic reservoirs. The cavities of large machine tools can be used as the reservoir. For example, the hollow structural members of equipment, such as the arms of front end loaders, may serve for both storage and fluid transmission. In other mobile pieces of equipment, the oil used for the hydraulic system also functions as the lubricating oil for the transmission and gear train. In those situations, the transmission/differential housing is, in a sense, serving as the system reservoir. See **Figure 8-6**.

8.1.4 Sizing the Reservoir

It is often stated that a hydraulic system reservoir should have a storage capacity three times the rated flow of the pump (3:1). Although this is a very general statement, it has proved to be appropriate, especially in stationary industrial systems with typical actuator sizes. In any case, it is a good starting point when determining the size of a reservoir in a new or modified system.

In actuality, many factors must be considered when determining reservoir size. First, the reservoir must be large enough to store more than the anticipated maximum volume of oil that will be returned to the reservoir during any part of the system cycle. An example is the fluid returned to the reservoir in a system containing a single-acting cylinder with a large bore and long stroke. Other special considerations should involve the basic reservoir functions discussed earlier in the chapter. For example, an increase in the capacity of the reservoir increases the amount of time the fluid is in

Used with permission of CNH America LLC

Figure 8-6. In some equipment using hydraulic power, machine cavities or transmission cases in conjunction with a constructed reservoir serve as the hydraulic system reservoir.

the tank. This ensures increased heat dissipation and separation of dirt or water.

Many times, the final selection of the reservoir size is a compromise influenced by space and weight limitation. Fixed installations, where weight is not a factor, often have a ratio of pump output to reservoir capacity above 3:1. On the other hand, mobile installations, where increased weight can be a problem, have ratios below that guideline.

8.2 Conductors

The individual component parts of a hydraulic system cannot function alone. They must operate as an integral part of a system. This is accomplished by moving fluid from the reservoir to the pump and on to the control valves and actuators. In most system designs, it is not practical to directly connect these components. This is because many of the components must operate machine members as well as interact with other hydraulic components in the system. *Conductors* ensure distribution of system fluid to all components, no matter how widely they are dispersed, **Figure 8-7**. The proper selection, installation, and maintenance of these conductors is of utmost importance to proper operation of a hydraulic system.

8.2.1 Inherent Design Requirements

A variety of different types of conductors and associated connectors and fittings are available for use in hydraulic systems. Pipe, tubing, and flexible hose are considered basic conductor categories. Of course, the items in each of these groups vary considerably in

Goodheart-Willcox Publisher

Figure 8-7. A wide variety of conductors, connectors, and fittings are available for use in hydraulic systems, as can be seen on this piece of earthmoving equipment. Selecting appropriate components requires familiarization with each of the types available.

appearance, construction, and application. However, a number of basic factors are related to items in all of the categories. Part of these factors are concerned with safe system operation, such as physical strength and shock resistance. Other factors deal with the efficiency of system operation, such as resistance to fluid flow and achieving correct fluid velocity. Still other factors deal with ease of installation and service.

Adequate strength

Conductor strength must be considered a critical factor when selecting a conductor for a system. The most obvious reason for this concern is system operating pressure. Strength must also be considered from the standpoint of pressure surges or shock caused by sudden, unexpected load changes during the operating cycle. In addition, other strength aspects that must be considered are mechanical stresses produced by thermal expansion, physical abuse, and even the weight of other system components or machine members the conductor is expected to support.

A conductor must have a continuous operating pressure rating equal to the expected system operating pressure plus an adequate safety factor. A factor of 4:1 is often stated as a minimum. The actual factor selected is based on experience with specific applications.

Safety factor is a margin of error used to provide protection from hydraulic shock pressure. *Shock pressure* is a momentary high-pressure surge that exceeds the system relief valve setting, resulting from machine-member or workload kinetic energy when a directional control valve is centered or reversed and the load suddenly stops or reverses. Shock pressures must be considered when selecting a conductor, as they are often higher than most circuit designers and machine operators realize. It is possible to measure shock pressures using electronic test instruments. These devices are fitted with appropriate accessories to convert the pressures into electrical pulses that can be displayed as a waveform on a screen or printed out for future analysis.

The pressure rating of all conductors depends on the tensile strength of the construction materials, conductor wall thickness, and inside diameter of the conductor. Formulas exist for calculating wall thickness for conductors when the required inside diameter and material tensile strength are known. However, in most practical system design situations, manufacturer data sheets are used to select the conductors that have adequate strength to safely function.

Low flow resistance

It is desirable to have the fluid in the lines of a system move with as little resistance to flow as possible. This requires conductors that are properly sized with an inside diameter large enough to allow the needed volume of fluid to move through the lines within

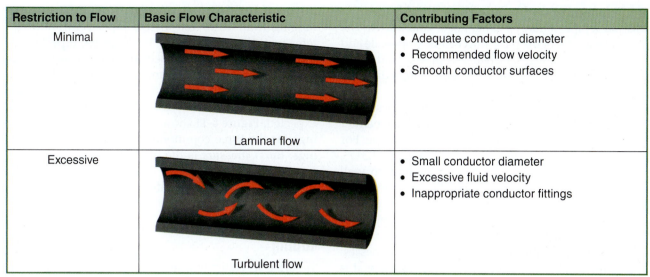

Restriction to Flow	Basic Flow Characteristic	Contributing Factors
Minimal	Laminar flow	• Adequate conductor diameter • Recommended flow velocity • Smooth conductor surfaces
Excessive	Turbulent flow	• Small conductor diameter • Excessive fluid velocity • Inappropriate conductor fittings

Figure 8-8. The inside diameter of a conductor must be large enough to move the fluid at a velocity no greater than the maximum recommended for the line. If this velocity is exceeded, fluid turbulence and power loss can result.

recommended velocity rates. Excessive velocity produces *fluid turbulence*, which increases flow resistance. See **Figure 8-8**. Smooth interior surfaces and gradual bends in conductors reduce the resistance to fluid flow in a system. This helps promote *laminar flow*, or straight-line flow. The selection of a fluid with the proper viscosity also helps ensure flow resistance is not too great in a system.

Resistance to flow in lines and fittings results in energy loses in a system. The energy being input into a system can be calculated using the fluid pressure and flow at the pump. As the fluid is forced through the system, the flow rate remains the same, but the pressure drops as resistance is encountered. This pressure drop is the result of the resistance to fluid flow caused by:

- Each square inch of conductor surface.
- Each bend or fitting encountered.
- Each orifice passed through.
- Level of turbulence in the fluid stream.
- Viscosity of the fluid.
- Resistance of the external load.

The pressure that is lost in the system due to all of these factors, except the external load, indicates the energy that has been lost in the system. This lost energy directly lowers the work output of the system. It also produces heat, which, in excessive amounts, can cause many system operating problems.

Installation and maintenance

To be effective, conductors must not only carry the fluid from component to component, but must be constructed in such a way as to allow economical installation and maintenance. Connectors and fittings must accommodate attachment of system components to the conductors. In actual use, the number of such units must be minimized. Each additional fitting placed in a system increases original cost and adds a point that may contribute to fluid flow resistance and turbulence.

Technicians who maintain fluid power systems often feel that planning for maintenance and service is not considered during new system design and construction. Components may be assembled in such a way as to make service difficult. However, once major maintenance is required, design variations can often be made that will allow a system to be more easily repaired during future service. An example of this is retrofitting components with flare fittings in place of the original threaded pipe connections.

8.2.2 Types of Conductors

A fluid power system can incorporate a number of different conductors to transmit fluid. The most common general classifications are pipe, tubing, and hose. A number of design variations, connectors, and fittings are available under each of these classifications. Also, some manufacturers classify manifolds as conductors, although these are often considered separate components. The conductor used depends on a number of factors, ranging from system operating pressure to initial cost.

Pipe

Pipe is a high-tensile-strength, rigid fluid conductor normally made from mild steel. It is normally installed by cutting and fitting various lengths into the desired configuration using threads cut directly on the pipe and threaded fittings. It is not intended to be bent.

Pipe used in hydraulic systems should be seamless, black pipe. It may be either hot or cold drawn. The internal surfaces of the pipe must be free from scale and any other contaminants. Galvanized pipe should *not* be used in a hydraulic system. The chemical ingredients in oil and its additives react with the zinc coating.

A *nominal sizing* system is used for pipes. With this system, the stated size, or nominal size, is neither the actual outside nor inside diameter of the pipe. For example, the outside diameter of a 1″ pipe is 1.315″, while the inside diameter varies depending on wall thickness. **Figure 8-9** includes all of the common nominal pipe sizes. It indicates the actual outside dimensions and the inside diameters of a number of the pipes rated by wall thickness.

Pipes are available in a variety of wall thicknesses, **Figure 8-10**. This is an important consideration as it is a key factor in establishing the safe maximum operating pressure of the conductor. The pipe with the thickest wall will withstand the greatest system pressure, if all other conditions are the same. Wall thickness of pipes is identified by a *schedule number* established by a standard of the American National Standards Institute (ANSI). This standard uses schedule numbers ranging from 10 to 160 to rate pipe. Hydraulic systems typically use schedule 40, 80, or 160 pipe. Schedule 40 pipe is generally used in lower-pressure parts of the systems, such as pump inlet lines and system return lines. Schedule 80 or 160 is used in the working lines, depending on the maximum system pressures expected. **Figure 8-11** shows typical burst pressure ratings for the three commonly used pipe schedule ratings. *Burst pressure* is the highest internal pressure a component can withstand without failure. Typically, the maximum recommended operating pressure is 25% of the burst pressure.

Tubing

Tubing is a relatively thin-walled, semirigid fluid conductor. It is available in a number of different materials, ranging from plastic to stainless steel. Tubing can be easily bent, which allows relatively easy conductor installation with a minimum of related fittings. The end result is a system with fluid conductors that provide proper flow characteristics and involve a minimum

Nominal Size (inches)	Outside Diameter (inches)	Inside Diameter (inches)			
		Schedule 40	Schedule 80	Schedule 160	Double Extra Heavy
1/4	0.540	0.364	0.302	—	—
3/8	0.675	0.493	0.423	—	—
1/2	0.840	0.622	0.546	0.466	0.252
3/4	1.050	0.824	0.742	0.618	0.434
1	1.315	1.049	0.957	0.815	0.599
1 1/4	1.660	1.380	1.278	1.160	0.896
1 1/2	1.900	1.610	1.500	1.338	1.100
2	2.375	2.067	1.939	1.689	1.503

Goodheart-Willcox Publisher

Figure 8-9. Pipe size is stated as a nominal dimension, which is neither the actual inside nor outside dimension of the conductor.

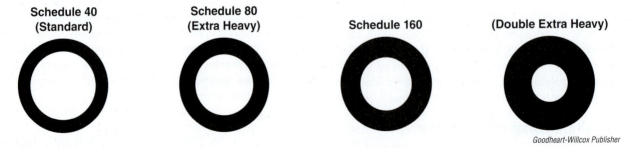

Schedule 40 (Standard)   Schedule 80 (Extra Heavy)   Schedule 160   (Double Extra Heavy)

Goodheart-Willcox Publisher

Figure 8-10. The wall thickness of a pipe is indicated by schedule numbers that are part of an American National Standards Institute (ANSI) specification.

Nominal Size (inches)	Burst Pressure (psi)			
	Schedule 40	**Schedule 80**	**Schedule 160**	**Double Extra Heavy**
1/4	16,000	22,000	—	—
3/8	13,500	19,000	—	—
1/2	13,200	17,500	21,000	35,000
3/4	11,000	15,000	21,000	30,000
1	10,000	13,600	19,000	27,000
1 1/4	8,400	11,500	15,000	23,000
1 1/2	7,600	10,500	14,800	21,000
2	6,500	9,100	14,500	19,000

Goodheart-Willcox Publisher

Figure 8-11. The maximum pressure at which a pipe can operate depends on the nominal diameter and schedule number classification. Burst pressures of the various pipe schedules vary between manufacturers.

amount of space while minimizing visual clutter. Care needs to be taken when selecting tubing to ensure the material is compatible with system fluid and the tubing is adequately strong to withstand system pressures. For example, copper tubing should *not* be used in a hydraulic system because of the chemical reaction between copper and the oil and its additives.

The size of tubing is indicated by measurement of the actual outside diameter. The inside diameter will vary depending on wall thickness. Most tubing is manufactured to the specifications of various standardizing organizations. The American National Standards Institute (ANSI), Society of Automotive Engineers (SAE), and American Iron and Steel Institute (AISI) all have standards covering various types of tubing. A generalization that can be made for any one nominal diameter of tubing is that the smaller the inside diameter, the greater the relative thickness of the wall. A thicker wall allows a higher system operating pressure. However, care must be taken to check the specifications of each type of tubing. Construction material and construction technique can produce widely varying specifications for any given size.

Lengths of tubing are bent to form a conductor between two components in a hydraulic system. The formed tubing is then attached to the components on either end by appropriate connectors and fittings. Care must be taken in designing these tubing sections. System vibration and thermal expansion and contraction must be considered. Both vibration and expansion/contraction can be absorbed by placing a bend in the tube between the attachment points, **Figure 8-12**. Tools are available to form these bends. The tools can be used to form most angles, although angles over 90° are difficult. Generally, 90° is the maximum bend angle. The tube tends to collapse at the bend, resulting in a reduced cross section that is not round. Tube inserts made from rubber or coiled wire are sometimes used to help prevent tube collapse during bending.

Used with permission of CNH America LLC

Figure 8-12. Tube can be safely bent to form angles up to 90°. This allows installations that contain fewer fittings than systems using pipe.

Hose

Hose is a flexible conductor made from a combination of different materials that are compatible with the system fluid and sufficiently strong to withstand system operating pressures. It is commonly used in systems to make connections where equipment parts tilt or swivel during operation or where severe vibration is encountered. See **Figure 8-13**. To achieve the combination of flexibility and strength, a hose has a flexible inner tube to conduct the fluid, a middle layer of reinforcing materials for strength, and an outer protective layer of tough material to withstand external abrasion and abuse.

The inner tube that conducts the fluid must be smooth and sufficiently flexible to freely bend, but strong enough to resist collapse, which would cause a reduction in cross-sectional area. This tube is usually made from synthetic-rubber compounds or thermoplastics. The middle reinforcement section is formed from braided wire, spiral-wound wire, or textile yarns that are either woven or braided. See **Figure 8-14**. Several layers may be used in this section, depending on the pressure rating of the hose. The outer section is a protective layer made from synthetic rubber, thermoplastic, or woven textile materials impregnated with rubber or thermoplastic. This outer material is formulated to resist the abuse of weather, abrasive dirt, and the chemical action of the system fluid or other materials from the environment. Additional plies of material are often used between layers to bind them together.

A wide variety of hoses are available. This makes selection difficult in many situations. Standardizing organizations, such as the Society of Automotive Engineers (SAE), provide guidelines that help when comparing the products. For example, SAE standard J517 includes 16 different hydraulic hose styles and gives dimensional and performance specifications covering

Hose

Figure 8-13. Flexible hydraulic hose is extensively used when fluid must be transmitted to actuators on movable machine members.

A

B

Sporlan Division, Parker Hannifin Corporation

Figure 8-14. Flexible hydraulic hose is made up of a minimum of three layers of materials. These include an inner layer for sealing, middle layer for strength, and outer layer for protection. A—Braided wire. B—Spiral-wound wire.

a full range of operating situations. Although these specifications are available, standardizing groups typically do not certify specific manufacturers or products. The standards are voluntary, but are very helpful when comparing products available from the many different hose manufacturers.

Manifolds

Manifolds are machined or fabricated machine elements used to distribute hydraulic system fluid. A manifold may be as simple as a larger-diameter pipe that serves as a common oil supply to several valves or actuators in a system. However, manifolds usually serve more-complex, multiple-valve situations and involve machined passages in blocks or multilayer subplates. These units eliminate pipe, tube, or hose between components. This allows a lighter, more-compact system.

8.2.3 Conductor Fittings for System Assembly

Constructing a hydraulic system requires a number of fittings to connect the conductors to the reservoir, pump, valves, actuators, and other component parts. The variety of fittings available from manufacturers for each of the various conductor types often makes selection confusing. However, selecting appropriate fittings is necessary to ensure reasonable initial construction cost, efficient system operation, and minimal ongoing maintenance.

Fittings serve as the means by which the ends of individual conductor sections are attached to other conductors or system components. Adapter fittings are often used between the conductor and the component to allow appropriate installation angles and clearances.

These additional fittings also facilitate disassembly and assembly during system service.

Applications that require frequent or rapid connecting and disconnecting of lines use *quick-disconnect couplings.* These fittings are commonly found on mobile equipment that uses a variety of interchangeable hydraulic implements. The couplings allow implements to be quickly changed with little or no fluid loss, **Figure 8-15**.

Quick-disconnect couplings contain an easily operated locking device and check valves. The locking device can usually be operated without tools. The check valves prevent fluid loss when the coupling parts are disconnected. See **Figure 8-16**.

Pipe

The sections of pipe used as fluid conductors in hydraulic systems are typically assembled using threads. Thread sizes are standard, even though the outside diameter of the pipe varies slightly. A pipe thread connection includes external threads cut on the pipe ends and internal threads cut into the openings of system components and pipe fittings. The threads are tapered 3/4″ per foot to ensure a positive seal at the connection.

A *Dryseal pipe thread* should be used, rather than standard pipe threads, when assembling pipe for use as a hydraulic system conductor. This thread design provides a tight seal that prevents the leakage encountered when standard pipe threads are placed under high system pressure. Standard threads tend to leak because of a continuous internal spiral clearance that exists in the threads. See **Figure 8-17**. *Spiral clearance* is the result of space between the crests and roots of the threads when assembled. This clearance exists in a spiral along the entire length of the thread, even when the threaded parts are properly fitted and tightened.

Quick-disconnect couplings

Goodheart-Willcox Publisher

Figure 8-15. Quick-disconnect couplings allow hydraulic lines to be quickly connected or disconnected. This allows, for example, implements to be quickly transferred on mobile farm and construction equipment.

**Connected
(Check Valves Open)**

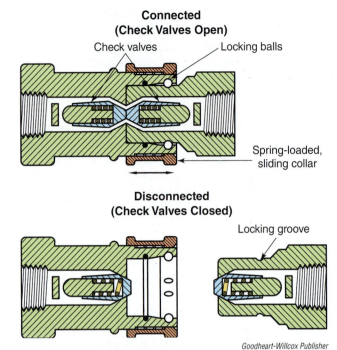

Check valves

Locking balls

Spring-loaded,
sliding collar

**Disconnected
(Check Valves Closed)**

Locking groove

Goodheart-Willcox Publisher

Figure 8-16. Easily operated locking devices and effective check valves allow quick-disconnect couplings to be quickly operated without spilling fluid.

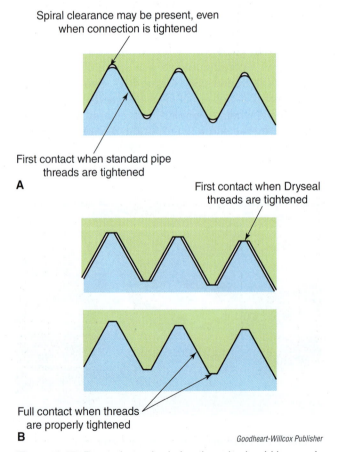

Spiral clearance may be present, even when connection is tightened

First contact when standard pipe threads are tightened

A

First contact when Dryseal threads are tightened

Full contact when threads are properly tightened

B

Goodheart-Willcox Publisher

Figure 8-17. Dryseal standard pipe threads should be used when constructing hydraulic system conductors from pipe. A—Standard pipe threads. B—Dryseal pipe threads.

A wide variety of pipe fittings are available to assist in constructing the conductors needed to carry the hydraulic fluid to system components. These include tees, elbows, couplings, unions, nipples, and several other shapes, **Figure 8-18**. These fittings should be made from steel or other material that will ensure a strength comparable to the schedule of the connecting pipe. All connections should be installed to permit easy removal and reassembly by common hand tools.

Union fittings allow easier connection of complex system lines. They provide a means of connecting lines using flanges that provide a positive seal. However, every attempt should be made to design pipe runs between components with a minimum of unions. It is critical to plan the system so that pipe, fittings, and system valves can be removed without dismantling large sections of the system or bending or otherwise damaging system parts.

Tubing

A wide variety of fittings are available for making connections to tubing. Manufacturers produce many different styles that are designed to provide the best means of attaching tubing to system components, **Figure 8-19**. Each must provide a proper fit and seal between the component and fitting and also between the tube and fitting.

Typically, the seal between the fitting and the system component is provided by either a tapered pipe thread or a straight thread with a separate seal. The separate seal may be an O-ring or a compression washer.

The seal between the fitting and the tube is provided by either a flared or compression connection. *Flared fittings* actually require physically flaring the end of the tube. A positive seal is provided by using a jamb nut and sleeve to force the flared tube end against a cone portion of the connection. In the United States, flared tube fittings are standardized at 37° and 45° angles. See **Figure 8-20**. Hydraulic systems typically use the 37° design.

A number of different designs can be found in *compression fittings* for tubing. Several manufacturers have patented designs. Typically, these include a fitting body, compression element such as an O-ring, *ferrule*, or proprietary compression sleeve, and compression nut, **Figure 8-21**. The end of the tube to be attached must be square, concentric, and free from burrs to ensure proper positioning and sealing. The tube end is slipped into the fitting until it rests against an internal shoulder that provides proper positioning. Tightening the compression nut squeezes the parts to secure the tube and provide a fluid-tight seal.

The tube may be attached to the fitting by swaging, brazing, or hard soldering. Swaging joins the tube

Figure 8-18. Several of the common threaded pipe fittings available for assembly of fluid conductor lines in hydraulic systems.

and fitting by squeezing the fitting onto the tubing. Brazing and hard soldering join the tube and fitting by using a low melting point, nonferrous metal to secure the joint without melting the base metals. The designs usually require split flanges or other devices to secure the fitting to the component. See **Figure 8-22**.

Goodheart-Willcox Publisher

Figure 8-19. Pipe threads or fittings with straight threads are used to attach hose fittings to hydraulic components. The seal on fittings with straight threads is obtained using an O-ring or a compression washer.

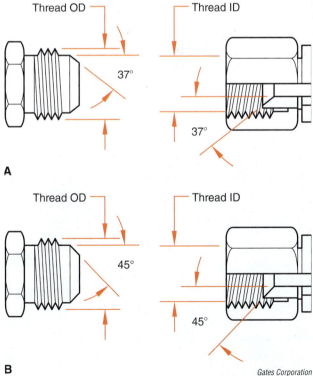

Gates Corporation

Figure 8-20. A 37° or 45° flare is used to provide a seal between components and tubing. A—Hydraulic systems typically have a 37° flare. B—A 45° flare may be found in other applications.

Hose

Flexible hose is attached to a *hose-end fitting* that includes an attachment mechanism designed to connect to other hydraulic lines, control valves, actuators, or other system components. This attachment mechanism or fitting may use pipe threads, 37° or 45° flare, flange, or other configuration. In some cases, the hose-end fitting is directly connected to system components. However, in most situations, an adapter fitting is used between the hose-end fitting and the component.

Hose-end fittings are available in permanently attached or reusable styles. A permanently attached unit consists of a nipple and socket on one end of the connector body and a threaded or other style of connection on the opposite end. The hose is forced over the nipple, filling the space between the nipple and socket. The metal of the socket is then swaged or crimped. This locks the connector over the hose and forms a permanent connection. See **Figure 8-23**.

These fittings are used on off-the-shelf hose assemblies. If proper equipment is available, an end user may apply these connections in the field.

Reusable hose-end fittings allow more flexibility in servicing hose assemblies. This is possible because these fittings can be attached to the hose using basic hand tools. They can also be removed from a conductor assembly that has failed and reused with a length of new hose. Reusable hose ends are generally classified as either screw-together or clamp types. See **Figure 8-24**.

The screw-together, reusable hose-end fitting consists of two parts. The first part is an outer shell fitted with internal threads that allow it to be threaded over the outer diameter of the hose. One end of this outer shell contains a smaller threaded opening through which a nipple is turned into the inner tube of the hose. This forces the hose to slightly expand, producing a leakproof seal between the hose and the fitting. The end of the insert opposite of the nipple contains the connection to be used for attaching the hose to system components.

Clamp type, reusable hose-end fittings typically consist of three parts. One part is an insert that is forced into the inner tube of the hose. This insert is constructed with a nipple on one end to aid in insertion. The other end of the insert contains the connections needed for attachment to system components. Two or four bolts are used to tighten two clamp pieces around the hose. This clamping action forces the hose onto the insert and produces a positive seal.

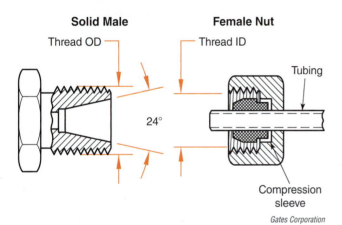

Gates Corporation

Figure 8-21. A number of different fittings depend on compression between the fitting and the tubing for a seal.

Goodheart-Willcox Publisher

Figure 8-22. Split-flange fittings that have been brazed or hard soldered to the tubing may be used to attach tubing to hydraulic components.

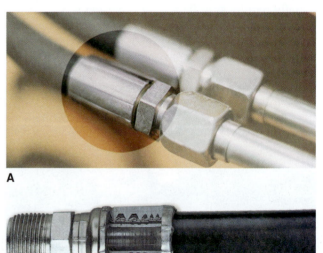

Gates Corporation

Figure 8-23. Permanently attached hose-end fittings are crimped or swaged onto the hose and cannot be removed. A—External view showing compression of the fitting socket. B—Cutaway showing hose materials held by the fitting nipple and compressed socket.

Adapter fittings

Adapter fittings are connectors used to eliminate twisting or excessively bending a conductor during installation. They also assist in disassembly and reassembly of the system during service. Adapter fittings are often used between pipe, tube fittings, or hose-end fittings and the component. A variety of connection styles are available for the ends of these adapters based on both American and international standards. See **Figure 8-25**.

The connection styles can be generally grouped into one of four categories:

- Category 1. Straight threads and a shoulder sealed with an O-ring or a metal compression washer.
- Category 2. Tapered pipe threads for both attachment and seal.
- Category 3. Flanged connection with the seal provided by an O-ring held in a groove.
- Category 4. Flared connections for the seal.

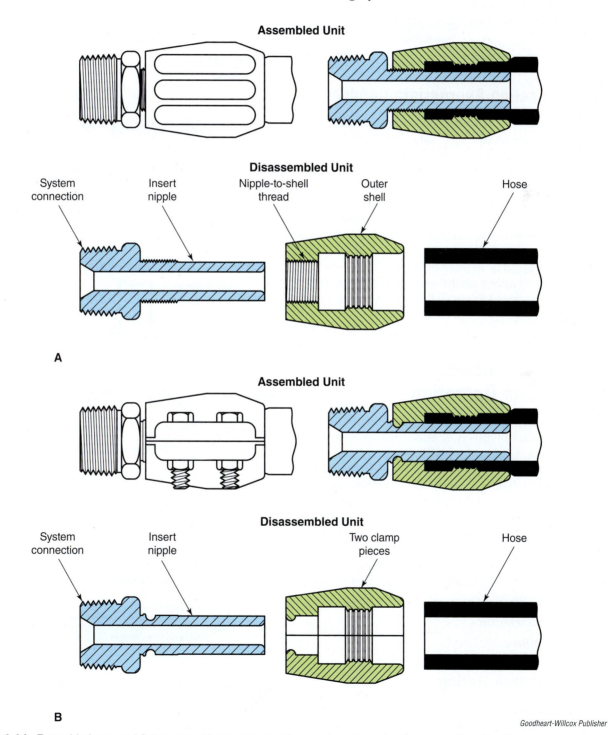

Goodheart-Willcox Publisher

Figure 8-24. Reusable hose-end fittings are either screw-together or clamp type. A—Screw together. B—Clamp.

Care must be taken when working with adapter fittings as several thread sizes and flare angles exist. Attempting to mate different threads or flare angles will cause problems.

Attachment methods from the first, second, or third categories are used with adapter fittings to connect major components. These components include pumps, valves, and actuators. Attachment methods from the first, second, or fourth categories are used when connecting an adapter to hose and tube-end fittings. In addition, adapters are available that form 45° or 90° elbows. Swivel fittings are also available to minimize hose twisting as a component moves or rotates during system operation.

8.2.4 Selecting Appropriate Conductors

There is no question that fluid conductors and their associated fittings are an essential part of any hydraulic system. Many factors must be carefully considered to obtain the most appropriate type and size of conductors and fittings. However, selection is often not as carefully considered as it should be. The result can be a system with conductors that cost more than necessary, contribute to inefficient system operation, fail prematurely, or are difficult to disconnect when system service is required. Many of these problems can be reduced or eliminated by selecting conductors only after a careful analysis of system operation, the environment in which the system operates, and manufacturer data sheets for conductors.

8.3 Analysis of Circuit and System Operation

Selecting a conductor requires an examination of the hydraulic circuit involved. It also involves examining the total hydraulic system and the mechanisms operated by the system. This complete analysis should not only take into consideration the obvious elements, such as system pressure and flow, but also include machine-related factors. Machine-related factors include such things as vibration and required machine member movement that directly affect conductor performance. The following factors need to be carefully considered when selecting and installing conductors of any type:

- Pressure rating.
- Flow characteristics.
- Temperature.

8.3.1 Pressure Rating

The recommended maximum operating pressure rating of a conductor is usually readily available in catalog

Category 1

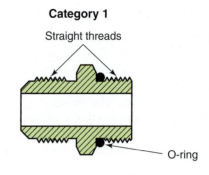

Straight threads

O-ring

Category 2

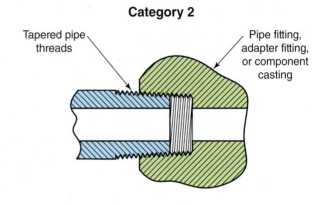

Tapered pipe threads

Pipe fitting, adapter fitting, or component casting

Category 3

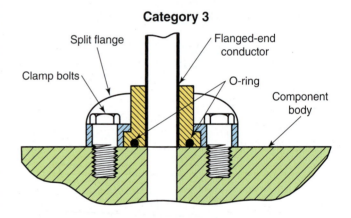

Split flange

Flanged-end conductor

Clamp bolts

O-ring

Component body

Category 4

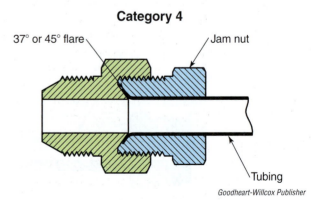

37° or 45° flare

Jam nut

Tubing

Figure 8-25. A variety of connection methods are used to attach fluid lines and components to provide sealing against fluid leaks while promoting easy assembly and disassembly.

materials distributed by the conductor manufacturer. These published ratings show the recommended operating pressure limits for conductors used in a continuously operating system. These recommended maximum pressures are typically no more than 25% of the burst pressure rating of the conductor. Care should be taken to avoid selecting a conductor with a recommended maximum operating pressure that is below anticipated system pressures.

Although every attempt should be made to eliminate operating situations that produce pressure surges, they need to be considered a possibility. Therefore, shock pressures also need to be considered when selecting a conductor for a system. If system pressure surges are frequent, the service life of system components will be considerably shortened. Conductor-related problems caused by pressure surges can be reduced by selecting a conductor that has a higher recommended maximum pressure rating than the normal working pressures in the system.

8.3.2 Flow Characteristics

A conductor must be large enough to accommodate the fluid flow required to operate the system. The inside diameter of the conductor establishes flow capacity. Flow capacity is also directly related to fluid velocity. The higher the fluid velocity, the higher the fluid flow rate for a given conductor. However, if the fluid velocity becomes too high, it can produce system operating problems. Different fluid velocities are recommended for different types of system lines and conductors. For example, the average fluid velocity in a pump inlet line should not exceed four feet per second (4 ft/sec). Average velocity in working lines should be held below 20 feet per second (20 ft/sec).

The velocity of the inlet line fluid is restricted to prevent the formation of a vacuum that is too far below normal atmospheric pressure. Excessively low pressure in an inlet line can result in pump cavitation. Cavitation can cause major pump damage. This problem does not relate to working lines, as they normally operate above atmospheric pressure. A typical working line pressure range is from slightly above atmospheric pressure to the maximum relief valve settings.

The problem with the movement of fluid through working lines is not pressure, but turbulence. Excessive fluid velocity produces turbulent flow, which causes resistance to fluid flow. This results in pressure losses in the system, which are converted into heat. The heat, in turn, results in increased system operating temperatures. The average fluid velocity in lines can be calculated using the formula:

$$V = \frac{Q}{A}$$

where:

V = Flow velocity

Q = Flow rate

A = Conductor cross-sectional area

Care needs to be taken to use quantities with the same units. This formula, for example, can use a flow rate of gallons per minute (gpm), but it may be useful to convert to inches per second (in^3/sec) if the conductor is also measured in inches. Cubic inches per second can be easily calculated by using the formula:

$$in^3/sec = \frac{gpm \times 231\ in^3}{60\ sec}$$

EXAMPLE 8-1

Fluid Velocity in a Conductor

Calculate the velocity of the fluid in a conductor with a 1/2″ inside diameter that is moving five gallons per minute.

1. Calculate the inside cross-sectional area of the conductor:

$$Area = \pi r^2$$
$$= 3.1416 \times .25″ \times .25″$$
$$= 0.196\ in^2$$

2. Calculate the flow rate in cubic inches per second (in^3/sec):

$$in^3/sec = \frac{5\ gpm \times 231\ in^3}{60\ seconds}$$
$$= 19.25\ in^3/sec$$

3. Calculate the average fluid velocity in the conductor:

$$Velocity = \frac{19.25\ in^3/sec}{0.196\ in^2\ seconds}$$
$$= 98.21\ in/sec$$

or

$$= 8.19\ ft/sec$$

8.3.3 Temperature

Although pipe and tubing are generally not affected by temperature, both ambient and system operating temperatures must be carefully considered when selecting hose, **Figure 8-26**. Very high or very low temperatures can adversely affect a hose. Extreme ambient temperatures can reduce hose service life by causing deterioration of the hose cover and reinforcement materials. Continuous system operating temperatures that are at or exceed the rated temperature of a hose will cause deterioration of both the hose inner tube and cover. This also results in a shortened service life.

Hose specifications should be carefully examined to ensure the hose selected for the system can tolerate the temperatures that will be encountered during system operation.

8.4 Interpreting Data Sheets from Manufacturers

A sizable number of companies produce products related to hydraulic fluid conductors. The *Fluid Power Handbook and Directory* lists product categories for pipe, tubing, and hose. For example, the Hydraulic Hose category identified in the directory lists the companies that produce hose. In addition to the product categories, the directory includes a considerable number of other categories related to conductors, such as fittings and seals. The companies publish a wide variety of catalog materials and technical data sheets that describe their products in detail.

The published materials usually provide a considerable amount of information about the products of a company. It can be difficult to obtain comparative information on similar products because no standard format is required by the industry. However, the

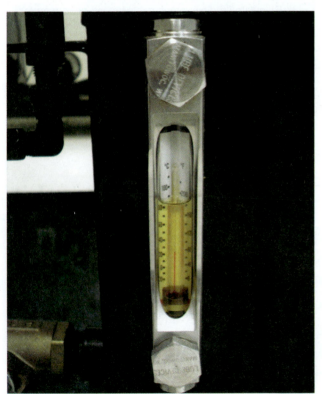

Used with permission of CNH America LLC

Figure 8-26. Temperature-sensing devices range from simple thermometers to elaborate temperature recording devices. System operating temperatures must be carefully considered when selecting hose.

materials do provide information that can be readily used to select items for a system. Common information includes:

- Construction details and typical applications.
- Dimensions.
- Acceptable operating temperatures and pressures.
- Flow rates for different conductor sizes and pressure drops.

Typically, the information provided for pipe and tubing relates to the material the item is made from and the finished dimensions. The information also specifies the rating of the product in relation to an SAE specification or one from another standardizing group.

Many companies produce hose with a wide variety of features not included in the specification of a rating organization. As such, the information provided on data sheets for hydraulic hose usually includes additional information not found on pipe and tubing data sheets. Hydraulic hose catalogs and data sheets often include construction details, typical applications, and acceptable operating temperature ranges. They also include materials and information related to specification numbers of standardizing organizations. Detailed information concerning the fittings required to connect conductors and specific system components are also available in company catalogs.

Care must be taken to carefully examine catalog data. The exact type of information supplied will vary from one manufacturer to another manufacturer. Hydraulic service personnel need to be reasonably proficient in interpreting catalog information. This proficiency will increase as you work with a variety of catalogs on the job.

8.5 Conductor Installation

When selecting conductors for hydraulic-powered equipment, installation factors must be considered in addition to pressure and flow requirements and temperature conditions. When examining the equipment, consider required component movement, conductor flexibility, and potential operating conditions involving vibration and abrasion. These factors influence the types of conductors selected and the methods used to install them in the equipment.

8.5.1 Pipe

When installing pipe as the conductor in a hydraulic system, the primary considerations are:

- Matching the centerline of the pipe to the centerline of the component connections.
- Determining the proper length of conductor.

Pipe sections should fit without being distorted or placed under tension. Both of these conditions will cause stress on the pipe and other components. This stress could result in material fatigue and lead to part failure. Careful planning must also be done to allow easy removal of components during system service. This may require the use of unions to provide an easy means of removing pipe, fittings, and components.

8.5.2 Tubing

Tubing is more expensive than pipe. However, it is more easily installed, lighter in weight, produces less flow resistance, and generally provides a neater installation. The number of fittings can be reduced in a tube installation by bending the tubing whenever possible. Manufacturer specifications indicate the minimum radius of bends for each type, diameter, and wall thickness of tube. Kinks, flattened spots, and wrinkles can be avoided when fabricating tube sections by using the proper bending equipment. Both hand-operated and power benders are available for making up tube sections.

Straight-line connections between components should be avoided in tube installations. These connections place excessive stress on the tube, tube fittings, and system components. The use of strategically placed bends provides connections that will perform very well in most installations.

Long lengths of tube should be supported by brackets or clamps to secure the conductor to the equipment. This is done to prevent fatigue caused by vibration or the weight of the tube and the hydraulic fluid it contains.

Care must also be taken when fabricating sections of tubing to ensure the completed tube sections are the correct length and the angles of the bends are appropriate. Tubing sections that are the incorrect length or do not contain the proper bend angles will cause stress on all parts involved. This occurs when they are sprung into position or forced into alignment when fittings are tightened.

8.5.3 Hose

Hose allows the transmission of hydraulic fluid between stationary and movable machine components. Although hoses are flexible to provide this connection, they must be installed according to manufacturer specifications to obtain maximum service life.

A wide variety of materials and manufacturing methods are used by hydraulic hose producers. This makes selection of a specific hose for an application more complicated. Basic factors to consider when selecting a hose for an application are:

- Pressure and flow ratings.
- Chemical resistance to the fluid used in the system.
- Resistance to external abrasion.
- Flexibility.

Once these factors have been established, specific installation factors must be considered.

Changes in system pressure must be considered when working with hose. As the pressure increases, the length of the hose tends to decrease and its diameter tends to increase. Always provide some slack in a hose during assembly to allow for length change. See **Figure 8-27**. When the hose passes through openings in machine members, be sure adequate clearance is provided for increases in hose diameter. Also, be careful to avoid any twists in hoses when the installation is complete. See **Figure 8-28**. Pressure tends to straighten the hose, which can actually loosen fittings. In addition,

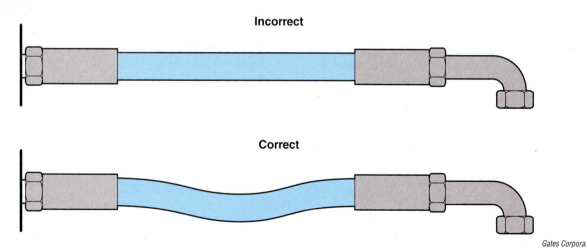

Incorrect

Correct

Gates Corporation

Figure 8-27. When installing hydraulic hose, always provide slack in lines. This counters the effect of the hose expanding in diameter and decreasing in length as system pressure increases.

the constant twisting caused by this tendency to straighten under pressure can result in separation of the hose reinforcement and lead to hose failure.

Hose fittings, elbows, and angle adapters should be used to relieve the strain on hose assemblies and maintain as neat and uncluttered of an appearance as possible. See **Figure 8-29**. Also, use standoff brackets wherever hoses must pass close to hot machine parts. These brackets hold the hoses away from the hot parts to allow air circulation. Direct contact with hot parts can shorten the service life of a hose.

Clamps should be used to support long hose runs. This is especially true when there is a possibility of abrasive damage from the hose being rubbed by moving machine members. Do *not* clamp high- and low-pressure hoses together. Also, do *not* clamp a hose at a bend. Pressure variations can cause hose movement resulting in damage to the hose.

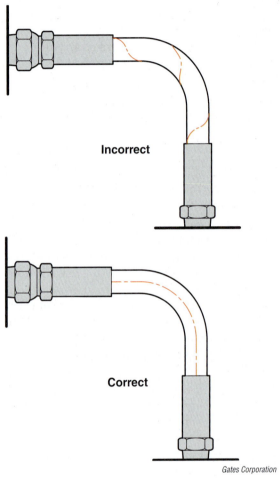

Gates Corporation

Figure 8-28. Be certain hoses are installed without twists. Hoses will tend to straighten (unwind) under pressure, placing considerable stress on the internal structure of the hose.

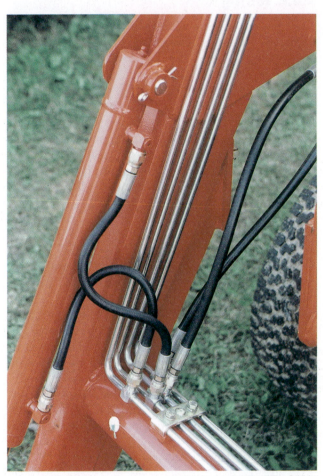

Goodheart-Willcox Publisher

Figure 8-29. When installing hose, the careful selection of fittings will produce a less-cluttered appearance, decrease hose stress, and reduce the amount of hose used for the installation.

Summary

- The reservoir of the hydraulic system not only provides storage for the fluid used in the system, but also helps dissipate excess heat, release trapped air, and separate both water and solid particles from the fluid.

- A typical hydraulic reservoir is a rectangular steel tank fitted with ports for easy cleaning, a vented breather to allow the exchange of air, and three lines that connect the reservoir to the system. These lines include the pump inlet line, a system return line, and a drain line.

- The reservoir must be large enough to store appropriate amounts of fluid to operate the system and to hold the fluid long enough to allow heat to dissipate and contaminants to separate. A hydraulic system reservoir should have a storage capacity three times the rated flow of the pump (3:1).

- Conductors must have adequate strength to withstand high system pressures, low flow resistance to ensure low energy loss during system operation, and a design that will allow economic installation and low maintenance costs.

- The maximum operating pressure rating of all conductors depends on the tensile strength of the construction materials, conductor wall thickness, and inside diameter of the conductor.

- Conductors should be properly sized, have smooth interior surfaces, and bend gradually to lower the risk of fluid turbulence and to minimize resistance to fluid flow. High flow resistance can result in lower work output and can produce unwanted heat.

- Pipes, tubing, and flexible hoses are the basic conductors of the system.

- Pipe is a high-tensile-strength rigid fluid conductor normally made from mild steel. Pipes with the thickest walls can withstand the highest system pressures.

- Tubing is a relatively thin-walled semirigid fluid conductor that can be bent and shaped into lines that provide good flow characteristics with minimal visual clutter.

- Hose is a flexible conductor consisting of a flexible inner tube to conduct the fluid, a middle layer of reinforcing material for strength, and an outer protective layer to withstand external abrasion and abuse.

- The type and amount of component movement, conductor flexibility, and the potential for vibrations and abrasion during operation are factors that should be considered when selecting and installing conductors for a system.

- Manufacturers' data sheets include the construction details, typical application, and technical data about pressure rating, flow capacity and temperature tolerance of each conductor.

Internet Resources

The following are some useful resources available on the Internet. Enter a company or organization name into a search engine to access its website. Explore the various areas of the sites to discover useful fluid power resources.

Dayco Products. Provides links to information on Dayco's line of hose, couplings/adapters, and crimpers. Also provides a link to the Dayco Hydraulic Catalog. This site provides the opportunity to examine one manufacturer's product specifications and recommendations for selection and installation.

Eaton Corporation. Lists details on one company's quick-disconnect coupling products. Coupling construction, selection information, and safety information are included.

Gates Corporation. Includes links to hydraulic hose catalogs, part interchange information, selection wizard, and other engineering tools.

Gates Corporation. Provides links to calculators that determine pressure losses for user-input data. These calculators are based on a group of assumptions and average conditions. Results should only be used as a guide.

HYDAC Technology Corporation. Includes descriptions of filters, coolers, fluid-level indicators, breathers, and temperature sensors.

JWF Technologies. Provides an extensive listing of hydraulic pipe and tube fittings. Includes product data sheets and an index illustrating each of the fittings.

Chapter Review

Answer the following questions using information in this chapter.

1. The component that is designed to hold the fluid *not* currently in use in the system is called the _____.
   A. baffle
   B. conductor
   C. reservoir
   D. pipe

2. *True or False?* Baffles are used in the interior of reservoirs to direct the movement of fluid and reduce heat, dirt, air, and water.

3. The three types of system fluid lines connected to a typical reservoir are inlet line, return line, and _____ return line.

4. Two reservoir configurations that are suitable for use with a pump that requires a positive pressure on the inlet line are L-shaped and _____.

5. *True or False?* The three basic categories of fluid conductors used in hydraulic systems are pipe, tubing, and adapters.

6. *True or False?* Shock pressure is short duration, high pressures in a hydraulic system resulting from control valve shifting and unexpected load changes.

7. What factors within the hydraulic system cause resistance to fluid flow in the system?
   A. Viscosity of the fluid.
   B. Conductor bends.
   C. Fluid turbulence.
   D. All of the above.

8. *True or False?* Pipe, as related to a hydraulic system, is a conductor made from high-tensile-strength mild steel used to create distribution systems.

9. *True or False?* American National Standards Institute (ANSI) provides a pipe wall thickness standard using schedule numbers ranging from 1 to 5.

10. The size of hydraulic system tubing is indicated by the actual _____ of the tube.
    A. thread size
    B. length
    C. outside diameter
    D. flow rate

11. When installing tubing in a hydraulic system, how can both vibration and thermal expansion/contraction be absorbed?
    A. Install smaller fittings.
    B. Place extra fittings in the lines to increase heat transfer.
    C. Place a bend in the tube between the attachment points.
    D. Increase flow rate through the line.

12. Of the three layers of materials used in the construction of most types of hydraulic hoses, what is the function of the middle layer?
    A. Reinforcement.
    B. Flexibility.
    C. Heat transfer.
    D. Increase hose rigidity.

13. Proper selection and installation of pipe, tube, or hose fittings in hydraulic systems ensure reasonable initial construction cost, efficient system operation, and _____.
    A. straight-line connections
    B. minimal ongoing maintenance cost
    C. easy disassembly and storage
    D. None of the above.

14. A(n) _____ pipe thread should be used when assembling pipe to prevent leakage when the joint is placed under high system pressure.
    A. Dryseal
    B. O-ring
    C. jam nut
    D. tapered

15. *True or False?* The two general hose-end fitting styles used with hydraulic flexible hose are permanently attached and reusable.

16. Which of the following is a situation that may result if conductors and fittings are *not* carefully selected for a given hydraulic system?
    A. Excessive initial cost.
    B. Inefficient system operation.
    C. Premature part-failure.
    D. All of the above.

17. Which of the following factors must be considered when selecting and installing conductors?
    A. Volumetric efficiency.
    B. Conductor color.
    C. Pressure rating.
    D. Flow characteristics.

18. It is typically recommended that the average fluid velocity in a hydraulic system working line should *not* exceed _____ feet per second.
    A. 20
    B. 30
    C. 60
    D. 100

19. *True or False?* Straight-line connections should be avoided when fitting tubing in order to avoid excessive stress on materials and fittings.

20. *True or False?* Low- and high-pressure hoses are often clamped together.

Apply and Analyze

1. Describe how you might modify a fluid storage and distribution system to reduce overheating of hydraulic fluid during operation.

2. If the pump flow rate in a simple hydraulic system is 10 gpm and the maximum system pressure is 3,000 psi, what size reservoir and conductors for pump inlet and working lines would you choose to install? Why?

3. What are the possible effects of replacing a pump inlet line with a conductor that has an inside cross-sectional area half the size of the original line? What information do you need to determine if your conductor size is adequate for system operation?

4. You are called to troubleshoot a pump that is very noisy when operating near capacity. The noise started after the system reservoir was replaced with a different model. What questions would you ask about reservoir design and installation to determine what is responsible for the change in system operation? List at least three questions you might ask and explain how the answers to these questions will assist you in your troubleshooting process.

Research and Development

1. Investigate the current trend toward use of smaller hydraulic reservoirs. Determine what is motivating this trend and the technological advances that may be supporting it.

2. Prepare a report that describes the fluid storage and distribution system of a specific model of skid loader. Your report should include specific information about reservoir design, conductor type and size, and conductor connections used, as well as a discussion of how these specific components affect the operation and capacity of the loader.

3. Research the design, composition, and capacity of hydraulic hose used in an extreme environment, such as extreme temperature environments or high system pressures. Be prepared to explain how the design and materials allow the hose to withstand the conditions in which it operates.

4. Design and perform an experiment to measure or illustrate how average fluid velocity changes as cross-sectional area of the conductor changes.

9 Actuators
Workhorses of the System

Hydraulic systems are capable of easily transmitting and applying power directly to the required work location. In addition, the systems have an advantage over other power-transmission methods because of the ability to readily provide both linear and rotary motion. Major factors relating to these capabilities are the actuators or work devices. Actuators range from large units that can generate huge force or torque outputs to miniature components that produce extremely low power. This chapter deals with basic cylinder and motor designs as well as specialized actuators, such as limited-rotation devices.

Learning Objectives

After completing this chapter, you will be able to:

- Describe the construction and operation of basic hydraulic cylinders, limited-rotation actuators, and motors.
- Compare the design and operation of various types of hydraulic cylinders and motors.
- Select appropriate cylinder design for mounting hydraulic cylinders and reducing hydraulic shock.
- Size hydraulic cylinders and motors correctly to meet system force and speed requirements.

- Interpret manufacturer specifications for hydraulic cylinders.
- Compare the design and operation of various types of hydraulic motors.
- Contrast the operation of fixed- and variable-displacement hydraulic motors.
- Describe the construction and operation of a basic hydrostatic transmission.

Key Terms

barrel
cam-type radial-piston motor
cap
clevis mount
closed-loop circuit
cushioning
cylinder
double-acting cylinder
double-rod-end cylinder
dynamic seal
effective area
effective piston area
fixed-centerline mount
fixed-displacement motor
fixed-noncenterline mount
head

hydrostatic drive
hydrostatic transmission
inertia
limited-rotation actuator
mill cylinders
motor
one-piece cylinder
open-loop circuit
orbiting-gerotor motor
parallel circuit
pinion gear
piston
pivoting-centerline mount
rack gear
ram
replenishment circuit

rod
screw motor
series circuit
side loading
single-acting cylinder
static seal
stator
telescoping cylinder
threaded-end cylinder
tie-rod cylinder
trunnion mount
universal joint
variable-displacement motor
wiper seal

9.1 Hydraulic Cylinders

A hydraulic *cylinder* is the hydraulic system component that converts fluid pressure and flow into linear mechanical force and movement. See **Figure 9--1**. Often called a linear actuator, in a sense it extracts energy from the hydraulic system fluid and converts it into force and mechanical movement. The force generated by the cylinder is determined by the area of the piston operating within the cylinder and the system operating pressure. The speed of movement is controlled by the effective volume of the cylinder per unit of travel and the rate of fluid flow entering the component. Effective volume is the volume of the cylinder minus the volume of the rod (if applicable). The maximum distance of movement is controlled by the physical length of the cylinder. Controlling system pressure and flow rates with even the most simple designs allows a wide range of cylinder forces and speeds.

9.1.1 Basic Cylinder

A basic cylinder consists of a movable element made up of a piston and piston rod assembly. The piston and piston rod assembly operate within a close-fitting cylindrical tube. One end of the tube is closed, while an opening in the opposite end allows the piston rod to extend outside of the component for attachment to machine members. Each cylinder end also contains a port connected to the system allowing fluid to be forced into and out of the cylinder. Seals are incorporated in the cylinder design to prevent fluid leakage around the piston. This increases cylinder efficiency and prevents the leakage of fluid to the exterior of the cylindrical tube. See **Figure 9-2**.

This design forms two sealed chambers inside of the cylinder that are separated by the piston. Forcing system fluid into the chamber between the closed end of the cylinder and the piston causes the piston and piston rod assembly to move, enlarging the volume of that chamber. The piston movement reduces the volume of the second chamber, forcing fluid out while extending the piston rod. Forcing fluid into the chamber on the opposite side of the piston reverses the process, causing the piston rod to retract. The piston rod is usually attached to the movable machine element. The body of the cylinder is typically anchored to a stationary machine member.

Each of the parts of a basic cylinder are identified by specific terms:

- Barrel.
- Head.
- Cap.
- Piston.
- Rod.
- Seals.
- Rod wipers.

These names are typically used in the fluid power industry. The next sections describe the materials and processes commonly used for the construction of these parts.

Goodheart-Willcox Publisher

Figure 9-1. The primary purpose of a cylinder is to convert fluid pressure and flow into linear mechanical movement.

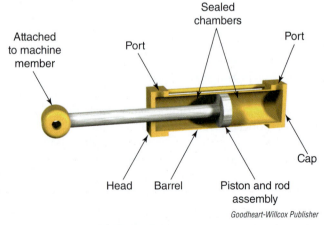

Goodheart-Willcox Publisher

Figure 9-2. Sealed chambers are formed on either side of the piston of a cylinder. Fluid pressure acting on the piston surfaces moves the piston and rod assembly.

Barrel

Figure 9-3 shows a parts list for a typical cylinder. The tube portion of a cylinder is called the *barrel*. It is made from metal tubing that has been machined to provide the desired bore diameter. The surface finish on the walls of the bore is obtained by honing. A proper finish produces a minimum of piston seal wear as the cylinder extends and retracts during operation. Some cylinder manufacturers chrome plate the barrel bore surface to increase the service life of the cylinder.

Wall thickness and the type of material used for construction of the barrel are two important factors that establish the maximum safe operating pressure of a cylinder. Manufacturers must consider the weight, strength, and bulk of materials in order to provide a cylinder that is as light and compact as possible, while providing adequate strength for safe operation at specified system operating pressures.

Head and cap

The ends of the barrel are closed by component parts called the head and the cap. The *head* closes the end through which the cylinder rod passes. The *cap* closes the opposite end, the end without the rod. The head is often called the rod end, while the cap is called the blind end. Seals are provided at both ends of the cylinder to prevent fluid leakage between the barrel and the head and cap.

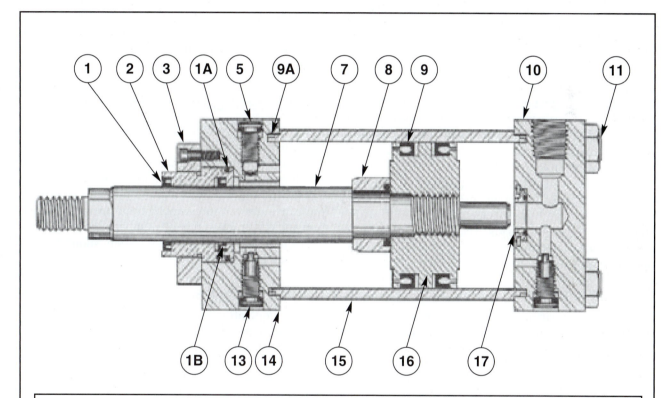

ITEM	DESCRIPTION	ITEM	DESCRIPTION
1, 1A, & 1B	ROD SEAL KIT **	10	REAR CAP
1, 1A, 1B, & 2	ROD GLAND KIT **	11	TIE RODS AND NUTS
3	RETAINER RING/PLATE	13	CUSHION NEEDLE ASSEMBLY
5	BALL CHECK ASSEMBLY	14	FRONT HEAD
7	PISTON ROD	15	CYLINDER TUBE (BARREL)
8	CUSHION SPUD W/ SEAL	16	PISTON
9 & 9A	PISTON SEAL KIT	17	CUSHION STAR ASSEMBLY

Yates Industries, Inc.

Figure 9-3. The component parts of a cylinder must withstand the expected maximum system operating pressure and provide a positive seal to prevent fluid leakage.

The hole that allows the cylinder rod to pass through the head contains a bearing surface to support the rod during extension and retraction. It also contains a seal to prevent leakage around the rod. The head and cap also contain the ports that allow system fluid to enter and exit the cylinder during extension and retraction. The cap is often used to attach the cylinder to a stationary machine member.

Piston and rod

The piston and rod are usually thought of as the working parts of the cylinder. The *piston* separates the cylinder into two, variable-size chambers. A positive seal is provided between the chambers by piston rings or other seals located between the piston and the bore surface of the barrel. Applying pressure on one side of the piston produces the force needed to either extend or retract the cylinder. Applying force on the opposite side of the piston reverses the cylinder.

The piston is machined to a specific diameter to accommodate the type of seal used by the manufacturer. The width of the piston is also carefully considered to provide an adequate bearing surface to withstand side loading as the cylinder extends. *Side loading* is force applied to the rod/piston assembly that is not parallel to the axis of the rod.

The *rod* is connected to the piston and transfers the force generated to act against the load. Rods are usually made from high-tensile-strength steel to withstand heavy operating loads. The surface of a rod is often chrome plated to extend service life under dirty operating conditions.

Seals and rod wipers

Leakage in any part of a hydraulic system results in reduced operating efficiency, reduced cleanliness, and increased safety hazards. Seals are incorporated in several locations in a basic cylinder to prevent internal and external leakage of system fluid, **Figure 9-4**. These locations include:

- Contact points between the barrel and the head and cap.
- Clearance between the piston and barrel.
- Clearance between the rod and the opening in the head.

Seals can be classified as static or dynamic. *Static seals* are materials compressed between two rigid, nonmoving parts. They may be flat, paper gaskets; molded synthetic O-rings; or a variety of other materials or forms. *Dynamic seals* function between parts that are moving relative to each other. These seals may be formed from natural material (such as leather), molded synthetic materials, or machined metal compression rings. These seals are subject to wear because at least one of the moving parts must move across the seal material.

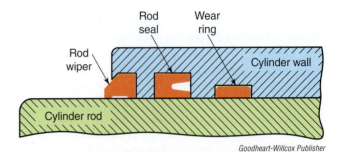

Goodheart-Willcox Publisher

Figure 9-4. A wide variety of seal designs are used to ensure leak-free cylinder operation.

The seal between the rod and the opening in the cylinder head is often made up of two parts. The first seal prevents external leaks. The second is a *wiper seal* or scraper that prevents water, dirt, or other contaminants from entering the system as the rod is drawn back into the cylinder during the retraction stroke. See **Figure 9-5**.

9.1.2 Classification of Cylinders

A number of variations from the basic cylinder design are available for use in fluid power systems. This variety is necessary to provide the type of linear movement and force application required by the many machines using fluid power. Two common methods to classify these cylinder variations are by basic operating principles and construction type. A number of other cylinder forms that do not fit into these classifications are also discussed in this section.

Operating principles

Cylinder operation may be classified as single acting or double acting. A *single-acting cylinder* exerts force only on extension or retraction. It depends on some outside force to complete the second movement. A *double-acting cylinder* can exert force on both extension and retraction. See **Figure 9-6**.

In a single-acting cylinder, only one side of the piston can be pressurized. The chamber on the opposite side of the piston is vented to the atmosphere. Some designs use pressure to extend the cylinder, while others apply pressurized fluid for retraction. In either case, the cylinder is returned to its initial position by a spring or gravity. A spring-return cylinder typically has a spring mounted inside of the vented side of the cylinder barrel. Returning the cylinder by gravity requires a vertically mounted cylinder or an external lever that allows the weight of the load to return the piston. A directional control valve directs fluid to the cylinder to allow force to be applied only in one direction. During cylinder return, the valve blocks the flow of fluid from the system pump, while energy from the compressed spring or gravity forces fluid back to the reservoir.

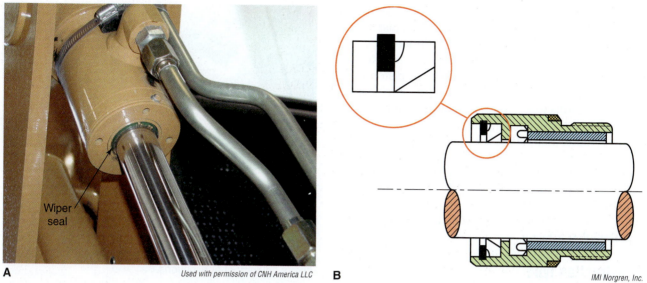

Figure 9-5. When a cylinder operates under dirty conditions, a wiper seal or scraper can be used to prevent contaminants from entering the system. A—A wiper seal on a cylinder. B—A detailed cross section of a wiper seal.

A　Used with permission of CNH America LLC　B　IMI Norgren, Inc.

Double-acting cylinders closely match the description of the basic cylinder covered in an earlier section of this chapter. They are the most commonly used of the linear actuators because they can generate force during both extension and retraction. A directional control valve used with these cylinders directs system pump output to the chamber on one side of the piston. At the same time, the valve also connects the chamber on the opposite side of the piston to the reservoir. The incoming fluid moves the piston and rod to generate cylinder rod movement to do work. At the same time,

fluid is forced from the opposite side of the piston back to the reservoir. Shifting the directional control valve reverses the connections. This reverses the fluid movement in the cylinder, causing the piston and rod to reverse direction.

The force that a double-acting cylinder can exert at a given system pressure differs between extension and retraction. This variation is the result of the difference between the piston area exposed to system pressure. See **Figure 9-7.** The piston surface exposed to system pressure during cylinder extension is the total area

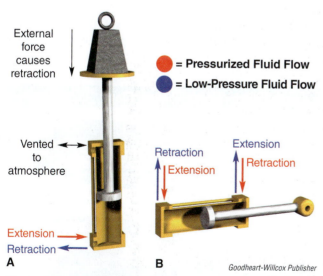

Figure 9-6. Cylinders may be classified as either double acting or single acting. A—A single-acting cylinder must use external force to return the piston. B—The piston in a double-acting cylinder can be returned by pressurizing the opposite side of the piston.

Goodheart-Willcox Publisher

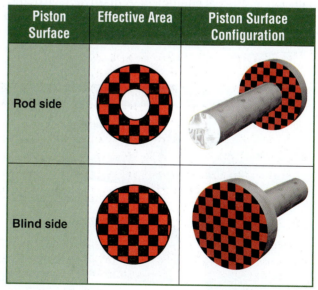

Goodheart-Willcox Publisher

Figure 9-7. The force a double-acting cylinder can generate varies between extension and retraction because of the effective area of the piston.

of the piston face. On the other hand, the piston area exposed during retraction is the area of the piston face *minus* the cross-sectional area of the cylinder rod. Force is calculated by multiplying piston area by system pressure. Therefore, the smaller effective piston area, which is the area exposed to system pressure, during retraction results in less force for any system pressure.

The speed of extension and retraction of a double-acting cylinder also differs for a specific pump output. This variation is the result of the difference in total volume between the rod end of the cylinder and the blind end without the rod. See **Figure 9-8**. The volume of the rod end is smaller because of the space taken up by the cylinder rod. This reduced volume results in a retraction time that is faster.

Construction type

The construction classification relates primarily to the method used to attach the cylinder head and cap to the barrel. Typically, manufacturers produce cylinders that can be classified as tie rod, mill, threaded end, or one piece. However, other variations may be found.

A *tie-rod cylinder* has the head and cap pieces secured on the barrel using external rods. The head and cap are usually square with holes drilled through the corner areas to facilitate placement of the rods. See **Figure 9-9**. Typically, this type of cylinder has a groove machined into the head and cap pieces to properly align them with the barrel. A static seal is also placed in the groove. Tightening the tie rods both compresses the seals, promoting leak-free operation, and forms the components into a sturdy actuator.

Tie-rod cylinder designs are used in a wide variety of fluid power applications. They are commonly found on heavy industrial equipment. A disadvantage of the design is the external placement of the tie rods. Since they are exposed, the rods may be damaged during

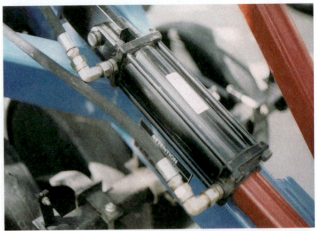

Goodheart-Willcox Publisher

Figure 9-9. Tie-rod cylinders are commonly used for mobile, heavy equipment applications.

machine operation. The irregular surfaces inherent with the external placement of the tie rods also increase the accumulation of dirt. In some applications, this may result in heat buildup.

Mill cylinders are constructed using a barrel formed of heavy-weight steel with heavy, continuous flanges welded to each end. The head and cap pieces are attached to these flanges with several short bolts. See **Figure 9-10**. A seal between the flanges and the head and cap is ensured by placing a gasket or some other sealing material between the parts. The mounting bolts compress the material to create a positive seal. This design was originally developed for applications in heavy industries, such as foundries and steel mills. These environments require a cylinder that can withstand external physical abuse, as well as operate under hot and dirty conditions.

Threaded-end cylinders have the head and cap attached to the barrel using machine threads.

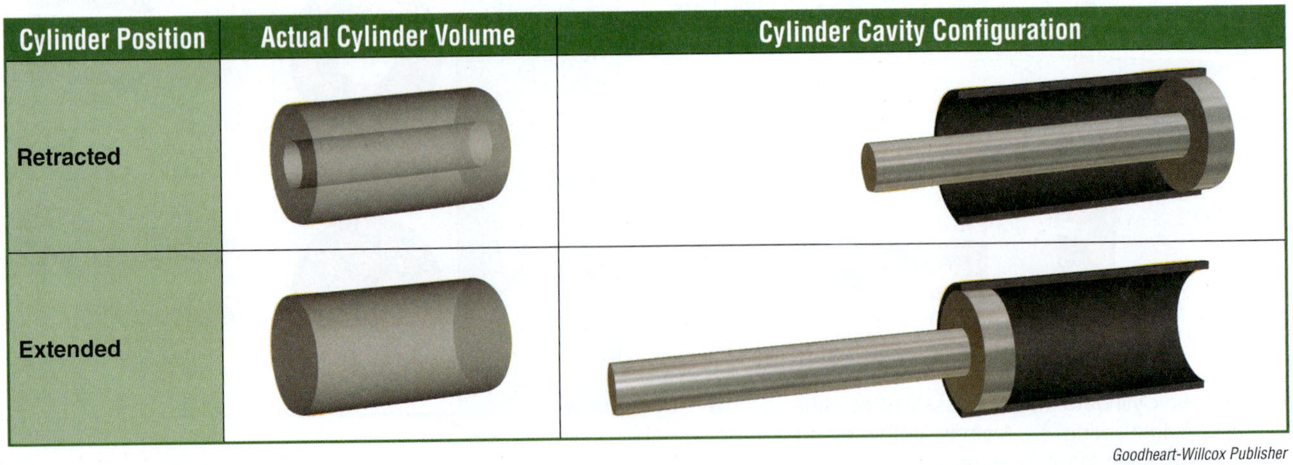

Cylinder Position	Actual Cylinder Volume	Cylinder Cavity Configuration
Retracted		
Extended		

Goodheart-Willcox Publisher

Figure 9-8. The speed of a double-acting cylinder varies between extension and retraction because of the difference in volume between the rod side and blind side of the piston.

Yates Industries, Inc.

Figure 9-10. Mill cylinders are designed using heavy materials in the barrel and other parts to withstand damage in situations where severe external abuse may be encountered.

Bailey International Corporation

Figure 9-11. Threaded-end cylinders provide streamlined exterior surfaces that are easier to keep clean.

The design typically uses internal threads on the head and cap and external threads on the barrel. However, the reverse is used by some manufacturers, as shown in **Figure 9-11**. This design normally produces a compact cylinder that is relatively light in weight. The uncluttered exterior surface is also more streamlined in appearance and easier to keep clean.

A *one-piece cylinder* is a cylinder in which the head and cap are permanently attached to the barrel. This may be done by welding the parts together or rolling the ends of the barrel into a groove machined in the head and cap pieces. See **Figure 9-12**. In both cases, the cylinder is considered a throwaway item as the unit cannot be disassembled for service. These cylinders are relatively inexpensive to produce. They are considered to be cost effective for circuits not requiring continuous operation.

Some manufacturers produce modifications of the threaded-end and one-piece cylinders. These modifications involve welding the cap to the barrel and providing some type of alternative design for the head to allow disassembly of the cylinder for service. These alternatives include threads or other mechanical designs that allow easy removal of the head. A modified snap ring design is often used in these situations.

Other cylinder forms

The linear actuators discussed to this point follow the design of the basic cylinder. Two of the characteristics of that design are differences in:

- Volume of the cylinder chambers per linear unit.
- Effective surface area the piston faces in each of the two chambers.

These characteristics result in both force and speed differences between extension and retraction of the cylinders. A double-rod-end cylinder eliminates these differences.

The *double-rod-end cylinder* has a rod attached to both faces of the piston. A head plate is attached to each end of the cylinder barrel. The heads allow the rods to extend from the cylinder. With this configuration, the effective surface area of the piston and the volume of the chambers per linear unit are the same on both sides of the piston. See **Figure 9-13**. These factors allow equal force for a given pressure during both extension and retraction. In addition, the speed of the cylinder is the same for extension and retraction for a given fluid flow. A disadvantage of the design, however, is that the

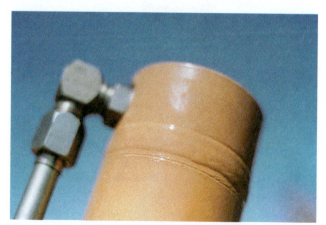

Goodheart-Willcox Publisher

Figure 9-12. This one-piece cylinder has the head and cap welded to the barrel. One-piece cylinders are relatively inexpensive, but cannot be disassembled for service.

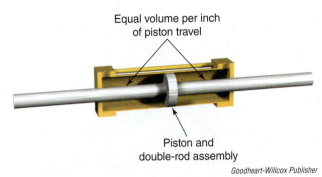

Equal volume per inch
of piston travel

Piston and
double-rod assembly

Goodheart-Willcox Publisher

Figure 9-13. The double-rod-end cylinder not only provides the possibility of two attachment points, but operates at the same extension and retraction speeds without individual flow control systems

overall length of the cylinder is increased by the length of the additional rod.

Another linear actuator that does not fit the form of the basic cylinder is the ram. The distinguishing feature of a *ram* is the diameter of the rod. The rod is the same diameter as the bore of the barrel with only sufficient clearance provided to allow rod movement and the placement of seals. The end of the rod serves as the surface against which the pressurized system fluid acts to generate the force of the actuator. No chamber is provided in the rod end to allow retraction using system pressure. Rams are, therefore, always single acting. Retraction is achieved using gravity acting on at least part of the load the ram was lifting. A common form of this actuator is the hand-operated hydraulic jack used for raising loads in many consumer and industrial applications, **Figure 9-14**.

In some fluid power applications, the cylinder must extend a long distance even though the mounting space is limited in the equipment. An example of this is the cylinder used to lift a dump box on trucks and other mobile equipment, **Figure 9-15**. These extended linear distances can be provided by telescoping cylinders. *Telescoping cylinders* are constructed of several nested tubes. See **Figure 9-16**. The largest of the tubes fits snuggly into the barrel of the cylinder. This first-stage tube forms the largest-diameter cylinder rod. Its inside bore acts as the barrel for another tube. This second tube is the second stage and has a slightly smaller rod diameter. This nesting continues as needed by the design. A solid rod is typically used for the last stage.

Goodheart-Willcox Publisher

Figure 9-14. A hand-operated hydraulic jack is a common example of a hydraulic ram.

Goodheart-Willcox Publisher

Figure 9-15. A telescoping cylinder is used to provide elevation on this carnival ride.

cylinder represents a considerable amount of energy. A direct impact can easily cause damage to the cylinder. *Cushioning* is commonly used to control the approach of the rod to the cap or head at the completion of its stroke. The most common cushioning devices are constructed as an integral part of a hydraulic cylinder.

A cushioning device consists of plunger cavities, check valves, and adjustable orifices located in the head and cap of the cylinder. Also, tapered plungers are located on both faces of the piston. See **Figure 9-17**. The diameter of each plunger cavity is only slightly larger than the largest diameter of the plungers. The cavities connect the cylinder chambers with the ports used to attach the cylinder to the hydraulic lines of the circuit. The check valves and adjustable orifices are placed in separate passageways that also connect the cylinder chambers and ports. See **Figure 9-18**.

A

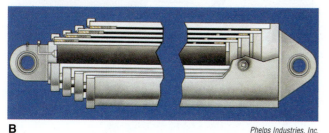

B

Phelps Industries, Inc.

Figure 9-16. A—A telescoping cylinder can extend several times its initial, closed length because of the nesting of the tubes that make up the rod. B—This cross section shows the nesting of a telescoping cylinder.

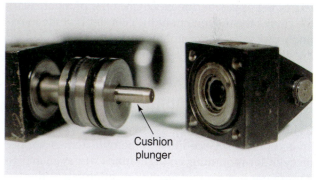

Goodheart-Willcox Publisher

Figure 9-17. Reducing the speed of the rod near the end of the stroke is accomplished by restricting fluid flow out of the cylinder. This is done with a cushioning device.

In a telescoping cylinder, special attention must be given to the seals between the rod and barrel surfaces. Attention must also be given to the mechanisms used to stop extension at the end of each stage of rod travel. These cylinders may contain five or more stages and exceed twenty feet in length when fully extended.

Telescoping cylinders are typically single acting. This is due to the physical relationship between the bores and the rods, which is much like a ram. Some special designs are produced that will allow the telescoping cylinder to be retracted using hydraulic pressure. However, the force produced is restricted to that needed to retract the cylinder, not move a load.

9.1.3 Cylinder Cushioning

When a rapidly moving piston and rod assembly approaches the end of its stroke, the assembly must be slowed to prevent the piston from directly impacting the cylinder cap or head. The combined momentum of the moving piston assembly and the fluid entering the

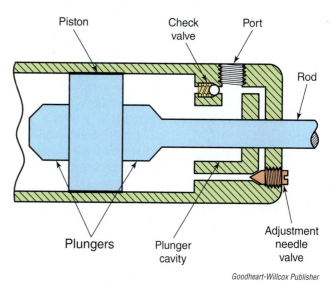

Goodheart-Willcox Publisher

Figure 9-18. The deceleration rate produced by a cylinder cushion is determined by the size and shape of the plunger and the cross-sectional area of the adjustable orifice controlled by the adjustment needle valve.

Near the end of the cylinder stroke, the end of the tapered plunger enters the plunger cavity. This begins the deceleration and cushioning process. As the plunger continues to move into the cavity, the clearance between the plunger and the cavity walls decreases. As a result, the amount of fluid that can pass out of the cylinder chamber is reduced. This, in turn, reduces the rate of movement of the cylinder rod. Once the largest diameter of the plunger has entered the cavity, the remainder of the fluid in the cylinder chamber must exit through the adjustable orifice (needle valve). The adjustable orifice allows a machine operator to tune the system to obtain the desired deceleration rate as the cylinder completes its movement.

When the directional control valve in the circuit is shifted to reverse cylinder direction, pump output is directed into the port passageway. The easiest route for this fluid to follow is through the check valve and into the cylinder chamber. This forces the piston and rod in the opposite direction. Once the plunger clears the cavity, the movement of the cylinder is at the maximum rate until the cushioning device is contacted on the opposite end of the cylinder.

9.1.4 Cylinder Mounting

Cylinders must be properly mounted in equipment if an acceptable service life is to be obtained. A variety of cylinder-mounting configurations are available to equipment designers. The configuration selected must be based on the expected cylinder loading. Loading includes forces parallel to the centerline of the piston rod and side loading caused by the movement of machine members. Mountings can be classified as fixed centerline, fixed noncenterline, and pivoting centerline. Variations of each of these classifications are discussed in the following sections.

Fixed centerline

A *fixed-centerline mount* aligns the centerline of the cylinder with the forces created by the system load. See **Figure 9-19**. This mounting style is strong, but cannot tolerate misalignment. Misalignment produces side loading on the cylinder. There are three mounting variations commonly available in this classification:

- Tie rod.
- Head-end flange.
- Cap-end flange.

In the tie-rod design, the tie rods that hold the head and cap onto the cylinder barrel are extended and used as bolts to mount the cylinder to the equipment. Depending on the requirements of the equipment, the cylinder may be mounted using the tie-rod bolts extending from either the head or the cap end.

In the head- and cap-end flange designs, a heavy metal plate is bolted to the head or cap end of the cylinder. See **Figure 9-20**. This flange is then mounted directly to the equipment at the desired location.

Fixed noncenterline

A *fixed-noncenterline mount* is attached to equipment by feet connected to the head and cap ends. However, the feet are not on the centerline of the cylinder. See **Figure 9-21**. The resulting off-center attachment position causes the forces generated during system operation to attempt to rotate the cylinder around the mounting points. This rotating force can apply undesirable stress to machine and actuator parts, causing service problems, unless cylinders are carefully sized for an application.

Goodheart-Willcox Publisher

Figure 9-19. Fixed-centerline cylinder mounting provides a strong mount, but one that does not stand up well to side loading.

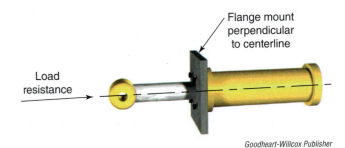

Goodheart-Willcox Publisher

Figure 9-20. A head-end flange cylinder mount is one type of a fixed-centerline cylinder mount.

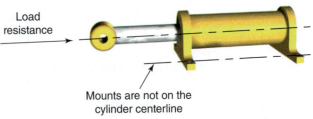

Goodheart-Willcox Publisher

Figure 9-21. Fixed-noncenterline cylinder mounting can cause stress to machine members because the force generated attempts to rotate the actuator around the mounting point.

The feet can be angle brackets bolted to the cylinder head and cap, lugs cast as part of or welded to the head and caps, or holes drilled and tapped into the head and cap. Each of these designs allows direct mounting of the cylinder using bolts or machine screws.

Pivoting centerline

Equipment often includes arms or other machine members that move in an arc as they perform their task. These machine elements are often powered by a cylinder with a *pivoting-centerline mount*, which allows the cylinder to pivot to follow the movement. Two mounting methods are used to allow the cylinder to follow this curved path. The lines of force caused by cylinder loading are concentric with the centerline of the cylinder in both of these mounts.

The first mounting method is the *clevis mount*. A clevis is a U-shaped shackle. A clevis mount includes a clevis on the cylinder cap. The cap is normally attached to the body of the machine. Another clevis is included on the end of the cylinder rod. This is attached to the movable machine member. A clevis mount is shown in **Figure 9-22**.

The second mounting method is the *trunnion mount*. This mount has pivot pins, or trunnions, mounted on the sides of the cylinder. See **Figure 9-23**. The trunnions may be positioned on head, cap, or at any point in between in order to obtain a desired pivoting action. The trunnions are attached to the body of the machine. A clevis or similar device is used to attach the cylinder rod to the movable machine member.

9.1.5 Cylinder Calculations

The selection of cylinders for a particular application requires that a series of calculations be completed. These calculations establish that the cylinders can perform at the levels indicated by system specifications using the system power unit and recommended pressure levels and flow rates. Calculations include:

- General cylinder extension/retraction force and speed rates.
- Speed, pressure, and flow rates needed to generate force and speeds for specific cylinders.

Force

The force a cylinder can generate depends on system pressure and the effective area of the piston. *Effective piston area* is defined as the surface area of the piston exposed to system pressure. See **Figure 9-24**. During the extension of a double-acting cylinder, the *effective area* is equal to the area of the piston on the side without the rod. During retraction, the effective area is the area of the piston *minus* the cross-sectional area of the piston rod.

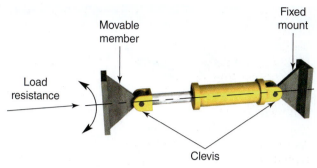

Goodheart-Willcox Publisher

Figure 9-22. Pivoting-centerline cylinder mounting allows the cylinder to follow a curved path as it applies force to a moving machine member. This is a clevis mount.

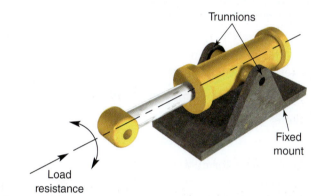

A   *Goodheart-Willcox Publisher*

B   *Yates Industries, Inc.*

Figure 9-23. A—A trunnion mount allows a cylinder to pivot if a machine member is required to move through an arc in its operation. B—This is a large cylinder with a trunnion mount.

End	Effective Area	Calculation
Blind end of cylinder	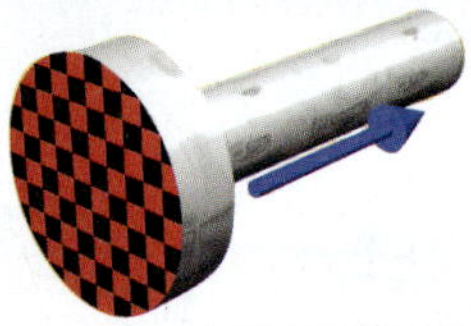	Effective area = Total area of piston face
Rod end of cylinder	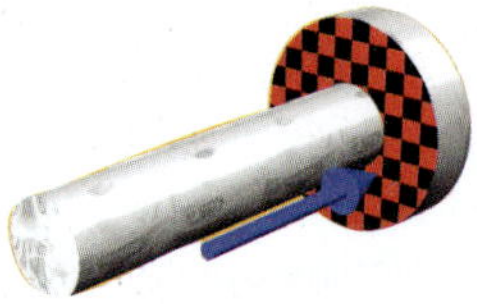	Effective area = Total area of piston face − Area of rod cross section

Goodheart-Willcox Publisher

Figure 9-24. An understanding of the effective area of the cylinder piston is critical when calculating cylinder force, speed, and required flow rates.

This formula is used to calculate the force of a cylinder on the extension stroke:

Extension Force = System Pressure × Piston Area

where:

Piston Area = $\pi \times radius^2$

A similar formula is used to calculate the force of the cylinder on the retraction stroke:

Retraction Force = System Pressure × (Piston Area − Rod Area)

EXAMPLE 9-1

Force Generated by a Cylinder

Determine the maximum force generated by a double-acting cylinder during extension and retraction. The cylinder has a piston area of 5 in² with a rod area of 1 in². The maximum system pressure is 500 psi.

Extension Force (lb) = System Pressure (psi) × Piston Area (in²)

= 500 psi × 5 in²

= 2,500 lb

Retraction Force (lb) = System Pressure (psi) × (Piston Area (in²) − Rod Area (in²))

= 500 psi × (5 in² − 1 in²)

= 500 psi × 4 in²

= 2,000 lb

A closely associated, additional calculation that is often desired is the system pressure needed to produce a required force while using a cylinder of a specific size. This calculation is accomplished by modifying the above formulas:

$$\text{System Pressure} = \frac{\text{Cylinder Force}}{\text{Effective Piston Area}}$$

EXAMPLE 9-2

System Pressures Required to Generate a Force

Determine the system pressure required to generate a force of 1,000 pounds during extension and retraction of a double-acting cylinder with a piston radius of 1″ and a rod radius of 0.25″.

1. Find the areas of the piston and rod.

Piston Area = $\pi \times radius^2$

= $\pi \times 1$ in²

= 3.14 in²

Rod Area = $\pi \times radius^2$

= $\pi \times 0.25^2$

= 0.19 in²

2. Find the system pressure required for the extension phase.

$$\text{System Pressure (psi)} = \frac{\text{Cylinder Force (lb)}}{\text{Piston Area (in}^2\text{)}}$$

$$= \frac{1{,}000 \text{ lb}}{3.14 \text{ in}^2}$$

= 318 psi

3. Find the system pressure required for the retraction phase.

$$\text{System Pressure (psi)} = \frac{\text{Cylinder Force (lb)}}{\text{Piston Area (in}^2\text{)} - \text{Rod Area (in}^2\text{)}}$$

$$= \frac{1{,}000 \text{ lb}}{3.14 \text{ in}^2 - 0.19 \text{ in}^2}$$

$$= \frac{1{,}000 \text{ lb}}{2.95 \text{ in}^2}$$

= 339 psi

Speed

The speed at which a cylinder rod extends or retracts depends on the volume displaced per inch of piston travel and the amount of fluid entering the cylinder. The effective area of the piston, as defined in the cylinder force calculations above, must be used to determine cylinder speed. The flow delivery rate, which determines the amount of fluid entering the cylinder, must be converted from gallons per minute (gpm) to cubic inches per minute (in³/min). The conversion factor used is 1 gal = 231 in³.

The extension speed of a cylinder can be calculated using this formula:

$$\text{Extension Speed (in/min)} = \frac{\text{Flow Rate (gpm)} \times 231 \text{ in}^3/\text{gal}}{\text{Piston Area (in}^2\text{)}}$$

The following formula can be used to calculate the speed of the cylinder on the retraction stroke:

$$\frac{\text{Retraction Speed}}{(\text{in/min})} = \frac{\text{Flow Rate (gpm)} \times 231 \text{ in}^3/\text{gal}}{\text{Piston Area (in}^2) - \text{Rod Area (in}^2)}$$

Another closely associated calculation is the system flow rate needed to produce a desired extension or retraction speed for a cylinder of a specific size. This calculation is accomplished by modifying the cylinder speed formulas:

$$\frac{\text{Flow Rate}}{(\text{gpm})} = \frac{\text{Piston Area (in}^2) \times \text{Cylinder Speed (in/min)}}{231 \text{ in}^3/\text{gal}}$$

EXAMPLE 9-3

Speed of Cylinder Extension and Retraction

A double-acting cylinder has a piston radius of 1" and a rod radius of 0.25". Determine the maximum extension and retraction rates for the cylinder if the system pump output is 5 gpm.

1. Find the areas of the piston and rod.

$$\text{Piston Area} = \pi \times \text{radius}^2$$
$$= \pi \times 1^2$$
$$= 3.14 \text{ in}^2$$

$$\text{Rod Area} = \pi \times \text{radius}^2$$
$$= \pi \times 0.25^2$$
$$= 0.19 \text{ in}^2$$

2. Find the maximum system speed for the extension phase.

$$\frac{\text{Extension Speed}}{(\text{in/min})} = \frac{\text{Flow Rate (gpm)} \times 231 \text{ in}^3/\text{gal}}{\text{Piston Area (in}^2)}$$
$$= \frac{5 \text{ gpm} \times 231 \text{ in}^3/\text{gal}}{3.14 \text{ in}^2}$$
$$= 368 \text{ in/min}$$

3. Find the maximum system speed for the retraction phase.

$$\frac{\text{Retraction Speed}}{(\text{psi})} = \frac{\text{Flow Rate (gpm)} \times 231 \text{ in}^3/\text{gal}}{(\text{Piston Area (in}^2) - \text{Rod Area (in}^2))}$$
$$= \frac{5 \text{ gpm} \times 231 \text{ in}^3/\text{gal}}{(3.14 \text{ in}^2 - 0.19 \text{ in}^2)}$$
$$= \frac{5 \text{ gpm} \times 231 \text{ in}^3/\text{gal}}{2.95 \text{ in}^2}$$
$$= 392 \text{ in/min}$$

9.1.6 Interpreting Cylinder Manufacturer Specifications

A large number of different companies produce hydraulic cylinders. The online *Fluid Power Manufacturers Directory* lists companies that manufacture various types of hydraulic cylinders. The products are grouped into categories to help in searching. With the amount of information available about this wide selection of cylinders, it is often difficult to interpret and compare the data. This section discusses cylinder specifications often included in manufacturer literature or data sheets.

Typical specifications

Manufacturers generally provide detailed specifications for each of their cylinder lines. The format for presenting the specifications varies considerably between manufacturers. However, the following factors are often readily available from company literature.

- **Duty service.** Light-, medium-, or heavy-duty based on the specifications of national standards groups.
- **Construction type.** Tie-rod, mill, threaded-end, or one-piece designs.
- **Pressure rating.** Depends on the bore size of the cylinder barrel.
- **Fluid rating.** Hydraulic oil or other specialized fluid that may require special materials for seals.
- **Acceptable operating temperature.** May range from below 0°F to over 165°F.
- **Bore size.** The diameter of the cylinder barrel bore may range from 1" to well over 8", depending on the cylinder series.
- **Piston rod diameter.** The rod diameter may range from 1/2" to over 6", depending on the cylinder series.
- **Mounting styles.** Most cylinder series provide a variety of mounting configurations to facilitate attachment to machines, including flange, lugs, clevis, trunnion, and so on.
- **Stroke.** Most manufacturers indicate that their cylinders are available in any practical stroke length.
- **Cushions.** Usually optional at either end of the cylinder.
- **Rod end configuration.** Several thread styles and sizes are considered standard, with several others available as a custom option.
- **Ports.** NPTF pipe threads are normally used, but other styles, such as O-rings, are acceptable.

Display of cylinder specifications

Manufacturer literature typically includes a series of descriptions, illustrations, and tables that contain the information required to select a cylinder to meet the specifications of an installation. **Figures 9-25** through **9-27** are materials taken from the cylinder section of a cylinder manufacturer's catalog. These materials are representative of the approaches used by companies to indicate the characteristics of their products. This information is critical when a machine designer is developing equipment. It is also important when service personnel are selecting cylinders to replace failed components or upgrading components to improve equipment performance.

The cylinder section of a typical catalog or manufacturer website includes dimension drawings of each basic cylinder design. Rather than showing specific sizes on the drawings, the dimensions are labeled with code letters that can be cross-referenced with dimensions shown on the tables associated with the drawings. These tables include information on all of the critical dimensions of a cylinder, including the exterior envelope, fittings for mounting, rod diameter and length, and rod end fittings. **Figure 9-25** illustrates a typical basic cylinder type and identifies critical dimensions. The table below the drawing contains the dimensions for each bore size available for this specific cylinder model. The drawings and tables provide the machine designer or technician with all of the dimensional information required for selecting a cylinder.

9.2 Limited-Rotation Hydraulic Actuators

Machine shafts and other components often require torque in situations where rotation is limited to one or two revolutions or less. If the demanded rotation is less than 180°, a hydraulic system may use a standard linear cylinder attached to a lever to produce the required motion and torque. If this arrangement cannot produce the desired rotation and torque or the components are too bulky for the space allowed in the machine design, a limited-rotation actuator may provide a desirable solution. A *limited-rotation actuator* provides restricted rotary or reciprocating rotary motion for lifting/lowering, opening/closing, or indexing operations. These devices are sometimes called torque motors.

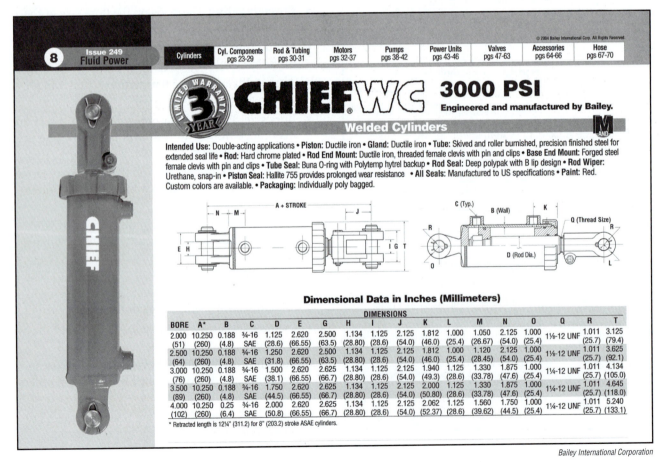

Bailey International Corporation

Figure 9-25. Manufacturer catalogs provide a wealth of information about the construction and size of their cylinder models.

BAILEY NO.	STROKE	ROD DIAMETER	RETRACT	COLUMN LOAD	PORT SIZE	PIN DIAMETER	SHIP WT.
				2" Bore • 3000 PSI			
286-001	4"	1⅛"	14¼"	Full PSI	SAE 8	1"	12
286-002	6"	1⅛"	16¼"	Full PSI	SAE 8	1"	15
286-003	8" ASAE	1⅛"	20¼"	Full PSI	SAE 8	1"	17
286-004	8"	1⅛"	18¼"	Full PSI	SAE 8	1"	16
286-005	10"	1⅛"	20¼"	Full PSI	SAE 8	1"	17
286-006	12"	1⅛"	22¼"	Full PSI	SAE 8	1"	19
286-007	14"	1⅛"	24¼"	Full PSI	SAE 8	1"	20
286-008	16"	1⅛"	26¼"	Full PSI	SAE 8	1"	21
286-009	18"	1⅛"	28¼"	8100 lbs.	SAE 8	1"	23
286-010	20"	1⅛"	30¼"	6880 lbs.	SAE 8	1"	26
286-011	24"	1⅛"	34¼"	5140 lbs.	SAE 8	1"	27
286-014	30"	1⅛"	40¼"	3550 lbs.	SAE 8	1"	31
286-015	36"	1⅛"	46¼"	2600 lbs.	SAE 8	1"	39
286-100		1⅛"		WC Packing Kit			
				2½" Bore • 3000 PSI			
286-016	4"	1¼"	14¼"	Full PSI	SAE 8	1"	17
286-017	6"	1¼"	16¼"	Full PSI	SAE 8	1"	18
286-018	8" ASAE	1¼"	20¼"	Full PSI	SAE 8	1"	17
286-019	8"	1¼"	18¼"	Full PSI	SAE 8	1"	20
286-020	10"	1¼"	20¼"	Full PSI	SAE 8	1"	21
286-021	12"	1¼"	22¼"	Full PSI	SAE 8	1"	23
286-022	14"	1¼"	24¼"	Full PSI	SAE 8	1"	25
286-023	16"	1¼"	26¼"	Full PSI	SAE 8	1"	27
286-024	18"	1¼"	28¼"	12500 lbs.	SAE 8	1"	28
286-025	20"	1¼"	30¼"	10610 lbs.	SAE 8	1"	29
286-026	24"	1¼"	34¼"	7920 lbs.	SAE 8	1"	33
286-027	30"	1¼"	40¼"	5470 lbs.	SAE 8	1"	38
286-028	36"	1¼"	46¼"	4000 lbs.	SAE 8	1"	46
286-101		1¼"		WC Packing Kit			

Bailey International Corporation

Figure 9-26. Manufacturer catalogs provide details covering cylinder bore, stroke, and other basic envelope dimensions.

Bailey International Corporation

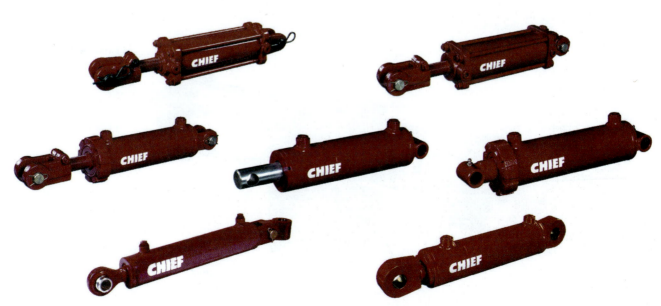

Figure 9-27. Many cylinder catalogs provide illustrations covering the configurations available for a cylinder rod end.

9.2.1 Design and Construction

A variety of designs are used in limited-rotation actuators. The three most common designs are:

- Rack-and-pinion.
- Vane.
- Helical shaft.

Other designs that may be found are units consisting of a cylinder, piston, connecting rod, and crankshaft or a cylinder, piston, chain, and sprocket. The three most common design types are discussed in the next sections.

Rack-and-pinion

A rack-and-pinion design consists of a rack gear and a pinion gear powered by two concentric cylinders. The *rack gear* is machined into the cylinder rod with the teeth along its length. The *pinion gear* is a round gear with teeth that mate with the rack. The pistons of the two cylinders are attached to the ends of the cylinder rod containing the rack gear. See **Figure 9-28**. The cylinder chambers are located on either end of the rack gear/piston assembly.

The housing also contains the pinion gear attached to a power-output shaft. The axis of the output shaft is perpendicular to the axis of the rack gear located on the cylinder rod. The housing holds the pinion gear in constant mesh with the rack gear.

Pressurizing the cylinder chamber on one end forces the rack gear/piston assembly to move. This movement turns the pinion gear, thus rotating the output shaft. Pressurizing the opposite cylinder chamber forces the rack gear and piston assembly in the opposite direction. This reverses the direction of rotation of the output shaft.

Rack-and-pinion units may incorporate one, two, or more cylinder/piston/rack sets. Each of these sets drives the pinion and output shaft. Their output force is combined to produce the desired torque. The rack-and-pinion design can also produce more than 360° of output shaft rotation. The amount of output shaft rotation depends on the length of travel of the piston and rack gear in relation to the diameter of the pinion gear.

Load-carrying capacity of a rack-and-pinion unit is determined by the strength of the housing and the load rating of the output-shaft bearings. This design adapts well to a wide range of loads. In fact, one manufacturer offers units that produce from under 50 inch-pounds of torque to units that produce several million inch-pounds of torque.

The torque output of a rack-and-pinion unit can be determined by this formula:

$$\text{Torque Output} = \text{Piston Area} \times \text{System Pressure} \times \text{Gear Pitch Radius}$$

The operating efficiency of rack-and-pinion units is approximately 90%. This efficiency plus the availability of units in a wide range of load-carrying capacities and torque outputs make this design desirable for limited-rotation applications.

Vane

The vane type of limited-rotation actuator consists of a barrel-shaped body, rotor, and two head pieces. The body contains a stationary barrier called the *stator*. The rotor consists of a vane attached to the output shaft. The rotor fits into the body and is held in place by bearings in the head pieces. The head pieces cover the ends of the body.

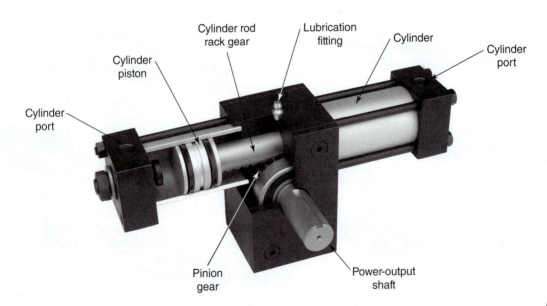

IMI Norgren, Inc.

Figure 9-28. A common type of limited-rotation actuator uses a rack-and-pinion to produce rotary motion.

The stator and rotor vane divide the cavity in the body into two tightly sealed chambers. See **Figure 9-29**. Forcing fluid into one of these chambers applies force to the rotor. This causes the output shaft to rotate. Forcing fluid into the other chamber causes the shaft to rotate in the opposite direction.

Units that use a single-vane design turn less than 360°. The exact amount below 360° of rotation depends on the width of the stator. A maximum rotation of approximately 280° is typical. Units are also available with rotors containing two vanes and two body stators. These double-vane units produce less rotation than single-vane designs. The maximum rotation of these designs is usually about 100° because of the width taken by the stators.

The operating efficiency of vane-type limited-rotation devices can be as high as 90%. Even with this high efficiency, the design tends to have some leakage around the vane seals. This slight leakage makes it difficult for the unit to hold position for an extended time. The design also requires external stops, especially in applications that require high-speed operations on high-inertia loads.

The torque output of vane-type limited-rotation devices can be as much as 750,000 inch-pounds. Torque output can be calculated using this formula:

$$\text{Torque Output} = \text{Vane Length} \times \text{Vane Width} \times \text{Vane Radial Distance} \times \text{System Pressure}$$

Helical piston and rod

The helical piston-and-rod design provides high torque output for a relatively small physical size. The unit consists of a cylindrical housing that contains a sleeve piston and the power output shaft. The open end of the sleeve piston is fitted with a nut containing helical grooves, similar to screw threads. The grooves in the nut match helical grooves machined into the surface of the output shaft. The sleeve piston is fitted into the housing using a design that allows linear movement, but not rotary movement. The output shaft is threaded through the helical nut of the sleeve piston. The output end of the shaft is mounted in a bearing attached to the housing of the unit. See **Figure 9-30**.

During operation, hydraulic pressure is applied to the sleeve piston, resulting in linear movement. Due to the helical grooves in the output shaft, it is forced to turn as the sleeve piston moves. The result is limited rotational movement of the output shaft. Shaft rotation is reversed when pressure is applied to the opposite side of the sleeve piston.

The total rotation of these designs is limited, but may be more than 360°. Manufacturers often have a number of stock and special-order rotation options available. The torque output depends on the system pressure, piston area, and angle of the helix machined into the rotating output shaft. Models are available in a range of torque capacities, from a maximum output of under 100 inch-pounds to those with a capacity of 4 million inch-pounds or more. Regardless of the maximum capacity, the design is considered to have high torque output in relation to its actual physical size and weight. However, average efficiency for this design is lower than the rack-and-pinion and vane units.

Single-Vane Design

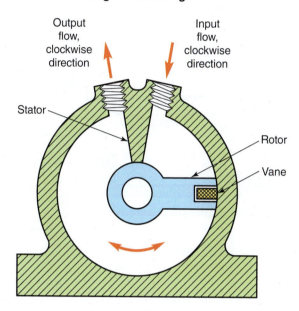

Dual-Vane Design

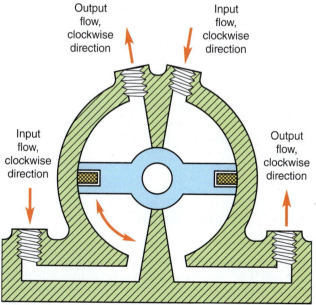

Goodheart-Willcox Publisher

Figure 9-29. The vane type of limited-rotation actuators are restricted to movement of less than one revolution because of their internal design. These designs may be single or dual vane.

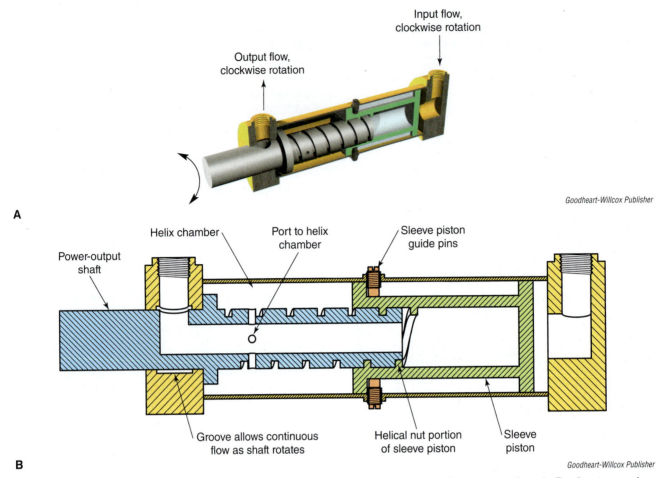

Input flow,
clockwise rotation

Output flow,
clockwise rotation

Goodheart-Willcox Publisher

A

Helix chamber   Port to helix
chamber   Sleeve piston
guide pins

Power-output
shaft

Groove allows continuous
flow as shaft rotates   Helical nut portion
of sleeve piston   Sleeve
piston

Goodheart-Willcox Publisher

B

Figure 9-30. A—A sleeve piston fitted with a helical nut forces the rod to turn as the piston moves linearly. B—A cutaway view of the assembly.

Efficiency ratings of only 80% are typical because of internal resistance relating to the movement of the piston and shaft.

9.2.2 Applications of Limited-Rotation Actuators

Limited-rotation actuators are used to perform a multitude of functions in a wide variety of industries. They are found in many types of manufacturing and assembly equipment, in material handling situations, on barges and ships, in steel mills and other heavy-metal industries, and in many other situations.

Several specific examples from manufacturing include the powering of indexing devices, opening and closing of clamps, and holding materials under tension. In addition, manufacturing-line robots often use limited-rotation actuators for positioning, **Figure 9-31**.

Material-handling applications include the switching of conveyor lines in warehouse and distribution facilities. On lift trucks, limited-rotation actuators can be used to operate specialized clamping and rotating devices for the handling of barrels or hard-to-grasp materials.

See **Figure 9-32**. In the steel industry, they are used to charge furnaces, tilt electric furnaces for pours, and handle large coils of the completed metal.

Applications in the marine industry include such basic functions such as opening and closing hatches and the operation of large valves. Limited-rotation actuators may also be involved in the operation of cargo-handling equipment, such as booms and hoists. In addition, some ships use these actuators for steering control.

9.3 Hydraulic Motors

A hydraulic **motor**, often called a rotary actuator, is the hydraulic system component that converts fluid pressure and flow into torque and rotational movement. See **Figure 9-33**. A motor, in a sense, extracts energy from the hydraulic system fluid and converts it into torque and mechanical motion. The amount of torque generated is determined by system pressure and the geometry and surface area of the internal parts of the motor. The speed of rotation is determined by the displacement

IMI Norgren, Inc.

Figure 9-31. Robots commonly use limited-rotation actuators for positioning.

Goodheart-Willcox Publisher

Figure 9-32. Material handling equipment often makes use of limited-rotation actuators for specialized clamping. Here, the scoop head can rotate about the arm's axis.

of the internal chambers of the motor and the rate of fluid flow entering the unit. The maximum speed and torque of a motor is controlled by:

- Efficiency of the design of the motor.
- Physical size of the internal elements involved in extracting the energy from the fluid.
- Fluid flow rate of the system.

Goodheart-Willcox Publisher

Figure 9-33. Although the appearance of a hydraulic motor is similar to a pump, the function is different. Motors use system flow to produce rotary motion and system pressure to generate torque.

A wide range of torque and rotational speeds can be obtained with hydraulic motors simply by varying system pressure and flow.

Hydraulic motors are often compared to hydraulic pumps because of their similar design. Even though motor and pump structures may be similar, a conceptual understanding of how motors operate is important. You need to understand motor operation in order to be successful when designing hydraulic circuits containing motors. Likewise, this understanding is essential when troubleshooting fluid power systems incorporating motors and motor circuits.

9.3.1 Conceptual Hydraulic Motor

Conceptually, a hydraulic motor consists of three component groups:

- Housing.
- Rotating internal parts.
- Power output shaft.

The housing contains inlet and outlet ports, an internal motor chamber designed to hold the internal rotating parts of the motor, bearings to support the power output shaft, and any fittings to plumb and mount the motor. The internal rotating parts of the motor consist of two sliding vanes fitted into slots machined into the outer diameter of a rotor. The rotor is securely mounted to the output shaft. The output shaft delivers the rotary power of the motor. The construction of a basic motor is shown in **Figure 9-34**. The mechanical parts of a motor require a design incorporating four basic concepts:

- The vanes must be equally spaced around the rotor to form chambers in the housing.
- The fluid moved by the system pump through the inlet port of the motor chamber must contact the surface of one side of the rotor vanes.
- The inlet-port side of the motor chamber must be sealed so fluid forced into the chamber cannot exit without forcing the vanes and rotor to turn the power output shaft.
- As the motor turns, an outlet port must allow fluid passing through the motor to be returned to the system reservoir at pressures close to atmospheric.

Figure 9-35 illustrates the operation of the conceptual motor. When system fluid is directed to the inlet port of the motor, system pressure begins to increase. This pressure increase results from any load caused by internal resistance in the motor and any external load on the motor output shaft. System pressure is applied to the surface of the inlet chamber, rotor, and surface of the vanes. Pascal's law indicates the pressure is equal on all of these surfaces. However, the pressure in the chamber connected to the outlet port is very close to atmospheric because it is connected to the line leading to the reservoir. The pressure difference across the vanes increases until the force on the inlet side of the vane overcomes the internal resistance and exterior load of the motor. When this occurs, the motor turns.

In summary, system fluid comes into the inlet chamber and applies pressure to the vane. This moves the rotor, which moves this chamber to a point where it connects to the output port. Fluid from the chamber is then returned to the reservoir. The chamber that was connected to the output port next moves to connect to the inlet port, thus continuing the cycle. The direction of rotation of the motor may be reversed by simply reversing the inlet and outlet connections to the system pump and reservoir.

9.3.2 Basic Motor Design and Operation

Several different designs of hydraulic motors are used in fluid power systems. The three most common, basic designs of hydraulic motors are gear, vane, and piston. The principles of the conceptual motor can be applied to each of these designs, even though their structure may vary considerably. Each design is capable of converting hydraulic system fluid pressure and flow into torque and mechanical rotary motion. This section discusses displacement and bearing load classifications;

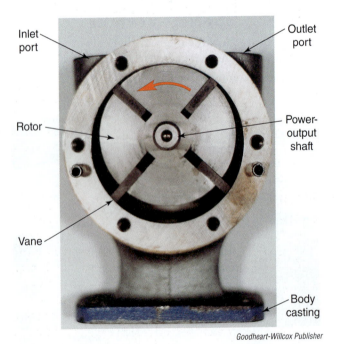

Inlet port

Outlet port

Rotor

Power-output shaft

Vane

Body casting

Goodheart-Willcox Publisher

Figure 9-34. The basic construction of a hydraulic motor requires internal rotating parts that use system pressure to generate torque, which is then applied through the output shaft of the unit.

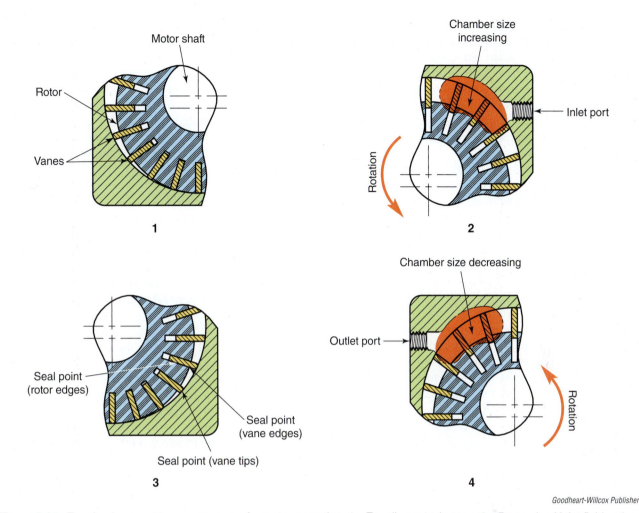

Figure 9-35. Four basic concepts are necessary for motor operation: 1—Equally spaced vanes, 2—Pressurized inlet fluid acting on the vanes, 3—Sealed chambers between the vanes, and 4—An outlet port returning low-pressure fluid to the reservoir.

Goodheart-Willcox Publisher

essential characteristics of various basic gear, vane, piston, and other motor designs; and construction and operational details.

Displacement and bearing load classification

Motors are classified by the control of the internal displacement of the unit and the type of side loading placed on the bearings of the power output shaft. These classifications are indicators of motor speed control and expected bearing performance.

Displacement may be either fixed or variable. In a *fixed-displacement motor*, the volume of fluid needed for each output shaft revolution is not adjustable. With this design, the internal motor geometry cannot be altered. Each motor revolution requires a fixed amount of fluid. In the *variable-displacement motor*, an operator can change the internal geometry of the motor using a handwheel or other mechanical device. Doing so either increases or decreases the displacement of the motor. This allows the operator to easily change the speed (revolutions per minute) of the motor.

Bearing load design may be classified as balanced or unbalanced. A motor classified as *balanced* incorporates a design that balances the load by exposing the opposing sides of the bearings to equal pressure. This reduces or eliminates side loading. The *unbalanced* designation means that the bearings of the output shaft are side loaded by the pressure difference between the inlet and outlet ports of the motor.

External-gear motors

External-gear hydraulic motors are generally considered the simplest of the basic motor types. Their cost is comparatively low. They are fixed-displacement units and pressure unbalanced. However, some designs incorporate a pressurized bearing-lubrication system that counteracts some of the bearing wear caused by pressure differences between the inlet and outlet ports.

The external-gear motor design consists of a housing that contains a number of cavities. These cavities allow system fluid into and out of the unit. The largest of the cavities contains a closely fitted, external-tooth drive gear attached to the motor output shaft. It also contains a meshed follower gear that turns freely on a fixed bearing shaft attached to the housing, **Figure 9-36**. The output shaft is supported on either end by bearings located in the housing. The drive and follower gears are closely meshed and tightly fit into the housing. There are minimal clearances between the sides of the gears and the housing and the tips of the gear teeth and the housing.

The primary housing cavity is oval, **Figure 9-37**. The end portions of the cavity match the diameter of the drive and follower gears. This primary cavity is further divided into a number of small, generally triangular cavities formed by the space between the gear teeth and the cavity wall. The middle portion of the primary cavity does not follow the curve of the gear diameters, rather it forms two larger, triangular, open cavities. The inlet and outlet ports of the motor are located in these cavities. Because of the precision fit between the gear teeth, gear sides, and walls of the housing, a minimal amount of fluid leakage occurs between all of these cavities. Separate pressure- or spring-loaded wear plates are used in some gear motor designs. These wear plates control gear end play and leakage around the sides of the gears.

When system fluid is directed into the inlet port of the motor, the fluid pressure begins to increase. This is an attempt to overcome the resistance to motor rotation caused by friction of the internal parts of the motor and the load on the output shaft. When the pressure difference between the inlet side and the outlet side is high enough to generate adequate force on the gear teeth,

the motor turns. As a result, torque is produced at the output shaft. The pressurized fluid entering the motor inlet cavity is carried in the small cavities formed by the space between the gear teeth to the outlet cavity. It is then returned to the system reservoir at near atmospheric pressure. The motor continues to turn as long as the force generated by system pressure can overcome the internal motor resistance and the shaft load. Reversing fluid flow though the motor reverses shaft rotation.

Internal-gear motors

Internal-gear hydraulic motors use a basic gerotor configuration with several component variations in their construction. These gear motors are fixed displacement and pressure unbalanced.

The construction of the gerotor involves a housing that contains a cavity fitted with three rotating elements. See **Figure 9-38**. The first element rotates, while using the walls of the housing cavity as a bearing surface. This element is a disk-like component containing specially shaped internal-gear teeth. The second element is a smaller diameter disk constructed with external-gear teeth of a matching design. The third element is the power output shaft. This is placed off center in the opening of the disk with the internal teeth. The gear with the external teeth is securely mounted on the output shaft. This gear has one less tooth than the gear with the internal teeth. As a result, the gear teeth can mesh and the shaft and gears can freely turn. The key to the motor's operation is the fact that the tips of the external teeth are always in contact with the

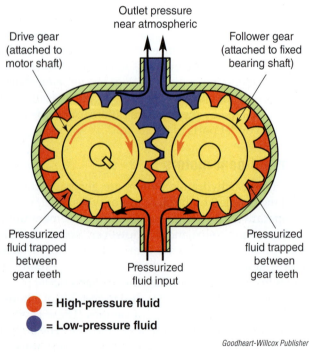

Drive gear
(attached to
motor shaft)

Outlet pressure
near atmospheric

Follower gear
(attached to fixed
bearing shaft)

Pressurized
fluid trapped
between
gear teeth

Pressurized
fluid input

Pressurized
fluid trapped
between
gear teeth

● = High-pressure fluid

● = Low-pressure fluid

Goodheart-Willcox Publisher

Figure 9-36. A basic external gear motor design.

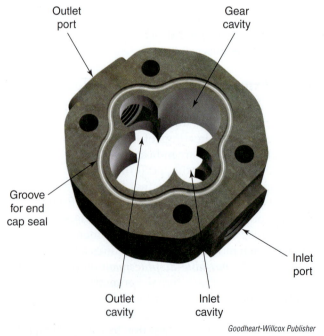

Outlet
port

Gear
cavity

Groove
for end
cap seal

Inlet
port

Outlet
cavity

Inlet
cavity

Goodheart-Willcox Publisher

Figure 9-37. The primary housing cavity for an external gear motor is oval.

surface of the internal teeth. This produces a series of individual chambers that increase and decrease in volume as the unit rotates.

The final design feature of the gerotor motor is the placement of an inlet and an outlet port. These crescent-shaped ports are located in an area of the housing that contacts the side of the rotating-gear elements. The leading edge of the inlet port is located at the point where the volume of each chamber begins to enlarge. The trailing edge of the inlet port is located where the chamber has reached its maximum volume. The leading edge of the outlet port is located where the chamber volume begins to decrease. The trailing edge of the outlet port is located near the point where the chamber volume is the smallest.

When system fluid is directed into the inlet port, the fluid pressure begins to increase to overcome internal motor friction and the external load. Several chambers formed by the clearance between the internal and external gear teeth are exposed to this pressure. Once the pressure increase has generated adequate force to rotate the motor, the system pump supplies the fluid needed to fill the motor chambers as they enlarge during motor rotation. As the motor rotates, the chambers connected to the outlet port decrease in volume. This pushes the spent fluid, which is near atmospheric pressure, into the system lines leading to the reservoir. The motor continues to rotate as long as the system continues to supply fluid and the pressure is sufficiently high to overcome the internal motor resistance and system load.

A variation of the gerotor motor is the *orbiting-gerotor motor*. In this design, the gear with the internal teeth is fixed and does not rotate. The gear with the external teeth rotates. The center point of this gear follows an elliptical path around the center point of the fixed gear element. Instead of crescent-shaped ports, this design has a rotating valve plate that turns at the same speed as the rotating inner gear. The openings in the valve plate allow the appropriate chambers to be pressurized as they are increasing in volume. It also connects the chambers to the reservoir when they are decreasing in size.

A design variation of orbiting-gerotor motors incorporates modifications to both gear elements. See **Figure 9-39**. The gear that normally has internal teeth instead has rollers. The teeth of the other gear are modified to mesh with the rollers. This results in a unit that basically functions the same as the gear tooth design. However, the rollers reduce internal motor friction. This, in turn, increases operating efficiency and reduces wear.

Vane motors

Vane hydraulic motors closely resemble the design of the conceptual motor discussed earlier in this chapter. However, an actual motor has a number of refinements to improve the operation of the unit. In addition, there are several design variations of vane motors that provide different performance characteristics.

The most basic vane motor is a fixed-displacement, pressure-unbalanced design. This configuration consists of three component groups:

- Housing.
- Rotating internal parts.
- Power output shaft.

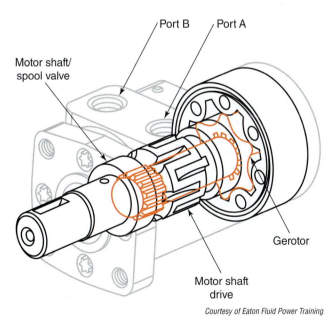

Courtesy of Eaton Fluid Power Training

Figure 9-38. The internal gear motor uses a gerotor design similar to the gerotor pump.

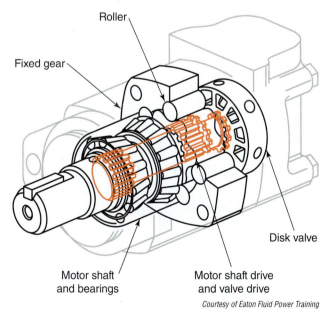

Courtesy of Eaton Fluid Power Training

Figure 9-39. A variation of the gerotor design is the orbiting gerotor. This design provides reduced wear by using rollers.

The housing contains an internal, circular motor chamber to hold the rotating internal parts, bearings to support the power output shaft, inlet and outlet ports, and any fittings needed to plumb and mount the motor. The internal rotating parts consist of a rotor and a number of sliding vanes fitted into machined slots evenly spaced around the outside diameter of the rotor. The tips of these vanes are held in contact with the chamber walls using springs, system pressure, or centrifugal force. They form a series of individual chambers. A spline fitting is typically used to securely mount the rotor to the motor output shaft. The output shaft, rotor, and vanes are located off-center in the housing. Crescent-shaped inlet and outlet ports are located in an area of the housing that contacts the side of the rotor and vanes. Construction details of a basic vane motor are shown in **Figure 9-40**.

When system fluid is directed to the inlet port, pressure begins to increase in the individual chambers exposed to the port. This increase is caused by the attempt of the system to overcome any resistance caused by internal friction in the motor and any external load. The increased system pressure is applied to the surface area of the inlet chamber, including the surfaces of the vanes. However, the pressure in the chambers connected to the outlet port is very close to atmospheric because of the connection to the reservoir line. The pressure difference between the vane surfaces in the inlet and outlet port areas increases until the force generated overcomes the internal resistance and exterior load, forcing the motor to turn. As the motor rotates, each chamber increases in volume during the time it is exposed to the inlet port. Those chambers exposed to the outlet port decrease in volume, forcing system fluid to return to the reservoir.

Variable-Displacement Vane Motor. A variation of the basic vane motor design is a variable-displacement configuration. This variation also has unbalanced loading of the bearings as system pressure increases. The design allows the displacement per revolution of the motor to be easily changed. This feature permits an operator to change motor speed without changing the system flow rate.

The variable-displacement feature involves a movable cam ring. The ring surrounds the rotor and vane assembly. The internal surface of the ring serves as the contact point for the ends of the vanes. The sides of the housing provide the walls of the pumping chambers. The ring can slide within the housing. Moving the center point of the ring in relation to the center point of the rotor changes the chamber sizes. If the center points are concentric, all of the chambers have the same volume. As the center points are offset, the chamber volumes vary. When the points are at maximum offset, the volume is the largest. At this maximum offset point, the motor will operate at the slowest speed with the highest torque potential for a given flow rate. As the center points move toward each other, the speed increases, but the torque capability is reduced.

Balanced-Vane Motor. Another common variation of the vane motor is often simply called the balanced-vane design. This motor has a fixed displacement and a configuration that places a balanced load on the shaft bearings. The design uses a cam ring with a slightly oblong chamber opening. In actuality, the internal shape of the ring consists of two overlapping circles of equal size with centers that are spaced slightly apart. The rotor and vane assembly is positioned in this oblong opening at the midpoint of a line extending between the center points of the overlapping circles. As a result, two identical, crescent-shaped chamber areas are created. Inlet and outlet ports are located in each of these chambers.

During motor operation, system fluid is forced into the chamber through the inlet ports connected to both of the crescent-shaped chambers. Fluid pressure increases to the point where sufficient force is applied to the motor vanes to overcome internal motor and load resistance. Once this pressure is reached, the motor turns. Fluid continues to fill the chambers between the vanes until maximum volume is reached. By then, the chamber has passed over the inlet port and is connected to the outlet. At the outlet, fluid is allowed to return to the system reservoir as the chambers decrease in volumes.

The bearings of this motor design have a balanced load placed on them during unit operation. The reason for this is the placement of the ports. The two

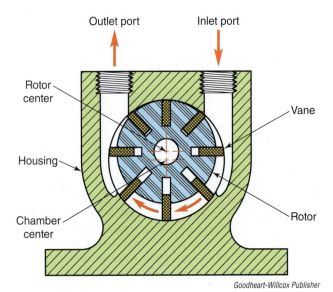

Goodheart-Willcox Publisher

Figure 9-40. The basic vane motor uses a design similar to a vane pump. System fluid entering the inlet port applies pressure to the vanes, which forces the motor shaft to turn.

inlet ports are located 180° from each other. The two outlet ports are also placed opposite each other. The 180° spacing causes the forces that load the bearings to counteract each other.

Piston motors

There are a variety of piston motor designs in use today. However, they can generally be grouped under the two broad design classifications: axial piston and radial piston. Both of these designs use a cylinder barrel to hold the cylinders and pistons. In axial-piston motors, the centerline of the cylinders and pistons is arranged parallel to the centerline of the barrel. In radial-piston motors, the cylinders and pistons radiate out perpendicular to the centerline of the barrel.

All of these designs use multiple pistons and cylinders and some form of connecting mechanism to force rotation of the power output shaft. The motor designs are available as both fixed and variable displacement units. Bearing loads are generally considered unbalanced. However, the variety of mechanical connections used between the cylinder/piston assemblies and output shaft make a specific designation of the bearing loading somewhat unclear.

The next discussion provides a general operating description that can be applied to the various piston motor designs. Details that apply to construction and specific operation for each design are discussed later in this section.

Pistons reciprocating in the cylinders are the source of the force required to rotate the power output shaft. The pistons are closely fit in the cylinder or use an additional sealing device, such as piston rings, to prevent leakage between the pistons and cylinder walls. All of the designs alternately connect the inlet port of the cylinder chambers above the pistons to the pressurized system fluid pump and the outlet port to return to the reservoir at low pressure.

When the cylinder chamber volume is the smallest, the chamber is connected to the inlet port. System fluid pressure increases until the force generated on top of the piston is adequate to overcome internal pump resistance and the load on the motor shaft. When the force is high enough to overcome the resistance, the piston moves. This turns the motor through a mechanical connection to the output shaft. The mechanical connection then returns the piston, reducing the volume of the cylinder chamber. This forces system fluid back to the reservoir through the outlet port under near-atmospheric pressure.

The three most common piston motor designs are inline, bent axis, and radial. Even though the inline and bent-axis motors vary considerably in mechanical appearance, they are classified as axial-piston units. **Axial-Piston Motor. Figure 9-41** illustrates the components and construction of an inline design. In this motor,

the pistons and cylinders are exactly inline with the power output shaft. The major parts of the motor are:

- Cylinder block.
- Pistons.
- Piston shoes and retainers.
- Power output shaft.
- Swash plate.
- Housing.

The cylinder block contains multiple cylinders. The cylinder block is attached to a shaft connected to the power output shaft via splines. The pistons are fitted into the cylinders with their exposed ends attached to shoes that contact the swash plate during motor rotation. The swash plate is stationary in the housing. It provides an angled surface that converts the linear force generated from pressure on the pistons into the torque required to turn the motor. The housing contains all of the parts, as well as the inlet and outlet ports and shaft bearings.

During operation, the inlet port supplies system fluid to a cylinder when the chamber volume is small. When pressure increases, force is applied to the piston. This is transmitted to the shoe resting against the declining angle of the swash plate. When the force is high enough to overcome internal motor resistance and output shaft load, the piston extends. This causes the shoe to slide along the declining angle of the swash plate. The movement causes the cylinder block and power output shaft to rotate. At the same time, the shoes on the pistons exposed to the output port are sliding up the inclining angle of the swash plate. This reduces cylinder chamber volume while moving system fluid back to the reservoir at low pressure.

The Oilgear Company

Figure 9-41. The inline-piston motor is one of the common axial-piston designs. The centerlines of all the pistons are parallel to the centerline of motor output shaft.

Inline piston motors may have a fixed or variable displacement. The swash plate angle determines the piston stroke length. Fixed-displacement motors have a swash plate that cannot be adjusted. However, the swash plate angle can be adjusted on variable-displacement units. The swash plate is mounted on a pivot so the angle can be changed by a hand-operated lever or wheel. Sophisticated electronic systems are also used to provide automatic sensing and swash plate angle adjustment. As the angle of the swash plate is changed, the piston stroke length changes, thus varying the displacement of the motor.

The bent-axis motor design uses a rotating cylinder barrel and piston assembly similar in appearance to the inline motor. However, the centerlines of the power output shaft and the cylinder barrel/piston assembly form an angle. This angle takes the place of a swash plate and allows the motor to operate, **Figure 9-42**.

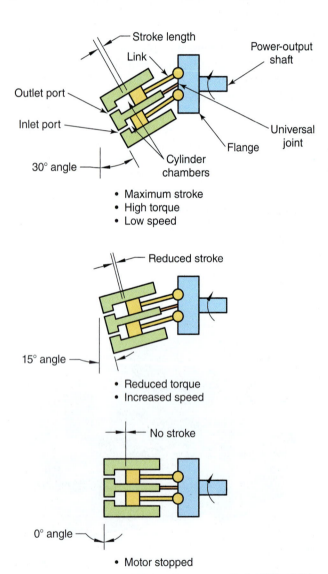

- Maximum stroke
- High torque
- Low speed

- Reduced torque
- Increased speed

- Motor stopped

Figure 9-42. Operation of a bent-axis piston motor.

The internal end of the power output shaft contains a flange fitted with links. A link is connected to the end of each piston in the cylinder barrel. Because of the angle, a *universal joint* is used to connect the power output shaft and the cylinder barrel. During operation, the cylinder barrel and pistons, universal joint linkage, and power output shaft turn as a unit. The angle between the cylinder barrel and output shaft causes the pistons to reciprocate as the motor turns. Inlet and outlet ports are used to connect the cylinder chambers to the system pump and reservoir.

The cylinder chambers aligned with the inlet port have less-than-maximum volume. This is because the pistons have been moved into the cylinder due to the geometry of the motor design. Fluid pressure in these chambers increases in an effort to overcome the resistance to motor rotation as fluid is pumped into the cylinders. As pressure increases, additional force is applied to the pistons, which transmit it to the power output shaft through the piston/flange links and the flange. When the force is high enough to overcome internal motor resistance and output shaft load, the pistons move. This increases the distance between the flange and the cylinder block. This linear piston movement causes the cylinder block and power output shaft to rotate as a unit because of the angle between the centerlines of those components. At the same time, the pistons in the cylinders exposed to the outlet ports are being pushed into the cylinders by the flange and link connections. This returns fluid to the system reservoir.

Bent-axis piston motors may have a fixed or variable displacement. The angle between the center of the cylinder barrel and the power output shaft determines the length of the piston stroke. The larger the angle, the longer the piston stroke and the higher the volume of fluid needed to turn the motor. Fixed-displacement motors have a fixed angle and, thus, a fixed volume. Variable-displacement units typically have a mechanical adjustment of the angle. The cylinder block is mounted on a pivot so the angle can be changed using a simple hand-operated lever or wheel. Sophisticated electronic systems can also be used to provide automatic sensing and angle adjustment.

Radial-Piston Motor. In a radial-piston motor, the pistons and cylinders are perpendicular to the centerline of the power output shaft. The major parts of the motor are:

- Cylinder block.
- Pistons.
- Reaction ring.
- Power output shaft.
- Pintle.
- Housing.

The cylinder block contains multiple cylinders. The cylinder block is attached to the power output shaft. The open end of the cylinders contact the pintle. The pintle contains the inlet and outlet ports. As the cylinder block rotates, the cylinders are alternately exposed to the inlet and outlet ports. The pistons are fitted into the cylinders with their ends extending to contact the reaction ring. The reaction ring is stationary in the housing. It is set off center to the centerline of the power output shaft. This offset configuration provides the angled surface needed to convert the linear force generated from pressure on the pistons into the torque required to turn the motor. See **Figure 9-43**.

The cylinder chambers with less-than-maximum volume are aligned with the inlet ports in the pintle. Fluid pressure in those chambers increases in an effort to overcome the resistance of the motor load. As pressure increases, force is applied to the pistons, which transmit it to the power output shaft. When the force is high enough to overcome the load, the pistons extend. This causes the piston ends to slide along the declining angle of the reaction ring. This movement is what causes the cylinder block and power output shaft to rotate. At the same time, the ends of the pistons exposed to the output port are sliding up the inclining angle of the ring. This reduces the cylinder chamber volume and moves system fluid back to the reservoir.

Another radial-piston motor design uses fixed-position cylinders. The major parts of the motor are a housing on which the cylinders are mounted, an eccentric that is fitted to the power output shaft, pistons, and connecting rods located between the pistons and surface of the eccentric. Fluid-transmission passages or lines connect each cylinder chamber to a rotating spool valve that directs system fluid into and out of the cylinders.

This motor begins the operating cycle when the rotating spool valve connects the cylinder chambers that are increasing in volume to the inlet line coming from the system pump. At this point, the connecting rods of these pistons are resting on the declining slope of the eccentric. When the force generated by the pressure on the pistons is high enough to overcome the resistance of the power output shaft, the pistons move. This forces the ends of the connecting rods down the slope of the eccentric. In turn, this action forces the output shaft to turn. At the same time, the connecting rods of the cylinders connected to the spool valve outlet are being forced up the inclining slope of the eccentric. This movement reduces cylinder chamber volume and pushes fluid back to the reservoir.

Both radial-piston motor designs may have a fixed or variable displacement. The components that determine the displacement are the reaction ring in the rotating-cylinder-block motor and the eccentric in the fixed-cylinder design. If the centerline of these components cannot be changed in relation to the centerline of the power output shaft, the motor has a fixed displacement. Changing the relationship of the centerlines of these components varies the piston stroke. This, in turn, changes the motor displacement. The reaction ring can be shifted with a simple mechanical device, such as a screw fitted with a handwheel. The eccentric is typically hydraulically shifted because of its internal location. Both motor designs, when having a variable displacement, can also use sophisticated electronic controls to provide maximum operating performance.

Other motor designs

Although gear, vane, and piston motors are by far the most common, several other designs will be encountered as you work in the fluid power area. Some of these motors can be considered general purpose, but others are designed for special applications. The screw- and cam-type radial-piston motors warrant a brief discussion.

The *screw motor* provides very quiet, vibration-free operation. The primary components of this motor are three meshed shafts with external screw threads contained in a housing. One of these shafts serves as the power output shaft of the motor, while the other two shafts are idler rotors. The threads of the idlers are meshed with the threads of the output shaft. This forms a series of isolated chambers between the housing and the shafts. An inlet port on one end of the housing allows system fluid to be delivered to the chambers. An outlet port on the other end allows the return of fluid to the reservoir. The pressure difference between the inlet and outlet ports acts on the thread areas of the shafts. This produces the torque that causes the motor output shaft to rotate.

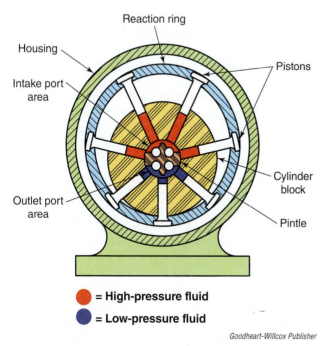

● = High-pressure fluid

● = Low-pressure fluid

Figure 9-43. Operation of a radial-piston motor.

Cam-type radial-piston motors are often used in applications where equipment drive wheels are directly mounted on the power output shaft. **Figure 9-44** shows one of the many designs of this type of motor. The motor consists of:

- Rotor shaft with an offset portion that serves as both a cam and a pintle.
- Cylinder block with pistons.
- Substantial external housing.

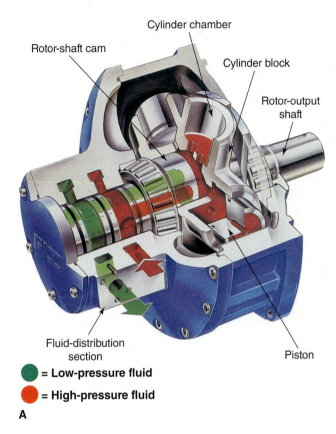

= Low-pressure fluid

= High-pressure fluid

A

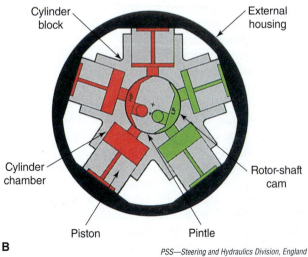

B

PSS—Steering and Hydraulics Division, England

Figure 9-44. Cam-type radial-piston motors are heavy-duty units often used as direct-drive wheel motors on heavy equipment.

Fluid enters the cylinder chambers through distribution lines in the rotor shaft. The lines end in the pintle designed into the cam of the shaft. Pressurized fluid enters the cylinder cavities. This pushes the pistons against flat surfaces machined in the housing. Several cylinders are exposed to high-pressure fluid at all times. This applies thrust to the cam, which results in rotation of the rotor shaft. The output is smooth, high-torque rotation over the entire speed range of the motor.

9.3.3 Basic Motor Circuits

Hydraulic motors are used in a variety of circuits. However, a number of these circuits are considered basic. The basic circuits often encountered when working with hydraulic motors are series and parallel, braking, and replenishment circuits. An understanding of these circuits will assist in understanding how motors can be used to produce efficient and safe hydraulic systems.

Series and parallel connections

Figure 9-45 illustrates motors connected in series. In a *series circuit*, the inlet line of the first motor is connected to the system pump. The inlet lines of any additional motors are connected to the outlet of the preceding motor. The outlet of the last motor in the series returns the fluid to the reservoir. Series circuits require motors that can withstand high outlet port pressures.

In a series circuit, the total system pressure is divided in proportion to the load placed on each motor. The pressure required in the inlet line of *Motor A* to develop the torque to move its load is added to the pressure required by each additional motor to develop the torque they need to turn. As long as the total pressure requirement is below the system relief valve setting, each motor will turn, even though individual motor loads may vary.

If the displacement of each motor is equal, the motors will operate at approximately the same speed, even if the individual loads are different. This is because the outlet of *Motor A* is directly connected to the inlet of *Motor B* with the same fluid passing through both motors. Any speed variations between equal-displacement motors is the result of internal leakage.

Two parallel-motor circuits are shown in **Figure 9-46**. In a *parallel circuit*, the inlet lines of the motors are connected to a common supply line coming from the pump. A simple form of a parallel circuit is shown in **Figure 9-46A**. In this circuit, the size of each motor and their loads must be balanced to provide consistent operating speed. Changing the load on either of the motors during system operation results in relative speed changes. **Figure 9-46B** shows a parallel circuit where pressure-compensated flow control valves are used to divide pump

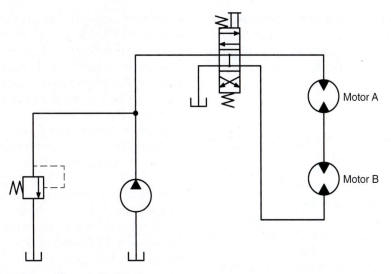

Figure 9-45. A circuit with two hydraulic motors in series.

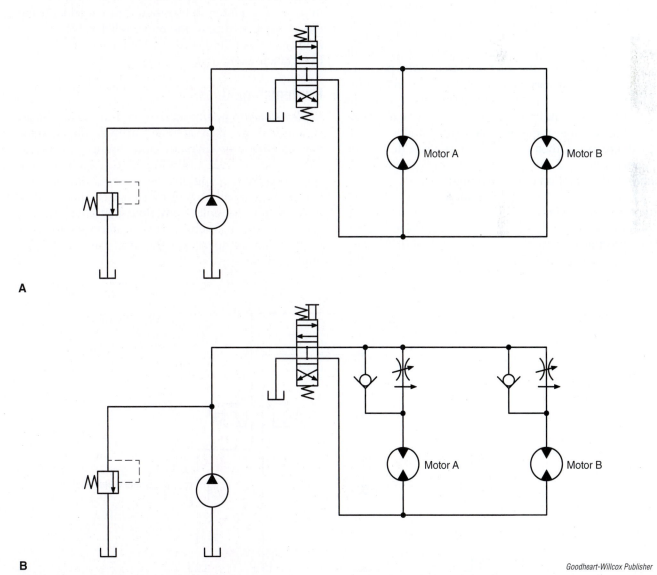

A

B

Figure 9-46. A circuit with two hydraulic motors in parallel. A—A basic parallel circuit. The motors and loads must be balanced. B—This parallel circuit provides adjustment for each motor branch in the circuit.

fluid output. This ensures uniform flow to the motors. A variety of design variations can be used with parallel connections to provide a range of individual flow and pressure settings needed to obtain the performance required from individual motors in a circuit.

Braking circuits

When hydraulic-motor-driven equipment involves a heavy rotating load, the inertia of that load must be considered in designing the system. *Inertia* is the tendency of an object to remain either in motion or at rest. A heavy, rotating object has a lot of inertia. Under these conditions, stopping the motor involves more than placing a directional control valve in the inlet or outlet lines of a hydraulic motor. This is because the inertia of the rotating load will tend to maintain the rotation of the hydraulic motor even after the directional control valve has been shifted to stop fluid flow from the pump. This continued application of rotational force causes the motor to act as a pump. It attempts to draw fluid in through the inlet port and move fluid through the outlet port to other sections of the circuit. The result can be cavitation in the inlet port area, if the pressure drops too low, or very high shock pressures in the outlet area, if the discharge fluid has no escape route. To prevent these conditions, a motor braking circuit can be used to allow a motor to be driven under normal load situations, then coast to a stop when desired or quickly stopped using hydraulic braking when a shorter deceleration time is required.

Figure 9-47 shows a basic motor braking circuit that uses a three-position directional control valve to provide the three operating conditions. Position *1* for the directional control valve is the normal drive position for the motor. Pump flow passes through the

motor and produces rotation until the load causes system pressure to increase to the point where the system relief valve functions. The relief valve then returns fluid to the reservoir and the motor stalls. When the directional control valve is shifted to the center position *2*, both sides of the motor and the pump are connected to the reservoir. This allows the motor and its load to coast to a stop.

However, when the directional control valve is in position *3*, the pump outlet is directly connected to the reservoir. This allows the pump output to be dumped without going through the motor. The outlet of the motor is also connected to the reservoir, but through the brake relief valve. The inlet of the motor remains connected to the pump outlet line. If there is adequate inertia from the load, the motor continues to rotate and begins to act as a pump. This draws fluid in from the pump line and forces it through the brake relief valve to the reservoir. When the directional control valve is in position *3*, motor deceleration is provided by using the pressure setting of the brake relief valve. Also, a ready supply of fluid is ensured in the motor inlet port to eliminate cavitation.

Replenishment

Replenishment circuits replace the fluid lost through system leakage. They are used in hydraulic systems that continuously circulate fluid through the motor and pump without returning it to a large-volume system reservoir. Typically, the reservoir of these systems is very small. A replenishment circuit provides makeup fluid to replace any fluid that is lost during system operation. These circuits are also used to make up the fluid dumped over the relief valve in motor-braking circuits.

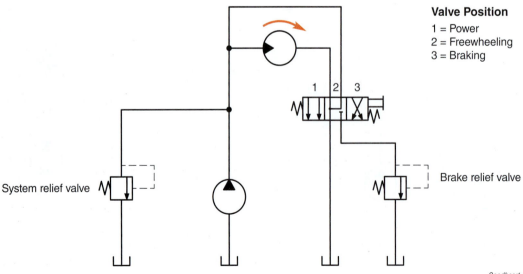

Figure 9-47. A basic braking circuit using a three-position directional control valve.

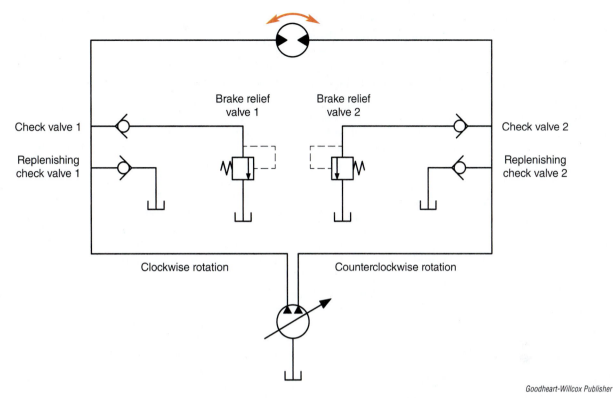

Figure 9-48. A replenishment circuit may be used to make up fluid that is lost during the operation of closed-loop hydraulic motor circuits.

Figure 9-48 illustrates a replenishment circuit. This circuit works in conjunction with a braking system for a bidirectional motor. A variable-displacement pump powers the circuit. The system maintains the volume of the hydraulic fluid in the line functioning as the connection between the low-pressure motor outlet and pump inlet. The relief valves are set to maintain the desired operating pressure in that line. The replenishing check valves allow fluid to be supplied to either side of the system, depending on the position of the variable-displacement system pump. The position of the system pump controls the direction of motor rotation.

9.4 Hydrostatic Drives

Designing a compact drive unit that allows easy and accurate control of output shaft power is a challenge for any machine designer. A relatively simple solution to this challenge has emerged in the form of hydrostatic

| A | *MDMA Equipment—Menomonie* | B | *Used with permission of CNH America LLC* |

Figure 9-49. Hydrostatic drives provide an easy means of transmitting and controlling power in consumer products as well as heavy industrial applications. A—A common lawn tractor with a hydrostatic drive. B—A skid steer with a hydrostatic drive.

drive devices, **Figure 9-49**. A *hydrostatic drive* is simply a drive unit where a pump provides fluid to drive a motor. Like other drive units, these devices are not prime movers. They depend on an electric motor, internal combustion engine, or some other device for power. However, hydrostatic drives efficiently transmit power. They also have configurations that allow them to easily adjust and accurately control output shaft speed, torque, horsepower, and direction of rotation. In addition, when compared to conventional transmissions, these units:

- Have a high power output-to-size ratio.
- May be stalled under full load with no internal damage.
- Accurately maintain speed under varying load conditions.
- Provide an almost infinite number of input/output speed ratios.

9.4.1 Conceptual Circuit Construction and Operation

Hydrostatic drives consist of basic components typical of other hydraulic motor circuits. These include a hydraulic pump that generates fluid flow, a hydraulic motor that produces the rotary motion, lines to transmit fluid between components, valves to control direction and pressure, and a reservoir to store fluid. These components may be arranged using either open-loop or closed-loop circuit configurations.

The *open-loop circuit* configuration has a layout typical of a basic hydraulic motor circuit. See **Figure 9-50**. The pump inlet is connected to the reservoir and the outlet to a directional control valve. The directional control valve is connected to the motor with the fluid returning from the motor directed to the reservoir. A relief valve controlling maximum system pressure is located between the pump and directional control valve. This design allows the motor to be stopped or reversed by shifting the directional control valve. Motor speed depends on the relationship between the displacement of the pump and motor. The reservoir serves to compensate for any fluid lost from system leakage.

The *closed-loop circuit* configuration uses a layout where the outlet of the pump is directly connected to the inlet of the motor. The outlet of the motor is plumbed to the inlet of the pump. The only control valve used in this configuration is a relief valve that sets the maximum pressure that may be developed in the system. A small replenishment pump and reservoir are used to make up any fluid lost during system operation. As a result, these systems are much more compact than the open circuit design. See **Figure 9-51**.

When the pump rotates in the closed circuit design, the system fluid is moved directly to the inlet of the motor. As the motor rotates, the fluid passing through the unit is returned directly to the inlet of the pump. The speed of the motor is established by the relationship between the displacement of the pump and motor. The replenishment pump replaces any fluid that leaks from the lines or other components.

9.4.2 Pump-Motor Arrangements

All hydrostatic drives involve a hydraulic pump and hydraulic motor that work together to transmit and control prime mover power. The pumps and motors may be either fixed or variable displacement. There are four possible pump/motor, fixed/variable combinations. The following discussion is based on controlling maximum circuit-operating pressure using a relief valve with fixed-displacement pumps and pressure compensation with variable-displacement pumps. The discussion also assumes the prime mover is running at a constant speed.

In the first pump/motor arrangement, both components have fixed displacements. This combination

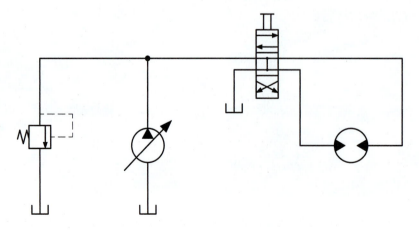

Figure 9-50. A circuit configuration typical of a basic, open-circuit hydrostatic drive.

results in a drive unit where the maximum horsepower, torque, and output shaft speed are fixed. These cannot be adjusted.

In the second pump/motor arrangement, the pump has a fixed displacement and the displacement of the motor is variable. A drive unit using this combination has a fixed maximum horsepower output, but torque and speed are variable. Output shaft rotation may be reversed if the variable-displacement motor is reversible. The need for a relief valve to control maximum system pressure lowers the efficiency of any drive unit using a fixed-displacement pump.

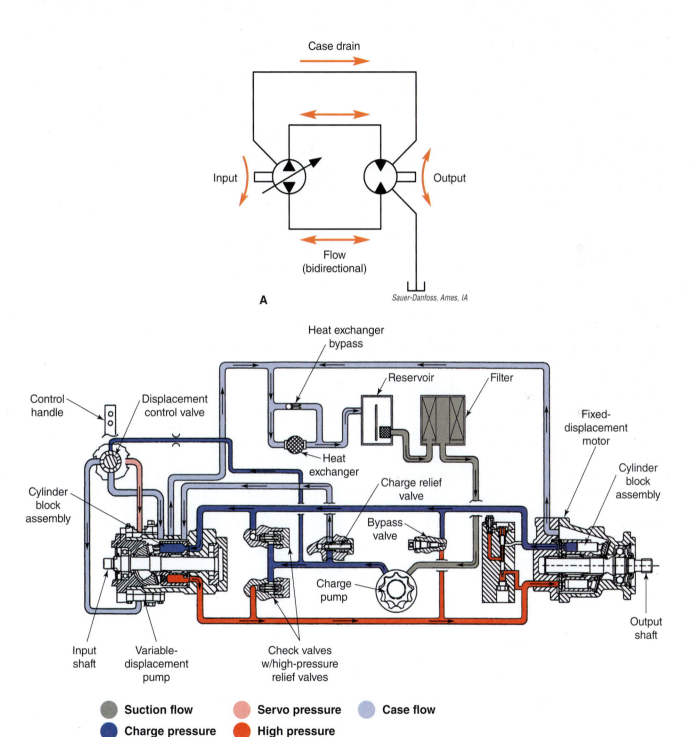

Figure 9-51. A circuit configuration typical of a basic, closed-circuit hydrostatic drive. A—Schematic with symbols. B—Pictorial schematic.

Used with permission of CNH America LLC

The third pump/motor arrangement has a drive unit with a variable-displacement pump and a fixed-displacement motor. This arrangement produces a hydrostatic drive having variable horsepower and speed capability. However, the torque output is constant. Output shaft rotation may be reversed if the design of the pump allows the flow to be reversed.

In the fourth pump/motor arrangement, the pump and motor have variable displacements. In this drive, the horsepower, torque, and output shaft speed and direction can be changed. This arrangement is the most versatile of the pump/motor combinations. It allows a wide selection of torque- and speed-to-power ratios.

9.4.3 Hydrostatic Transmissions

A hydrostatic drive is typically considered a *hydrostatic transmission* when the pump, motor, or both have a variable displacement. Without the variable feature, the drives can be compared to a basic gear train that simply transmits power from the prime mover to the load. With the variable feature, hydrostatic transmissions can be manually or automatically adjusted to control torque, speed, and power.

Two general types of construction are used with hydrostatic transmissions: integral and nonintegral. Integral construction combines all of the transmission parts into a single housing. This includes the pump, motor, fluid lines, valving, and accessories. In addition, the housing may contain axles and mountings for the prime mover and machine parts. This type construction is generally used in transmissions in lower-horsepower devices. In nonintegral construction, the pump is attached to the prime mover and the motor is attached to the workload. Hoses or tubing assemblies connect the pump, motor, and accessories.

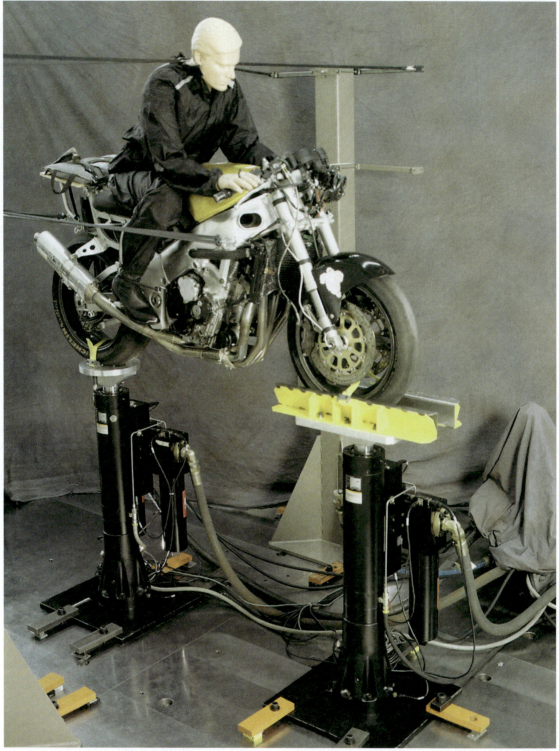

(MTS Systems Corporation)

This is a testing apparatus that makes use of fluid power. The two actuators are used to simulate changes in the road as the motorcycle is operated. Make note of the various types of lines and connections used in this setup.

Summary

- Actuators are the components used in a hydraulic system to convert fluid pressure into mechanical force and movement.

- Cylinders are linear actuators that convert fluid flow into linear mechanical motion. Cylinders consist of a piston and piston rod assembly moving within a close fitting cylindrical tube called a barrel. Forcing system fluid into the sealed volume of the barrel drives the piston and rod assembly to move, expanding the chamber volume and performing the work of the system.

- Cylinders are typically classified by operating principle or by construction type.

- Cylinder operation may be classified as single acting or double acting. Single-acting cylinders exert force on either extension or retraction and require an outside force to complete the second motion. Double-acting cylinders generate force during both extension and retraction. A directional control valve is used to alternately direct fluid to opposite sides of the cylinder to produce movement.

- Cylinder classification by construction type depends primarily on the method used to attach the cylinder head and cap to the barrel. Designations include tie-rod, mill, threaded-end, or one-piece construction.

- A variety of mounting configurations are used to attach the cylinder body and rod end to machinery. Mountings are classified as fixed centerline, fixed noncenterline, and pivoting centerline. Expected cylinder loading is the major factor in the selection of the mounting style.

- Hydraulic cylinder manufacturers provide detailed specifications concerning the construction, physical size, and load capacity of their products. This information includes basic factors such as bore, stroke, and pressure rating, as well as service rating, rod-end configurations, and dimensioned drawings of cylinder types.

- Hydraulic motors are rotary actuators that convert fluid pressure and flow into torque and rotational movement.

- All basic hydraulic motors consist of a housing, rotating internal parts, and a power output shaft. System fluid enters the housing, applies pressure to the rotating internal parts, which move the power output shaft applying torque to rotate a load.

- The displacement of a hydraulic motor indicates the volume of fluid needed to turn the output shaft one revolution. Motors can have fixed- or variable-displacement depending on whether the internal geometry of the motor chambers can be changed.

- Hydraulic motors may be classified as unbalanced or balanced depending on the type of load applied to the bearings of the output shaft. If side loading of the shaft bearings has been eliminated, the motor is classified as balanced.

- Most hydraulic motors are constructed using gears, vanes, or pistons as the primary internal components producing the rotating motion.

- Hydraulic motors may be incorporated into circuits using series or parallel connections. In series connections, the total system pressure is divided in proportion to the load placed on each motor. In parallel connections, care must be taken to balance the size of each motor and its load to provide constant operating speed.

- Hydrostatic drive systems use combinations of fixed- and variable-displacement pump/motor arrangements to provide effective transmission of power and the easy adjustment and control of output shaft speed, torque, horsepower, and direction of rotation. Hydrostatic drives are typically considered hydrostatic transmissions when both the pump and motor of the unit are variable displacement.

Internet Resources

The following are some useful resources available on the Internet. Enter a company or organization name into a search engine to access its website. Explore the various areas of the sites to discover useful fluid power resources.

Eaton Corporation. Product literature library on light-duty hydrostatic transmissions designed for applications such as lawn tractors. Includes general descriptions, features, benefits, and performance data.

Helac Corporation. Provides details, specifications, and applications for hydraulic rotary actuators manufactured by the company.

Hydraulics & Pneumatics Magazine. Provides information on a number of circuits basic to effective system operation. Actuator braking, replenishment, and series/parallel connections are covered in detail.

Industrial Quick Search Manufacturer Directory. Provides a variety of information on hydraulic motors. Includes links to other manufacturer sites that feature small motors; large, industrial motors; and mobile units.

Parker Hannifin Corporation. Catalog of hydraulic rotary actuators manufactured by the company. Provides product information, applications, and a selection guide.

Peninsular Cylinder Company. Homepage of a fluid power cylinder manufacturer. Explore the hydraulic cylinder section of the site, which provides details of cylinder features.

Chapter Review

Answer the following questions using information in this chapter.

1. The system component that converts fluid pressure and flow into linear force and movement is called a _____.
   A. cylinder
   B. vane
   C. trunnion
   D. rack and pinion

2. The tube portion of a fluid power cylinder is called the _____.
   A. actuator
   B. rotor
   C. barrel
   D. vane

3. The component used to prevent contaminants from entering a cylinder as the rod is drawn back into the cylinder during the retraction stroke is called a _____.
   A. static seal
   B. flange
   C. cleaner
   D. wiper seal

4. An actuator that can exert force during both extension and retraction strokes is called a(n) _____ cylinder.
   A. dual service
   B. double-acting
   C. extension-retraction
   D. pivoting centerline

5. A cylinder that has the head and cap pieces secured on the barrel using external rods is called a _____ cylinder.
   A. double-mounted
   B. tie-rod
   C. trunnion
   D. twin-mounted

6. The _____ is the same diameter as the bore of the actuator barrel in a hydraulic ram.

7. Which of the following is a basic configuration used to mount cylinders to equipment?
   A. Fixed centerline.
   B. Fixed noncenterline.
   C. Pivoting centerline.
   D. All of the above.

8. *True or False?* During retraction, the effective area of the piston of a double-acting cylinder is the cross-sectional area of the piston plus the cross-sectional area of the piston rod.

9. *True or False?* The speed at which a cylinder rod extends or retracts depends on the volume displaced per inch of piston travel and the amount of fluid entering the cylinder.

10. Which of the following is a common design of limited-rotation actuators?
    A. Reciprocating plunger.
    B. Vane.
    C. Rotating cylinder.
    D. All of the above.

11. The three conceptual component groups that make up any hydraulic motor are _____.
    A. rotor, vanes, and eccentric
    B. housing, rotating internal parts, and power output shaft
    C. housing, reciprocating internal parts, and power input shaft
    D. rotating internal parts, power input shaft, and power output shaft

12. When the internal geometry of a hydraulic motor can be changed to vary output shaft speed, it is classified as a _____.
    A. variable-output motor
    B. variable-speed motor
    C. variable-displacement motor
    D. variable-volume motor

13. What are the three most common, basic designs of hydraulic motors?
    A. Gear, vane, and piston.
    B. Piston, gear, and cam-type.
    C. Parallel, nonparallel, and vane.
    D. All of the above.

14. Some external-gear motor designs control leakage around the ends of the gears by the use of spring-loaded _____.
    A. swash plates
    B. reaction rings
    C. rotating valve plates
    D. wear plates

15. *True or False?* The stationary-gerotor design of an internal-gear motor uses a fixed gear with internal teeth and a rotating valve plate.

16. To vary the displacement of a variable-displacement vane motor, a movable _____ is used to change the size of the pumping chambers.

17. *True or False?* The balanced-vane motor configuration places a balanced load on the bearings of the motor.

18. Which of the following is a broad design classification of piston motors?
    A. Radial piston.
    B. Vane.
    C. Internal gear.
    D. All of the above.

19. In the inline piston motor, the _____ angle determines the piston stroke length.
    A. wear plate
    B. swash plate
    C. bearing
    D. reaction ring

20. *True or False?* In the rotating-cylinder block design of a radial-piston motor, an eccentric is used to convert the reciprocating motion of the pistons to the rotary motion of the power output shaft.

21. *True or False?* Cam-type radial-piston motors are often used in applications where equipment drive wheels are directly mounted on the power output shaft of the motor.

22. *True or False?* In a nonparallel circuit, the sizes of the motors and their loads must be balanced to provide consistent operation of the motors.

23. The tendency of an object to remain either in motion or at rest is known as _____.

24. *True or False?* The open-loop type of hydrostatic drive uses a configuration in which the outlet of the pump is directly connected to the inlet of the motor and the outlet of the motor is directly connected to the inlet of the pump.

25. Which of the following is a possible pump/motor arrangement that may be used with a hydrostatic system?
    A. The pump has a fixed displacement and the motor a variable displacement.
    B. The pump has a variable displacement and the motor a fixed displacement.
    C. Both pump and motor have variable displacement.
    D. All of the above.

Apply and Analyze

1. A hydraulic system must move a load linearly both forward and back; the speed and output force must be the same in each direction. What kind of actuator should be installed in the system? Why?

2. List three specific examples in which a fixed-centerline mounted cylinder would be the best design choice. Explain why this mounting design is the best choice.

3. A 10″ diameter ram can operate up to 2,000 psi.
    A. What is the maximum output force generated during extension?
    B. If 15 gpm is delivered by the system pump, how fast will the ram extend?

4. A 500 lb load must be moved by a double-acting cylinder.
    A. What pressure is needed to retract the load if a cylinder with 6″ diameter piston and a 3″ diameter rod is used?
    B. How does the system pressure change if the same cylinder is used during extension?
    C. If the load must be extended at a rate of 2 in/sec, what pump flow rate is required?

5. When choosing a limited-rotation actuator for use in a machine tooling application, what specific information about the application is needed to make a choice between the three most common designs? List three criteria and explain how each would help make the decision.

6. A simple hydraulic circuit consists of a loaded gear motor driven by fluid supplied by a pump.
   A. What changes in system operation will occur if the load on the motor is doubled?
   B. How would replacing the gear motor with a variable-displacement vane motor affect the capacity and operation of the system?
   C. If a second motor with the same displacement is connected to the outlet of the first motor, how will system operation be affected?

Research and Development

1. Choose one example of a hydraulic cylinder currently available for purchase. Prepare a poster that shows the basic cylinder design, describes cylinder components, lists operating capacity, and explains the applications for which it is best suited.

2. Log splitters use hydraulic cylinders to apply the force required to split logs. Prepare a short report that compares the hydraulic cylinders used in three different log splitters. Include an explanation of how the choice of cylinder affects the capacity, operation, and cost of the log splitter.

3. Investigate one type of dynamic hydraulic seal used in a cylinder. Prepare a fact sheet that describes the composition, material properties, seal design, manufacturing process, and the applications in which cylinders with this seal are used.

4. Design and conduct an experiment that illustrates how vane size or gear pitch diameter affects torque created by an actuator.

5. Create a chart that compares the design and performance profiles of the different types of hydraulic motors.

6. Find an example of equipment that uses a hydrostatic drive. Investigate the operating characteristics and capacities of the pumps and motors used and prepare a report on how the chosen pump/motor arrangement affects the operation of the equipment.

10 Controlling the System
Pressure, Direction, and Flow

The operation of a hydraulic system requires the control of fluid contained in the system. This control is provided by a number of specialized valves located between the pump and actuators of the system. These valves provide control of fluid pressure, flow direction, and flow volume. A wide variety of valve designs are available to provide desired control variations in each of the areas. Understanding the primary function of these valve types allows the system users the opportunity to obtain maximum performance from the system. This chapter provides in-depth information on basic valve design and operation in each of the control areas.

Learning Objectives

After completing this chapter, you will be able to:

- Explain the function of each of the three general types of control valves used in hydraulic systems.
- Compare the design and operation of direct-operated and balancing-piston pressure control valves.
- Describe the function of the various types of pressure control valves used in hydraulic systems.
- Compare the design and operation of shutoff, check, three-way, and four-way directional control valves.

- Describe the characteristics of the various spool configurations used in three-position directional control valves.
- List and compare the methods used to position control spools in valves.
- Compare the design and operation of noncompensated and compensated flow control valves.
- Explain the effect fluid temperature and pressure variations have on the operation of flow control valves.

Key Terms

balancing-piston valve	four-way valve	pressure control valve
ball valve	full-flow pressure	pressure override
brake valve	gate valve	pressure-reducing valve
bypass flow control valve	globe valve	pump unloading control
check valve	hydraulic pressure fuse	relief valve
compound relief valve	internal force	restriction check valve
control element	land	restrictor flow control valve
control valve	needle valve	safety valve
counterbalance valve	normally closed	sequence valve
cracking pressure	normally open	shutoff valve
detent	normal operating position	spool
directional control valve	orifice	spool valve
direct-operated valve	pilot pressure	temperature compensation
external force	platen	three-way valve
fixed orifice	poppet valve	valve body
flow control valve	pressure compensation	

10.1 Primary Control Functions in a Hydraulic System

Control valves allow hydraulic systems to produce the type of motion and level of force required to complete the functions expected during operation. Control components are designed to produce desired system operation by sensing and controlling system pressure, directing fluid flow, and controlling fluid flow rate. Each of these three control factors requires valves especially designed to produce the type of system operation desired by the equipment manufacturer, **Figure 10-1**. Pressure control valves, directional control valves, and flow control valves all fall into the broad category of control valves.

The primary function of *pressure control valves* in a hydraulic system is to protect the overall system from damage from excessive pressure buildup by limiting maximum operating pressure, **Figure 10-2**. Pressure control valves are also used to prevent the initial movement of actuators by blocking fluid flow into a portion of the circuit until a predetermined pressure level is reached. Another function of these valves is to limit the pressure in a selected section of the system to a level that is lower than the maximum operating pressure of the total system.

Directional control valves direct the flow of fluid in the conductors of the circuit to establish the movement of system actuators, **Figure 10-3**. These valves include a number of configurations that control the direction of fluid to the ports of the various actuators, allowing cylinders to extend and retract and motors to rotate clockwise or counterclockwise. Directional control valves may also limit fluid flow through a conductor to only one direction. In addition, another variation of directional control valves allows automatic switching between two fluid sources when one source fails.

Figure 10-1. A complex hydraulic system uses a variety of valves to produce desired actuator direction, speed, and force. In this image, hand-operated, multi-positional directional control valves mobilize the cylinder controlling the movement of the equipment.

Goodheart-Willcox Publisher

Figure 10-2. One of the primary purposes of pressure control valves is to limit the maximum pressure in a system.

Used with permission of CNH America LLC

Figure 10-3. Directional control valves are very evident to equipment operators. These valves start and stop actuators and control their direction. At various points in an operation sequence, directional control valves can become flow control, speed control, or even shutoff mechanisms.

The fluid flow rate in the lines leading to an actuator controls the operating speed of that cylinder or motor. Flow rate in a hydraulic circuit can be varied by using a variable-displacement pump. However, *flow control valves* must be used to provide varying flow rates with a fixed-displacement pump or when several actuators operate at different speeds, **Figure 10-4.** The amount of fluid passing through any flow control valve depends on the pressure drop across the controlling orifice of that valve. In a system where this pressure drop does not vary, a noncompensated flow control valve can be used. However, in systems where loads vary, the result is a variation in pressure drop across the orifice. This causes variations in fluid flow and actuator speed. To eliminate these variations, a pressure-compensating device must be added to the valve. This device automatically maintains a constant pressure drop across the orifice, producing a constant flow rate at any given valve setting.

These brief descriptions suggest the complexity of providing the fluid control necessary to obtain the desired operation of a hydraulic system. The remainder of this chapter provides a foundation in the design and operation of control valves. Knowing the function and operation of each of the control components is critical. A solid foundation relating to control components and their placement in a circuit allows a fluid power professional to improve the design and operation of those components, the related circuits, and the machines in which the components operate.

Flow control valve

Shutoff valve

Pump

Goodheart-Willcox Publisher

Figure 10-4. Flow control valves are not usually evident to machine operators. These valves are not adjusted during each machine cycle.

10.2 Basic Structure and Features of Control Valves

The internal design and the external appearance of control valves varies considerably, depending on the manufacturer. Even though the descriptions of the valves are very similar in the advertising literature, the features and performance of the valves can vary quite a bit. However, the valves incorporate many common structural component parts and operating principles. A variety of these factors are discussed in this section.

10.2.1 Valve Body Construction and Function

The *valve body* serves as the holder of all the working elements of the valve. The body may be a special casting with a unique exterior shape or simply standard, round or rectangular metal stock cut to the proper length. See **Figure 10-5**. During valve manufacture, the internal fluid passageways and bores are machined into the raw body. These passageways are needed for the springs and internal moving control parts. Manufacture also includes machining inlet, outlet, and control port fittings. These fittings allow the valve to be easily assembled into the hydraulic circuit. The body also typically includes fittings for mounting the valve on the equipment using the hydraulic system.

10.2.2 Fluid Control Elements

All of the various control valves operate by regulating system fluid movement in some form. Therefore, all of these valves must have *control elements* designed to allow, direct, meter, or stop the flow of fluid. In hydraulic valves, these elements typically take the form of a fixed orifice, needle valve, spool or piston,

Sauer-Danfoss, Ames, IA

Figure 10-5. The body of a control valve may be manufactured using a casting especially made for the component, as shown here, or it may be machined from standard metal bar stock.

poppet valve, or sliding plate. These elements may be used as stand-alone units or in various combinations to obtain the performance desired from the valve.

A *fixed orifice* is a small hole machined into the valve body or a piece of metal stock inserted into the valve body. The hole serves to control fluid flow rate through a passageway. See **Figure 10-6**. It is fixed because the size of the hole cannot be changed. When the pressure drop across the fixed orifice is constant, the flow rate is constant. However, varying the pressure drop causes the flow rate to change. Fixed orifices can be used as basic flow control valves or as part of the control mechanism of complex flow or pressure control valves.

A *needle valve* is basically a fixed orifice with the end of an adjustable screw aligned with the center of the orifice. A taper is machined on the end of the screw allowing the cross-sectional area of the orifice to be varied by turning the screw in or out of the orifice. Needle valves are used as simple, adjustable flow control valves. They are also used as a means of adjusting the control mechanisms of a variety of other complex control valves.

A *spool* is a cylindrical metal piece precision-fitted into a bore machined into the valve body, **Figure 10-7**. The spool has grooves machined into its surfaces in various configurations. The clearance provided by these grooves between the spool and bore allows system fluid to move though the valve. Sliding the spool back and forth in the bore aligns the grooves with fluid passageways in the valve body. This directs the movement of fluid through the valve. Spools are commonly used in directional control valves to route fluid to system actuators. However, spool variations can be found

in other valve types, such as pressure reducing, flow divider, and servo valves.

A variety of piston designs are used in control valves. Like spools, pistons operate in bores in which the pistons are precision-fitted. Unlike spools, however, pistons are generally used to operate other valve parts in order to obtain the valve performance desired. They are commonly used to sense pressure and operate other valve elements or control fluid flow. For example, pistons can be used to sense pressure differences in adjoining valve chambers. When a predetermined pressure difference is reached, the unequal forces exerted on the opposite sides of the piston cause

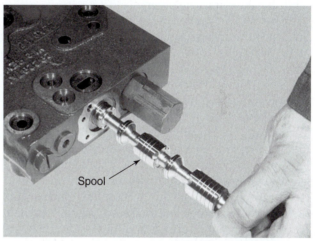

Used with permission of CNH America LLC

Figure 10-7. A spool is a cylindrical metal piece precision-fitted in a bore machined into a valve body. They are commonly used to control fluid flow direction in directional control valves.

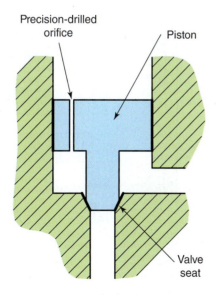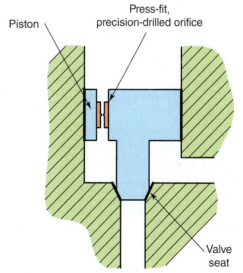

Goodheart-Willcox Publisher

Figure 10-6. In many situations, a fixed orifice is used to control fluid flow in hydraulic valves. Orifices may be a precision hole drilled through the casting of the valve body to connect two passageways or separate parts threaded or press fit into the casting.

it to move. This movement operates other valve components to allow or stop fluid flow or other action.

Poppet valves similar to the designs in **Figure 10-8** may be found in hydraulic systems. Although there are many variations, they contain a valve seat located in the body and a valve face on a movable *poppet*. The poppet may have the appearance of a piston precision-fitted in a bore with the seat machined into a surface on the valve body, as shown in **Figure 10-8A**. Forcing the face and seat together creates a seal that blocks fluid flow. Moving the face and seat apart allows fluid flow through the valve. These valves may also be constructed similar to the valves found in internal combustion engines, as shown in **Figure 10-8B**. This style is commonly used to allow fluid flow in one direction while blocking it in the other.

The sliding plate design uses two precision-machined, flat plates held in close contact with each other. See **Figure 10-9**. A series of channels and ports are machined into the surfaces of the plates. Moving the plates in relation to each other exposes the channels and ports in various combinations. This allows the hydraulic fluid to be directed in different flow patterns. The plates are typically held together with a center bolt that allows the plates to slide over each other as they are rotated in relation to each other. This design is not common, but it is used in one type of directional control valve.

Further details about the function and characteristics of the various elements are discussed later in this chapter. It is important to understand these valve elements. The operation of the various control valves will be easier to understand if you are familiar with these concepts.

Goodheart-Willcox Publisher

Figure 10-9. Sliding plate valves use precision-machined surfaces on two plates that oppose each other. Channels in the plates direct fluid flow through the valve.

10.2.3 Positioning of Control Valve Internal Elements

When a hydraulic control valve is operating in a system, the internal valve elements produce the desired valve action. Internal and external forces are used to position these valve elements for the desired operation. *Internal force* is generated by internal valve components or by fluid pressure that exists in the internal passageways of the valve. *External force* is applied to the elements from sources external to the valve. Several different techniques are used to produce these forces.

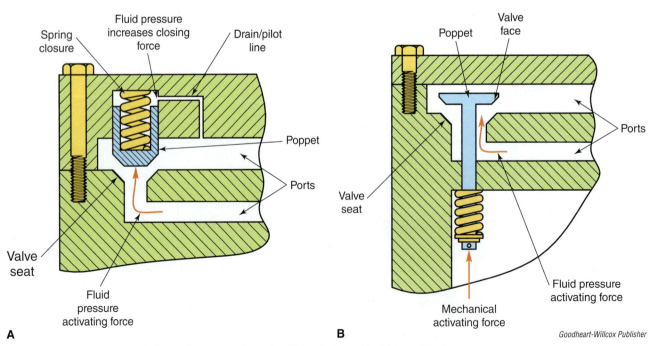

Figure 10-8. There are variations of poppet valves. A—Piston in bore. B—Valve with stem.

Goodheart-Willcox Publisher

Internal forces

Springs and pilot pressure typically provide the internal forces used to position the valve elements. Springs of various weights are often used in conjunction with valve components such as spools, pistons, or poppets. The springs are fitted into the bore that holds these components or in a separate chamber located at the end of the bore. Springs typically hold these valve elements in what is considered the *normal* position of the valve. This is usually the open or closed position. However, certain valves may have another position as the normal position.

Pilot pressure may be thought of as the pressure used to produce the force needed to move the internal elements of valves to obtain a desired valve operation. System fluid pressure that exists in the internal passageways of a valve is often used as pilot pressure to shift internal elements of the valve. For example, the fluid pressure in the valve interior acts on the end surface of a piston or poppet to generate a force. The piston or poppet will not move until the force is high enough to overcome the resistance that is holding it in place. This holding resistance can be the result of friction, spring action, or pressure acting on another surface of the component. Pressure and flow control valves often use internal pilot pressures to produce their operating characteristics.

External forces

Manual force, pilot pressure, and electromagnets are external forces typically used to position internal elements of a control valve. Manual force is applied to the valve elements by a machine operator physically moving a handle or turning a threaded adjustment screw. Manual force can also be applied by levers or ramps attached to machine members that automatically position a valve element as the machine moves through an operating cycle.

The source of pilot pressure for operating control valves can be external as well as internal. The basic operating principles of internally operated pilots also applies to externally operated pilots. The primary difference is the location of the pressure sensing. When a valve uses external pilot pressure, it senses pressure at a remote location in the circuit. In most situations, the pressure used is from a location beyond a second valve in the circuit. This, of course, requires the use of a separate pilot-pressure line extending from the valve to the sensing point.

Electromagnetic force may be used to operate internal elements in a control valve. Electricity is used to generate the electromagnetic force. This may involve a simple on/off switch that a machine operator controls. The switch is used to apply electrical power to a solenoid that shifts the spool in a directional control valve.

Electrical control can range from this very simple arrangement to servo valves that externally sense position or velocity of a machine component and position control valve elements to maintain desired operating characteristics.

10.2.4 Draining Internal Leakage

Gaskets, O-rings, and compression packings are commonly used to prevent external leakage in control valves. However, in most hydraulic control valves, a precision fit is used to prevent excessive internal leakage. The fluid leakage that occurs around the internal parts is caused by the clearance between the parts and the pressure differences between the various chambers. The limited leakage that occurs between the bore and a piston or spool, for example, is desirable as it serves to lubricate the parts.

However, internal leakage can cause a valve to malfunction. This occurs when the chamber into which the leakage passes is not properly drained. If the fluid cannot escape from the chamber, there is the possibility of fluid buildup. This will cause an increase in pressure, which can result in the valve elements not properly functioning. Valve failure does not usually occur. When it does, the valve will not properly operate. It may not respond when an operator attempts to adjust a pressure or flow setting.

Valves may be internally or externally drained to remove the fluid that results from internal leakage. See **Figure 10-10**. Internal drains connect the valve chambers that need to be drained to the normal outlet of the valve. Valves with outlet or return lines that are *never* subjected to system operating pressure can be internally drained. External drains connect the valve chambers that need to be drained to an external line connected to a low-back-pressure return line leading to the system reservoir. If a valve does not have a separate return line and its outlet line is subjected to system operating pressure, it must be externally drained.

10.2.5 Identifying Normal Valve Position

Control valves are often classified by their *normal operating position*. This refers to the position the internal elements assume when the system is shut down. Although it is difficult to classify all control valves in relation to their normal operating position, the classification can be useful when analyzing the operation of a hydraulic component or circuit.

Typically, pressure control valves are either fully open or fully closed when a hydraulic system is shut down. A *normally open* valve shifts to a fully open position when the system is shut down. A *normally closed*

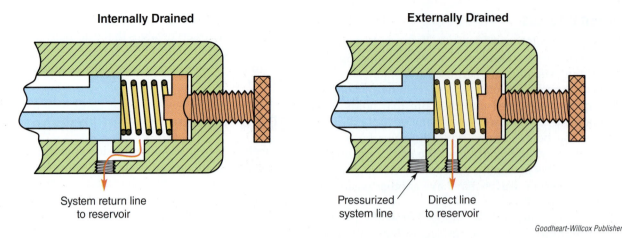

Internally Drained

Externally Drained

System return line
to reservoir

Pressurized
system line

Direct line
to reservoir

Figure 10-10. Hydraulic fluid leakage around the internal moving parts of valves serves to lubricate valve parts. This fluid must be returned to the reservoir to eliminate back pressure that will cause a valve to malfunction.

valve shifts to a fully closed position when the system is shut down. See **Figure 10-11**. When the system is operating, the internal pressure control elements shift to provide the desired system performance. Depending on the valve setting and the load on the system, the position of the internal elements may range from fully open to fully closed during system operation. A valve's normal position classification relates only to the position when the valve is not subjected to system load.

It may be more difficult to classify the normal position of directional control valves. A primary reason for this is the fact that directional control valves may be simple on/off valves or may be more complex valves having two, three, or more positions. See **Figure 10-12**. Another complicating element is the valves may hold a position until they are physically shifted, even though the system has been shut down. Being aware of the variations of directional control valves and how they react when shifted or when the system is shut down can be helpful when analyzing system operation.

The normal position of flow control valves is also not as easily identified as pressure control valves. Flow control valves range from manually operated needle

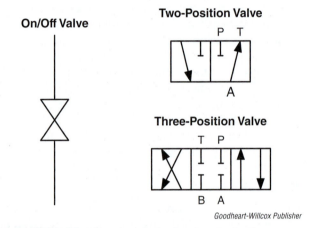

On/Off Valve

Two-Position Valve

Three-Position Valve

Figure 10-12. Directional control valves are typically classified by the number of internal operating positions in which the valve may function. Normally, these are identified by graphic symbols as on/off, two position, and three position. More than three positions are possible.

valves to complex pressure-compensated valves. However, in general terms, these valves may be considered normally open even though the size of the opening varies in some designs. Manually operated valves are set by a machine operator. The opening remains fixed when cycled through a variety of loads during operation or when the system is turned off. Pressure-compensated valves also remain open, although the opening varies in size to provide a uniform flow rate as the load on the system changes.

10.3 Valve Operation and Springs, Fluid Pressure, and Fluid Flow

Springs, pressure, and flow are fundamental to the operation of hydraulic system control valves. Pressure acting against the surface of internal valve components

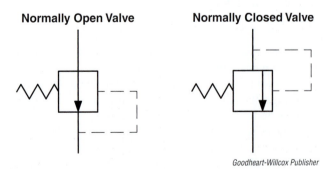

Normally Open Valve

Normally Closed Valve

Figure 10-11. Pressure control valves are typically classified by the position the internal parts of the valve assume when the system is not operating. These positions are identified as normally open or normally closed.

produces the force needed to compress a spring to open, close, or adjust many of these valves. Likewise, basic fluid-flow concepts are used to produce conditions that allow a valve to function within design specifications. Awareness of spring, pressure, and flow concepts is necessary to fully understand the operation of valves and, in some cases, their accessories.

10.3.1 Springs and Their Use in Valves

Springs can take many different forms in hydraulic valves. They can be constructed using leaf, coil, or spiral forms and operate through expansion, compression, or torsion. They are used in hydraulic control valves to perform a number of functions, **Figure 10-13**. The three basic functions are:

- Moving the spool of a directional control valve back to its normal position after an operator has released the shifting force.
- Establishing the maximum pressure at which a pressure control valve operates.
- Serving as a biasing device in pressure-compensated flow control valves to maintain a constant pressure drop across a metering orifice.

One basic spring characteristic that influences the operation of hydraulic valves is the change in tension as a spring is activated. As a spring is compressed, stretched, or twisted, the tension increase per unit of movement is not equal. The most uniform tension change occurs in the middle movement range of the spring. Manufacturers of hydraulic components must take this concept into consideration to produce valves that maintain the most consistent pressure, flow, or movement control.

10.3.2 Fluid Pressure Used as a Valve Operator

Fluid pressure acts on internal surfaces in actuators to generate a force. This is generally understood to be the function of hydraulic cylinders and motors. Pressure is also used in several additional ways in a hydraulic circuit. One area where pressure is critical is the operation and control of system valves. Pressure is used to perform a number of tasks, including these four basic tasks:

- Opening a control valve to limit the maximum pressure in a system.
- Closing a control valve to provide a reduced pressure in a small part of a system.
- Remotely operating a control valve to promote effective operation or increased operator safety.
- Operating a compensating device to provide a source of uniform fluid flow.

A normally closed pressure control valve can serve as an example of how pressure is used in valve operation. In practice, these valves do not remain closed until the selected operating pressure is reached and then suddenly open to prevent excessive pressure. Rather, they gradually open due to the characteristics of a spring. The maximum operating pressure is set by adjusting the tension on a spring that holds a poppet or ball against a seat. This prevents fluid from passing through the valve. The maximum operating pressure of the valve is directly related to the amount of spring tension. If the spring tension setting is low, a low maximum pressure results. A high maximum pressure requires a spring tension that is much greater.

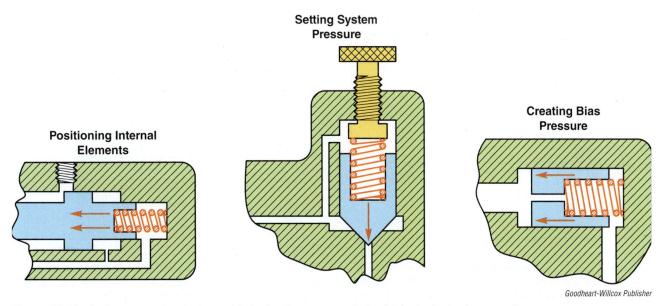

Goodheart-Willcox Publisher

Figure 10-13. Springs are extensively used in hydraulic components to obtain desired valve operation.

When the load on the system actuators produces a pressure lower than the setting of the valve, the spring tension keeps the poppet or ball seated. Thus, the valve is closed. As the load increases on the system actuators, the pressure increases. As the pressure increases, the force generated eventually approaches the force produced by the spring tension. Due to the characteristics of springs discussed earlier, the poppet or ball actually begins to unseat at a pressure below the maximum pressure setting. This allows a small volume of fluid to pass through the valve. As the pressure continues to increase, the fluid flow though the valve increases until the pressure is at the maximum setting and the valve is passing the maximum volume of fluid.

The pressure at which the valve just opens is called the *cracking pressure*. The pressure at which all of the fluid is passing through the valve is termed the *full-flow pressure*. The difference between the cracking pressure and the full-flow pressure is referred to as the *pressure override*. See **Figure 10-14**.

10.3.3 Fluid Flow Used in Valve Operation

Controlling fluid flow is typically considered the means used to vary the operating speed of actuators. For example, the higher the flow rate entering an actuator, the faster the actuator extends or turns. However, fluid flow can also be used in control valves to establish pressure differences that allow the valve to produce desired operating characteristics.

The pressure difference is produced by using an orifice to restrict flow from one chamber of the valve to another. Moving the fluid through the orifice requires a small amount of energy. This lost energy results in a pressure difference between the upstream and downstream sides of the opening. The higher the flow rate through a fixed-size orifice, the greater the pressure difference between the two sides. When used with a balancing piston and a biasing spring, this concept is commonly used in both pressure and flow control valves. See **Figure 10-15**.

10.4 Pressure Control Devices

Devices that sense and control pressure in a hydraulic system make up one of the three basic groups of control valves. Five different types of pressure-sensing valves fall into the general classification of pressure control valves, making this component group the largest of the control valve groups. This group includes valves that:

- Control the maximum pressure allowed in a system.
- Automatically operate system actuators in a selected sequence.
- Restrain movement of a load held by a system actuator.
- Hold pressure in an isolated sector of a circuit, while allowing the system pump to unload fluid output to the reservoir at near-zero pressure.
- Maintain a reduced pressure in one section of a hydraulic circuit, while the remainder of the circuit functions at full system pressure.

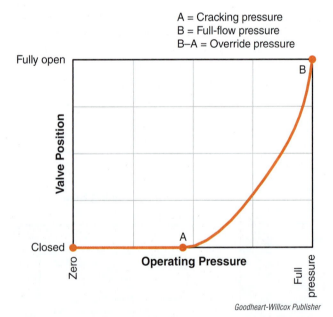

A = Cracking pressure
B = Full-flow pressure
B–A = Override pressure

Goodheart-Willcox Publisher

Figure 10-14. Fluid pressure acting on the surfaces of the internal components of pressure control valves is a key factor in the operation of a hydraulic system. Three pressure operating terms are key to the understanding of valve operation: cracking pressure, full-flow pressure, and override pressure.

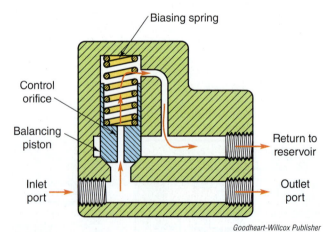

Goodheart-Willcox Publisher

Figure 10-15. An orifice is often used to restrict flow between two areas in control valves to produce the pressure difference needed to move a valve element.

These valves may be further separated based on two different internal modes of operation: direct operation or balancing-piston operation. *Direct-operated valves* use heavy, internal springs to directly establish the operating pressure of the valve. *Balancing-piston valves* use lightweight springs combined with pressure acting on the surface area of pistons or spools to operate the internal mechanism of the valve. Initial cost and efficiency of operation are factors that must be considered when selecting the design to use in a hydraulic system. Balancing-piston designs usually cost more but operate more efficiently.

Manufacturers produce a number of valve designs. However, under close examination, most pressure control valves can be classified using the above groupings. The following material relates to a number of valves that are used in the fluid power field today. The discussion is structured around the five groups of pressure control valves and the two internal modes of valve operation. The concepts on basic valve structure and features covered earlier will be related to specific valve types and other, more-specialized characteristics will be discussed.

10.4.1 System Maximum Pressure Control

Pressure control valves and devices that control maximum system pressure are referred to as relief valves, safety valves, and hydraulic fuses. *Relief valves* are used to prevent system pressure from exceeding the desired maximum, normal operating pressure. This operating pressure is established by the pressure needed to perform required machine operations. Normal operating pressure is usually well below the level where circuit, prime mover, machine, or product damage will occur. Safety valves and hydraulic pressure fuses are incorporated into hydraulic systems to prevent damage resulting from excessive system pressure caused by relief valve malfunctions. They are set at a pressure higher than the system relief valve and their design ensures a minimal failure rate.

Relief valves and safety valves remain closed until the pressure setting is reached. Then they open, dumping fluid to the reservoir to prevent further pressure increases. When system pressure drops below the pressure setting, the valves close. This allows continuing fluid flow to operate the system actuators.

Hydraulic pressure fuses, on the other hand, rupture when the pressure setting is reached. The fluid is dumped to the reservoir to prevent further pressure increases. The pressure-sensitive fuse material is often called a rupture disk. The rupture disk must be replaced after system pressure exceeds the pressure setting and the disk ruptures.

Relief valves

Relief valves are normally closed valves that open as system pressure approaches the maximum, normal operating pressure. They are included in systems using fixed-displacement pumps to prevent the system pressure from exceeding a preselected maximum operating level. Various manufacturers produce a variety of designs to meet the needs of the many hydraulic systems in operation today. However, relief valves can generally be identified based on one of the two internal modes of operation.

Direct-Operated Relief Valve. The most basic of the operating modes is the direct-operated type. It uses the pressure generated by the resistance to fluid movement in the actuators to directly compress a spring and open the valve. This action maintains a selected maximum system pressure by allowing fluid to return to the reservoir.

Figure 10-16 shows the internal structure of a typical direct-operated relief valve. These valves consist of:

- Body containing passageways with pressure and reservoir ports.
- Ball or poppet and a seat.
- Heavy spring.
- Spring-adjustment mechanism.

The body may be a casting specially designed for the valve or it may be made from bar stock cut to length with the internal passageways and ports machined into the block.

When the assembled valve is operating in a system, the pressure port is connected to the main operating portion of the circuit. The reservoir port is connected to a line directly leading to the system reservoir. The operating pressure of the valve is set by adjusting spring tension using the spring-adjustment mechanism. The spring holds the ball or poppet against the valve seat.

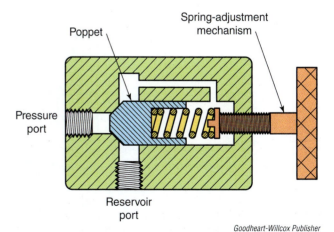

Goodheart-Willcox Publisher

Figure 10-16. The internal structure of a typical direct-operated relief valve. These valves depend on spring tension to establish maximum system pressure.

During system operation, the pressure in the system increases as the load on the system actuators increases. System pressure directly acts on the end surface of the ball or poppet, generating a force that attempts to compress the spring. When the force is high enough, the spring begins to compress and the ball or poppet is forced off the seat. A portion of the system fluid is then returned to the reservoir through the valve.

As actuator resistance increases, the fluid pressure increases. This generates more force on the ball or poppet, which moves farther off the seat as the spring is compressed more. The result is an increase in flow through the relief valve until the actuators stall and all pump flow is returned to the reservoir through the valve.

Balancing-Piston Relief Valve. The relief valves in the second design group use a balancing-piston mode of operation. These valves are commonly called *compound relief valves*. They have internal designs that are more complex. Compound relief valves commonly use a pilot stage and a balancing piston for operation. Pressure differences are used to operate internal components to dump the fluid required to prevent the system from exceeding the selected maximum operating pressure.

Figure 10-17 shows the internal structure of a typical compound relief valve. These valves consist of a pilot section and a balancing-piston section combined in a single valve. The pilot section is a small, direct-operated relief valve that controls the position of a balancing piston. The balancing piston is a primary part of the second section. The two sections work together to produce a valve that is generally more efficient and quieter than direct-operated designs.

The pilot section is at the top of the valve. The balancing-piston section is located between the pilot section and the chamber connected to the hydraulic system. The balancing piston is a critical operating part of this section. It typically includes:

- Poppet valve attached to the system side of the balancing piston.
- Cavity on the pilot side of the piston that contains a small balancing spring.
- Metering orifice that connects the pilot section and hydraulic system chamber.

The exact design of these features varies between valve manufacturers.

The face of the poppet valve attached to the balancing piston rests against a machined seat in the valve body. The balancing spring located on the opposite side of the balancing piston keeps the valve face in contact with the seat. This contact prevents fluid flow back through the relief valve until a selected system operating pressure is reached.

The metering orifice plays an especially critical part in the operation of this type of valve. It is located in a passageway that connects the pilot-valve side and system side of the balancing piston. The primary function of the orifice is to create a pressure difference between the pilot and system sides of the balancing piston.

Compound Relief Valve Operation. During hydraulic system operation, the direct-operated relief valve in the pilot section of the compound relief valve indirectly establishes maximum system pressure. Adjusting the spring tension of the direct-operated valve controls the maximum pressure possible in the chamber between the pilot section and the balancing piston. The balancing-piston

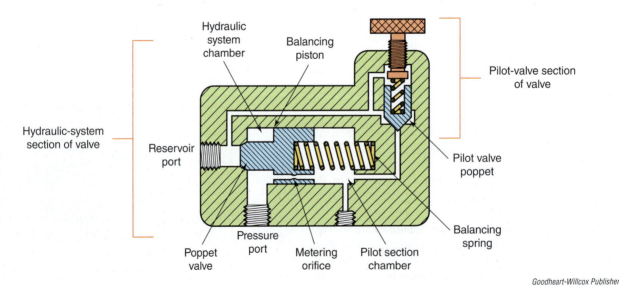

Figure 10-17. The internal structure of a typical compound relief valve. These valves depend on a pilot section to produce an internal pressure differential that moves the balancing piston, which establishes maximum system pressure.

section of the valve then responds to the pilot valve setting to prevent system pressure from exceeding the selected maximum level. Refer to **Figure 10-18**.

The following conditions exist in the valve when the load on the system is constant and the pressure required for actuator operation is below the valve setting:

- The small poppet of the pilot valve is seated as the pressure in the pilot-section chamber is below the setting of the valve.

- The pressure in the pilot-section chamber and the hydraulic-system chamber are equal because they are connected by the metering orifice passageway and no fluid is moving through the orifice and the pilot valve.

- No fluid is being passed to the reservoir around the poppet valve attached to the balancing piston as the pressure is equal on either side of the piston, which allows the tension of the balancing spring to keep the poppet seated.

As the load on the system actuator increases, the valve adjusts to begin the process of controlling maximum system pressure. When the actuator load requires a pressure in the system approaching the valve setting:

- The small poppet of the pilot valve begins to open, allowing a small volume of system fluid from the pilot-section chamber to return to the reservoir.

- Fluid from the hydraulic-system chamber begins to flow through the metering orifice passageway replacing the fluid flowing through the pilot valve.

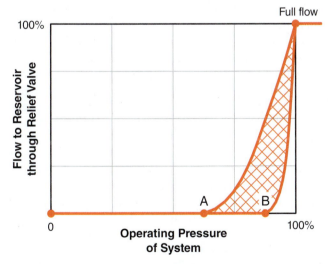

A = Cracking pressure of direct-operated valve

B = Cracking pressure of pilot-operated valve

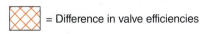 = Difference in valve efficiencies

Goodheart-Willcox Publisher

Figure 10-18. Compound pressure control valves are more efficient than direct-operated valves. This graph indicates cracking and full-flow pressures for typical valves.

- A slight pressure difference develops between the fluid in the pilot-section chamber and the hydraulic-system chamber because of the resistance to fluid flow through the metering orifice.

- The poppet valve attached to the balancing piston remains seated as the force generated by the tension of the balancing spring is still higher than the force difference generated on the balancing piston by the pressure difference between the pilot and system chambers.

Further increasing the load on the system actuator causes the pressure in the system to increase until the valve must begin dumping fluid to the reservoir through the poppet valve attached to the balancing piston. This prevents the pressure from exceeding the preset system level. As the actuator load increases:

- The pressure in the pilot-valve chamber increases causing the pilot valve to open further, which increases fluid flow from the pilot chamber to the reservoir.

- This pressure increase also increases flow through the metering orifice passageway, resulting in an increased pressure difference between the pilot and hydraulic-system chambers.

- The increased pressure difference causes the forces acting on the balancing piston to become more unequal.

- The difference between these forces eventually overcomes the force generated by the tension of the balancing spring.

- As a result, the poppet valve attached to the balancing piston opens.

- When the poppet valve attached to the balancing piston opens, sufficient fluid is dumped to the reservoir to prevent system pressure from exceeding the preset level.

Safety valves

If the relief valve fails, system pressure will exceed the maximum desired system operating pressure. This type of system failure could cause damage to system components, damage to materials being processed by the system, and stalling of the prime mover. *Safety valves* are placed in hydraulic systems to protect the system from damage if the system relief valve should fail to open at the selected pressure. See **Figure 10-19**.

The structure and operation of a safety valve is usually the same or only slightly different from a direct-operated relief valve. The primary differences are the valve:

- Has been placed in the system as a backup safety device.

- Uses a direct-operated, spring-loaded design.

- Typically is set at a pressure 25% above the system relief valve.

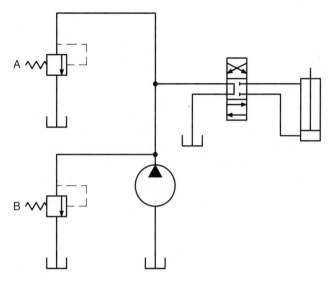

A. Safety valve: Direct-operated valve with cracking pressure set above required system operating pressure.
B. System relief valve: Set at required system operating pressure.

Goodheart-Willcox Publisher

Figure 10-19. Safety valves are direct-operated pressure control valves used in circuits to act as a backup to prevent excessive system pressure if the system relief valve should fail.

These three factors provide a hydraulic circuit with a highly reliable device that is set at a pressure slightly above the desired system pressure, but not sufficiently high as to cause system damage.

Hydraulic pressure fuses

Hydraulic pressure fuses are safety devices that do not reseat after discharge. The device is equipped with a disk that ruptures when the predetermined pressure is reached. See **Figure 10-20.** A hydraulic pressure fuse

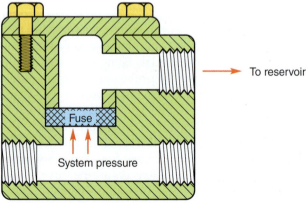

Goodheart-Willcox Publisher

Figure 10-20. Hydraulic fuses use a disk that ruptures if system operating pressure exceeds their rating due to relief and safety valve failures.

is similar in function to fuses in an electrical system in that the circuit ceases to function when the fuse "blows."

Fuses can be incorporated in a hydraulic circuit to ensure that a predetermined maximum pressure is not exceeded. The device is placed in a line connecting the reservoir and the circuit section that needs positive-pressure protection. After the disk ruptures, all system fluid is diverted to the reservoir and system pressure drops to near zero. System pressure is only high enough to force the fluid through the lines and back to the reservoir after the disk has failed. The system must be shut down to replace the disk.

The term hydraulic pressure fuse is also applied to a variety of devices that sense sudden changes in system pressure or flow. These devices automatically close when a sudden change is sensed. This blocks flow to the protected section of the system to prevent fluid loss or other problems.

10.4.2 Actuator Sequence Control

The need to sequence the operation of multiple actuators or sections of a circuit is a fairly common requirement. This need can be met using direct-operated pressure control valves to limit fluid flow to an initial portion of a circuit until a predetermined operation is completed. Once this first operation is completed, system pressure increases until the control valve opens. This allows fluid flow to activate a second portion of the circuit, which then performs the next operation in the sequence. The fact that these valves are pressure activated requires that each step in the sequence increases the system pressure to a higher level to initiate the next step in the process. Valves that control this type of movement are called *sequence valves.*

The simple clamp-and-drill hydraulic circuit shown in **Figure 10-21** illustrates the concept of sequencing. The clamp (*Cylinder 1*) must secure the workpiece before the drilling operation (*Cylinder 2*) can begin. This is achieved by using a sequence valve to block flow to the actuator that advances the drill. When the machine operator activates the circuit, hydraulic fluid flows only to the clamp actuator. This is because the sequence valve is closed, blocking flow to the drill actuator. When the clamp closes, pressure increases in the initial portion of the circuit to the level of the sequence valve setting. This causes the sequence valve to open, which allows flow to the drill actuator. The drill actuator then advances and completes the operation. Maximum system pressure is determined by the setting of the relief valve. This setting depends on the pressure required to provide effective operation of the sequence valve as well as the clamping and drilling force requirements.

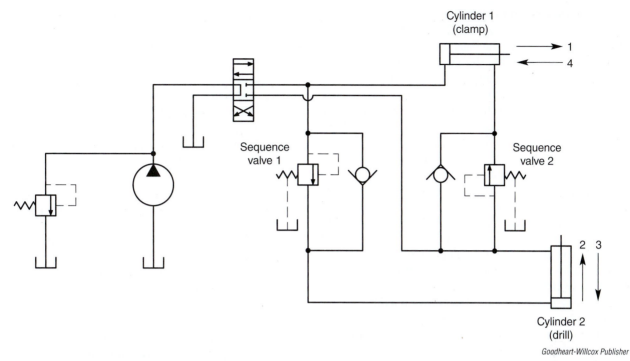

Goodheart-Willcox Publisher

Figure 10-21. A clamp-and-drill hydraulic circuit is an example of sequence control. The clamp portion of the circuit functions before the drill is activated. The drill withdraws before the workpiece is unclamped.

Figure 10-22 shows a cutaway of a typical sequence valve. The operating concepts of these valves are similar, even though design variations exist between manufacturers. These valves consist of a body containing passageways and ports, which are attached to the initial- and secondary-movement portions of the circuit; a spool precision fit in a bore; a spring and spring-adjustment mechanism; and passageways allowing pilot and drain functions. The body may be a casting especially designed for the valve or it may be machined from bar stock.

Sequence valves are basically normally closed valves that block the movement of fluid between the initial and secondary ports until a preset pressure is reached.

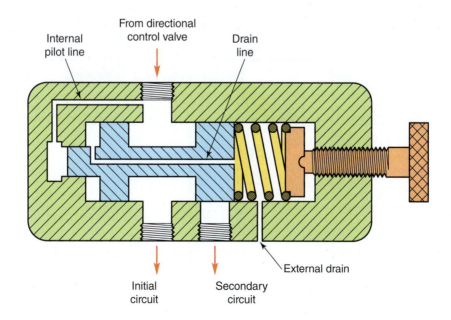

Goodheart-Willcox Publisher

Figure 10-22. Sequence valves are normally closed valves that block the movement of fluid between the initial and secondary portions of a hydraulic circuit.

The pressure needed to open the valve is set by adjusting the tension of the spring that holds the spool in the closed position. In the basic clamp-and-drill circuit shown in **Figure 10-21**, the sequence valve remains closed until the clamp actuator has closed the clamping device. Once the clamp has closed, increased resistance to fluid flow causes system pressure to increase. The valve uses a pilot line and chamber to sense this pressure and apply force to the end of the spool opposite of the spring-adjustment mechanism. When the force is high enough, the spring begins to compress and the spool slides within the bore. This allows system fluid to move into the second portion of the circuit, starting the drilling operation. As the drill actuator encounters greater resistance, system pressure increases and generates more force on the pilot end of the spool. The increased force causes the spool to continue to slide in the bore until the passageway between the two sides of the valve is fully open. This allows maximum flow through the valve.

In sequence valves, the chamber containing the spring and adjustment mechanism must be externally drained. This is required as both the initial and secondary sides of sequence valves are subject to full system pressure at the same time during some phases of circuit operation. If an internal drain is used, the pilot and drain chambers will be exposed to equal pressure. This will generate equal forces on the ends of the spool, allowing the spring to move the spool into the closed position. A circuit malfunction results.

Sequence valves are normally closed valves that do not allow reverse fluid flow without the addition of a check valve. **Figure 10-23** illustrates a sequence valve containing an integral check valve. This is a common method used to allow reverse flow to bypass the valve. Sequence valves are also available with balancing-piston operation.

10.4.3 Restrained Movement Control

Restrained movement control refers to the control of actuator movement in a hydraulic circuit by placing a pressure control valve in the outlet line leading from an actuator. These normally closed, direct-operated valves are used to restrain the movement of both cylinders and hydraulic motors. The valves provide both increased hydraulic system control and operator safety, **Figure 10-24**. Typically, the valves are called *counterbalance valves* when used with cylinders and *brake valves* when the actuator is a motor.

These valves are very similar in construction to the sequence valves discussed in the previous section.

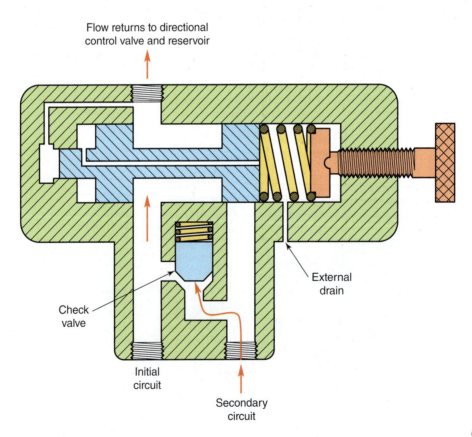

Flow returns to directional control valve and reservoir

Check valve

External drain

Initial circuit

Secondary circuit

Figure 10-23. Check valves are commonly incorporated in sequence valves to allow reverse flow through the system to bypass the valve.

Goodheart-Willcox Publisher

Figure 10-24. Counterbalance valves are often used in the actuator portion of load-lifting equipment. They restrain the uncontrolled drop of a load due to equipment failure or improper operation.

The primary differences are found in the pilot and drain connections and their applications in a hydraulic system. Internal, external, or multiple pilots may be used in either counterbalance or braking valves, depending on the system demands. Internal drain connections are appropriate for most circuits. Like sequence valves, these valves are normally closed. This requires a check valve to allow reverse flow to freely move around the valve. The check valve may be an integral part of the valve or a separate component. See **Figure 10-25**.

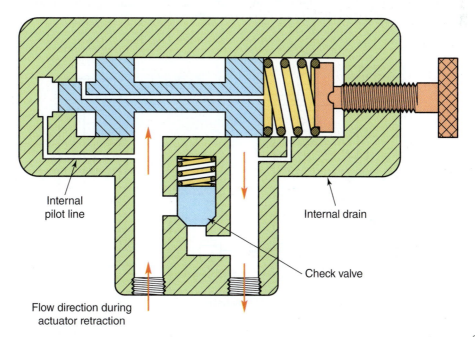

Internal pilot line

Internal drain

Check valve

Flow direction during actuator retraction

Goodheart-Willcox Publisher

Figure 10-25. The internal structure of a counterbalance pressure control valve. These valves include either an internal or external pilot and, in most cases, an integral check valve. In this example, the check valve allows the free movement of fluid around the valve during actuator extension.

Counterbalance valves

The simple circuit shown in **Figure 10-26** illustrates the location of a counterbalance valve in a hydraulic press. Variations of this circuit can be found in mobile equipment, such as a cherry picker, or other equipment used in manufacturing operations. In this example, when the press is raised, fluid in the rod end of the cylinder and the line leading to the counterbalance valve is trapped. Even if the system pump is not operating, this trapped fluid is pressurized because of gravity acting on the weight of the press *platen* and any load. This pressure is equal to the weight of the platen and the load divided by the surface area of the blind side of the piston.

If the cracking pressure of the counterbalance valve is set at a pressure higher than that generated by this weight, the load will not drop even if the directional control valve is shifted. In order to lower the load, the system pump needs to be operating with the directional control valve shifted to extend the cylinder. The pressure increase caused by the fluid coming from the pump added to that generated by the weight of the platen and load is sufficient to open the counterbalance valve, extend the cylinder, and lower the press platen.

However, without the counterbalance valve in place, the platen and load will drop if the directional control valve is accidentally shifted, even without the pump operating. Gravity acting on the platen and load causes them to move downward, extending the cylinder. This movement creates a negative pressure in the blind end of the cylinder. The negative pressure draws fluid out of the reservoir to fill the blind end, while

fluid in the rod end is returned to the tank. The platen and load may fall uncontrolled, causing damage or operator injury.

> **CAUTION**
>
> Service personnel must be aware of circuit designs that may allow unexpected actuator movement, even without system pump operation.

Brake valves

The hydraulic motor circuit shown in **Figure 10-27** illustrates the location of a brake valve in a circuit. Brake valves are variations of normally closed counterbalance valves. They are designed for use in hydraulic motor circuits. Their specific purpose is to prevent an overrunning load from turning the motor at a higher speed than desired.

Brake valves use two pilot lines to sense pressure at the inlet, or remote, side and the outlet, or load, side of the motor. The pilot lines are connected to two chambers that generate force on the end of the valve spool that is opposite of the control spring. **Figure 10-28** shows the construction of the valve. Notice the location and difference in size of the pistons in the two pilot chambers. The valve spring is used to set the pressure of the braking load.

In the following example, the braking load for the valve is set at 500 psi. The pistons in the inlet and load pilot chambers have an area ratio of 5:1 (inlet to load).

When the system is started and 100 psi is reached in the motor inlet pilot line, the braking valve opens and the motor turns. The motor continues to turn,

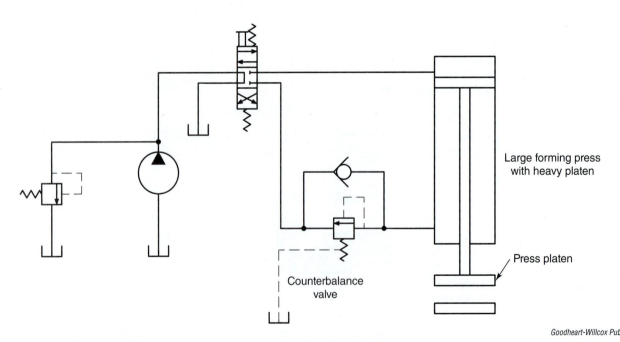

Large forming press
with heavy platen

Press platen

Counterbalance
valve

Goodheart-Willcox Publisher

Figure 10-26. Placing a counterbalance valve in the outlet line of a cylinder prevents the uncontrolled drop of a load, even if the directional control valve is improperly shifted or a line fails between the directional control and counterbalance valves.

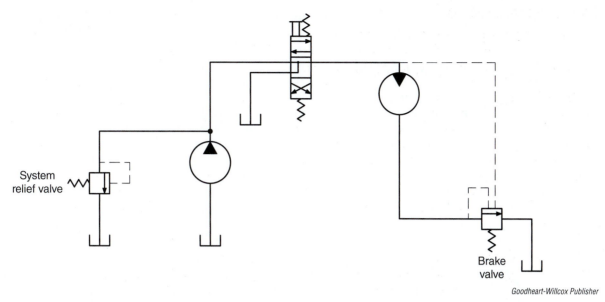

Goodheart-Willcox Publisher

Figure 10-27. Placing a brake valve in the outlet line of a hydraulic motor prevents overrunning loads from turning the actuator at a higher-than-desired speed.

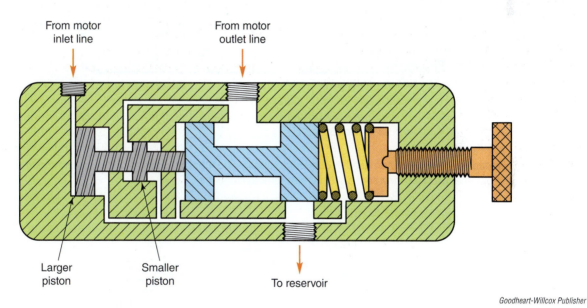

Goodheart-Willcox Publisher

Figure 10-28. The structure of a brake valve. The braking load is determined by the tension of the spring and the ratio of the piston areas in the inlet and load pilot chambers.

with the actual line inlet pressure varying to meet the motor-load demand.

If the motor load begins to run away, motor speed increases to the point where fluid requirements match or exceed the pump output. The result is inlet line pressure drops below the 100 psi needed to keep the valve open. This pressure drop in the inlet pilot allows the valve spring to close the brake valve, restricting flow out of the motor.

The rotational momentum of the motor and load momentarily turns the motor into a pump, causing a pressure increase in the motor outlet line. Because of the difference in areas between the two pilot pistons, the outlet line pressure must reach 500 psi before the load pilot opens the valve. Once open, fluid is allowed to again pass to the reservoir from the motor outlet.

The 500 psi back-pressure load has a braking effect on the overrunning speed of the motor. This continues until the motor speed has dropped to a level where pressure in the motor inlet line again reaches 100 psi. At that point, the motor has been slowed to the desired operating speed and continues normal operation until another overrunning-load condition occurs.

10.4.4 Pump Unloading Control

Operating a hydraulic system pump at full flow and pressure levels results in maximum energy and maintenance costs. This is wasteful of resources, especially when many systems do not require the pump to supply full-flow output under maximum pressure on a continuous basis. Stopping the system prime mover and pump when fluid flow is unneeded is not usually a practical solution. This is also expensive and shortens the service life of equipment. One technique that is

often used is a design called a *pump unloading control*. This design uses a pressure control valve to monitor a direct, low-pressure flow path between the pump and the reservoir. See **Figure 10-29**. Unloading valves are commonly used in circuit designs using dual pumps or accumulators.

Unloading valves are classified as normally closed valves as they prevent the return of fluid to the reservoir until a desired operating pressure is achieved. They must also have an external pilot as the operating mechanism must sense pressure at a remote location in the system. See **Figure 10-30**. These valves are available using a direct-operated design, similar to the sequence and counterbalance valves discussed earlier. They may also

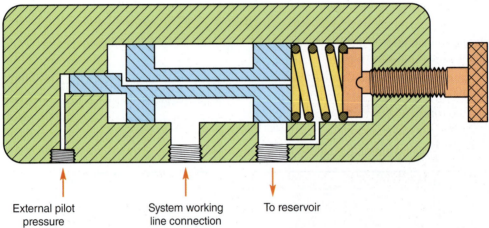

External pilot pressure   System working line connection   To reservoir

Goodheart-Willcox Publisher

Figure 10-29. Unloading valves are often used in dual-pump applications that provide maximum system operating efficiency. These valves use external pilots as they sense pressure at remote locations.

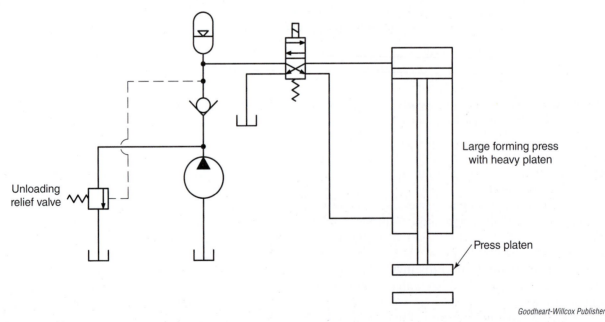

Unloading relief valve

Large forming press with heavy platen

Press platen

Goodheart-Willcox Publisher

Figure 10-30. Unloading relief valves are often used in circuits where accumulators supply additional fluid volume or hold an actuator under pressure for extended periods of time.

be more complex, unloading relief valves following a variation of the design found in compound relief valves.

Direct-operated unloading valve

The high-low pump circuit shown in **Figure 10-31** illustrates one application of pump unloading control that uses a direct-operated unloading valve. The circuit has two pumps, a check valve, a relief valve, and the direct-operated unloading valve. It provides high-volume, low-pressure fluid for the initial operation of a circuit and then automatically switches to low-volume, high-pressure output in the second stage of operation. During the second stage, the unloading valve dumps the output of the largest pump to the reservoir at near-zero pressure to reduce energy consumption and heat buildup.

When the system starts, the low-volume pump moves fluid directly to the system. The flow from the high-volume pump passes through the check valve before blending with the low-volume pump output. When the actuator in the system encounters resistance, system pressure begins to increase. The external pilot line of the unloading valve senses system line pressure near the low-volume pump. When that pressure reaches the cracking pressure of the unloading valve, flow from the high-volume pump begins to flow to the reservoir. As the system actuator resistance increases, the pressure continues to increase until the unloading

valve is fully open. At that point, all of the output of the high-volume pump is passing though the unloading valve to the reservoir.

As flow from the low-volume pump attempts to reach the low-pressure area of the unloading valve, it closes the check valve. With the check valve closed, the low-volume pump continues to supply fluid to the system actuator until the pressure setting of the system relief valve is reached. The unloading valve remains open because of the high pressure in the pilot line. The high-volume pump continues to return all fluid output to the reservoir, as long as system pressure is above the pressure setting of the unloading valve.

Unloading relief valve

When greater sensitivity is needed to provide pump unloading control, a valve with the expanded design features of a balancing piston is used. **Figure 10-32** shows the construction of an unloading relief valve used for such situations. This valve is very similar in basic construction to a compound relief valve. However, additional features are incorporated into the valve:

- Integral, spring-loaded check valve in the outlet line of the valve stops reverse fluid flow.

- Plunger mechanism in the pilot section of the valve assists in operating the balancing piston.

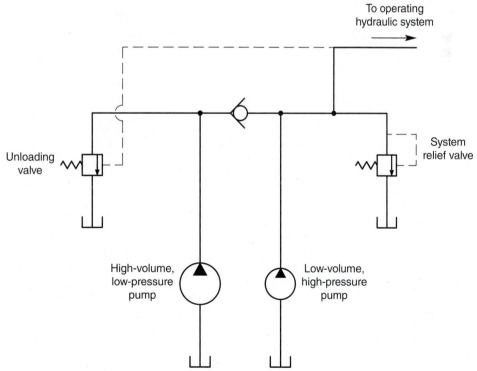

Goodheart-Willcox Publisher

Figure 10-31. A high-low pump circuit can use a direct-operated unloading valve to deliver two different flow rates under two different pressures. Maximum system pressure is controlled by the relief valve located in the low-volume-pump section of the circuit.

The addition of these two elements provides a valve that is compact and has the unloading feature desirable in accumulator and other circuits.

Figure 10-33 shows a basic accumulator circuit using an unloading relief valve. This diagram and the internal structure shown in **Figure 10-32** are used to explain the operation of the valve.

The integral, spring-loaded check valve is located in the passageway between the inlet and outlet ports of the valve. It is in close proximity to the pilot section of the unloading valve. This check valve is used to stop reverse fluid flow from the actuator portion of the system. A plunger mechanism is located in the pilot section of the valve. It is designed to assist in unseating and holding open the poppet of the small, direct-operated relief valve in the pilot section. The plunger is activated by pressure from an internal pilot sensing pressure in the passageway between the check valve and outlet port.

In the circuit shown in **Figure 10-33**, the directional control valve is centered, blocking flow to the system. When this occurs, the pump supplies fluid to the accumulator through the built-in check valve. As the accumulator approaches its fluid-storage capacity, pressure in the system begins to rise. The small, direct-operated relief valve in the pilot section of the valve senses this pressure increase through the orifice in the balancing piston of the unloading valve. At the same time, the pilot line leading from the plunger mechanism is sensing pressure in the passageway between the check valve and the outlet port of the valve. As system pressure increases, the force generated on the poppet of the

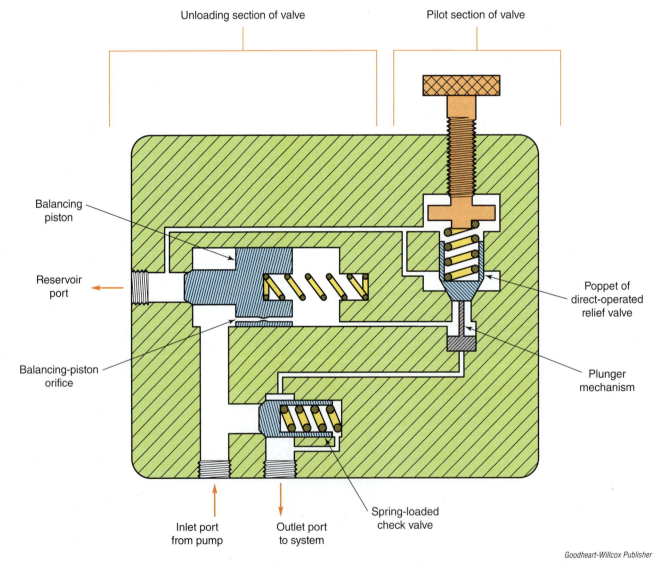

Unloading section of valve

Pilot section of valve

Balancing piston

Reservoir port

Balancing-piston orifice

Poppet of direct-operated relief valve

Plunger mechanism

Spring-loaded check valve

Inlet port from pump

Outlet port to system

Figure 10-32. An unloading relief valve provides effective control of a single-pump unloading system. This is achieved by using a plunger mechanism to assist in the operation of the direct-operated relief valve located in the pilot section of the valve.

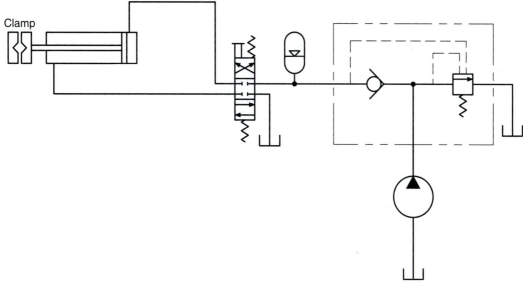

Figure 10-33. This is an accumulator circuit using an unloading relief valve. This combination of components allows the pump to operate at near-zero pressure while the accumulator maintains a high holding pressure in the system.

small, direct-operated relief valve begins to compress the spring of that valve. Pilot pressure in the plunger mechanism also generates force that is transmitted to the end of the poppet by a rod extending from the end of the plunger. These two forces combine to unseat the poppet. This begins the process of shifting the balancing piston to unload pump output directly to the reservoir.

As the balancing piston shifts, allowing pump output to pass to the reservoir, the accumulator begins to discharge. This reverses flow as the fluid attempts to reach the low-pressure area of the reservoir. The spring-loaded check valve immediately closes to block the reverse flow. This traps pressurized fluid in the accumulator and the lines leading to the check valve and directional control valve. The plunger mechanism continues to hold the poppet of the small, direct-operated relief valve open because of high pilot pressure sensed in the passageway between the check valve and the outlet of the valve. Under these conditions, the unloading valve fully opens and dumps pump output to the reservoir at minimal back pressure. This produces considerable energy savings.

Once the directional control valve is shifted to activate system actuators, the accumulator discharges fluid to operate the system. This results in a system pressure drop that eventually allows the poppet in the small, direct-operated valve to close. The unloading valve then closes and the process is repeated. The amount of pressure drop required to reset the valve depends on the difference in the effective area of the plunger in the plunger mechanism and the area of the poppet in the small, direct-operated relief valve. For

example, one manufacturer designs valves of this type to reseat at 85% of the maximum pressure setting.

10.4.5 Reduced Pressure Control

Some circuits require that a branch of the circuit operate at a maximum pressure lower than the setting of the system relief valve. This lower-pressure branch may involve a single actuator or a more extensive portion of the circuit. The reduced pressure is usually required to prevent damage to machine tooling or the materials being handled. Reduced pressure control is accomplished by using a *pressure-reducing valve* in the circuit.

Figure 10-34 illustrates a circuit using a pressure-reducing valve. In this circuit, the two cylinders require different operating pressures. The maximum pressure for *Cylinder 1* is set by the system relief valve, while the pressure for *Cylinder 2* is controlled by the pressure-reducing valve. The pressure-reducing valve:

- Controls the maximum pressure that can be obtained in the designated branch circuit.

- Allows passage of the fluid required to operate the actuator to complete the task involved in the branch circuit.

Pressure-reducing valves are normally open valves and require an external drain. They typically have an internal pilot that senses pressure on the outlet-port side of the valve. These valves are available using direct or balancing-piston operation.

Figure 10-35 shows the basic internal construction of a pressure-reducing valve using the balancing-piston

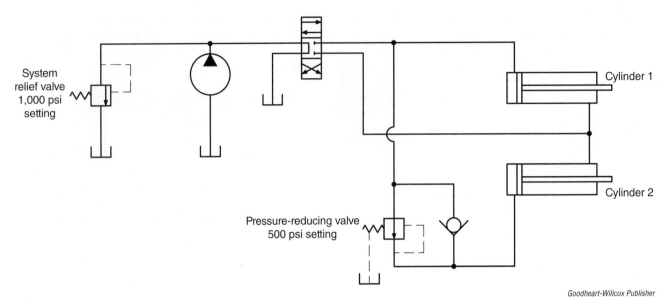

Figure 10-34. Pressure-reducing valves allow a portion of a hydraulic circuit to operate at a pressure below the normal operating pressure of the circuit. In this circuit, the maximum operating pressure for Cylinder 1 is set by the system relief valve, while the maximum for Cylinder 2 is established by a pressure-reducing valve.

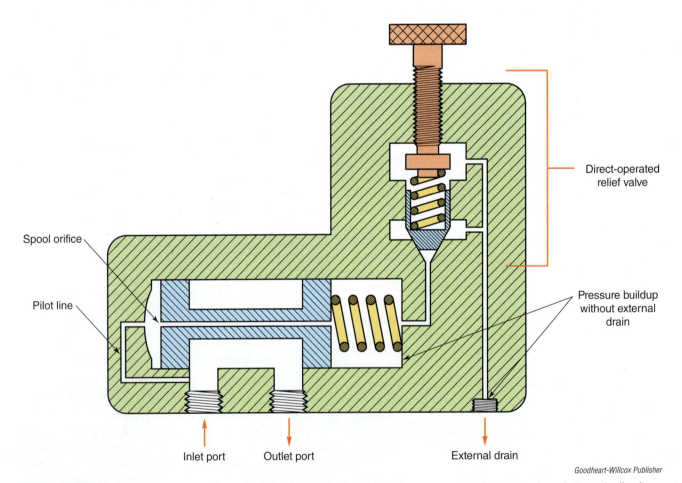

Figure 10-35. The internal structure of a typical pressure-reducing valve. These are normally open valves that gradually close to limit the maximum pressure in a portion of the circuit.

mode of operation. The primary parts of the valve are a small, direct-operated relief valve and a balancing-spool assembly. The relief valve sets the reduced pressure level of the pressure-reducing valve. The spool assembly responds to the relief valve setting and controls the pressure in the outlet port area by controlling the fluid flow allowed into the outlet area.

The desired reduced-pressure level is set by adjusting the spring tension in the small relief valve. During system startup, the poppet of the relief valve is seated and the spring in the spool assembly shifts the spool to the fully open position. An orifice in the center of the spool allows fluid to pass from the pilot line to the chamber connected to the relief valve. This orifice allows both ends of the spool to be subjected to equal pressure.

As the pressure at the outlet of the valve increases, the pressure on the relief valve poppet increases an equal amount due to the connection through the spool orifice. When the pressure on the poppet reaches the valve setting, the poppet unseats and flow begins through the valve drain.

The flow through the spool orifice causes a pressure drop between the pilot line and the spool-spring chamber. As the pressure increases from the pilot line, the flow through the spool orifice increases, causing the pressure difference to further increase. When this pressure difference becomes sufficiently large, the pilot pressure acting on the pilot end of the spool generates a force adequate to compress the spool spring. This force begins to close the valve. As a result, fluid flow from the valve outlet is restricted and the pressure increase downstream from the valve is limited to the desired reduced pressure.

When a valve of this design is operating at the full, reduced pressure, there is continual fluid flow through the drain line. Any resistance to the flow through the drain reduces valve efficiency. In situations where return flow through the valve is required, a built-in check valve is recommended.

10.5 Directional Control Devices

Devices that control the direction of fluid flow in a hydraulic system make up the second of the three basic groups of control valves. These valves can be divided into four general classifications according to the routing of fluid through their passageways, **Figure 10-36**.

- *Shutoff valves,* often called two-way valves, have a single passageway between two ports allowing fluid to move in either direction.
- *Check valves* have a fluid route between two ports, but generally allow fluid to move through the valve in only one direction.

- *Three-way valves* have two fluid paths used alternately between three ports to direct fluid to and from a single-acting actuator or circuit segment.
- *Four-way valves* have two fluid paths used simultaneously between four ports to direct fluid to and from double-acting actuators or circuit segments.

Manufacturers produce a wide variety of directional control valve designs. However, under close examination, most of these valves can be classified using the above groupings. The following material relates to the most common of the directional control valves used in the fluid power field today.

10.5.1 Shutoff Valves

The primary purpose for shutoff valves is to block fluid flow in a system line. They are commonly used to isolate a component or a circuit branch during maintenance. These valves have a single internal passageway connecting two ports. Designs include globe, gate, ball, spool, and needle valves. All of these valves are simple in design. Although these valve designs can be used to throttle (regulate) fluid flow, they are generally considered inefficient as flow control devices. They should be treated as valves that function best in the fully open or fully closed position, except needle valves.

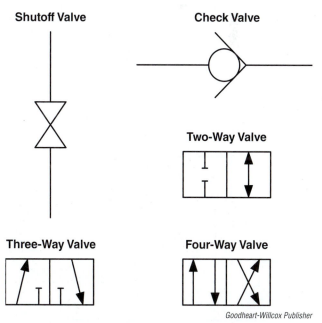

Goodheart-Willcox Publisher

Figure 10-36. Directional control valves are classified under one of four general groups: shutoff valves, check valves, three-way valves, and four-way valves. These graphic symbols are used to indicate each valve type.

Globe valves

Globe valves are very commonly used in water, air, and hydraulic systems. **Figure 10-37** shows the internal structure of these valves. The passageway through the valve is slightly offset with a seat located at the offset. This offset produces fluid turbulence, which results in increased pressure drops at higher flow rates. A globe or disk with a machined surface that matches the seat is mounted on a threaded shaft attached to an external handwheel. Turning the handwheel allows an operator to adjust the valve opening from fully closed to fully open.

Gate valves

The internal structure of a *gate valve* provides a straight line between the inlet and outlet ports. See **Figure 10-38**. The device that opens and closes the valve is a rectangular piece, called the gate, mounted on a threaded shaft attached to an external handwheel. The gate is fitted into a slot that is perpendicular to the valve passageway. Turning the handwheel allows the operator to slide the gate to a point where it fully blocks the passageway to stop fluid movement. Turning the handwheel in the opposite direction can completely retract the gate from the passageway. This allows unrestricted movement of fluid through the valve.

Ball valves

Ball valves consist of a valve body with a straight-line passageway connecting the inlet and outlet ports. The valve mechanism is a spherical piece attached to a shaft. See **Figure 10-39**. The ball contains a precision-machined hole that is the same size as the passageway through

the valve body. When the valve parts are assembled, the shaft holds the hole through the ball in alignment with the passageway through the valve body. A handwheel or other handle on the shaft allows an operator to rotate the ball. Turning the ball 1/4 turn (90°) moves the valve

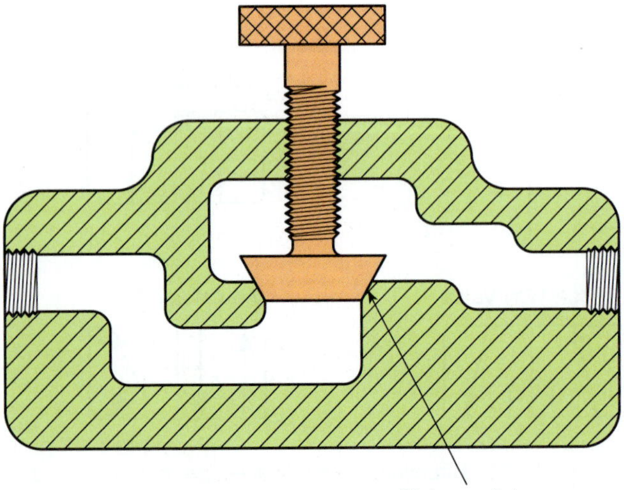

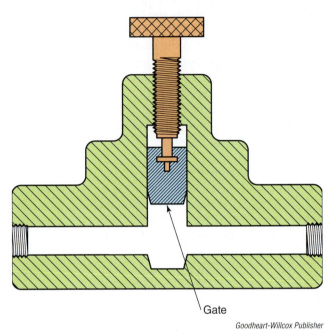

Goodheart-Willcox Publisher

Figure 10-38. The internal structure and flow path through a basic gate valve. When fully open, this valve produces little flow resistance.

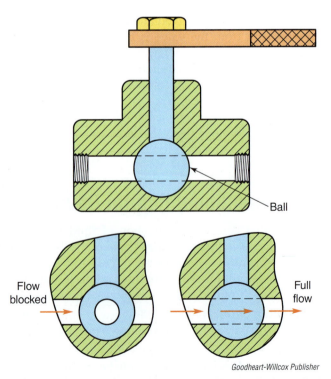

Goodheart-Willcox Publisher

Figure 10-39. The internal structure and flow path through a basic ball valve. Only a 1/4 turn of the handle is required to adjust this valve from fully closed to fully open.

Goodheart-Willcox Publisher

Figure 10-37. The internal structure and flow path through a basic globe valve. This valve design produces high resistance to fluid flow.

from the fully open to the fully closed position. When the valve is in the fully open position, there is unrestricted movement of fluid through the valve.

Spool valves

A *spool valve* consists of a valve body with a precision-machined bore and passageways connecting inlet and outlet ports. A precision-fit spool is placed in the bore to provide a means by which the inlet and outlet ports can be connected and disconnected. **Figure 10-40** shows the internal structure of a spool valve. Different spool configurations and shifting methods may be used to provide different operating characteristics. These valves can be designed to function with minimal flow resistance, resulting in a low pressure drop across the valve even at maximum flow capacity.

Needle valves

Needle valves are very commonly used in fluid power systems as both shutoff and metering devices.

Figure 10-41 shows the internal structure of these valves. The passageway through the valve contains an orifice. This orifice may be simply a hole machined in the valve body or a separate part that can be easily replaced. The needle has a tapered point and is mounted on a shaft with fine threads attached to an external handwheel. Turning the handwheel allows an operator to adjust the valve opening from fully closed to fully open.

This type of valve produces fluid turbulence, resulting in increased pressure drops at higher flow rates. In addition, the small diameter of the orifice produces a pressure drop as the fluid is metered through the needle valve.

> **NOTE**
> A more detailed discussion of using a needle valve as a metering device is found in the Flow Control Devices section of this chapter.

10.5.2 Check Valves

A standard design of check valve has the primary purpose of allowing free flow of fluid through a line in one direction, while preventing reverse flow in that line. More complex check valve designs also are available that include restriction and pilot-operated units. Both of these versions allow free flow in one direction, but may also allow reverse flow in selected circuit operating situations.

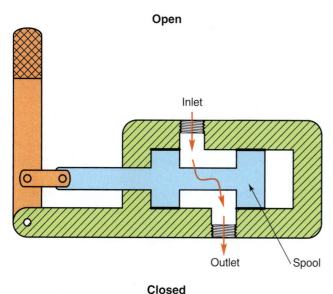
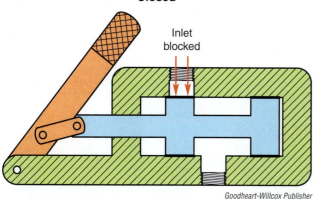

Figure 10-40. The internal structure and flow path through a basic spool valve. Linear movement of the valve handle provides rapid adjustment of this valve from fully open to closed.

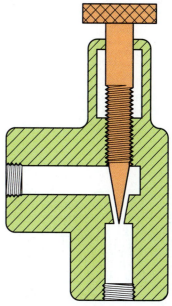

Goodheart-Willcox Publisher

Figure 10-41. The internal structure and flow path through a basic needle valve. This design is commonly used for both shutoff and metering of fluid in hydraulic circuits.

Depending on their design, check valves may be incorporated in hydraulic systems to provide a number of different functions.

- Allowing the free flow of fluid through a hydraulic system line in only one direction.
- Allowing fluid to bypass selected components during the return portion of a system cycle.
- Isolating components or system sections to prevent fluid loss during system maintenance.

- Providing fluid flow resistance to maintain adequate pilot pressure for activating a system.
- Providing restricted reverse flow from a large actuator to allow decompression or speed control.
- Selectively allowing unrestricted reverse flow to pass through the valve by using a pilot system to open the valve.
- Providing a bypass to return fluid to the reservoir if a filter becomes clogged.

Standard check valves

The construction of typical, standard, one-way check valves is shown in **Figure 10-42**. Both inline and right-angle valve designs are shown. These designs essentially operate the same. The design selected for a particular circuit is based on the amount of space available for installation and service or, at times, only the preference of the machine designer. These valves are also commonly included as an integral part of pressure, flow, and other directional control valves as well as some actuators where flow must be limited to only one direction.

A standard check valve consists of a valve body containing a passageway connecting inlet and outlet ports. The passageway between the ports contains a one-way valve that blocks fluid flow in one direction, while allowing free flow in the opposite direction. A ball or poppet is the most common movable part of these one-way valves. However, a hinged flapper is sometimes used. See **Figure 10-43**. The ball, poppet, or flapper is generally held against a valve seat by a spring to ensure a positive valve seal. Valve designs that do not use a spring rely only on fluid flow to seat the movable part and block the passageway.

When fluid enters the inlet of a standard, spring-loaded check valve, a resistance to flow is encountered. This resistance is based on the strength of the spring holding the valve shut. Typically, a 5 psi pressure drop across the movable part is required to compress the spring and unseat the movable part. This allows the free passage of fluid through the check valve. In contrast, when fluid enters the outlet of the check valve in

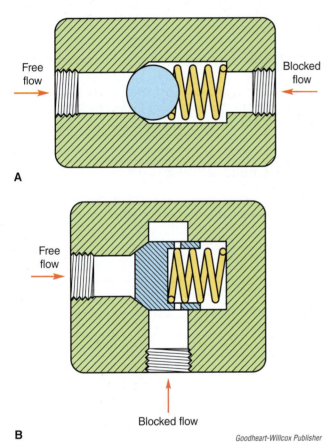

A

B

Goodheart-Willcox Publisher

Figure 10-42. Internal structure of typical check valves. The selection of configuration is based on available space and expected service requirements. A—Inline. B—Right angle.

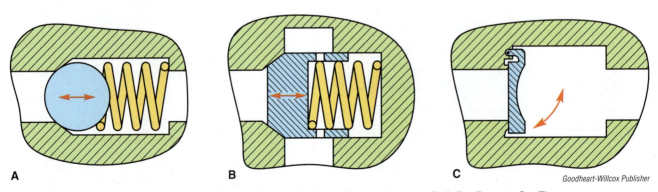

A   B   C

Goodheart-Willcox Publisher

Figure 10-43. Different internal sealing elements are available in check valves. A—Ball. B—Poppet. C—Flapper.

an attempt to reverse flow through the system line, the spring seats the movable part to block fluid flow. The blocked line increases resistance to the flow of fluid. This causes the pressure in the line to increase and the one-way valve is seated even more firmly. This prevents any reverse fluid flow through the check valve.

Some check valve designs do not use a spring to seat the movable part. This can be a desirable feature in some systems because a pressure drop is not required to unseat the movable part. The movable part is seated by gravity or by the initial fluid movement when the fluid flow attempts to reverse direction. If the valve depends on gravity to close, it must be mounted in a near-vertical position to ensure proper operation. If fluid flow alone seats the movable element, care needs to be taken to match the flow required to close the valve to the expected reverse flow.

Restriction check valves

There are specialized check valves designed to allow free flow in one direction and restricted or metered flow in the opposite direction. These valves are called *restriction check valves*. **Figure 10-44** illustrates a typical valve of this type. These valves are basically a standard, poppet-type check valve with a metering orifice in the poppet.

When fluid enters the valve at the inlet port, the resistance of the spring holding the poppet closed causes pressure to increase in the valve. The pressure increases until the check valve opens to allow free flow of fluid to the system. When fluid flow attempts to reverse direction, the poppet seats to block flow. However, the orifice in the poppet allows a measured amount of fluid to return through the valve. The exact amount of return flow is determined by the size of the opening. This is specified when selecting the valve for an application.

Restriction check valves may be found in large forming presses where they are used to control the lowering speed of the large cylinders that move the platen. They are also used to decompress large hydraulic systems where the sudden release of high system pressure can cause undesirable shock to components.

Pilot-operated check valves

Pilot-operated check valves have all of the features of standard check valves. However, they may be opened to allow reverse flow through the valve. In some designs, the pilot feature may also be used to prevent flow through the valve in either direction.

Figure 10-45A illustrates a check valve with a pilot that can be used to open the poppet to allow reverse flow through the valve. When pilot pressure is not applied, the valve operates as a standard check valve, limiting fluid flow to only one direction. When pilot pressure is applied, the pilot piston unseats the ball against the force generated by the valve spring and the system pressure on the ball. As long as sufficient pilot pressure is applied, the valve remains open to allow free flow in either direction.

Figure 10-45B illustrates another version of a pilot-operated check valve. It is used to prevent the movement of fluid through the valve in either direction. In this way, it acts as a shutoff valve. When the pilot section of this valve is activated, the poppet is forced against the seat to prevent fluid flow in either direction.

The pressure required to energize the pilot in pilot-operated check valves depends on the force needed to unseat the movable part to allow reverse flow or to close it and hold it seated. When opening the valve to allow reverse flow, the pilot must overcome the force generated by the strength of the spring and system pressure acting against the area of the movable part. When acting as a shutoff valve, the pilot pressure acting on the movable part must produce a force greater than the force generated by system pressure attempting to unseat it.

10.5.3 Three-Way Valves

Three-way valves are directional control valves used to operate rams and single-acting cylinders. They provide a means to extend the actuator using hydraulic system pressure, while using the weight of the external system load, a spring return in the cylinder, or the weight of a machine member to power the return stroke.

Three-way valves have three external ports with connections to the system pump, one port of the cylinder, and a system line leading to the reservoir. Shifting the valve in one direction connects the pump to one side of the cylinder. This allows system fluid to extend the cylinder. In this shifted position, the port leading to the reservoir is blocked. Shifting the valve in the opposite direction blocks the pump port and connects the cylinder port to the reservoir. External forces acting on

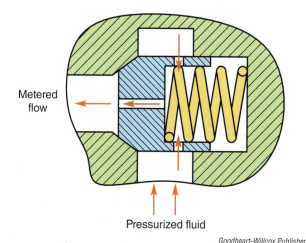

Metered flow

Pressurized fluid

Goodheart-Willcox Publisher

Figure 10-44. A restriction check valve has an orifice machined into the movable sealing element. This orifice allows a measured reverse flow through the valve.

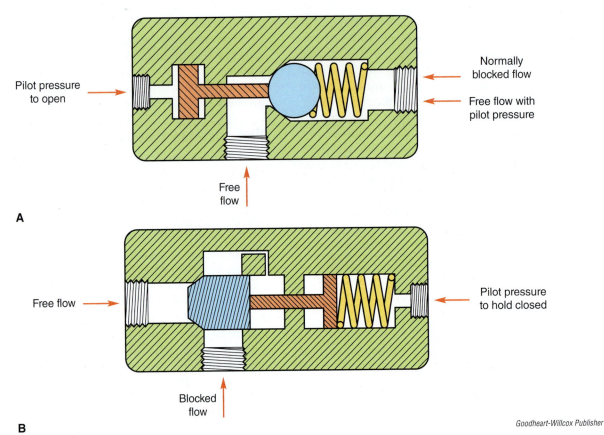

Pilot pressure to open

Normally blocked flow

Free flow with pilot pressure

Free flow

A

Free flow

Pilot pressure to hold closed

Blocked flow

B

Figure 10-45. Pilot-operated check valves. A—Pilot pressure to open the valve against system pressure. B—Pilot pressure to hold the valve shut against normal reverse flow.

the cylinder causes it to retract. This forces the fluid contained in the cylinder to flow to the reservoir. See **Figure 10-46**.

Figure 10-47 shows the structure of a basic three-way valve. The valve consists of a valve body containing a precision-machined bore. The bore is connected to pump, actuator, and reservoir ports. Fitted in the

bore is a sliding spool designed to direct the flow of fluid through the valve. A clearance machined into the spool allows ports to be connected to produce the desired fluid passageway through the valve.

The system actuator is activated by moving the spool so the pump and actuator ports are connected through the clearance on the spool. At the same time,

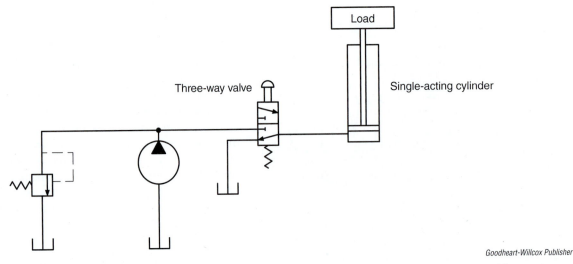

Load

Three-way valve

Single-acting cylinder

Figure 10-46. Three-way valves provide pressurized fluid to power one side of a single-acting cylinder. External force returns the actuator to its original position when the valve is shifted to the starting position.

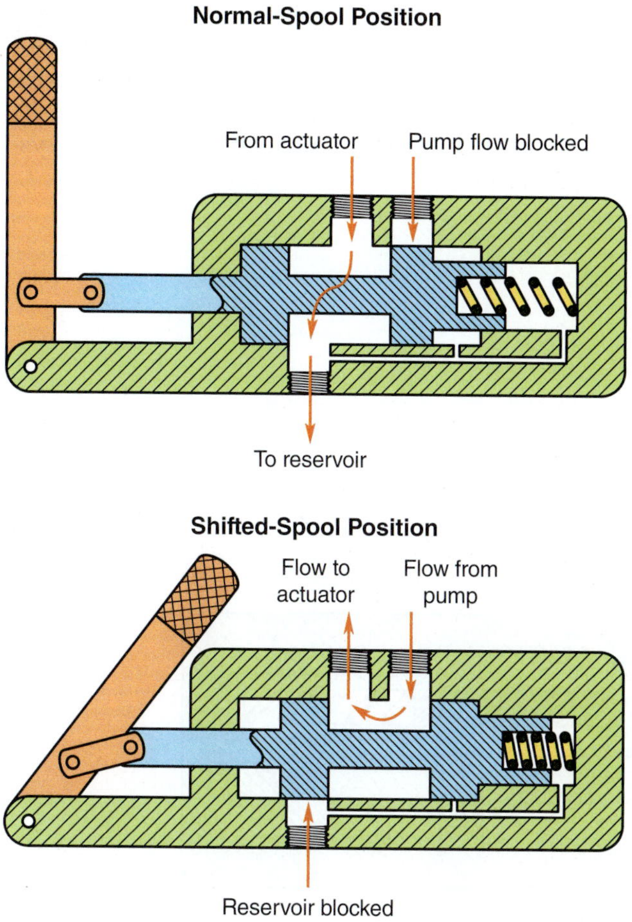

Normal-Spool Position

From actuator   Pump flow blocked

To reservoir

Shifted-Spool Position

Flow to actuator   Flow from pump

Reservoir blocked

Goodheart-Willcox Publisher

Figure 10-47. Three-way directional control valves have three ports, providing connections to the pump, actuator, and reservoir.

the reservoir port is blocked by one of the full-diameter sections of the spool. These sections are called *lands*.

Shifting the spool in the opposite direction connects the actuator and reservoir ports. The pump port is blocked by a spool land. Under this condition, the fluid from the actuator flows thorough the valve to the reservoir. This allows the cylinder to retract while the pump output flow is returned to the reservoir through the system relief valve. The retraction speed of the actuator depends on the force produced by the external factors acting on the single-acting cylinder.

Located on either end of the spool is a spring and an actuating mechanism, such as a solenoid or lever. The actuating mechanism is selected to best fit the valve application. The spring and mechanism allow easy and accurate control of the spool position in relation to the valve ports.

Three-way valves may be constructed as either normally open or normally closed. A valve is considered normally open if it connects the pump to the actuator when the actuating mechanism is not activated. If the valve connects the pump to the reservoir in the normal position, it is normally closed.

10.5.4 Four-Way Valves

Four-way valves are used in circuits with hydraulic motors or double-acting cylinders. They provide a means of using system pressure to both extend and retract the cylinder. See **Figure 10-48.** Likewise, the valve design allows hydraulic motors to be powered in either rotation direction.

Four-way valves have four external ports with connections to the system pump, inlet and outlet lines of the actuator, and a system line leading to the reservoir. Shifting the valve in one direction connects the pump to the cylinder, allowing system fluid to extend the cylinder. In this shifted position, the valve also connects the opposite end of the cylinder to the reservoir. This allows fluid from that chamber to return to storage with minimal flow resistance. Shifting the valve in the opposite direction connects the pump to the rod end of the cylinder and the blind end of the cylinder to the reservoir. This forces the cylinder to retract. A comparable combination of connections allow a hydraulic motor to be powered in a clockwise or counterclockwise direction.

Two-position, four-way valve

A basic two-position, four-way valve causes the actuator to move in one direction while in the first position and to reverse the operating direction when shifted into the second position. This valve consists of a valve body containing a precision-machined bore connected to internal passageways leading to the pump, reservoir, and two actuator ports. The ports are often identified as *P* for the pump port, *T* for the reservoir port, and *A* and *B* for the actuator ports. Fitted in the bore is a sliding spool designed to direct the flow of fluid through the valve. Clearances machined into the spool allow the ports to be connected to produce the desired fluid movement through the valve.

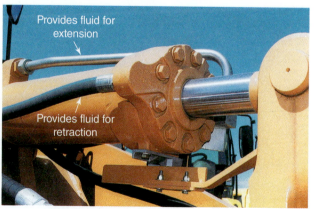

Provides fluid for extension

Provides fluid for retraction

Goodheart-Willcox Publisher

Figure 10-48. Four-way directional control valves provide pressurized fluid to power double-acting cylinders during both extension and retraction.

Shifting the valve to the first position connects the pump port (P) to the actuator port (A) leading to the blind-end of the system cylinder. See **Figure 10-49A**. Fluid from the pump enters the valve and passes through the internal passageways and clearance on the spool. The fluid is directed through the actuator port (A) and enters the blind end of the cylinder. As the cylinder extends, fluid from the rod end is forced back to the directional control valve through the actuator port (B). This internally directs fluid to port (T), which returns the fluid to the system reservoir.

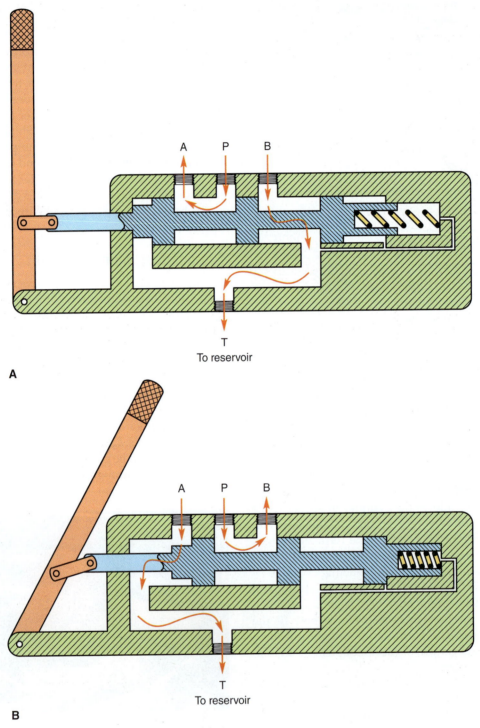

Figure 10-49. A two-position, four-way directional control valve showing the relationship of the ports. A—Fluid is directed to port A. Port B is open to the reservoir. B—Fluid is directed to port B. Port A is open to the reservoir.

Shifting the valve spool to the second position connects the pump to the rod end of the cylinder. See **Figure 10-49B**. This supplies the fluid needed to retract the actuator, while routing the fluid returning from the blind end of the cylinder to the reservoir.

Four-way valve center position

Three-position valve design variations provide four-way directional valves with a third control that goes beyond the simple extension and retraction of cylinders and rotational direction of motors. This third control position is located between the two extreme positions described in the previous section. See **Figure 10-50**. It is often called the center position. This position is sometimes referred to as the intermediate position. In some specialized industries, such as the mobile equipment field, two or more of these positions are added to provide desired operating characteristics.

The spool configuration and internal passageways in the valve body provide the operating characteristics desired for the valve center position. The location of the valve body passageways, the spacing of the grooves and lands on the spool, or passageways machined into the center of the spool are used to produce the desired operating characteristics.

Several center position configurations are available from valve manufacturers. However, this discussion is limited to five commonly available center design options. These include closed, open, tandem, floating,

and regenerative centers. The symbols of these center positions are shown in **Figure 10-51**.

The closed-center configuration provides a center position where the pump (*P*), reservoir (*T*), and actuator ports (*A* and *B*) are all closed. With the pump port closed, the pump output is returned to the reservoir over the system relief valve at full system pressure. With both actuator ports blocked, the actuator is held in a locked position. Because the system is operating at relief-valve pressure, the full capacity of the pump may be used to operate another actuator in the system.

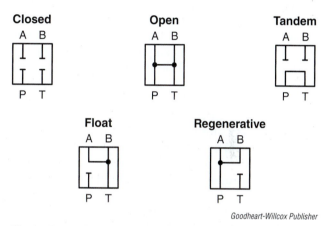

Goodheart-Willcox Publisher

Figure 10-51. Many center design options are available for three-position directional control valves. However, the five center-design options shown here are widely available.

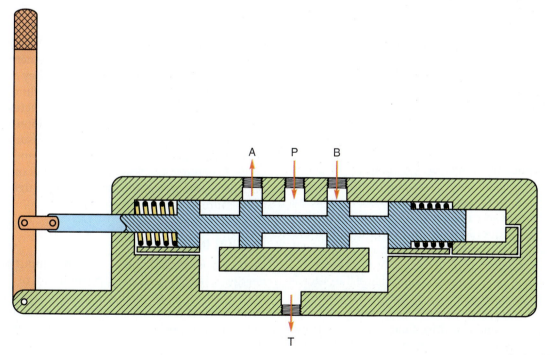

Goodheart-Willcox Publisher

Figure 10-50. A three-position, four-way directional control valve. The two extreme positions of the valve direct fluid the same as the two-position valve shown in **Figure 10-48**. The center position shown here provides a third alternative, which blocks all ports of the valve.

If the system is using multiple actuators with closed-center, four-way valves, simultaneous operation is possible, but system response time is reduced. Another characteristic of this center position is the performance of the system as the valve is moved from either of the activated positions. Shifting the valve from an activated position to the center position abruptly blocks fluid flow. This may produce high shock pressures in the circuit. Under some operating and load conditions, this may lead to component damage.

The open-center configuration connects the pump (P), reservoir (T), and actuator ports (A and B). This interconnection allows the pump to return fluid to the reservoir at minimal pressure while permitting fluid to freely move into and out of either side of the actuator. This configuration minimizes energy consumption by the prime mover when the system is idle. However, it eliminates the possibility of the pump output operating other system actuators when the directional control valve is centered. Also, with both sides of the actuator connected to the pump and reservoir, the actuator is free to float in response to external forces applied by the load.

The tandem-center configuration connects the pump (P) and reservoir (T) ports and blocks both actuator ports (A and B). This results in the pump output being transferred to the reservoir at minimal pressure when the system is idle, while the actuator is locked in place. This allows minimal energy consumption by the prime mover while the system is idle. In addition, two or more of these valves may be used in series by connecting the reservoir and pump ports of alternate valves. In a series configuration, a single valve may be operated without diminished performance. However, simultaneously operating valves modifies system performance.

In the floating-center configuration, the pump port (P) is blocked while the actuator ports (A and B) are connected to the reservoir port (T). The blocked pump port allows the full capacity of the pump to be used in other system segments. Since both actuator ports are connected to the reservoir, the actuator is free to float in response to external loads.

The regenerative-center design connects the pump port (P) to the actuator ports (A and B), while the reservoir port (T) is blocked. This configuration is used in situations where a rapid advance is required for a double-acting cylinder. When the directional control valve is shifted into this center position, both actuator ports are pressurized. Because of differences in piston areas, the force generated by system pressure acting on the blind side of the piston is greater than the force generated by the same pressure acting on the rod side of the piston. This greater force causes the cylinder to extend and drive fluid from the rod side. The fluid is returned to the valve where it combines with fluid coming from the pump. This increased flow is then routed back to the blind end of the cylinder, causing an increase in the extension rate.

As you can see, the center position of a four-way directional control valve has considerable influence on the performance of a hydraulic system. The center position may direct fluid flow through the directional control valve or control the way in which actuators respond to system startup, idling, and stopping. Care must be taken to select the configuration that is the most energy efficient for the application, provides the control desired to slow and stop an actuator, and provides the performance characteristics desired while the system is idle.

10.5.5 Actuation Methods for Directional Control Valves

Directional control valves, more than any other general valve type, use a variety of methods to shift the valve. The actuation methods range from simple, self actuation to electrical control that may be more complex, but extremely accurate. These actuation methods can be grouped in five general operating categories:

- Fluid-flow actuation.
- Manual operation.
- Mechanical operation.
- Pilot operation.
- Electrical operation.

In some situations, two of these methods may be used in combination to produce the desired results.

Fluid-flow actuation

With fluid-flow actuation, the valve is automatically opened or closed by the internal movement of fluid. Basic check valves are flow actuated as they are opened and closed by fluid movement. System fluid flow acting against the ball or poppet typically actuates these valves. If the check valve includes an internal spring to keep the valve closed, a small pressure increase is needed to unseat the ball or poppet and open the valve. When reverse flow occurs, the ball or poppet is initially seated by the spring, but is more tightly sealed as pressure increases in the system.

Manual operation

Manual operation involves the direct input of human energy. A handwheel, lever, pushbutton, or foot pedal may be directly attached to the valve. The operator's hands or feet are used to move the actuating device to the desired position. This aligns the internal mechanisms of the valve to produce actuator movement.

Manual controls are used on all types of shutoff, three-way, and four-way directional control valves. For example, a handwheel is typically attached to a threaded shaft used to open a needle, globe, or gate shutoff valve. Also, various designs of lever controls can be found on some spool-type shutoff valves and most three-way and four-way valves. In addition, pushbuttons are sometimes used with spool-type shutoff valves. Foot pedals may be used in special situations to shift a spool-type shutoff valve and three-way and four-way directional control valves.

Mechanical operation

Mechanical operation is often used when the opening and closing of a directional control valve must occur at a specific position in actuator travel. An example is a valve actuated using a roller on the end of a shaft connected to the valve spool. See **Figure 10-52**. The roller follows a cam positioned on a moving machine

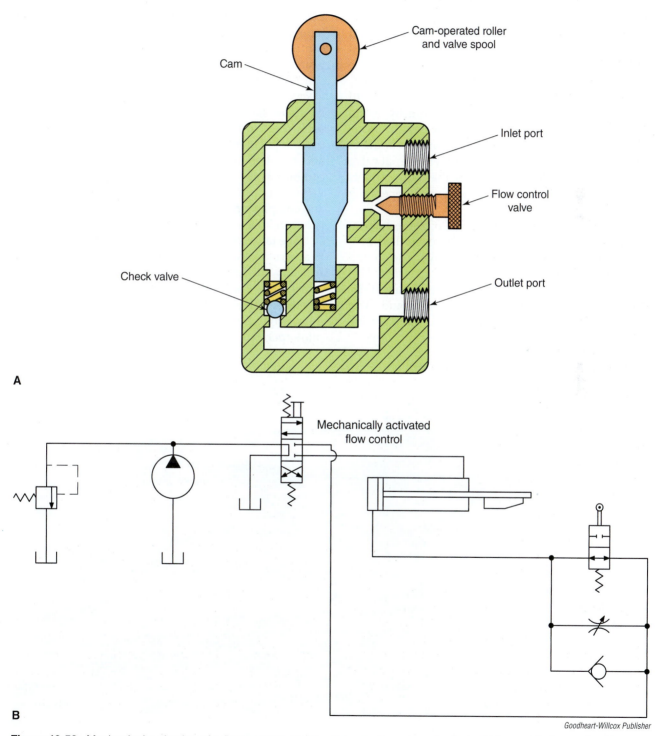

Goodheart-Willcox Publisher

Figure 10-52. Mechanical activation of a flow control valve is often used to automatically control a rapid advance/feed sequence of an actuator in machine tool systems. A—Cutaway of a valve with roller activation. B—The valve in a circuit.

member or the actuator being controlled. The shape of the cam causes the valve to open and close to produce the desired sequence of valve operation. Other combinations of mechanisms using levers, rams, and cams can also be used to shift the internal parts of directional control valves.

Pilot operation

The internal mechanisms of some directional control valves are shifted using system fluid pressure as the source of power. This technique is called piloting. It is accomplished by using the pressure of the system fluid to produce the force that shifts the valve into the desired position. The exact way in which the pressure is applied depends on the design of the component. In some units, the pressurized fluid may directly act on the end lands of the valve spool. Other designs incorporate a separate pilot piston. This piston is in direct contact with the end of the spool or other internal valve parts. The fluid transmission lines that power pilots are usually small in comparison to the main operating lines of the system.

The system fluid that powers the pilot mechanism of a directional control valve is often controlled by a second, smaller directional control valve. The second valve is either manually or electrically activated. This configuration is called staged control. See **Figure 10-53**. The smaller size of both the pilot control line and valve allows more flexibility in regard to their location on the equipment.

Staged control of pilot controlled valves has two advantages. First, the power-assisted shifting of valves with high flow capacities reduces the physical effort required of a machine operator. Second, it allows operator controls to be smaller, less-expensive pilot lines and control valves. The operator controls can also be located in remote locations.

Electrical operation

Electrical control of hydraulic components is very common in both industrial and mobile equipment. These control systems can range from basic to very sophisticated. The most basic of these systems use relatively simple solenoids operated by mechanically activated electrical switches. The solenoids shift the internal elements of standard hydraulic control valves. In more sophisticated forms, proportional solenoids may be used to position control valves with extreme accuracy and repeatability. These valves often are designed to combine both directional and flow control in a single valve to reduce the physical size of systems. The design and complex operation of proportional solenoid valves is beyond the scope of this text.

The term solenoid valve is, however, often applied to any control valve using electrical power for control. Actually, the term usually deals only with the means of actuating the valve. Care should be taken to classify these valves as electrically operated to distinguish them from flow-actuating, manual, mechanical, or pilot-operated valves.

Pilot control valve

System directional control valve

Courtesy of Eaton Fluid Power Training

Figure 10-53. System fluid is used to shift the spool of pilot-operated directional control valves. Manual or electrical power is commonly used to operate the pilot section of these valves. The resulting fluid movement then shifts the primary valve.

Solenoids, however, can be a very important part of the control of hydraulic valves, especially directional control valves. As shown in **Figure 10-54**, a solenoid is basically a linear electric motor that consists of a metal frame, wire coil, and plunger. The doughnut-shaped wire coil is placed inside of the metal frame and the plunger slides through the opening of the wire coil. When electricity is passed through the wire coil, a strong magnetic field is created. The lines of force of the magnetic field pull the plunger to the center of the field. Plunger movement is used to shift the internal elements of the valve to obtain desired system operation.

The shifting force generated by a solenoid is relatively small. This restricts the size of the valve that can be directly shifted by such a unit. A design that is often used to overcome this disadvantage is a valve that is solenoid controlled and pilot operated. These valves involve a solenoid-activated, smaller-capacity directional control valve that activates the pilot section of a larger-capacity valve. This two-stage operation is very effective.

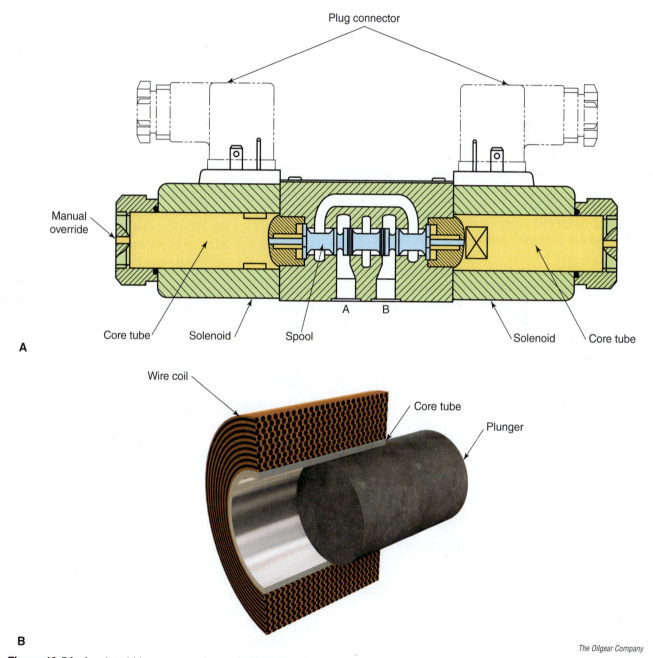

The Oilgear Company

Figure 10-54. A solenoid is a commonly used actuator on hydraulic valves. It is a linear motor that shifts the valve spool by using a magnetic field to center the metal plunger located in the center of the wire coil. A—A solenoid controlled valve. B—Cross section of a solenoid.

10.5.6 Holding Position in Multiple-Position Directional Control Valves

In an operating hydraulic circuit, it is often desirable to have the spool of a multiple-position directional control valve held in a selected position until the machine completes an operating cycle. At that point, the valve may return to a predetermined normal position. A relatively simple way to provide this capability is through the use of detents or springs.

Detents are locking devices used to keep the spool of a directional control valve in a selected position. **Figure 10-55** shows a typical detent. It consists of notches machined into a rod on the end of the valve spool. A spring-loaded ball mounted in a hole in the valve body rests against the rod. When the spool is shifted, the ball encounters one of the notches at some point in the spool travel. The spring forces the ball into the notch, which holds the spool at that position. The holding power required of the detent is determined by the valve designer based on circuit operating characteristics. When a detent is engaged, the actuation force can be removed as the detent will hold the valve in position until an additional actuation force is applied. Detents can be used to hold a valve in the centered or either shifted position.

Springs can be used to automatically position a valve spool in a normal or restart position, **Figure 10-56**. The springs are located on the ends of the spool. For example, in a two-position, three-way valve, a single spring can be used to close the valve as soon as the actuation force is removed. This blocks the pump connection and opens the reservoir connection, allowing the actuator to drain to the reservoir. In three-position valves, springs on either end of the spool shift the spool into the center position whenever the actuation force is removed. The circuit then functions according to the type of center incorporated into the spool design.

Manufacturers produce a variety of detent and spring positioning combinations. Some valves may use only a detent or only springs. Others may use a combination to produce the desired characteristics. For example, one relatively common combination is a three-position, four-way valve with a detent and is spring centered. The detent holds the valve in the position to extend the cylinder. Once the cylinder is fully extended, the increased pressure in the spring chamber generates additional force on the end of the spool. This force added to the force of the spring overcomes the holding power of the detent and the valve automatically centers itself. The system then goes into the mode of operation designed into the valve center.

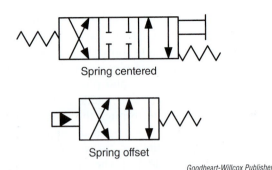

Spring centered

Spring offset

Goodheart-Willcox Publisher

Figure 10-56. Springs located on the ends of the spool are often used to hold a directional control valve in a selected position. Springs can hold the valve in a centered position or offset it in either direction.

10.6 Flow Control Devices

Devices that control the rate of fluid flow in a hydraulic system make up the third of the three basic groups of control valves. These valves can be divided into two general types according to the method used to deliver fluid flow to a system actuator:

- Restrictor.
- Bypass control.

Both designs produce the desired rate of actuator operating speed by controlling the volume of fluid allowed to reach the actuator. Both designs also depend on having a flow volume coming from the system pump that is larger than necessary to operate the actuator at the desired speed. However, the designs vary in the way they achieve the desired control.

Restrictor flow control valves limit the flow passing through the valve. They force excess pump output fluid to return to the reservoir through the system pressure relief valve. See **Figure 10-57**. In comparison,

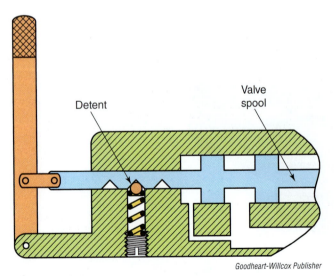

Detent

Valve spool

Goodheart-Willcox Publisher

Figure 10-55. Detents are used to hold a multiple position valve in a selected position. Detents typically consist of a spring-loaded ball that drops into a notch cut into a movable member attached to the valve spool.

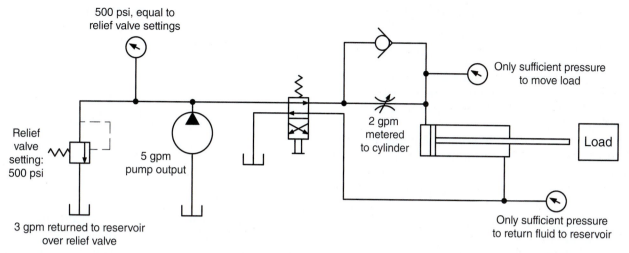

Figure 10-57. This circuit contains a restrictor flow control valve that meters fluid flow to the actuator with excess pump output returned to the reservoir through the relief valve at maximum system pressure.

bypass flow control valves use an integral control port to return excess fluid directly to the reservoir. See **Figure 10-58.** Both restrictor and bypass valve designs can use a variety of devices to maintain a controlled flow rate even when inlet and outlet pressures and fluid temperatures vary.

Manufacturers produce a variety of flow control valve designs. However, under close examination, most of these valves can be classified using the above two groupings. The following material relates to the most common of the flow control valves used in the fluid power field today.

10.6.1 Fluid Mechanic Concepts Basic to Flow Control Devices

Flow control devices are designed to accurately limit the volume of fluid entering an actuator to control its rate of movement. This is necessary to provide accurate

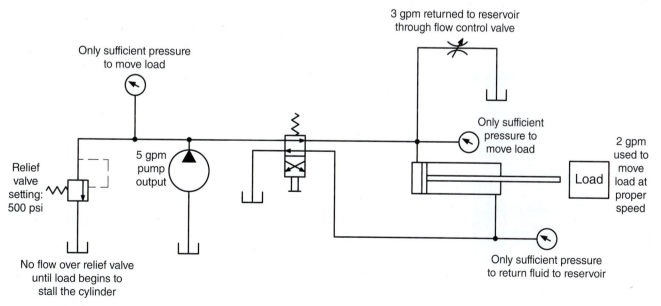

Figure 10-58. This circuit contains a bypass flow control valve that meters fluid flow to the actuator with excess pump output returned to the reservoir at a pressure determined by the load on the actuator. Maximum operating pressure is controlled by the setting of the system relief valve.

control of cylinder velocity and hydraulic motor rotation. Accurate tool cutting speeds, grinding speeds, lifting rates for material handling equipment, and a multitude of other applications depend on accurate flow control. Although the outward appearance, complexity, and accuracy of various flow control valves vary, their conceptual operation may be traced to the operation of a basic orifice.

In its simplest form, an *orifice* is nothing more than a calibrated hole machined into a plug that is fitted into a line transporting fluid. See **Figure 10-59**. The line may be a pipe, hose, tube, or passageway machined into a valve body or component part. When fluid is forced through the line, the orifice restricts the rate at which fluid flows. The fluid flow rate through an orifice depends on:

- Area of the orifice opening.
- Pressure difference between the inlet and outlet sides of the orifice.
- Fluid viscosity, which depends on the temperature of the fluid.

Changing any of these three variables changes the rate of fluid flow through the orifice. It is easy to maintain the cross-sectional area of a valve orifice. It is more difficult to maintain a constant pressure drop across an orifice or a consistent fluid temperature. Because of these variable conditions, flow in an operating section of a hydraulic system may be difficult to accurately predict or control.

Books dealing with fluid mechanics show a number of mathematical formulas relating to calculating flow through orifices. The formulas typically make use of velocity, contraction, and discharge coefficients to simplify calculations. An example of a formula dealing with fluid discharge through an orifice is:

$$Q_A = A_O \times C_D \times 2g \times H$$

where:

Q_A = Actual quantity of flow

A_O = Cross-sectional area of the orifice

C_D = Coefficient of discharge

g = Gravity

H = Head

Formulas such as this are typically used during the design phase of valve development. Once a valve design reaches the prototype stage, laboratory test results are used to develop charts and tables that illustrate the performance of the valve under operating conditions. If a new design does not perform as expected during the initial tests, further theoretical development is done to produce a valve that operates within the stated specifications.

10.6.2 Restrictor-Type, Noncompensated Flow Control Valves

Restrictor-type, noncompensated flow control valves are the simplest of the restrictor flow control devices. They can provide a consistent flow rate in a section of a circuit as long as actuator loads and system fluid temperature remain constant. Varying system loads result in varying pressure drops across the valve and, in turn, varying flow rates. Varying system operating temperatures produces viscosity changes in system fluid that also result in varying flow rates. Restrictor-type, noncompensated flow control valves are available as units with either fixed or adjustable flow rates.

Fixed-flow-rate devices

Fixed-flow-rate flow control devices typically have a simple orifice, such as described earlier in the chapter. This type of device is not generally available from hydraulic component manufacturers for use in aftermarket applications. However, they can be easily machined from metal stock and fitted into system distribution lines to produce the desired reduced flow rate in a system section.

Fixed-flow-rate orifices also play an important part in the internal design of a number of hydraulic components. They are used in pressure control valves to create pressure drops for the mechanisms that operate the compound feature of these valves. They are also used in the mechanisms that operate pressure-compensated flow control valves.

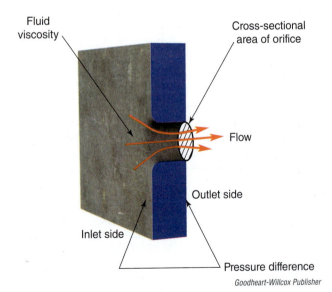

Fluid viscosity

Cross-sectional area of orifice

Flow

Outlet side

Inlet side

Pressure difference

Goodheart-Willcox Publisher

Figure 10-59. Flow through an orifice is determined by the cross-sectional area of the opening, the pressure drop between the inlet and outlet sides, and the viscosity of the fluid.

Adjustable-flow-rate devices

The simplest of the adjustable-flow-rate flow control valves is the needle valve. **Figure 10-60** shows a cutaway view of a typical needle valve. These valves are available in a variety of shapes and sizes. The valve consists of a valve body containing inlet and outlet ports, a metering orifice, and a tapered needle with a threaded stem for adjusting the effective opening through the orifice.

The threaded stem of the needle is fitted with a screwdriver slot or some form of handwheel for adjusting the effective area of the orifice. Turning the needle into the orifice reduces the effective area of the opening.

This decreases the flow rate for a given pressure drop across the orifice. The angle of the taper on the needle and the fineness of the threads on the stem determine the precision of valve adjustment. However, it must always be kept in mind that the flow through this valve will fluctuate for any needle setting if the pressure drop changes across the orifice.

A check valve is often an integral part of a needle valve. See **Figure 10-61**. This produces a compact valve that provides metered fluid flow in one direction and free flow in the opposite direction.

Valve performance

Figure 10-62 illustrates a basic hydraulic circuit using a restrictor-type, noncompensated flow control valve. The flow control valve is placed between the blind end of the cylinder and the directional control valve. In this example

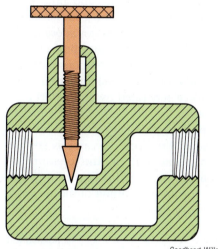

Goodheart-Willcox Publisher

Figure 10-60. A needle valve is a common adjustable-flow-rate flow control device. The valve consists of an orifice fitted with a tapered needle that can be easily adjusted to vary the cross-sectional area of the orifice opening.

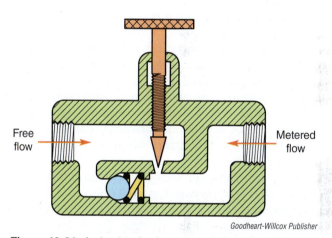

Free flow → ← Metered flow

Goodheart-Willcox Publisher

Figure 10-61. A check valve is commonly incorporated into a needle valve to allow reverse flow to pass unrestricted.

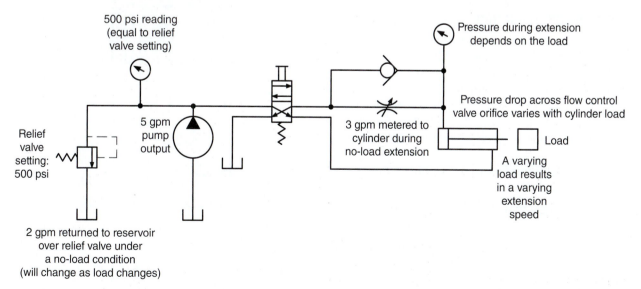

500 psi reading (equal to relief valve setting)

Pressure during extension depends on the load

5 gpm pump output

Relief valve setting: 500 psi

Pressure drop across flow control valve orifice varies with cylinder load

3 gpm metered to cylinder during no-load extension

Load

A varying load results in a varying extension speed

2 gpm returned to reservoir over relief valve under a no-load condition (will change as load changes)

Goodheart-Willcox Publisher

Figure 10-62. A basic hydraulic circuit using a restrictor-type, noncompensated flow control valve.

circuit, the pump has an output of 5 gpm, the relief valve is set at 500 psi, and the flow control valve is set at 3 gpm. When the cylinder is extending under a no-load condition, the flow control valve delivers 3 gpm to the cylinder. The extra 2 gpm of pump output is forced through the relief valve. The pressure drop across the flow control valve is 500 psi because of the relief valve setting and no load on the cylinder. However, when the cylinder is loaded, the pressure drop across the flow control valve decreases. This is because the relief valve setting remains the same, while the pressure in the blind end of the cylinder increases due to the cylinder load. The result is a cylinder speed that varies as the load varies.

10.6.3 Restrictor-Type, Compensated Flow Control Valves

Hydraulic system actuators must often perform at consistent speeds under varying load and operating conditions. This requires a flow control valve that can provide a consistent pressure drop across its metering orifice. In addition, the valve must provide a consistent flow rate even as the fluid viscosity varies due to changes in system fluid operating temperature. Valves are available to control both of these factors. These are compensated valves that can maintain a consistent flow rate over a range of constantly changing pressures and system fluid temperatures.

Manufacturers have a number of designs that can be used to maintain a consistent pressure drop across the metering orifice. Various devices are also used to compensate for variations in the flow rate as viscosity varies due to fluid temperature changes. The following sections discuss pressure- and temperature-compensating devices as well as designs used to provide fixed and adjustable flow rates for compensated control valves.

Pressure compensation

A valve with *pressure compensation* contains a mechanism that maintains a constant pressure difference between the inlet and outlet passageways. This mechanism is called a pressure compensator. The flow control device is typically a fixed orifice or adjustable needle valve in restrictor-type flow control valves. A pressure-compensated valve also includes a compensator that consists of:

- Spool fitted into a precision-machined bore in the valve body.
- Biasing spring that establishes a preselected pressure drop across the flow control device.
- Two pilot lines connecting chambers at the ends of the spool to the inlet and outlet lines of the flow control device.

Figure 10-63 shows a basic pressure-compensated flow control valve.

When the hydraulic system is not operating, the force of the biasing spring slides the spool of the compensating device to allow full flow past the spool. In this position, the inlet and outlet ports of the valve are directly connected through the compensating spool area and the flow control device. One pilot line connects the outlet port of the flow control device to the end of the compensator spool containing the biasing spring. The second pilot connects the opposite end of the spool to the chamber formed between the compensating spool and flow control device. During compensator operation, this compensation chamber reflects the pressure in the passageway between the flow control device and the pressure compensator.

Compensator operation is based on forces generated by the biasing spring and fluid pressure acting on the end surfaces of the compensator spool. The force pushing the compensator spool to restrict flow is established by the area of the spool end multiplied by the fluid pressure in the compensation chamber. This force is counteracted by the force of the biasing spring plus the force generated by outlet port pressure acting on the area of the spool end with the biasing spring.

In an operating system, when a pressure-compensated flow control valve is set at a flow rate less than pump output, the pressure between the valve compensator and the flow control device quickly increases toward the setting of the system relief valve.

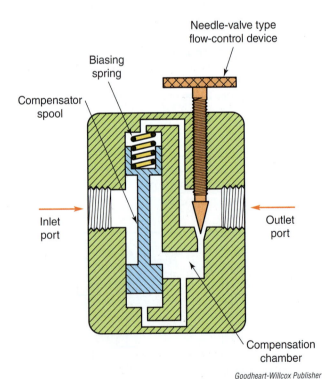

Goodheart-Willcox Publisher

Figure 10-63. The internal design of a simple, pressure-compensated flow control valve. Note the component parts and their relationship to the inlet and outlet ports of the valve.

The compensation chamber between the compensator spool and flow control device is exposed to this increasing pressure. The pilot line leading from the chamber also exposes one end of the compensator spool to this changing pressure.

When the pressure transmitted through the pilot line generates a force on the end of the spool greater than the force generated by the bias spring and the pilot pressure acting on the opposite end of the spool, the biasing spring is compressed and the spool slides to restrict flow. The spool movement reduces the size of the compensator passageway between the valve inlet port and the flow control device. This reduces flow and limits the pressure increase on the inlet side of the flow control device and the pilot line leading from this chamber to the end of the compensator spool. The result is a consistent pressure drop across the flow control device based on the force generated by the biasing spring.

When the load on the system actuator varies, the pressure on the outlet side of the flow control device also varies. This causes the pilot pressure on the end of the compensator spool with the biasing spring to vary. These pressure variations cause the forces acting on the ends of the compensator spool to slide the spool back and forth. See **Figure 10-64**. This slight spool movement varies the size of the compensator passageway opening. When the forces resulting from pilot pressures are equalized, the size of the opening is stabilized. The stabilized opening allows the passage of sufficient fluid to maintain the selected pressure difference between the inlet and outlet of the flow control device, while maintaining the flow rate selected to operate the system actuator(s). The result is a stable pressure drop across the flow control device that produces the selected rate of fluid flow. It must be understood that under varying system load conditions, the compensator spool and orifice size are constantly varying to maintain the desired flow.

Temperature compensation

As the temperature of a fluid changes, so does its viscosity. When viscosity changes, the rate of fluid flow through lines and valves changes. In hydraulic systems where varying loads and operating conditions are encountered, the operating temperature of the system can vary considerably. Where accurate flow control rates are critical, it is especially important to have a means to compensate for changes in fluid temperature. Without *temperature compensation*, accurate flow control is not possible. Controlling changes in flow rate due to changes in fluid viscosity caused by fluid temperature variations is accomplished using either a:

- Specially designed, sharp-edged orifice.
- Heat-sensitive metal rod for the automatic adjustment of the opening in an orifice.

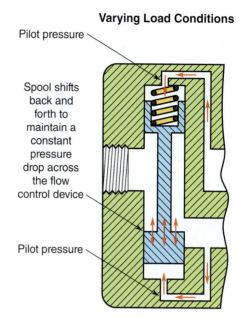

Varying Load Conditions

Pilot pressure

Spool shifts back and forth to maintain a constant pressure drop across the flow control device

Pilot pressure

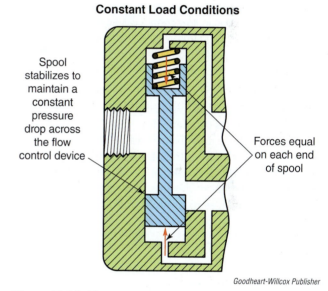

Constant Load Conditions

Spool stabilizes to maintain a constant pressure drop across the flow control device

Forces equal on each end of spool

Goodheart-Willcox Publisher

Figure 10-64. The operation of a compensator maintains a constant pressure drop across the flow control device in a pressure-compensated flow control valve.

The sharp-edged orifice is considered viscosity insensitive (unaffected by viscosity). It does not physically adjust valve components to obtain a uniform flow, **Figure 10-65**. The theory of the design is that the orifice minimizes changes in flow rates even when temperature and viscosity changes occur in system fluid. These orifices are difficult to design and manufacture. However, some technical data sheets claim accuracies within 2%, even under substantial viscosity changes.

The heat-sensitive metal rod changes dimensions as the fluid temperature changes. The expansion rate and the geometry of the other mechanism parts must correlate to the viscosity changes that occur in the fluid. The design must be accurate so the device can modify

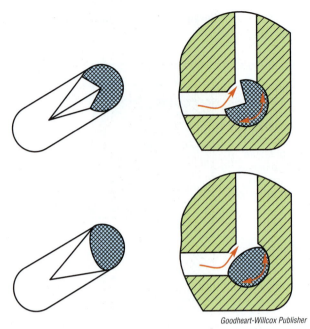

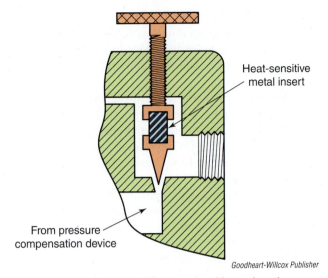

From pressure
compensation device

Heat-sensitive
metal insert

Goodheart-Willcox Publisher

Figure 10-66. A heat-sensitive metal rod in conjunction with a needle can be used as a temperature-compensating device.

Goodheart-Willcox Publisher

Figure 10-65. Manufacturers often use specially designed, sharp-edged orifices in flow control valves to greatly reduce flow variations resulting from viscosity changes produced by system temperature fluctuations. Two styles of hardware using this general concept are shown here.

the orifice size as fluid viscosity varies with temperature to maintain a constant flow rate. The orifice used with this design may be a traditional donut orifice with the heat-sensitive metal positioning a needlelike device for the desired control. See **Figure 10-66**. Other manufacturers have designs with specially shaped flow control devices to provide uniform flow.

Valve performance

Figure 10-67 illustrates a basic hydraulic circuit using a restrictor-type, compensated flow control valve. The flow control valve is placed between the blind end of the cylinder and the directional control valve. The pump has an output of 5 gpm, the relief valve is set at 500 psi, the flow control setting is 3 gpm, and the biasing spring tension requires 50 psi to overcome. This example circuit uses the same component layout as the noncompensated valve circuit shown in **Figure 10-62**. The difference is a pressure-compensated flow control valve is substituted for the noncompensated valve.

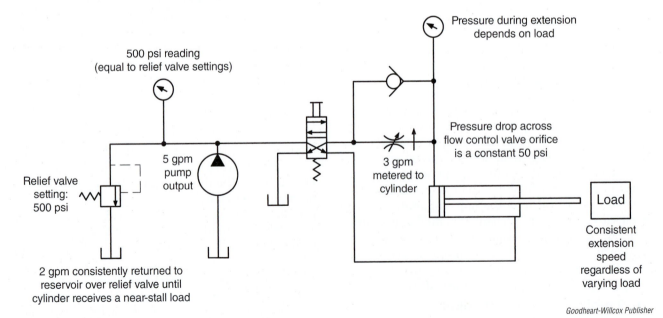

500 psi reading
(equal to relief valve settings)

Pressure during extension
depends on load

Pressure drop across
flow control valve orifice
is a constant 50 psi

5 gpm
pump
output

3 gpm
metered to
cylinder

Load

Consistent
extension
speed
regardless of
varying load

Relief valve
setting:
500 psi

2 gpm consistently returned to
reservoir over relief valve until
cylinder receives a near-stall load

Goodheart-Willcox Publisher

Figure 10-67. Pressure and flow conditions in a hydraulic circuit using a basic restrictor-type, compensated flow control valve.

In this circuit, when the cylinder is extending under a no-load condition, the flow control valve delivers 3 gpm to the cylinder. The extra 2 gpm of pump output is forced through the relief valve. This is the same as in the noncompensated circuit. The potential pressure drop across the flow control device is a consistent 50 psi. This is established by the tension of the biasing spring. When the cylinder is subjected to varying loads, the pressure drop across the flow control device continues to remain constant. The flow rate, therefore, remains constant as long as the valve is operated within specifications. The result is a constant actuator speed. This is possible because the pressure compensator built into the valve maintains a constant pressure drop between the inlet and outlet sides of the internal flow control device.

10.6.4 Bypass-Type Flow Control Valves

Bypass-type flow control valves provide accurate flow to actuators by maintaining a consistent pressure drop across the flow control device in the valve while directing any excess flow from the pump to the reservoir. A relief valve is incorporated in these circuits, but it only functions when the resistance from an actuator load is sufficient to increase system pressure to the valve cracking pressure.

Bypass valves are often referred to as three-port flow control valves as they have inlet, outlet, and bypass ports. In some applications, the bypass port is connected to a line leading to an additional function. More commonly, however, it is directly connected to the reservoir. In most models, the bypass connection cannot:

- Be pressurized.
- Effectively deliver fluid to another system branch.
- Accommodate reverse fluid flow.

Valve construction and operation

A bypass flow control valve consists of a valve body; flow control device; pressure-compensating device; and inlet, outlet, and bypass ports. See **Figure 10-68**. The flow control device in this valve is typically the same type of fixed orifice or adjustable needle valve used in restrictor-type flow control valves. The pressure compensator in this valve is similar to the compensator used in restrictor-type valves. Both types of valves are designed to provide a uniform pressure drop across the opening of the flow control device. However, in a bypass valve, this device dumps fluid to the bypass port rather than restricting flow through a compensator passageway.

The compensator consists of a spool fitted into a precision-machined bore in the valve body and a biasing spring that establishes the preselected pressure

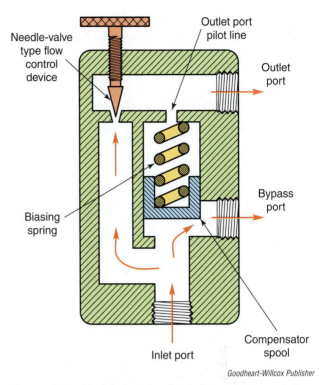

Goodheart-Willcox Publisher

Figure 10-68. The design of a simple bypass flow control valve. Note the internal components and their relationship to the inlet, outlet, and bypass ports of the valve.

drop across the flow control device. A pilot line connects the end of the spool with the biasing spring to the outlet port of the valve. The inlet and bypass ports of the valve are connected via a passageway in the bore in which the compensator spool operates. The flow control device is located in a passageway between the inlet and outlet ports of the valve.

As in the restrictor-type valve, compensator operation is based on the biasing spring tension and forces generated by fluid pressure acting on the ends of the spool. In this compensating device, the force sliding the spool to restrict flow to the bypass port is established by the area of the spool end multiplied by the pressure of the fluid in the outlet port plus the force generated by the biasing spring. This force is counteracted by the force generated by inlet port pressure of the valve acting on the end of the spool facing the valve inlet.

When the hydraulic system is not operating, the force generated by the bias spring slides the spool of the compensating device to block the passageway between the bypass port and inlet port of the valve. Once the system is operating, the pump delivers fluid to the actuator through the flow control device.

As an example, assume the pump has a fixed displacement of 10 gpm, the flow control device is set to 5 gpm, the tension of the biasing spring requires a system pressure of 50 psi to function, and there is no load on the actuator connected to the valve output port. Refer to **Figure 10-69** as you read the following description.

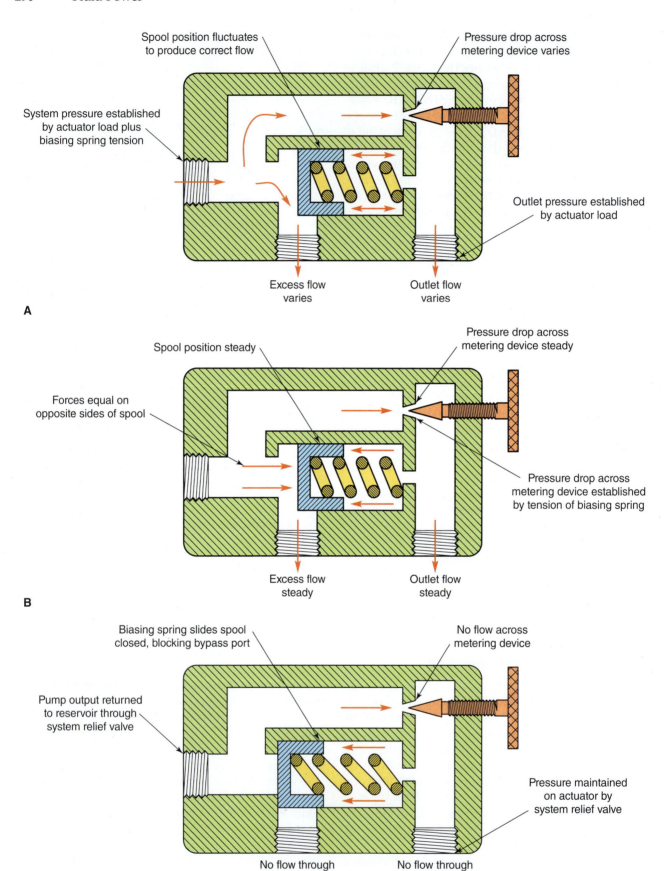

Spool position fluctuates to produce correct flow

Pressure drop across metering device varies

System pressure established by actuator load plus biasing spring tension

Outlet pressure established by actuator load

Excess flow varies

Outlet flow varies

A

Spool position steady

Pressure drop across metering device steady

Forces equal on opposite sides of spool

Pressure drop across metering device established by tension of biasing spring

Excess flow steady

Outlet flow steady

B

Biasing spring slides spool closed, blocking bypass port

No flow across metering device

Pump output returned to reservoir through system relief valve

Pressure maintained on actuator by system relief valve

No flow through bypass port

No flow through outlet port

C

Goodheart-Willcox Publisher

Figure 10-69. The operation of a bypass pressure-compensated flow control valve during three stages of system operation. A—Increasing or decreasing actuator load. B—Steady actuator load. C—Stalled actuator.

When the system is started, the pump immediately directs 10 gpm of fluid to the valve. Part of the fluid moves through the flow control device and outlet port on the way to the actuator. The remaining fluid enters the passageway leading to the compensator where it encounters the end of the spool. With no outlet to this passageway, the pressure of the fluid quickly increases to the 50 psi required to shift the compensator spool. When the 50 psi pressure is reached, the spool begins to shift. This exposes the passageway to the bypass port. At this point, the flow control device is passing 5 gpm to the actuator, while the other 5 gpm of pump output is returned to the system reservoir through the bypass port. The maximum inlet port pressure is approximately 50 psi, reflecting the activation pressure of the compensator. The pressure at the valve outlet port is near zero, as there is no load on the actuator.

When the actuator load increases, the pressure in the outlet port also increases. The pilot line connecting the end of the spool with the biasing spring exposes the diameter of the spool to the increased pressure. This increased pressure generates an increased force on the spool, moving it to decrease the size of the opening to the bypass port. This reduced opening increases the resistance to fluid flow through the opening, which, in turn, increases the fluid pressure in the valve inlet port.

The inlet port pressure may momentarily fluctuate but stabilizes at a pressure required to move the load plus overcome the tension of the biasing spring. Using the example pressures given above, the inlet pressure stabilizes at 50 psi above whatever outlet pressure is required to move the load. The flow through the flow control device stabilizes at 5 gpm because of the consistent 50 psi pressure drop between the inlet and outlet of the valve.

Valve performance

Figure 10-70 illustrates a basic hydraulic circuit using a bypass-type flow control valve. The flow control valve is placed between the blind end of the cylinder and the directional control valve. In this example, the pump has an output of 5 gpm, the relief valve is set at 500 psi, the flow control valve is set at 3 gpm, and the biasing spring requires 50 psi to overcome the tension. This example circuit uses basically the same component layout as the noncompensated and compensated restrictor valve circuits shown in **Figures 10-62** and **10-67**. The primary difference is the bypass port connection that leads directly to the system reservoir.

In this circuit, when the cylinder is extending under a no-load condition, the flow control valve delivers 3 gpm to the cylinder. The extra 2 gpm of pump output is returned to the system reservoir through the bypass port of the valve. Under a no-load condition, the total pressure drop across the flow control valve is approximately the 50 psi needed to overcome the biasing spring tension, rather than the 500 psi setting of the relief valve. When the cylinder is subjected to an increased load, the pressure in the inlet and outlet port areas increases to produce the force needed to move the load. However, the pressure compensator maintains a 50 psi pressure drop between the ports. This allows the flow control device of the valve to maintain the selected fluid flow rate. Therefore, a constant actuator speed is maintained, even under varying actuator load conditions.

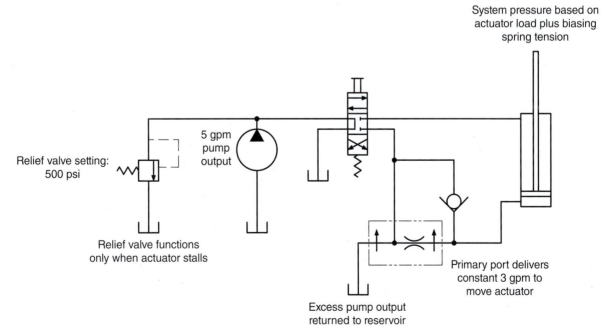

Goodheart-Willcox Publisher

Figure 10-70. Pressure and flow conditions in a hydraulic circuit using a bypass-type, pressure-compensated flow control valve.

The bypass-type of flow control is more efficient than the restrictor type. The surplus pump fluid returned to the reservoir is at a pressure below the relief valve setting. This reduces heat generation, which prolongs fluid life and lowers energy consumption. However, care must be taken when designing and operating a circuit of this type. The pressure required to operate the actuator must be accurately determined. In order for a bypass flow control valve to properly function, the cracking pressure of the system relief valve must be higher than the required actuator operating pressure plus the pressure needed to operate the compensator. If this is not true, the system will not operate properly.

10.6.5 Flow Divider Valves

Flow divider valves are used to divide one fluid supply between two circuit branches or subsystems. The valves are often used in mobile equipment where pump output flow varies with vehicle engine speed. Many of these valves can also be used to combine the flow from two different circuits. Both priority and proportional designs are available.

Priority divider valves

In hydraulic equipment with circuit subsystems, it is sometimes necessary to be certain a preset flow level is always available to a vital actuator. This may be related to the function of an important step in the machine operation or a safety issue in case of unexpected flow demands or a sudden pressure drop in the remainder of the circuit. A priority divider valve provides flow to one port before providing flow to a second port, **Figure 10-71**. These valves are often used in mobile equipment where pump output is controlled by engine speed. Because of safety issues and the absolute need

for easy maneuverability, the power steering unit of mobile equipment is often operated by priority flow coming from such divider valves.

Figure 10-72 shows the basic structure of a priority-type flow divider valve. The valve consists of a body containing a machined bore, pump inlet port, priority and secondary outlet ports, spool, metering orifice, and biasing spring. The priority and secondary ports are positioned so that sliding the spool slightly in either direction changes the size of the openings between the bore and the ports. Moving the spool in one direction reduces the size of the opening to the priority port and enlarges the size of the opening leading to the secondary port. Movement in the opposite direction causes the opposite size changes. The metering orifice located in the center of the spool establishes the priority flow rate

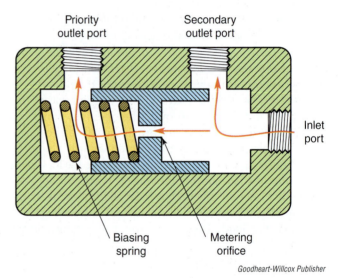

Figure 10-72. The internal structure of a priority-type flow divider valve.

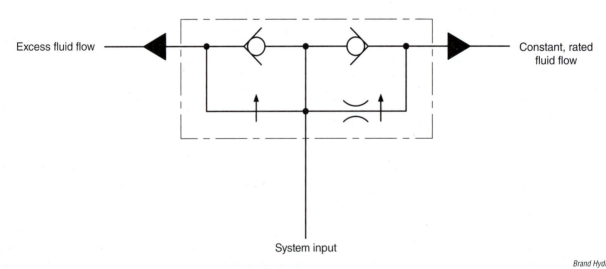

Figure 10-71. Symbol for a nonadjustable, priority flow divider valve. The check valves allow uncontrolled reverse flow through the valve.

of the valve. The biasing spring maintains a constant pressure drop across the orifice. This maintains a consistent flow rate to the priority port.

When the system is not operating, the biasing spring slides the spool toward the secondary port. This fully exposes the priority port and fully closes the secondary port. When the system pump begins to operate, fluid is supplied to the inlet port. Fluid flows through the metering orifice and out the priority port to the subsystem requiring preferential flow. As the flow increases, the pressure drop across the metering orifice increases. This generates a force on the spool that causes it to compress the biasing spring and move toward the priority port. As the spool moves, the size of the opening to the priority port is reduced. Also, the opening to the secondary port begins to be exposed.

The spool rapidly shifts back and forth until the forces generated by system pressure on the ends of the spool and the tension of the biasing spring are in equilibrium. The pressure drop across the metering orifice then produces the priority flow rate through the orifice. Any additional fluid coming from the pump flows out the secondary outlet port to operate other elements of the system.

Proportional divider valves

A proportional divider valve splits input port flow into two proportional output flows. These flows are directed to individual circuit subsystems or actuators. The ratio between the output flows may be fixed or variable. Proportions ranging from a 50:50 ratio to a 95:5 ratio are available from various valve manufacturers. A ratio of 50:50 is most common. See **Figure 10-73**.

A basic proportional-type flow divider valve consists of a body containing a machined bore, pump inlet port, and two outlet ports. Internally, these valves typically contain a spool fitted with control orifices leading to control chambers. During operation, valve inlet flow passes through the control orifices, control chambers, variable metering orifices, and on to the system through the outlet ports of the valve. The spool stays centered in the bore as long as the pressure generated by the load on each circuit subsystem is equal.

Increasing pressure in one of the control chambers produces an imbalance in the forces holding the spool in equilibrium. The spool rapidly moves back and forth to adjust the variable metering orifices until the pressures in the control chambers are again equalized. The result is an equal pressure drop across the control orifices. This produces the proportional flow.

Valve performance

Figure 10-74 illustrates a basic hydraulic circuit using a priority flow divider valve. The divider valve is placed between the pump and the circuit subsystems. In this example, the pump has a maximum output of 10 gpm and the relief valve is set at 500 psi. The priority flow setting of the valve is 2 gpm.

The priority divider valve delivers a maximum of 2 gpm to the subsystem connected to the priority port (the motor). Any excess flow, up to 8 gpm, is diverted to the second subsystem (the cylinder). The actuator in the cylinder subsystem operates on any excess flow until system pressure increases to the system relief valve setting. At that point, any flow over 2 gpm is returned to the reservoir through the system relief valve and the actuator in the second subsystem stops. The subsystem with the motor continues to receive the 2 gpm priority flow, as long as pump speed is sufficient to produce that volume. The spring-loaded valve

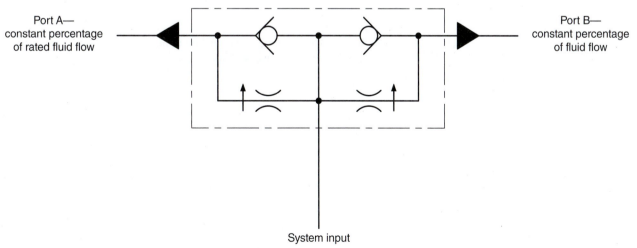

Port A—
constant percentage
of rated fluid flow

Port B—
constant percentage
of fluid flow

System input

Brand Hydraulics

Figure 10-73. Symbol for a nonadjustable, proportional flow divider valve. Ports A and B receive a predetermined percentage of the total system input based on the characteristics of the orifice and spring.

in the motor subsystem serves as a relief valve to set the maximum operating pressure of the motor.

The subsystem with the motor represents a steering mechanism on mobile equipment. As you can see, the steering system has priority over the second subsystem. That subsystem may be used to extend a load, for example.

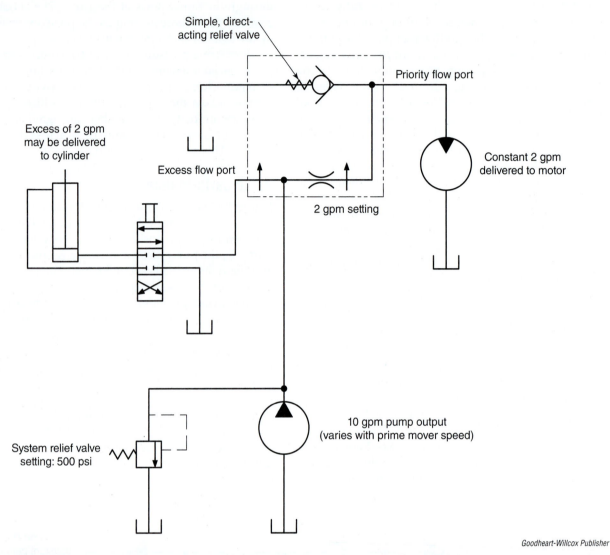

Figure 10-74. In this circuit, the priority outlet of the divider valve discharges 2 gpm. The remaining pump output goes to the second circuit. If pump delivery drops below 2 gpm, all fluid output is delivered through the priority port.

Summary

- Pressure control, directional control, and flow control valves are the three basic types of control valves used in the structure of hydraulic circuits.
- A valve body serves as the holder of the working elements of a hydraulic control valve. The working elements may include fixed orifices, needle valves, spools, poppets, and sliding plates.
- Pressure control valves control system maximum pressure, actuator sequence, restraining movement, pump unloading, and reducing pressure.
- Pressure control valves can be separated based on modes of operation: direct-operated or balancing-piston operation. Direct-operated valves use heavy, internal springs to directly establish the operating pressure of the valve. Balancing-piston valves use lightweight springs combined with pressure acting on the surface area of pistons or spools to operate the internal mechanism of the valve.
- Maximum system pressure control devices, which include relief valves, safety valves, and hydraulic fuses, are normally closed valves that open when system pressure approaches the desired maximum operating pressure.
- Directional control devices allow a system operator to control the direction of fluid flow in the system. Controlling fluid flow direction allows a machine operator to start, stop, and control the direction of actuator movement.
- Directional control devices can be classified as shutoff, check, three-way, and four-way valves.
- Shutoff valves block fluid flow through a hydraulic system line. Globe, gate, ball, spool, and needle valve designs can be used depending on the system application.
- Check valves primarily allow free flow through a conductor in one direction while preventing reverse flow.
- Three-way directional control valves use a sliding spool in a bored valve body to direct system fluid providing pressure to extend rams and single-acting cylinders. The actuator is returned to its original position by an external force, such as the system load or an internal spring built into the actuator.
- Four-way directional control valves provide a means to use system pressure to power cylinders during both extension and retraction, as well as hydraulic motors in either direction of rotation. The internal structure of the valve allows four valve ports to be alternately connected to the supply line.
- Four-way directional control valves may be two- or three-position valves. In two-position valves, the first position operates the actuator in one direction, while the second position reverses the direction. In three-position valves, a center position is added that provides additional distinct operating characteristics.
- The most common center positions for three-position valves are closed, open, tandem, floating, and regenerative. These position configurations not only affect directional control characteristics, but also overall system efficiency.
- Flow control devices limit the volume of fluid entering an actuator to control its operating speed.
- Flow control valve operation is based on the mechanics of fluid flow through an orifice. Orifice area, pressure differences between inlet and outlet sides of the orifice, fluid viscosity, and fluid temperature are among the factors that determine fluid flow rate.
- Compensated flow control valves, which maintain a constant pressure difference across the metering orifice, automatically adjust for fluid pressure variations and produce a consistent flow rate under varying load. Some models also compensate for fluid temperature and viscosity changes. The flow rate through noncompensated flow control valves can vary as the load or fluid viscosity changes.

Internet Resources

The following are some useful resources available on the Internet. Enter a company or organization name into a search engine to access its website. Explore the various areas of the sites to discover useful fluid power resources.

Brand Hydraulics. Provides a review of priority and proportional flow divider/combiner valves. Includes descriptions of valve features and general operating information. Also provides information on pressure, directional, and other flow control valves.

Dynex/Rivett, Inc. Describes one series of pressure control valves produced by the manufacturer. Includes specifications typically available, such as pressure, flow rates, function, and installation.

Eaton Corporation. Presents product information on the variety of industrial valves available from the manufacturer. Includes information on the markets that use various valve types.

Hydraulics & Pneumatics Magazine. Presents a two-part article on flow control valves. The article covers design and operation. Symbols are shown for each of the valve types discussed, as well as circuit schematics.

Prince Manufacturing Corporation. Provides descriptions of two lines of directional stack valves. Includes catalog pages, parts manuals, and instruction sheets. This material provides insight into the structure and operation of stack valves.

Chapter Review

Answer the following questions using information in this chapter.

1. *True or False?* The only purpose of control valves in a hydraulic system is to regulate the level of actuator motion.

2. *True or False?* The primary purpose of pressure control valves in a hydraulic system is to prevent damage to other system valves.

3. *True or False?* The valve body serves as the holder of the working elements of the valve.

4. A precision-machined hole in a valve body or component part used to control fluid flow rates is called a(n) _____.
   A. pilot
   B. compensator
   C. orifice
   D. flow divider

5. The two sources of internal force typically used to position control valve elements are _____.
   A. springs and pilot pressure
   B. orifices and internal pressure
   C. balanced orifice and pistons
   D. mini-valves and pressure overrides

6. *True or False?* It is *not* important to remove internal leaks in control valves.

7. *True or False?* When a valve is classified as normally open or normally closed, it is only in relation to the position of the internal elements when the system is shut down.

8. *True or False?* One basic task that fluid pressure performs in the operation of some control valves is closing the valve.

9. The difference between cracking and full-flow pressure of a valve is called _____.
   A. pressure override
   B. pilot pressure
   C. compensating pressure
   D. safety pressure

10. *True or False?* Balancing-piston control valve designs usually cost more, but operate more efficiently than direct-operated valves.

11. The two types of pressure control valves that are used to prevent damage to hydraulic systems in cases where the system relief valve fails are the safety valve and the _____.
    A. sequence valve
    B. brake valve
    C. unloading valve
    D. hydraulic fuse

12. *True or False?* The primary function of the metering orifice in the pilot section of a compound relief valve is to create a pressure difference between the pilot and system sides of the balancing piston.

13. Safety valves are typically set at a pressure _____ above the setting of the primary system relief valve.
    A. 10%
    B. 25%
    C. 35%
    D. 55%

14. Multiple operating steps actuated in a preselected order can be obtained in a hydraulic circuit by the use of _____.
    A. safety valves
    B. ball valves
    C. pilot valves
    D. sequence valves

15. *True or False?* Unloading valves are commonly operated by pressure created by actuator load and pump operation.

16. *True or False?* Unloading pressure control valves are commonly found in hydraulic circuits that use accumulators.

17. *True or False?* A pressure-reducing valve is normally open and must be externally drained.

18. A(n) _____ valve is used primarily to block fluid flow through a hydraulic system line.

19. _____ may be adjusted from the fully open position to the fully closed position by turning the control handle only 90° (1/4 turn).
   A. Safety valves
   B. Ball valves
   C. Pilot valves
   D. Sequence valves

20. The most common application of a check valve in a hydraulic system is to _____.
   A. adjust flow by 10 percent increments
   B. shut off flow completely
   C. allow free flow in one direction while blocking the reverse direction of flow
   D. restrict flow in the pump metering line

21. Three-way directional control valves have _____.
   A. one port connected only to the pump
   B. two ports connected to the pump and cylinder
   C. three ports connected to the pump, cylinder, and reservoir
   D. four ports connected the same as a four-way valve

22. *True or False?* A four-way directional control valve has five ports.

23. *True or False?* The center position in a four-way directional control valve provides a control feature beyond cylinder extension and retraction.

24. Which type of a three-position, four-way directional control valve has a middle position with the pump, reservoir, and actuator ports connected?
   A. Open-center.
   B. Closed.
   C. Tandem.
   D. All of the above.

25. The _____ method of directional control activation uses system pressure, which may be remotely sensed, to shift the valve.

26. In some directional control valves, the spool is held in a selected position by a locking device called a _____.
   A. detent
   B. plunger
   C. poppet
   D. gate

27. The two general operating designs of flow control valves are _____.
   A. flow divider and compensator
   B. biasing and metering
   C. priority port and sharp-edged orifice
   D. restrictor and bypass

28. In an operating fluid power system, the _____ of the fluid will change as the operating temperature of the system changes.
   A. viscosity
   B. schedule number
   C. turbulence
   D. pour point

29. The simplest of the adjustable flow control valves is a _____.
   A. relief valve
   B. needle valve
   C. gate valve
   D. ball valve

30. The mechanism that maintains a constant pressure difference between the inlet and outlet passageways of a flow control valve is called a _____.
   A. sharp-edged orifice
   B. priority flow divider
   C. pressure compensator
   D. bypass flow control

31. Two ways to accomplish temperature compensation in a flow control valve is a heat-sensitive insert in a needle valve or a _____.
   A. sharp-edged orifice
   B. priority flow divider
   C. pressure compensator
   D. bypass flow control

32. *True or False?* During system operation, when actuator pressure requirements are below the cracking pressure of the relief valve, a bypass-type flow control valve directs any excess pump output flow directly to the reservoir via a port on the valve.

33. Which of the following is a limitation that must be observed when incorporating most bypass-type flow control valves into hydraulic systems?
   A. Bypass line cannot be pressurized.
   B. Bypass line cannot effectively deliver fluid to another line.
   C. No reverse fluid flow.
   D. All of the above.

34. *True or False?* Proportional flow divider valves are available from a ratio of 50:50. to as high as 95:5.

35. A system pump is delivering 10 gpm to the inlet port of a priority-type flow divider valve. What amount of fluid is delivered to the secondary ports when the priority outlet is rated at 2 gpm?
   A. 2 gpm
   B. 4 gpm
   C. 6 gpm
   D. 8 gpm

Apply and Analyze

1. **Figure 10-11** shows the symbols for normally open and normally closed valves. Explain how these valves are activated and describe the conditions that result in the opening and closing of these valves.

2. A direct-operated pressure relief valve needs to be replaced in a hydraulic system.
   A. Predict the effects of the following choices of replacement valve on the system operation.
      - A direct-operated relief valve with a lower cracking pressure and greater override pressure.
      - A compound-pilot-operated relief valve.
   B. Which would you choose and why? List possible benefits and drawbacks of your choice.

3. **Figure 10-17** shows the internal structure of a typical compound relief valve in a closed position. Draw the internal structure of the valve as it would appear just before the poppet valve opens when maximum system pressure is reached. On your drawing, show the following:
   - Relative pressure at the reservoir port, the pressure port, and in the pilot section chamber.
   - The fluid flow paths.

4. Compare and contrast the design and operation of the counterbalance and brake valves shown in **Figures 10-25** and **10-28**. Could a counterbalance valve be used as a brake valve? Why or why not?

5. Describe the sequence of events that occur as the directional control valve in the circuit shown in **Figure 10-30** is shifted from one position to the other.

6. If the directional control valve in the circuit shown in **Figure 10-30** were replaced with a three-position, four-way directional control valve, what center design would you choose to use? Why? What impact would this change have on circuit operation?

7. For the circuit shown in **Figure 10-52**:
   A. Identify each of the five valves in the circuit and briefly describe their operation.
   B. Describe the sequence of events that occur as the three-position directional control valve is shifted through all three possible positions.

8. Compare and contrast the design and operation of a bypass flow control valve and a restrictor-type, noncompensated flow control valve.

9. Draw a hydraulic circuit in which a hydraulic motor operates at a specific rpm even when subjected to changing loads or variations in pump output.
   A. Explain the operation of the circuit and discuss your choice of components.
   B. How might changes in fluid temperature affect the operation of your circuit?

Research and Development

1. Design and conduct an experiment that illustrates the relationships between flow quantity, head, and cross-sectional area of an orifice. Create a poster to share your results.

2. Research the process by which hydraulic pressure relief valves are tested and certified. Present an overview to your class.

3. Find a schematic of a forklift's hydraulic circuit. Identify the components that are used to control the lifting system and be prepared to explain how each component contributes to the forklift's safe operation.

4. Investigate the history of directional control valves. Write a report that highlights the major steps in the development of current valves.

5. Research different designs of check valves and prepare a presentation that describes one type in detail, including basic design, materials, operation, capacity, and typical applications in which the check valve is used.

Equipment shaping the final contour of tunnels in mining and construction applications.

Accumulators

Pressure, Flow, and Shock Control Assistance

The noncompressible nature of a liquid provides an excellent means of energy transmission in a hydraulic system. However, that characteristic limits several other factors also required for effective hydraulic system performance. These factors include maintaining fluid pressure over a long period, compensating for intermittent flow rates that exceed the capacity of the system pump, and absorbing sudden pressure surges or system shock. Accumulators using weights, springs, or a gas can provide these added performance factors without affecting the noncompressible nature of the hydraulic fluid. This chapter provides the information needed to understand the design and operating characteristics of these units.

Learning Objectives

After completing this chapter, you will be able to:

- Explain the four basic functions of accumulators in hydraulic systems.
- Describe the design of weight-, spring-, and gas-loaded accumulators.
- Compare the basic operating principles of weight-, spring-, and gas-loaded accumulators.
- Compare the construction and performance of piston-, diaphragm-, and bladder-type gas-loaded accumulators.
- List and describe the typical uses of accumulators in hydraulic systems.
- Select the proper type and size of an accumulator for a circuit.
- List the safety factors that must be considered when working with accumulators.

Key Terms

accumulator
adiabatic
gas-charged accumulator
isothermal

precharging
pressure rating
pressure surge
rated capacity

shock pressure
spring-loaded accumulator
usable volume
weight-loaded accumulator

11.1 Basic Functions of an Accumulator

A simple hydraulic system that is made up of a prime mover, pump, lines, control valves, actuators, and a liquid hydraulic fluid cannot *store* energy. The prime mover must continually operate to keep the system energized. When the prime mover is operating, fluid flow is created. This, in turn, powers the actuator to move a load. Stopping energy input from the prime mover stops system operation and system pressure drops to zero. The noncompressible nature of liquid causes this type of system response. The noncompressible aspect of a liquid is a positive feature desirable for hydraulic systems. However, there may be times when it is desirable to store energy for short periods during some phases of system operation.

Accumulators are the fluid power components that absorb and store energy in a system, **Figure 11-1**. The stored energy is then used to:

- Maintain system pressure.
- Produce limited flow to expand pump output.
- Supply fluid to operate system subcircuits when the prime mover and pump are not functioning.

Accumulators are also used to prevent damage to the system by absorbing the energy of sudden system pressure increases caused by pressure surges. *Pressure surges* result from sudden changes to actuator loads or the rapid closing of a system valve.

Accumulators store energy using one of three methods:

- Lifting a weight.
- Compressing a spring.
- Compressing a gas.

Gravity acting on the lifted weight produces pressure on the system fluid. Likewise, the tension of the compressed spring maintains fluid pressure in spring-loaded units. Gas-type accumulators store energy using the compressible nature of gases. In each design, the stored energy maintains pressure on the noncompressible hydraulic fluid for later use in the system.

11.2 Accumulator Design and Operation

Accumulators use weights, springs or compressed gas to provide the energy storage needed to perform their function. New materials and applications of hydraulic systems have somewhat changed the appearance of accumulators over the years and may have expanded their tasks, but the basic concepts of operation remain

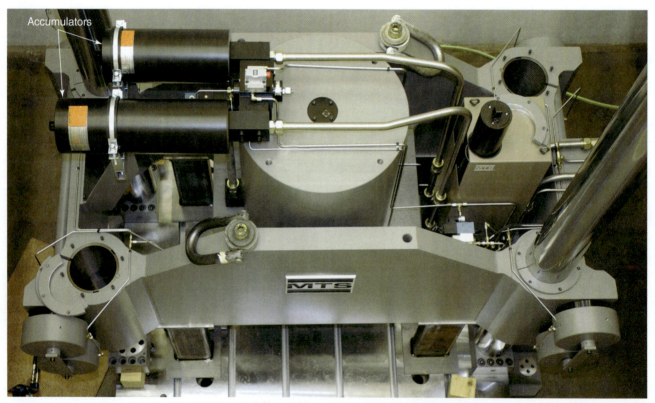

MTS Systems Corporation

Figure 11-1. These accumulators allow the system to apply the noncompressible characteristics of liquids while having the ability to store energy and smooth flow and pressure fluctuations during critical system operations.

the same. This section groups accumulators under the three energy storage methods of weight loaded, spring loaded, and gas charged.

11.2.1 Weight-Loaded Accumulators

A *weight-loaded accumulator* can be described as a large, vertically-mounted cylinder with a heavy weight attached to the end of the cylinder ram. See **Figure 11-2**. The blind end of the cylinder is plumbed to the hydraulic system. The weight is lifted when force generated by system pressure acting on the blind end of the ram is greater than the force generated by gravity acting on the weight. This stores energy in the accumulator. Likewise, the ram lowers when the system pressure drops below that needed to support the weight.

Weight-loaded accumulators are generally very large units. This is due to the weight needed to provide the pressure. They are typically used in central hydraulic systems powering applications such as heavy, stationary presses. In these applications, constant pressure and large fluid volumes are necessary. Weight-loaded accumulators may be found in large, forming-press operations. However, they are seldom, if ever, used today because of their massive size and weight.

A weight-loaded accumulator is unique as it is the only type of accumulator that can deliver its entire fluid volume at a near-constant pressure. However, an undesirable operating characteristic sometimes encountered with this design is high shock pressures. *Shock pressure* is the result of the inertia of the heavy weight when a control valve suddenly closes during the time the accumulator is delivering fluid to the system.

11.2.2 Spring-Loaded Accumulators

A *spring-loaded accumulator* uses one or more springs to store energy. The unit can be described as a cylinder with a sliding piston that divides the cylinder into two chambers. See **Figure 11-3**. One of the chambers is plumbed to the hydraulic system. The other chamber contains a coil spring that exerts a force on the piston to push it toward the end of the cylinder connected to the hydraulic system. Some designs have a screw adjustment on the spring end that allows adjustment of spring tension.

As the pressure in the system increases, it acts against the hydraulic-system side of the piston. This generates a force that slides the piston toward the spring end of the cylinder. As the coil spring is compressed, it stores energy. When the system pressure decreases, the spring extends. This pushes fluid back into the system.

The performance of a spring-loaded accumulator depends on the characteristics of the spring. The rate of spring compression per unit of pressure increase depends on the compression characteristics of the spring. The rate of compression is not the same over the entire length of the spring. The result is an accumulator that does not provide a constant rate of either fluid intake or discharge per unit of pressure change.

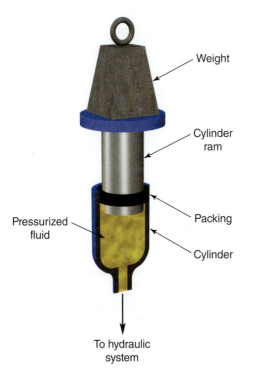

Goodheart-Willcox Publisher

Figure 11-2. Weight-loaded accumulators use gravity acting on heavy weights to store energy and smooth pressure surges.

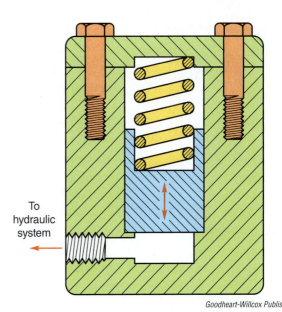

Goodheart-Willcox Publisher

Figure 11-3. Spring-loaded accumulators typically have a coil spring that stores energy, which can be used to enhance system flow and pressure.

An advantage of this design is that it can be mounted in any position. However, the design is not popular as it is difficult to adjust and not considered practical for systems using large quantities of fluid.

11.2.3 Gas-Charged Accumulators

Gas-charged accumulators use a compressed, inert gas for energy storage. These units are the most common type of accumulator in the hydraulic field. Design variations include piston, bladder, and diaphragm types. See **Figure 11-4**. The type indicates the method used to separate the gas from the liquid hydraulic fluid. This separation prevents entrainment of the gas in the hydraulic fluid during system operation. Physically separating the gas and liquid allows the units to be operated in multiple positions. It also allows the gas to be precharged to a specific pressure to obtain the desired operating characteristics.

Precharging is applying a pressure on the gas side of the accumulator after the pressure of the hydraulic side has been reduced to 0 psi. This is not to be confused with charging the accumulator with hydraulic fluid pressure. Only an inert (nonreactive) gas should be used to precharge any gas-charged accumulator. Dry nitrogen is the gas normally recommended for use in these units. Air should *not* be used. There is the possibility of fire or explosion when air and oil vapor are exposed to system pressures.

Design and basic construction

Piston, bladder, and diaphragm accumulators use basically the same operating theory. However, their appearance and construction varies considerably. Each design involves a body that is sufficiently strong to withstand the rated operating pressure of the unit. The piston, bladder, or diaphragm separates the internal volume of the body into two chambers. One of these chambers, the system-port chamber, is plumbed to the hydraulic system. The other chamber is the isolated gas chamber and contains a gas valve. Through this valve, the isolated gas chamber is precharged to a pressure that will produce the desired system operation.

A piston accumulator is basically a cylindrical body containing a free-sliding piston. **Figure 11-5** shows the

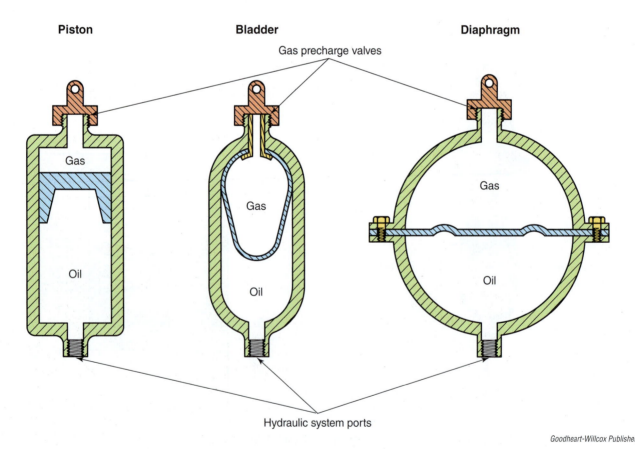

Goodheart-Willcox Publisher

Figure 11-4. Gas-charged accumulators use a compressed gas as the basis of their operation. They are available in three styles: piston, bladder, and diaphragm.

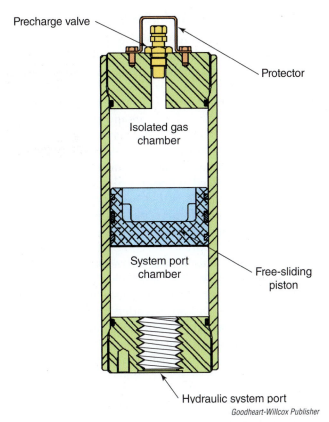

Figure 11-5. A piston-type gas-charged accumulator uses a free sliding piston in a cylinder to form separate chambers for the gas and the liquid.

structure of a typical piston unit. The piston and piston seals divide the cylinder cavity into the two chambers. Fittings on either end of the body allow the unit to be plumbed to the hydraulic system and a gas source.

The body of a bladder accumulator is a cylindrical metal shell with domed ends. One end of the shell contains a small hole for a gas valve. A larger hole is located on the other end, which allows insertion of the bladder. It also contains the system port connection with a poppet valve. The poppet valve prevents the precharge pressure in the bladder from pushing part of the bladder into the system line. The bladder is constructed of a synthetic elastomer and fitted with a gas valve on one end. See **Figure 11-6**.

A diaphragm accumulator is basically two metal hemispheres separated by a diaphragm made from a synthetic elastomer. The metal body sections are bolted together and form separate chambers on either side of the diaphragm. One of these chambers is fitted with the gas valve for precharging the accumulator. The other chamber is plumbed to the hydraulic system.

The elastomer used in the construction of bladders and diaphragms must be compatible with the fluid used in the hydraulic system. Manufacturers provide a selection of materials suitable for mineral oil, water-glycol, phosphate ester, and other fluids.

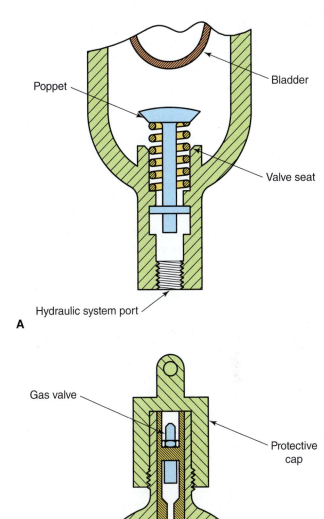

Figure 11-6. A—When a bladder-type accumulator is fully discharged, a poppet valve is often used to prevent the bladder from being extruded into the connecting line. B—The bladder is charged with an inert gas through a valve that also positions the bladder in the accumulator shell.

Operation of gas-charged accumulators

The operation of the three types of gas-charged accumulators is very similar. The following discussion begins with the piston accumulator, followed by the bladder and diaphragm types. This approach demonstrates the construction and operating similarities of the three designs.

Precharging a piston-type, gas-charged accumulator forces the piston to the system-port end of the cylinder. This minimizes the volume of the system-port chamber. During system operation, the hydraulic fluid in the system-port chamber is subjected to actual system operating pressure. This pressure acts on the surface area of the piston. The generated force attempts to slide the piston toward the gas-charged end of the unit. However, the gas in the isolated gas chamber generates an opposing force. When the force generated by system pressure becomes higher than the opposing force generated by the gas precharge pressure, the piston begins to slide toward the gas-chamber end of the unit. This movement reduces the volume of the isolated gas chamber. As a result, the gas pressure increases to match system pressure. As gas-chamber volume decreases, the volume of the system-port chamber increases. Hydraulic fluid enters the chamber and is stored under pressure. This filling process continues until the maximum system pressure established by the relief valve is reached.

Precharging a bladder-type, gas-charged accumulator expands the bladder to fill the accumulator shell. The poppet valve located at the system port closes to protect the bladder from damage that may occur if the bladder is forced into the port. During system operation, the pressurized system fluid enters the accumulator shell through the system port. There, it begins applying pressure to the bladder. When system pressure exceeds the precharge pressure of the gas inside of the bladder, the bladder begins to compress. Reducing the volume of the bladder increases the gas pressure to match system pressure.

As in a piston-type accumulator, fluid and energy are stored in a bladder accumulator whenever system operating pressure exceeds the precharge pressure. Whenever pressure is decreasing in the system, the higher gas pressure inside of the bladder expands the bladder. The expanding bladder fills the accumulator shell and forces fluid back into the system. This condition continues until the expanding bladder closes the poppet valve, blocking flow out of the unit.

The operation of the diaphragm accumulator is very similar to the piston and bladder designs. The flexible diaphragm is used to separate the system-port chamber and isolated gas chamber. The diaphragm moves to change the volume of the chambers. This provides basically the same operation as the piston and bladder designs. A metal disk is often attached to the center of the diaphragm to prevent extrusion of the diaphragm into the port as the unit discharges.

NOTE

A general rule to remember is that an accumulator begins to store fluid and energy when system operating pressure exceeds the gas precharge pressure. After that point is reached, an accumulator is storing fluid and energy as system pressure is increasing and dispensing fluid and energy as system pressure decreases.

Nature of compressed gases

The gas used in an accumulator responds to pressure, temperature, and volume changes in the same way as other gases. You may wish to review the discussion of the *ideal gas laws* in Chapter 4, *Basic Physical Principles*. Understanding these laws will help in understanding the way an accumulator functions in a hydraulic system.

The terms isothermal and adiabatic are often used in explaining the operation of accumulators. These terms relate to temperature and heat. An *isothermal* process is completed without a temperature change, although heat may be added or removed during the process. An *adiabatic* process is completed without heat gained from or lost to surrounding materials. These principles apply to the compression and expansion of a gas. Examining the operation of an accumulator should illustrate the importance of these factors.

An isothermal condition exists when an accumulator is filled over an extended period of time. In this situation, the temperature of the accumulator gas remains constant. Any heat generated from the compression of the gas is transferred to the surrounding materials. In this case, the volumes of the system-port chamber and isolated gas chamber can produce the volume of hydraulic fluid predicted by applying the ideal gas laws.

However, rapid filling of the accumulator can produce an adiabatic situation. Heat transfer cannot happen quickly enough to maintain a constant gas temperature. In turn, the pressure of the gas is increased beyond that predicted by the ideal gas laws. This increase in gas pressure increases gas volume beyond that produced by the isothermal process. The increased gas volume in the isolated gas chamber is reflected in a similar reduction in the volume of the system-port chamber. The end result is a reduction in the amount of pressurized hydraulic fluid available from the accumulator.

Figure 11-7 compares the differences in volume and pressure during isothermal and adiabatic compression of a gas. The area between the resulting curves is caused by the retention of heat in the gas. In actual practice, most accumulators function somewhere between the two curves. However, having a good understanding of these characteristics can be helpful in understanding manufacturer specifications and troubleshooting accumulator circuit performance.

11.3 Functions of Accumulator Circuits

An accumulator is the component that allows hydraulic systems to maintain the advantages of a noncompressible liquid while using weights, springs, or a gas

Hydraulic System Pressure

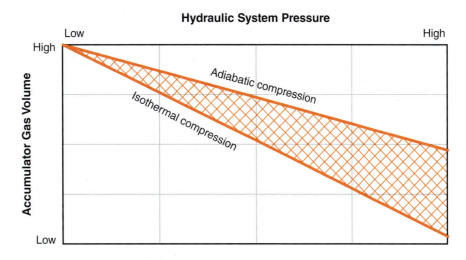

Goodheart-Willcox Publisher

Figure 11-7. The final volume of identical volumes of a gas compressed under isothermal and adiabatic conditions. The heat retained during adiabatic compression results in a larger volume of gas when an identical pressure is maintained in each test condition.

to enhance system operation. The selective use of these elements can increase the efficiency of a hydraulic system by:

- Providing an easy means to store energy.
- Controlling pressure and flow variations.
- Maintaining system pressure.
- Supplementing pump output flow.

This section examines each of these functions and provides basic examples.

11.3.1 Storing Energy

In a number of situations, a hydraulic system must be able to store energy for a period of time in order to increase system performance or efficiency. Even though a liquid can be slightly compressed, it is generally treated as incompressible. However, a lifted weight, compressed spring, or compressed gas can provide a means to maintain a significant amount of potential energy. Using one of these devices, an accumulator is able to effectively act as an energy-storage unit in a hydraulic system.

In one sense, all accumulator applications involve the storage of energy. This is simply because any accumulator is storing energy as it fills. It does not matter if that potential energy is in the form of a raised weight, compressed spring, or compressed gas. The energy is there and may be called on to perform a task.

The simple hydraulic motor circuit shown in **Figure 11-8** illustrates how an accumulator can be used to store energy for extended periods. The hydraulic motor may be a starting motor for an internal combustion engine that powers the pump for hydraulic equipment on mobile equipment. The engine drives

the pump to supply hydraulic fluid to all aspects of the system, including filling the accumulator. When the engine is stopped, flow from the accumulator closes the check valve. This traps pressurized fluid in the accumulator section of the circuit. The energy needed to restart the engine is held in the accumulator as a pressurized fluid. When engine operation is again required, the directional control valve is shifted to direct fluid flow to the hydraulic motor. The motor then spins the engine crankshaft to start the engine.

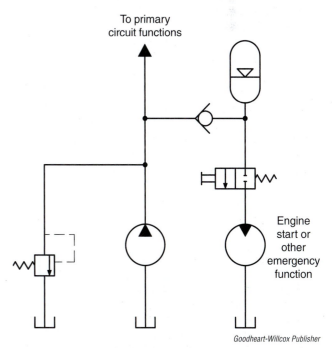

Goodheart-Willcox Publisher

Figure 11-8. A simple hydraulic circuit containing an accumulator that stores energy to be used later to operate a motor when the hydraulic power unit is not operating.

11.3.2 Dampening Pressure and Flow Variations

Under varying load conditions, certain hydraulic components and circuits can produce substantial pressure and flow variations. These variations can be great enough to:

- Reduce the efficiency of a system.
- Cause damage to system components.
- Be a safety issue that could harm machine operators or other workers in the area.

Extreme pressure increases can be caused when an actuator encounters a heavy load or when a control valve suddenly shifts to limit or totally block fluid flow. These situations may briefly produce pressures that are several times higher than the system relief valve setting. In addition, some pump designs inherently produce a pulsating flow. Under normal operating conditions, these flow variations are not critical. However, as system pressures increase, they may reduce system performance and shorten the service life of the pump and other components.

Accumulators can help reduce the above system problems. The effectiveness of an accumulator in controlling these problems involves careful sizing, precharging, and placement of the accumulator in the circuit. **Figure 11-9** shows the placement of an accumulator in a circuit to reduce shock pressures when a closed-center directional control valve is shifted to the center position. This placement also reduces shock when the cylinder reaches the end of its stroke in

either direction. High pump flow rates even in simple circuits such as this may require a carefully placed accumulator to prevent high shock pressures.

11.3.3 Maintaining System Pressure

The operation of many hydraulic systems involves a holding phase. During this phase, pressure must be maintained for an extended period of time with little or no actuator movement. Continuously operating the pump under full system pressure in this situation consumes an unnecessary amount of energy. However, without a small amount of fluid being added to the system, internal leakage in components may quickly cause the pressure to drop below the desired level.

Closely associated with this type of circuit design are changes in system pressure resulting from temperature variations. Isolating the fluid in the actuator area of a system for long periods of time causes the fluid to be affected by ambient temperature. As the fluid temperature changes, the fluid volume changes. This, in turn, alters the fluid pressure. Raising the temperature increases the pressure, while decreasing it reduces the pressure.

A properly sized and precharged accumulator can resupply the actuator in a holding circuit with the volume of fluid needed to maintain the desired system pressure. **Figure 11-10** shows a simple holding circuit used with a press. The accumulator is filled as the press closes and system pressure increases to the setting of the system unloading relief valve. After the unloading valve opens, the accumulator maintains the pressure in the actuator and resupplies any fluid lost around

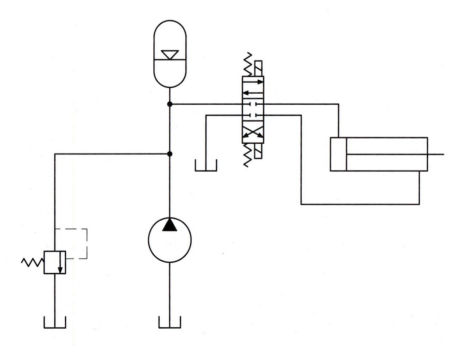

Figure 11-9. A hydraulic circuit designed to reduce shock pressures in the system when a directional control valve suddenly closes or an actuator suddenly encounters a heavy load.

the piston. This condition exists until sufficient fluid has leaked around the piston to drop the pressure to the point where the unloading valve closes. Then, the pump again supplies fluid to the system.

The same circuit can be used as an example of maintaining pressure in a system where ambient temperature is a factor. When the fluid in the press cylinder and the isolated lines cools, the accumulator supplies any fluid needed to maintain the holding pressure. Likewise, the accumulator absorbs any increase in fluid volume caused by an increase in fluid temperature due to a high ambient temperature.

11.3.4 Supplementing Pump Output

Another common problem with hydraulic systems is the intermittent need of high fluid-flow rates. A straightforward approach to satisfying this need is to size the pump so it adequately delivers the required flow throughout the entire operation of the system. However, the result is higher-than-needed initial hardware cost and operating cost related to both equipment service and energy consumption.

A solution that is often used is placing an accumulator in the circuit. The accumulator is filled during the low-flow-need portion of the operating cycle of the machine. The stored fluid and the fluid output of the pump are then combined during the high-flow-need

portion of the operating cycle. The combined fluid output provides the high flow rate needed to efficiently operate the machine.

This type of accumulator application is often found in die casting and plastic injection molding machines. These machines require high fluid-flow rates to rapidly inject the metal or plastic into the mold. However, the high-flow-rate demand is for only a small portion of the total machine cycle. **Figure 11-11** is a simple example of a circuit that uses an accumulator to provide a higher-than-pump output flow for both cylinder extension and retraction.

11.4 Selecting and Sizing Accumulators

The variety of accumulator uses and the performance variations of each design make a single selection and sizing procedure impractical, **Figure 11-12**. Most manufacturers provide detailed data concerning the accumulator models they produce. Most companies also provide suggested procedures to follow for the selection and installation of their products in various types of applications. The discussion in this section is limited to topics such as storage capacity, pressure ratings, and the usable volume of accumulators.

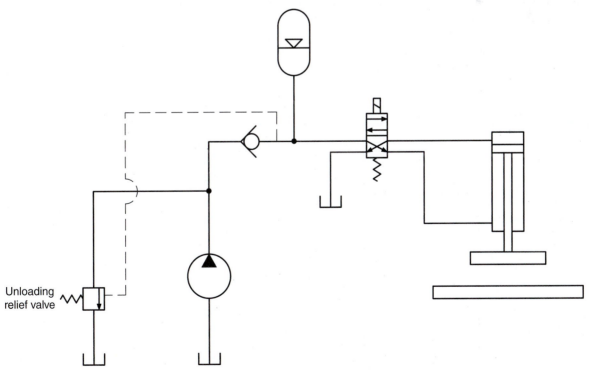

Goodheart-Willcox Publisher

Figure 11-10. A simplified circuit for a hydraulic press that must maintain a high pressure for an extended period of time. Unloading the system power unit significantly reduces energy consumption.

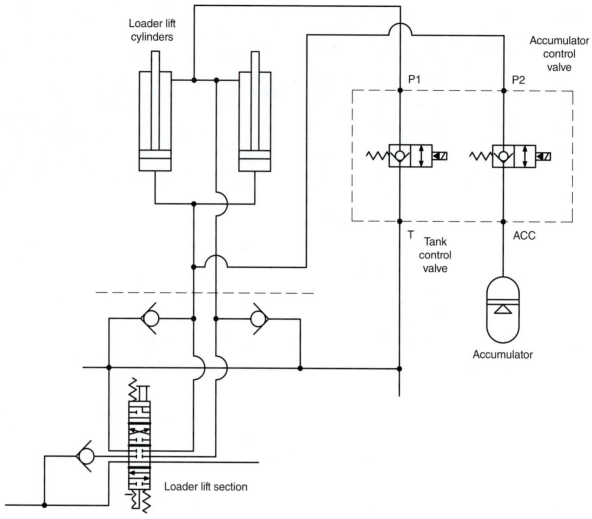

Figure 11-11. A portion of a circuit showing an accumulator used to supplement pump output. Adequate time to fill the accumulator must be provided when planning the cycle time of these circuits.

11.4.1 Rated Capacity and Pressure Ratings

The *rated capacity* of gas-charged accumulators is based on the gas volume when all liquid has been expelled. The actual value of liquid that can be stored is well below this rated capacity. The actual volume will vary somewhat between manufacturers. Rated capacities range from under 1 pint to over 200 gallons. However, some companies produce models that have even higher capacities.

Accumulators using the diaphragm design generally have the most limited capacity. Typically, the capacity of these designs ranges from about 5 cubic inches to approximately 1 gallon. Accumulators using the bladder design typically have capacities ranging from 10 cubic inches to over 40 gallons. Piston-type accumulators are available in an even wider range of capacities. Capacities range from under 10 cubic inches up to 150 gallons. This size range means piston accumulators vary in size from a small unit that is approximately 2 1/2″ in diameter and 8″ long to a large unit that is 16″ in diameter, 20′ in length, and weighing 3700 pounds dry.

Accumulators are also available in a number of pressure ratings. The *pressure rating* is based on the maximum safe operating pressure for the unit, as determined by the manufacturer. This varies from a maximum operating pressure of 2,000 psi to over 10,000 psi. One major manufacturer rates their units at either 3,000 psi or 5,000 psi. Other companies manufacture units rated at 2,000 psi; 2,500 psi; 3,000 psi; 5,000 psi; and 10,000 psi. These rated pressures are maximum system operating pressures, *not* accumulator precharge pressures. Generally, the precharge pressure should not exceed 50% of the rated pressure. For example, an accumulator rated at 3,000 psi should be precharged to no more than 1,500 psi.

11.4.2 Usable Operating Volume

The rated capacity of an accumulator does not indicate the usable volume of hydraulic fluid the unit can supply in a specific system application. The *usable volume* is the volume of hydraulic fluid the unit can supply in a specific system application. It is determined for a specific accumulator installation by three pressure levels:

- Maximum operating pressure, which is usually determined by the system relief valve setting.
- Minimum operating pressure established by the lowest level of pressure needed to perform the task involved.
- Precharge pressure, applied weight, or spring characteristic.

These pressures affect the amount of fluid that is stored in the unit and, therefore, the amount that can be discharged during the system task supported by the accumulator.

Most manufacturers recommend a precharge pressure that is 100 psi below the minimum operating pressure required to perform the task. This pressure difference allows the compressed gas of the accumulator to maintain at least a slight pressure on the fluid throughout the task.

Figure 11-12. Accumulators are available in a wide range of capacities. Stated capacity indicates the maximum volume of fluid that can be held in the unit, rather than usable operating volume.

Assume that the task to be supported requires 5 gallons of usable, pressurized hydraulic fluid from an accumulator. The pressure range is from a maximum system pressure of 2,000 psi to a minimum operating pressure of 1,600 psi. The first step is to select an accumulator model that can store 5 gallons of fluid between the two pressure levels. This selection can be easily done by using tables available from manufacturer catalogs. Next, after installation in the system, the selected accumulator needs to be precharged to 1,500 psi. This follows the general recommendations of precharge pressure 100 psi less than the minimum operating pressure. The following series of events then occurs in the system during the charging and discharging of the accumulator.

1. Once the hydraulic system pump is operating, the pressure in the system increases to the precharge level of the accumulator gas.
2. The accumulator begins to fill and system pressure begins to climb toward the setting of the relief valve.
3. As system pressure increases, the gas in the accumulator is compressed to match the pressure of the hydraulic fluid.
4. As the accumulator continues to fill, system pressure increases to the setting of the relief valve.
5. The system pump can be stopped.
6. The compressed gas of the accumulator forces fluid out of the accumulator to perform the designated task in the system. The accumulator provides the 5 gallons of usable volume of fluid before dropping to the minimum pressure of 1,600 psi.
7. The system pump again needs to be operated to refill the accumulator and repeat the cycle.

11.5 Setup and Maintenance of Accumulators

This section discusses general factors that can increase the operating efficiency of systems containing accumulators. The safe operation of hydraulic-powered and hydraulic-controlled equipment containing accumulators is also discussed.

CAUTION

Hydraulic accumulators are pressurized containers that need to be handled with a great deal of care. Only individuals with specific training should install or repair accumulators.

11.5.1 Mounting Position

The best mounting position for any style of accumulator is vertical with the hydraulic connection facing downward. See **Figure 11-13**. This reduces collection in the accumulator of contaminants from the hydraulic fluid. It also promotes the even loading of internal parts. The weight-type accumulator must *always* be vertically positioned because of the use of gravity. The two other designs can function in vertical, horizontal, or inclined positions.

Horizontal mounting tends to decrease the service life of the bladder in bladder-type accumulators. This is because of increased contact with the shell as the bladder expands and contracts. This contact increases in units used in systems with high cycle rates. It tends to be more critical in units that have a large storage capacity.

11.5.2 Precharging

The precharge pressure is critical to the proper operation of the accumulator. The correct precharge is necessary for the accumulator to store energy, dampen system flow and pressure variations, maintain system pressure, or supplement fluid output of the pump.

MTS Systems Corporation

Figure 11-13. The preferred mounting position for accumulators is vertical. This position provides the least stress on the internal components of the units.

The precharge pressure depends on the function of the accumulator in the hydraulic system. The example in the previous section recommended a precharge of 100 psi below the required minimum operating pressure. However, when the accumulator is being used as a shock or surge suppressor, a precharge of 100 psi *over* normal system operating pressure is often recommended. Consult the accumulator manufacturer to establish the initial precharge pressure for the application at hand. Adjustments can then be made to that pressure as needed for maximum system efficiency.

Accumulator manufacturers recommend a regular check of the gas precharge pressure. A loss in precharge pressure reduces system efficiency. In some situations, pressure loss can also cause damage to the accumulator bladder, diaphragm, or piston seals.

A charging and gauging unit is shown in **Figure 11-14**. This unit is used to precharge the accumulator with dry nitrogen. It is also used to check the existing precharge pressure in an installed unit. The charging/gauging units typically include the gauges, valves, hoses, and fittings to connect the accumulator gas valve to a nitrogen bottle, **Figure 11-15**. A typical gauge is shown in **Figure 11-16**.

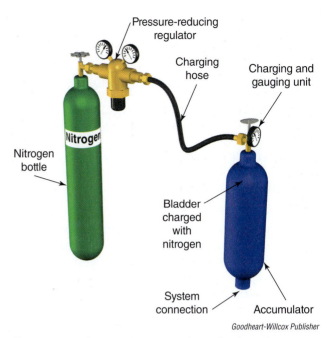

Goodheart-Willcox Publisher

Figure 11-15. Precharging a gas-charged accumulator requires pressurized nitrogen and an appropriate charging and gauging unit.

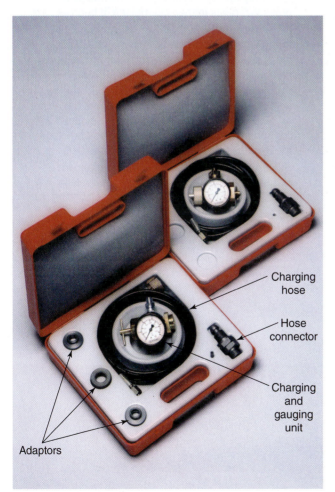

HYDAC Technology Corporation

Figure 11-14. The components of a charging and gauging unit.

HYDAC Technology Corporation

Figure 11-16. A typical gauge from a charging and gauging unit.

CAUTION

The procedure suggested by the equipment manufacturer should be carefully followed to ensure accuracy and personal safety during either gauging or charging. A gas-charged accumulator should only be charged after the hydraulic system has been reduced to 0 psi.

11.5.3 Working Safely with Accumulators

Whenever a hydraulic system includes an accumulator, the system should be treated with special respect. A charged accumulator has the ability to supply pressurized hydraulic fluid even when the system may appear to be powered down. The training of machine operators must include this type of information. In addition, signs should be prominently displayed to remind operators of the proper procedure for discharging the accumulator and locking out the system when shutting down machines. This process is a critical safety issue to prevent both injury to personnel and damage to equipment.

Safety bleed valves may be recommended by the accumulator manufacturer. These valves ensure an accumulator is discharged when the electrical power is shut off in the system. Electricity to the solenoid of the bleed valve is cut at the same time as the electricity to the motor of the hydraulic power unit. The spring in the bleed valve then shifts the valve to connect the accumulator directly to the reservoir. The pressurized hydraulic fluid remaining in the accumulator bleeds off to the reservoir. A flow control valve in the bleed-off line limits the rate of discharge flow from the accumulator to prevent an excessively high flow rate in the line to the reservoir.

CAUTION

Gas-charged accumulators are the most popular designs used today in both industrial and mobile hydraulic applications. The following is a list of safety factors that must be followed to provide a safe working environment when using gas-charged units.

- Only individuals with specific training should install accumulators or perform repair work.
- Never fill an accumulator with oxygen. An explosion may result if a mixture of oxygen and oil is pressurized.
- Never fill an accumulator with air. Three factors make the use of air in an accumulator undesirable. First, there is a risk of explosion as the oxygen in the air is compressed in the system. Second, the moisture contained in the air can condense, rusting system parts and reducing hydraulic fluid life. Third, oxygen in the air causes oxidation of the hydraulic oil.
- Always fill an accumulator with an inert gas, such as dry nitrogen. Using dry nitrogen eliminates the above problems.
- Never exceed the precharge pressure recommended by the accumulator manufacturer.
- Use soapy water to check for exterior gas leaks. Apply soapy water to all fittings and seams exposed to pressurized gas in the accumulator. If bubbles form, there is a leak.
- Internal gas leaks in the system may be identified by the appearance of bubbles in the reservoir.
- To remove all accumulator-related pressure from the hydraulic oil, first shut down the pump. Then, cycle an actuator in the accumulator portion of the circuit to relieve oil pressure. If the circuit is equipped with a bleeder valve, simply open that valve and bleed the hydraulic system pressure to 0 psi.
- Be certain to remove all gas pressure from an accumulator before removing it from the system.

Summary

- Accumulators are the fluid power components that absorb and store energy in the system. Energy is stored in an accumulator by lifting a weight, compressing a spring, or compressing a gas.

- Weight-loaded accumulators are vertically mounted cylinder-like units that store energy by lifting heavy weights mounted on the cylinder rod. Gravity then acts to pressurize the system fluid, allowing the entire volume of the fluid in the accumulator to be delivered at a near-constant pressure.

- Spring-loaded accumulators store energy by compressing a spring. The compressed spring maintains pressure on the system fluid and extends as system pressure decreases to push fluid back into system. The characteristics of the spring determine the rate of fluid intake and discharge.

- Gas-charged accumulators use compressed inert gas to store energy. As the system fluid pressure increases, the gas in the accumulator is compressed as hydraulic fluid moves into the accumulator; as the system pressure decreases, the gas expands to maintain system pressure or to push fluid back into the system lines. Gas may also be precharged prior to system operation.

- Pistons, bladders, and diaphragms are used in gas-charged accumulators to separate liquid and gas, preventing the gas from being absorbed by the liquid or entering the liquid carrying lines of the system.

- Piston-type accumulators are a cylinder with a free sliding piston that separates the cylinder into a gas chamber and a liquid chamber.

- Bladder-type accumulators use a flexible bladder fitted into a metal shell. The gas is stored in the bladder while the hydraulic fluid is held in the space between the shell and the bladder.

- A diaphragm-type accumulator is constructed of two metal hemispheres separated by a flexible diaphragm. Hydraulic fluid fills one of the hemispheres while the second holds the gas.

- Four basic functions of accumulators in circuits are: providing a means to store energy, controlling pressure and flow variations in a system, maintaining system pressure and supplementing pump output flow.

- Rapid pressure increases or fluctuations in flow can cause hydraulic component damage. Control of these problems requires careful sizing, precharging, and placement of the accumulator in a hydraulic circuit.

- Accumulators are available in a wide range of capacities (under 1 pint to over 200 gallons) and pressure ratings (typically 2,000 psi to over 10,000 psi).

- Most manufacturers provide suggested procedures to follow for the selection and installation of accumulators in various types of applications. Storage capacity, pressure ratings, and the usable volume should be considered when selecting an accumulator.

- Because charged accumulators can supply pressurized fluid even when the system appears to be powered down, special care must be taken to remind operators of the procedures to discharge accumulators. Safety bleed valves may also be recommended.

Internet Resources

The following are some useful resources available on the Internet. Enter a company or organization name into a search engine to access its website. Explore the various areas of the sites to discover useful fluid power resources.

Eaton Corporation. Provides descriptions of bladder, diaphragm, and gas-filled accumulators.

OLAER Fawcett Christie, Ltd. Maintenance and precharging instructions for gas-charged, hydraulic accumulators. Includes part drawings and information on maintenance, disassembly, cleaning, inspecting, and reassembly. Note the company's safety warnings related to accumulators.

HYDAC Technology Corporation. Provides information on manufacturing methods, types of designs, and specifications of selected accumulator models. Applications are also shown, along with the forms the company suggests to be used when collecting information needed to properly size an accumulator.

HydraulicSupermarket.com. Provides formulas recommended for calculating accumulator gas precharge pressure and accumulator size for various applications.

PACCAR, Inc.—Winch Division. Service bulletin from an end user that includes procedures for servicing/recharging gas-filled accumulators.

Parker Hannifin Corporation. Describes the use of safety blocks to protect, isolate, and discharge a hydraulic accumulator.

Chapter Review

Answer the following questions using information in this chapter.

1. An accumulator permits _____ to be absorbed and stored in a hydraulic system.

2. Accumulators use weights, _____, and compressed gas to provide the energy storage needed to perform their function in a hydraulic circuit.
   A. springs
   B. packings
   C. water
   D. All of the above.

3. _____-loaded accumulators use the force of gravity to allow the storage of energy in a hydraulic system.
   A. Weight
   B. Spring
   C. Water
   D. Gas

4. *True or False?* Weight-loaded accumulators are generally very small sized units.

5. Compressed, _____ is the most common type of gas used in gas-charged accumulators.
   A. inert gas
   B. atmospheric air
   C. hydraulic fluid
   D. active gas

6. Which of the following is a gas-charged accumulator design used in hydraulic systems?
   A. Piston.
   B. Bladder.
   C. Diaphragm.
   D. All of the above.

7. Which of the following can be used in gas-charged accumulators?
   A. Air.
   B. Oil vapor.
   C. Dry nitrogen.
   D. All of the above.

8. The body of a bladder-type accumulator consists of a cylindrical _____ shell.

9. In a bladder-type, gas-charged accumulator, the _____ valve prevents the precharge pressure in the bladder from pushing part of the bladder into the system line.

10. In a gas-charged accumulator, the volume of the isolated gas chamber _____ as the volume of the system-port chamber increases.

11. A(n) _____ process is one that is completed without heat gained from or lost to surrounding materials.

12. The volume of fluid available from an accumulator is greatest if the unit is charged and discharged as a(n) _____ process.

13. Which of the following is a result that may occur if pressure and flow variations are great enough in a system under varying load conditions?
    A. Increased efficiency of a system.
    B. Strengthen system components.
    C. A safety issue that could harm workers.
    D. All of the above.

14. *True or False?* The rated capacity of various gas-charged accumulators is under 1 pint to over 200 gallons.

15. _____ is used to establish the usable volume of an accumulator.
    A. Maximum operating pressure
    B. Minimum operating pressure
    C. Precharge pressure
    D. All of the above.

16. The best mounting position for any style of accumulator is _____.
    A. horizontal with the hydraulic connection facing downward
    B. vertical with the hydraulic connection facing downward
    C. vertical with the hydraulic connection facing upward
    D. Any position is appropriate.

17. *True or False?* The weight-type accumulator design *must* be mounted in a horizontal position.

18. *True or False?* Precharging a gas-charged accumulator should be done only after the hydraulic system pressure has been reduced to 0 psi.

19. *True or False?* Air is the most used precharge gas in gas-charged accumulators.

20. *True or False?* Use soapy water to check for exterior gas leaks at fittings exposed to pressurized gas in accumulator circuits.

Apply and Analyze

1. For each of the three basic types of accumulator (weight-loaded, spring-loaded, gas-charged), draw a graph showing how hydraulic pressure in the accumulator changes with time for the energy storage circuit shown in **Figure 11-8**. For all three types of accumulators, assume and label the following events:
   - Rapid initial pressurization of the accumulator with little chance for heat dissipation.
   - The engine is turned off. Flow from the accumulator closes the check valve.
   - Accumulator portion of the circuit reaches thermal equilibrium with the surroundings.
   - The directional control valve is shifted to direct fluid flow to start the hydraulic motor.
2. What design would you choose for an application that requires a horizontally mounted accumulator with a usable operating volume of 30 gallons? Why?
3. Explain why recommended precharge pressures are higher than normal system operating pressures when the accumulator is being used as a shock or surge suppressor.

Research and Development

1. Investigate the historical development of accumulators and create a time line to illustrate the critical events leading to current designs.
2. Find a circuit schematic for an application that uses an accumulator. Determine the function of the accumulator in the circuit, draw out the flow of high-pressure fluid through the circuit, and list the series of events that occur in the system during the charging and discharging of the accumulator.
3. Examine a specific model of gas-charged accumulator. Make a poster showing the overall design, components, and capacities. Explain how the accumulator functions in a typical application.
4. Research accumulators used in hydraulic hybrid vehicles. Prepare a presentation that compares hydraulic hybrid vehicle design, operating characteristics, efficiency, and cost with electric hybrid vehicles.

12 Conditioning System Fluid

Filtration and Temperature Control

Hydraulic systems operate in a wide variety of workplaces and under various load conditions. These conditions subject the system to dust, dirt, moisture, and heat. Each of these elements can be harmful to system components. This can reduce system efficiency and lead to major component damage. Although a clean, cool surrounding is ideal for system operation and service life, that condition is not possible in most industrial situations. This chapter provides information on the various filtration devices and heat exchanger equipment used to condition system fluid so it is clean and at a desired operating temperature.

Learning Objectives

After completing this chapter, you will be able to:

- Identify the typical contaminants found in hydraulic system fluid and describe the source of each.
- Explain the source of the energy responsible for increasing fluid temperature during hydraulic system operation.
- Describe how reservoir design can be used to reduce fluid contamination.
- List and compare the various filter media used for hydraulic fluid filtration.
- Describe the filtration rating methods used with hydraulic filters and strainers.

- Compare the characteristics of the various filter locations and circuits that may be used in hydraulic systems.
- Explain the function of heat exchangers in a hydraulic system.
- Describe the design and structure of heat exchangers commonly used in hydraulic systems.
- Identify the factors that must be considered when determining the need for a heat exchanger in a hydraulic system.

Key Terms

absolute rating	filter-element-condition indicator	nominal rating
absorbent filter		off-line filtration
adsorbent filter	filtration system	proportional filtration
bonnet	finned conductor	radiator
brazed-plate heat exchanger	fluid conditioning	shell-and-tube heat exchanger
bypass valve	full-flow filtration	suction filter
cleanout	heat exchanger	sump strainer
contaminant	immersion heat exchanger	surface-type filter
depth-type filter	micron	varnish

12.1 Need for Fluid Conditioning

Unlike the other major areas of hydraulic systems, fluid-conditioning equipment and circuits do not directly contribute to the development and control of system power output. However, these elements must be considered critical to the efficient and cost-effective operation of a system. These elements, in a sense, provide critical services to the components of the system and the fluid that is the lifeblood of the system.

Fluid conditioning provides the system with fluid that is clean and at an acceptable operating temperature. Without effective methods to provide these conditions, the operating efficiency and life of any hydraulic system is greatly reduced. It is often claimed that 75% of hydraulic system failures are a direct result of fluid contamination. Any system failure results in substantial increases in operating costs. This is due to system maintenance and the loss of production during the equipment downtime.

12.2 Contaminants and Their Sources

A hydraulic system must be protected from a wide variety of contaminants. *Contaminants* are particles or liquids that should not be in the system. They can reduce the operating efficiency or increase the rate of wear of component parts. Contaminants include those:

- Built into the components during manufacture.
- Entering the system during operation.
- Resulting from component wear.
- Resulting from the breakdown of the hydraulic fluid.

The processes used to produce equipment can result in a variety of manufacturing-, fabrication-, and shipping-related contaminants. These contaminants can include materials such as:

- Grains of sand from sand molds and cores used to cast the body.
- Rust chips and paint overspray related to cleaning and painting the product.
- Dust and water picked up during shipment of the assembled equipment to the final installation location.

It is the responsibility of the *filtration system* to remove contaminants. In new or reconditioned equipment, the filtration system initially needs more frequent attention to ensure the above built-in contaminants are removed as quickly and efficiently as possible.

Contaminants are also introduced into a hydraulic system during normal system operation, **Figure 12-1**. The cylinder rod of linear actuators may be exposed to dust, dirt, and liquid contaminants. On cylinder

Reprinted courtesy of Caterpillar, Inc.

Figure 12-1. External conditions expose a hydraulic system to dirt and water that can cause rapid deterioration of system fluid and components.

retraction, these may be introduced into the system. Also, dust and moisture may enter through the reservoir breather. Foreign materials may also enter a system during the disassembly of components for service and the routine refilling of the reservoir. Service personnel should always be careful to use clean work surfaces and containers.

An operating hydraulic system also internally produces contaminants during normal operation. These include particles released due to wear on bearings. Packing and seals also contribute contaminants as they wear.

In systems exposed to humid conditions, condensation may introduce water into the system during periods of shutdown. This contributes to the breakdown of the fluid and the formation of sludge and acids. Sludge is a viscous, often gel-like residue that forms when water and other contaminants interact with the hydraulic fluid. Although neither sludge nor acid is abrasive, they both contribute to system service problems.

12.3 Effects of Contamination

Hydraulic fluid contamination results in reduced system efficiency and physical harm to system components. These are caused by the reduced ability of the fluid to:

- Provide lubrication of moving parts.
- Seal the clearances between moving parts.
- Transfer heat.
- Transfer energy in the system.

Dirt particles reduce the ability of the fluid to provide lubrication. Reduced lubrication leads to excessive wear of component parts, **Figure 12-2**. In turn, bearing and spool clearances are increased. Excessive clearances cause internal component leakage and can lead to bearing failures in pumps and motors.

Sludge accelerates the build-up of residue on the surface of parts. It also leads to the gradual clogging of valve passageways. Acids attack metallic surfaces. This results in pitting and an increased rate of wear on bearing surfaces.

12.4 System Operating Temperature

System operating temperature is a major concern when designing and operating hydraulic systems. Systems function under a wide range of ambient temperatures. Systems also typically generate considerable heat during operation. This heat is a reflection of the relatively poor efficiency of these systems. Whenever the relief valve of a system functions, efficiency is reduced. The energy used to overcome the resistance of the valve is turned into heat. When all of this is considered, even well-designed systems convert 20% or more of horsepower input into heat. The heat is absorbed by the hydraulic fluid. Once started, this causes the temperature of the system to quickly increase.

System operating temperatures are usually monitored by measuring the temperature of the hydraulic fluid in the reservoir. In most systems, reservoir-oil temperature ranging from 110–140°F is considered desirable. Extreme system operating temperatures, either high or low, are undesirable in any hydraulic system. One manufacturer indicates that the temperature should not exceed 145°F for even a short period of time without special considerations for fluids and seal materials.

High hydraulic oil temperatures can cause the system oil to break down. This leads to the formation of *varnish.* This material adheres to internal surfaces of components and clogs orifices, contributes to the formation of acid, and promotes the formation of sludge. Also, as temperature increases, the viscosity of the oil decreases. Excessive oil temperature may reduce viscosity to the point where the lowered lubricating qualities lead to increased bearing surface wear and decreased component service life. Excessive system operating temperature also shortens the life of system seals. This leads to increased fluid leakage and increased overall maintenance costs.

Low hydraulic oil temperatures present another set of problems that can adversely affect system operation. Oil temperature lower than the recommended level causes larger-than-desired pressure drops and sluggish system operation. The result is an inefficient system. Fluid in equipment exposed to cold weather can have enough of a reduction in fluidity that major problems result, **Figure 12-3**. For example, highly viscous oil in the pump inlet line can cause cavitation. This leads to serious damage to internal pump parts. When ambient temperatures are low, a system should be operated under a no-load condition until the oil temperature approaches the recommended operating level.

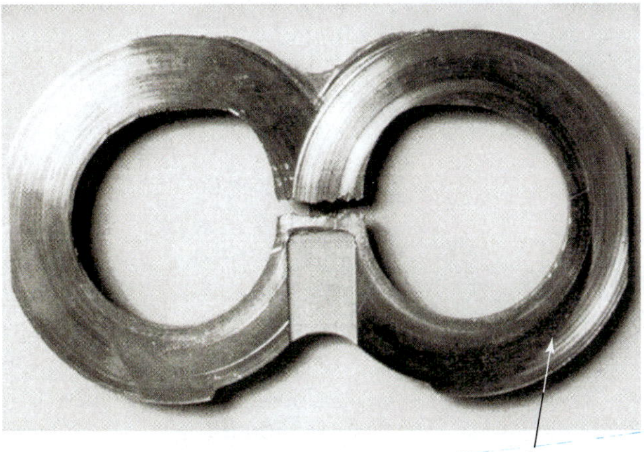

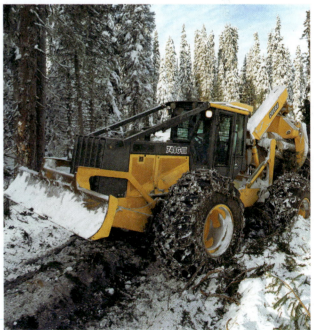

Deere & Company

Figure 12-3. Low system operating temperatures cause a special set of service problems, including pump cavitation during system startup.

Excessive wear

Schroeder Industries LLC

Figure 12-2. Dirt particles and acid in hydraulic fluids can cause rapid wear of bearing surfaces.

Hydraulic systems need to be designed to balance the system heat input with the rate of heat removed by conduction, radiation, or convection. A properly designed system should provide the means to balance this input and output of heat. If a hydraulic system is operating at a temperature higher or lower than recommended, maximum equipment performance will not be achieved.

12.5 Controlling and Removing Contaminants

The best route to follow to ensure a clean operating system is preventing contaminants from entering a hydraulic system. However, because of the many extreme operating conditions in which systems must operate, this solution is not easily achieved. It is possible to reduce contamination if designers, machine operators, and maintenance personnel are aware of the sources of contaminants and how they enter a system. Using this information, each of these individuals can help reduce contamination and improve performance and service life of systems.

The designer must have adequate information concerning the conditions in which a machine must function. With this information, designs can provide adequate seals and baffles to reduce the entrance of dust, dirt, and water. Machine operators need to be aware of the sources of contaminants. Situations that exceed the ability of the system to block the entrance of

contaminants must be avoided or promptly reported to supervisory personnel. Maintenance staff also needs to be alert to sources of contamination. These sources should be eliminated before a replacement component is placed in service.

This section examines the components and basic circuits that deal with the control of contamination. This includes terminology, design factors, operating theory, installation, and service.

12.5.1 Contamination Control and the Reservoir

The reservoir is not only the fluid storage unit of a hydraulic system; it is considered a basic contamination-control component. The reservoir can make a large contribution to a clean and efficient operating system. Careful reservoir design and construction provide features that prevent the entrance of dust, dirt, and water into the system. Features can also be provided to help remove contaminants that do gain access to the system. See **Figure 12-4**. Review the reservoir portion of Chapter 8 for information on design features and additional points relating to the role of the reservoir.

Reservoirs contain internal baffles. Baffles provide a long route for the fluid to travel between the return line and pump inlet line. This provides the longest possible time for dirt particles to settle, water to separate, and entrained air to escape.

Breather tubes allow air to flow into and out of the reservoir. This air movement occurs because of fluid

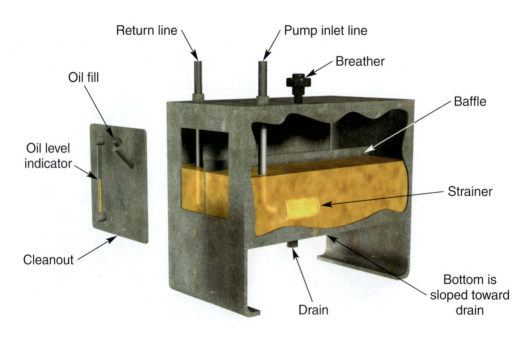

Return line Pump inlet line Breather Oil fill Baffle Oil level indicator Strainer Cleanout Drain Bottom is sloped toward drain

Goodheart-Willcox Publisher

Figure 12-4. The reservoir serves as the primary fluid storage unit of the system, but also acts as a basic fluid conditioning component. It helps remove dirt, water, entrained air, and heat from the system.

level changes that result from actuator operation. Filters must be included on breather tubes to clean the air.

Sealing material must be provided around inlet, outlet, and drain lines and the filler cap. This material prevents the entrance of dirt and water into the interior of the reservoir.

The bottom of the reservoir should be sloped or dish shaped. This ensures the best possible drainage when draining and cleaning the system. *Cleanouts* allow access to the interior of the reservoir for cleaning, **Figure 12-5**. They should be provided to help maintenance personnel in thorough servicing of the power unit. These features and the lines entering or leaving a reservoir should be placed for easy access. The placement of the reservoir in the overall scheme of the system often makes it difficult to gain access to these features. In order to encourage routine cleaning, well-laid-out reservoir features are essential.

12.5.2 Strainers and Filters

Filters and strainers are the devices normally associated with the cleaning of hydraulic fluid, **Figure 12-6**. The system fluid is forced to pass through the device. The device contains a porous material that traps insoluble materials contained in the fluid. However, the difference between a filter and a strainer is not precise. The National Fluid Power Association defines a strainer as a coarse filter. This indicates the device can remove only larger particles. However, some manufacturers may have strainers that can remove small particles.

Regardless of the term used to describe the device, it should be selected on its ability to maintain a specified level of fluid cleanliness. The device must be able to remove contaminants above a given size, provide a reasonable service life, and allow easy service to minimize system downtime while changing filter elements, **Figure 12-7**.

Goodheart-Willcox Publisher

Figure 12-5. A reservoir should have a means for draining and internal cleaning. In small systems, it often serves as the primary fluid conditioning unit.

Schroeder Industries LLC

Figure 12-6. Hydraulic system filters and strainers are available in many shapes and sizes to provide required system filtration.

Atlas Copco

Figure 12-7. A filter or strainer should provide adequate filtration and allow for easy service.

Filtering methods and materials

Filters and strainers can be classified by the two different methods of filtration used to remove contaminants: surface type and depth type. *Surface-type filters* provide surfaces containing numerous holes to allow the passage of fluid. Particles larger than the holes are caught and held in the filter. See **Figure 12-8**. On the other hand, *depth-type filters* use a mass of porous material that provides many different routes for the fluid to follow from the inlet to the outlet of the filter. Particles are removed by being trapped in one of the many flow routes in the porous material. See **Figure 12-9**.

Surface-type filters are made from a fine wire mesh, accordion-pleated paper, stacked metal disks, shaped metal ribbon, or molded cellulose. Depth-type filters use materials such as cotton, wood pulp, or paper to form the mass needed to provide the numerous long routes through the unit that serve to trap the unwanted contaminants.

Depth-type filters are also classified as either absorbent or adsorbent. *Absorbent filters* mechanically remove contaminants by trapping solid dirt particles, water, and water-soluble contaminants suspended in the hydraulic fluid. *Adsorbent filters* trap these same impurities, but also use chemical treatment to remove contaminants. Charcoal, chemically treated paper, and other materials are used in this process. Adsorbent filters are not typically used in hydraulic systems as they may remove desirable additives from the system fluid.

Level of filtration

Filtration ratings indicate how small of a particle the filter will remove. The most common measurement scale used to indicate the level of filtration is the micron. One *micron* is equal to .000039". **Figure 12-10** shows the relative size of common particles in microns. As shown in the figure, one division on the micron scale represents an extremely small size. For example, it takes 25 microns to equal .001", while the smallest particle that can be seen by the human eye without magnification is approximately 40 microns.

The National Fluid Power Association recommends a minimum filtration level of 25 microns for all hydraulic systems. However, many equipment manufacturers specify a filtration level exceeding this. A higher level is recommended in systems containing components manufactured to close tolerances or that require sensitive responses to system pressure and actuator position. Always follow the manufacturer's filtration recommendations to ensure maximum performance and service life of system components.

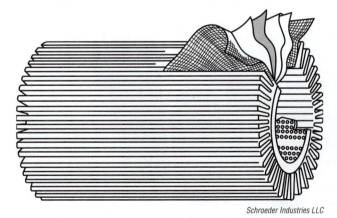

Schroeder Industries LLC

Figure 12-8. Surface-type filters depend on single or multiple filtering surfaces containing numerous holes of a specific size to catch and hold contaminants.

Donaldson Company, Inc.

Figure 12-9. An electron microscope view of the numerous routes through the material used in a depth-type filter element.

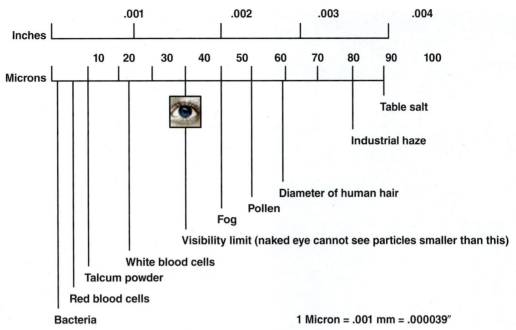

Schroeder Industries LLC

Figure 12-10. A comparison of common materials using the micron and inch units of measure.

The ratings found on filters may include an absolute and a nominal figure. The *absolute rating* states the largest pore opening. All particles larger than the indicated size will be removed by the filter. This rating is used with both surface- and depth-type filters. The rating is more meaningful with surface-type filters because their construction yields a more uniform pore size than that found in depth-type filters. The *nominal rating* typically indicates the average pore size of depth-type filters. This rating does not provide a sound indicator of the size of materials that may pass through the filter. It does *not* guarantee removal of all contaminants larger than the stated average pore size.

> **NOTE**
>
> Absolute and nominal filter ratings should be used only to identify the average and largest pore size in a filter element. Always consult with specific filter manufacturers and the component manufacturers when selecting a filter.

Filter housings

The filter style and the placement of the filter determine the way a filter is incorporated into the system. The simplest approach is demonstrated by the sump strainer, **Figure 12-11**. This filter is attached to the end of the pump inlet line that is submerged in the reservoir. One end of the strainer contains a threaded collar that directly attaches to the inlet line of the pump. Other designs include housings with inlet and outlet ports that allow the unit to be built into the pump inlet

line outside of the reservoir or into a system working or return line. These housings can include a:

- Bolt-on metal bowl that holds a separate filter element.
- Threaded fitting and sealing surface that allows spin-on filter cartridges to be easily attached to the housing. See **Figure 12-12**.

Bypass valves and indicators of filter element condition are also included in some housings.

The *bypass valve* is used to allow fluid flow past the filter if the filter becomes plugged, **Figure 12-13**. When flow resistance through the filter exceeds a preset level, the bypass valve opens. This allows fluid to move through the filter housing without passing through the filter. The bypass feature prevents:

- Pump cavitation when a filter in an inlet line is plugged.
- Rupture of the filter when it is placed in a working or return line.

Zinga Industries, Inc.

Figure 12-11. Sump strainers have a simple housing that attaches directly to the end of the pump inlet line.

Goodheart-Willcox Publisher

Figure 12-12. Spin-on filters have cartridges that may be quickly and easily changed during system maintenance.

Filter-element-condition indicators are often incorporated in a filter housing to show the pressure drop between the inlet and outlet sides of the filter, **Figure 12-14**. The scales on these indicators usually show the degree of filter contamination. They are used to prompt service personnel to replace filters on a timely schedule.

Filter housings are subjected to various pressures, depending on their placement in the system. Care should be taken to select housings that are compatible with the location. Pump inlet line pressures are normally slightly below atmospheric pressures, while return line ratings of 500 psi or lower are usually considered adequate. Working line filter placement requires units that can withstand pressures encountered during circuit operation. It should be remembered that, in some circuits, pressures higher than the relief valve setting can be experienced.

12.5.3 Locating Filtering Devices in a System

Using fluid filtration devices in a hydraulic system is generally considered important in obtaining the maximum system operating efficiency and service life. However, filters must be appropriately placed in a circuit to obtain their maximum performance. The three general locations for filters are:

- Pump inlet line.
- System working lines between the pump and actuators.
- Return and drain lines between components and the reservoir.

Zinga Industries, Inc.

Figure 12-13. A filter bypass valve allows system fluid to flow past the filter when contaminants clog the element and the pressure drop across the element reaches a predetermined level.

Zinga Industries, Inc.

Figure 12-14. Filter-element-condition indicators provide an easy means to check the pressure drop across a filter element.

Figure 12-15 illustrates the three basic locations in which filters may be placed to ensure the appropriate filtration of system fluid.

Pump inlet line

Two different types of filters may be used in the pump inlet line. A *sump strainer* is located inside of the reservoir on the submerged end of the pump inlet line. It is a coarse filter designed to protect the pump from larger dirt particles that have entered the reservoir. This type of filter is relatively inexpensive, as it does not

have a complex housing. The lack of a filter-element-condition indicator and the fact that the filter is located below fluid level makes service somewhat difficult. This often results in inconsistent servicing of the filter.

A *suction filter* is located in the inlet line, but on the outside of the reservoir. Suction filters provide finer filtration than sump strainers. This better protects the pump from dirt in the reservoir. Unlike a sump strainer, these filters can include a filter-element-condition indicator. Most suction filters are easily serviced without disassembling major system parts. However, care must

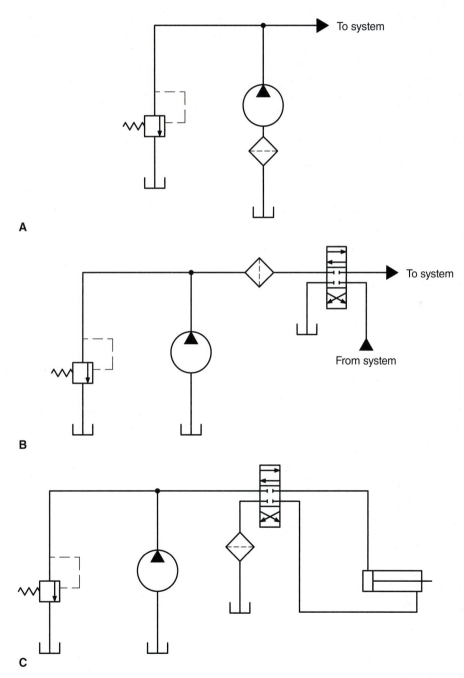

Figure 12-15. The most common locations of filters in a hydraulic system are shown here. A—Pump inlet line. B—System working line. C—System return line.

be taken when selecting a suction filter. Improper sizing can result in starving the pump, which can cause cavitation damage to internal pump parts.

System working lines

Filters in working lines remove contaminants that have:

- Passed through the pump from the reservoir.
- Been produced by the pump or other components upstream from the filter.

Placement may be between the pump and the first system component or between other system components further downstream in the system. These filters must be designed to withstand the normal working pressure of the system. If they are placed between an actuator and the directional control valve, they must also be designed to withstand reverse fluid flow.

Placing filters in working lines allows specific components to be protected from contaminants produced by other upstream components. Another advantage is the filtration level of filters used in these locations may be finer than other filters. This is possible as full system pressure is available to move the fluid through the filter element.

Return lines

Return line filtration removes contaminants generated in the system before returning the fluid to the reservoir. These contaminants may range from materials that enter on cylinder rod retraction to those generated by wear on bearing surfaces, seals, and rod scrapers.

The traditional position for a return line filter is in the return line connecting the directional control valve return port and the reservoir. **Figure 12-16** illustrates several additional locations where these filters may be placed in a system. These locations include a pump case drain line, the relief valve return line, and the return line of a bypass flow control valve. During typical system operation, these return line filter locations are not subjected to the maximum operating pressure of the system. This reduces the need for a high-pressure filter housing, which reduces component cost.

The flow rate in return lines varies in relation to the operating phases of a system. For example, the

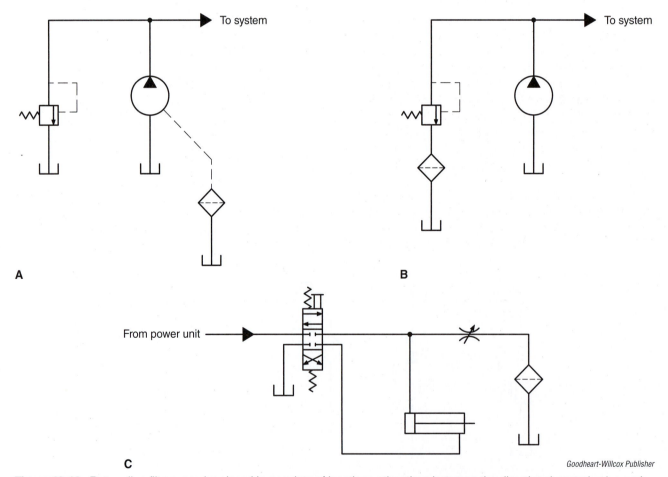

Figure 12-16. Return-line filters may be placed in a variety of locations other than between the directional control valve and reservoir. The locations shown have specific advantages and disadvantages in relation to filtration and service. A—Pump case drain. B—Relief valve return. C—Bypass flow control.

fluid being returned to the reservoir may exceed pump output. This may happen if the rod diameter is large in relation to cylinder bore. Careful analysis of return line flow must be done to ensure the selection of a filter with an adequate flow rating. An inadequately sized filter will produce excessive backpressure in the line. This reduces system operating efficiency and the effectiveness of the filter.

12.5.4 Routing Fluid Flow for Filtration

Three routing methods are used to direct system fluid through system filters:

- Proportional filtration.
- Full-flow filtration.
- Off-line filtration.

Variations of these methods are also used, such as duplex filters. Duplex filters provide two or more filters in a single housing. A control valve provides an easy switching method allowing the fluid flow route to be alternated through the various filters. An advantage of this variation is the filter not currently in use can be changed without shutting down the system.

Proportional filtration directs a measured amount of system fluid through a filter. **Figure 12-17** shows a circuit diagram of a simple circuit using proportional filtering. In this circuit, a line leading from a system working line contains a metering valve and the filter. Part of the pump output passes through the filter and is returned directly to the reservoir. Filtration is continuous. The theory is that eventually all system fluid will pass through the filter and be cleaned. This method has become less popular as system tolerances have increased, which requires more positive, finer filtration of system fluid.

Full-flow filtration filters all of the fluid circulated by the system pump. The location of a filter in the system and the level of filtration varies by system design. However, all fluid pumped into the system is filtered before returning to the reservoir. In some systems, filters are placed in several locations. This results in portions of the fluid being filtered more than once as it is circulated. An example of this is a system using a sump strainer and a return line filter. This provides double filtration.

The use of full-flow filters has become more critical with the increased precision of hydraulic component parts. This increased filtration requires bypass valves to prevent excessive pressure drops across the filter as it becomes blocked by contaminants. Excessive pressure drops in the pump inlet line can cause pump cavitation. An overall loss of system efficiency occurs when excessive pressure drops occur at other filter locations.

Off-line filtration consists of a separate circulating pump and filtering system, **Figure 12-18**. This type of filtration is sometimes called recirculating, kidney-loop, or side-stream filtration. It should be considered

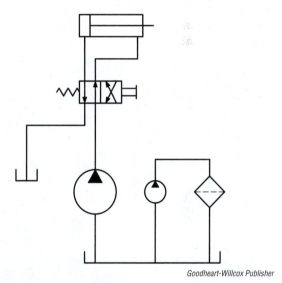

Goodheart-Willcox Publisher

Figure 12-18. An off-line filtration system involves a separate pump that continually circulates reservoir fluid through a filter.

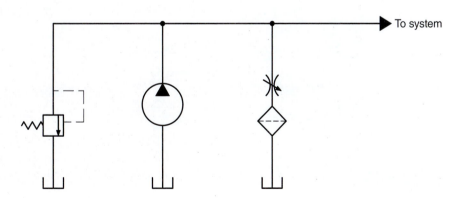

Goodheart-Willcox Publisher

Figure 12-17. This diagram illustrates one method of proportional filtration in which a percentage of total pump output is diverted through a filter on a continuous basis.

a subsystem that continuously pumps fluid out of the reservoir, through a filter, and then back to the reservoir. As a general rule, the circulating pump should have an output flow rated at 10% of the reservoir volume.

Off-line filtration systems can maintain a clean system if pump flow and filter size are matched to the operating hydraulic system. It is relatively easy to retrofit an existing system with an off-line filtration system. Also, they can be easily serviced without shutting down the overall system.

Figure 12-19 shows a portable, off-line system. These can easily be moved from point to point. They can be used to provide filtration in emergencies.

12.6 Filter System Maintenance

A carefully developed and supported maintenance plan is basic to the performance of industrial equipment and systems. A critical part of that plan is maintenance of the hydraulic filter system. The following basic items relating to filter maintenance need to be considered to ensure maximum performance and service life from hydraulic equipment.

- Set up and carefully follow a maintenance schedule for replacing filters and draining system fluid.
- Be certain that the fluid used in the system is the type recommended by the component and system manufacturer.
- Be certain that all reservoir cleanouts, breather cap filters, and filler caps are properly fitted and sealed.
- Store reserve system fluid in clean, tightly sealed containers.
- When filling the reservoir, be certain only clean containers, hoses, and funnels are used.
- Carefully handle filters. A bent or punctured filter element means the filter cannot perform its assigned task.

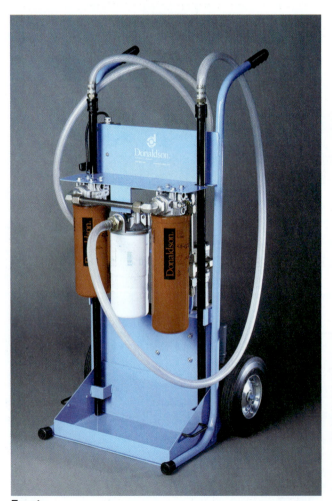

Front

Back

Donaldson Company, Inc.

Figure 12-19. Portable off-line filtering systems are available. These systems contain a circulating pump and a filter that can be used to boost filtration in a system on a temporary basis.

12.7 Controlling Fluid Temperature

System temperatures that are either higher or lower than the recommended temperature affect system performance. Extreme variations can cause major problems with both system performance and the service life of components. Lower-than-recommended temperatures increase fluid viscosity. This reduces the lubricating ability of the fluid, causes sluggish system operation, and may cause pump cavitation. High temperatures reduce fluid viscosity. This reduces lubricating ability and causes increased deterioration of seals, decomposition of the fluid, and buildup of varnish on component surfaces.

12.7.1 Function of Heat Exchangers

A *heat exchanger* in a hydraulic system adds heat to or subtracts heat from the system fluid to keep the fluid temperature within the desired range. In many hydraulic systems, ambient conditions and the heat-transfer characteristics of the system components maintain the desired operating temperature of the system fluid.

Of course, this is ideal and not possible in all systems. In the real world of hydraulic system operation, the atmospheric temperature can be too high or too low. In addition, increased system loads result in an

HYDAC Technology Corporation

Figure 12-20. A radiator heat exchanger is similar to automotive radiators. They are used in some hydraulic systems to maintain fluid temperatures within the recommended range.

increase in average system pressure. This varies heat input into the system. Even the internal or external cleanliness of the system can cause variations in system operating temperatures.

A well-designed hydraulic system will usually perform at a level that results in acceptable fluid operating temperatures. However, the temperature may be out of the suggested range for short periods of time. The real need for a heat exchanger occurs when extremes of fluid temperature threaten the service life of a system.

12.7.2 Design and Structure of Commonly Used Heat Exchangers

A variety of heat exchanger designs are used to cool or heat hydraulic system fluid. Some of these designs can cool or heat the fluid. Other designs are limited to just cooling or just heating. Air-cooled radiators and finned conductors are normally limited to cooling. Shell-and-tube units use water to absorb and dispense heat. They can either cool or heat system fluid. Immersion units located in the reservoir can be used to either cool or heat system fluid.

Radiators and finned conductors

Air-cooled radiators and finned conductors are closely related. Their basic operation involves hydraulic fluid passing through a tube with metal fins on its exterior surface. The fins increase the area of the heat-transfer surface.

Radiators are constructed of a series of small, fluid-carrying lines that pass through a common series of metal fins attached to their exterior surface. All of these lines end in common chambers on either end of the unit, **Figure 12-20**. In contrast, *finned conductors* are usually single, metal transmission lines fitted with a series of metal fins firmly attached to the exterior wall. In both designs, heat from the hot system fluid is transferred to the walls of the lines. The heat is conducted through the walls of the line and into the cooling fins. This heat is then dumped into the air surrounding the fins.

Still air around the metal cooling fins and, in finned conductors, laminar fluid flow in the hydraulic lines can reduce operating efficiency. Some manufacturers place devices inside of the transmission lines in finned conductors to cause turbulent fluid flow. The resulting increased heat transfer rate more than justifies the increased pressure drop caused by the higher fluid turbulence.

A fan is commonly used to move air over the fluid tubes and thin metal fins of the radiator. This increased air movement increases the efficiency of the heat transfer. However, it still must be kept in mind that air is not as effective of a heat conductor as water, oil, and metal.

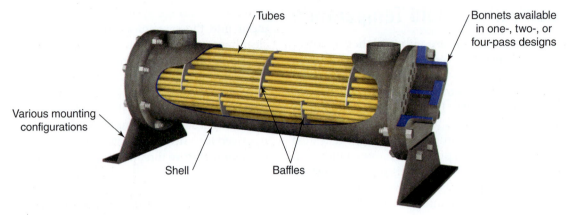

Tubes

Bonnets available
in one-, two-, or
four-pass designs

Various mounting
configurations

Shell   Baffles

Figure 12-21. Shell-and-tube heat exchangers use water to cool or heat hydraulic system fluid.

Shell-and-tube and brazed-plate heat exchangers

Both shell-and-tube and brazed-plate heat exchangers have the ability to cool or warm the hydraulic fluid. The shell-and-tube design has been used for many years in the hydraulic field as well as in other heat exchange applications. The brazed-plate heat exchanger is considered a newer design.

Shell-and-tube heat exchangers are constructed of a bundle of tubes enclosed in a metal shell. The ends of the tubes extend into common chambers cast into the bonnets. See **Figure 12-21.** The *bonnets* are caps fitted on either end of the metal shell. During operation, water flows inside the tubes as the hydraulic fluid flows around the tubes, **Figure 12-22.**

Shell-and-tube heat exchangers are available in one-, two-, and four-pass designs. The term *pass* refers to the number of times the water flows the length of the exchanger between entering the inlet port and being discharged at the outlet port, **Figure 12-23.** Increasing the

number of fluid passes through an exchanger increases the amount of heat transfer. Multiple passes are made possible by dividers located in the bonnets that route the flow of cooling water through only a portion of the tubes during each pass, **Figure 12-24.**

When these units are used to cool the hydraulic fluid, cold water is pumped through the tubes.

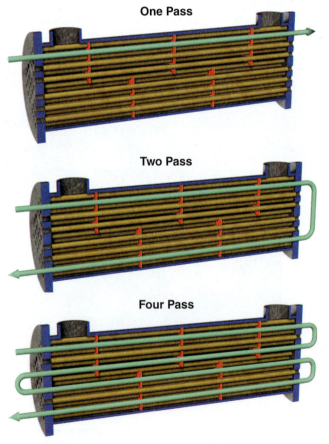

One Pass

Two Pass

Four Pass

Figure 12-23. The number of times the water flows through the heat exchanger determines whether the exchanger is a one-, two-, or four-pass design.

Baffles

● = Shell-Side Fluid

● = Tube-Side Fluid

Figure 12-22. Baffles are used to route the hydraulic fluid through the shell to maximize contact with the tubes carrying the cooling/heating water.

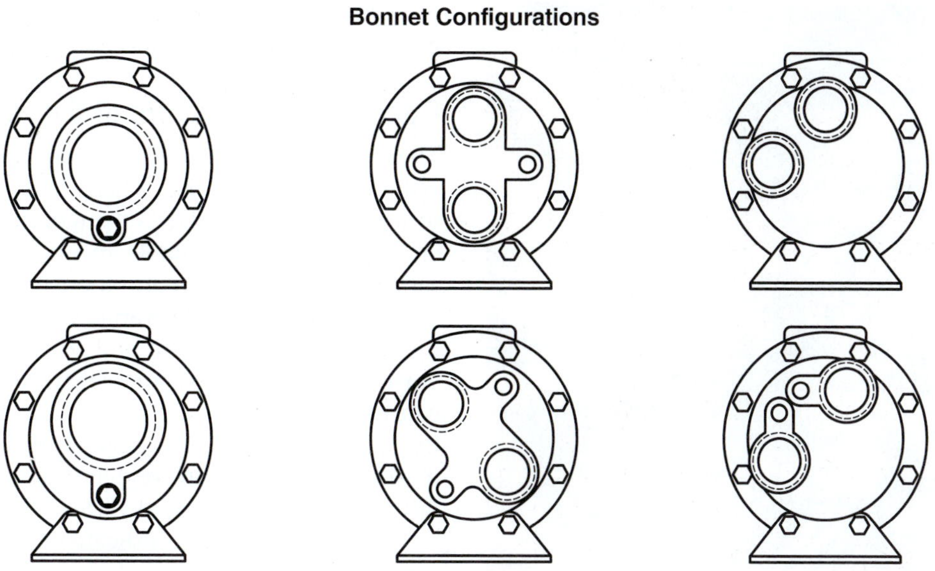

Bonnet Configurations

One pass   Two pass   Four pass

Figure 12-24. Shell-and-tube heat exchangers are available in single- or multiple-pass designs.

Heat from the system fluid is transferred to the walls of the tubes. It then moves through the walls of the tubes by conduction and is transferred to the cooling water. The water carries the heat out of the system to maintain the desired system operating temperature. When used to warm the system fluid, heated water is circulated through the tubes. The heat transfer process is reversed and the temperature of the hydraulic fluid is raised to the desired level.

The *brazed-plate heat exchanger* is constructed of a number of stainless steel plates. Each plate is stamped with a corrugated pattern. Thin, spacer sheets of copper or nickel separate the plates. This stack of plates and spacers is fitted with endplates and inlet and outlet fittings. The assembly is brazed in a high-temperature oven. The end result is a very compact, efficient, single-pass heat exchanger, **Figure 12-25**. The water and hydraulic fluid are directed through alternate channels formed between the plates. The operating principles of this unit and the shell-and-tube design are very similar.

Immersion heat exchangers

Immersion heat exchangers are typically located in the reservoir. They can be electric heating elements or tubes and pipes carrying water, **Figure 12-26**. In large installations, steam may be used instead of water. The units can be used to either cool or warm the hydraulic fluid.

Electric units heat the hydraulic fluid much like the heating elements in household water heaters. The elements are thermostatically controlled to prevent overheating. Overheating increases fluid oxidation and may result in a fire.

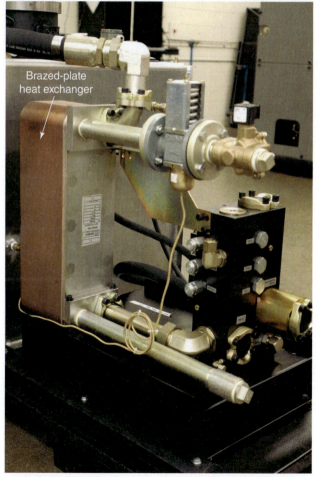

Brazed-plate heat exchanger

MTS Systems Corporation

Figure 12-25. Brazed-plate heat exchangers are very compact units providing heating or cooling with minimal pressure drop.

Kim Hotstart Manufacturing Company

Figure 12-26. Immersion-type electric heaters with thermostatic control are often used in mobile equipment.

The units using immersed tubes or pipes operate similar to the shell-and-tube exchangers. However, the total reservoir fluid volume is involved, rather than the confined space of the exchanger shell cavity.

Mobile equipment sometimes has an immersion heater to warm the reservoir oil before startup during cold weather operation. Placing the heater close to the pump inlet can prevent pump cavitation. This extends the service line of the pump.

12.7.3 Determining the Need for a Heat Exchanger

There are a number of basic factors that can be used to determine the need for adding a heat exchanger to an existing system. Key factors include:

- Operating temperature of the reservoir oil.
- Sequence, cycle time, and load of the system.
- Maintenance history.

Care must be taken to thoroughly consider these factors and other possible solutions. Installing an unnecessary unit or one that is too large adds to equipment costs and may produce unwanted results.

Operating temperature of the reservoir oil

The operating temperature of the reservoir oil is a critical indicator to use when determining the need to install a system heat exchanger. Most manufacturers indicate the acceptable reservoir oil temperature range is 110–140°F in a continuously operating system. Temperatures above this range can shorten the service life of the hydraulic fluid and seals. The formation of varnish on the internal surfaces of components and in control orifices is also accelerated. Temperatures below this range may cause sluggish system operation, formation of sludge from condensation, and pump cavitation due to the increased viscosity of the oil.

When observations indicate the oil temperature is continuously above or below the recommended range, further study is warranted. Always look for a simple solution first. For example, the problem may be solved by a simple redesign of the reservoir that increases its ability to transfer heat to the atmosphere. However, retrofitting the system with some type of heat exchanger may be the most cost-effective solution.

Sequence, cycle time, and load of the system

The sequence, cycle time, and load of the system are important factors that must be considered when looking for the source of excessive heat. Plot the sequence of operation, the time spent in each phase of the cycle, and the pressure encountered during each of the phases. This information will provide important information on how and when heat is generated during system operation.

For example, suppose a fixed-displacement pump is operating at full relief valve pressure during an extended holding phase. As a result, excessive heat is being generated. This may justify upgrading the pump to a compensated, variable-displacement pump. Upgrading the pump could eliminate the need for a heat exchanger while reducing energy consumption and overall system maintenance cost.

Maintenance history

The maintenance history should include a record of seal failures, evidence of rapid oxidation of the hydraulic fluid, and indications of varnish buildup. Examination of maintenance records can supply considerable information to indicate the need for the installation of a heat exchanger. These records may be particularly helpful when reservoir fluid temperatures are within specifications, but at the high end of the range. Records of excessive maintenance in the indicated areas may justify the cost of a heat exchanger.

Summary

- Fluid conditioning devices and circuits provide the system with fluid that is clean and maintained at an acceptable operating temperature, ensuring efficient system operation.

- Hydraulic system contaminants include those built into the system during manufacture, those entering the system during operation, and those resulting from component wear or the breakdown of the hydraulic fluid. Contaminants can reduce the ability of the fluid to perform key functions and may result in damage to system components.

- Hydraulic systems generate considerable heat during operation due to the general poor efficiency of the system. Hydraulic system designs need to balance the rate of system heat input with the rate of heat removed by conduction, radiation, and other cooling methods. Extreme variations in fluid temperature can cause major problems with both the performance of the system and the service life of component parts.

- The hydraulic system reservoir is designed to allow fluid contaminants to settle out or separate from the fluid and heat to be exchanged with the surrounding environment.

- Strainers and filters trap insoluble material contained in hydraulic fluid.

- Filters can be classified as surface- or depth-type. Surface-type filters provide a single surface containing numerous holes to trap contaminants. Depth-type filters use a mass of porous material to provide numerous routes that trap the particles.

- The common measurement scale used to indicate the level of filtration is the micron, which is .000039". A 25-micron filtration level is the recommended minimum for most systems.

- Absolute and nominal ratings are available for both surface and depth filters. The absolute rating suggests that all particles larger than the stated size will be removed. The nominal rating indicates the average pore size of a depth filter, but does not guarantee removal of all larger-sized particles.

- Filter housing style and the ability of the unit to withstand system operating pressure is dependent on the location of the filter in the circuit.

- Filters may be located in pump inlet lines, system working lines, return lines, and drain lines.

- Heat exchangers may be incorporated in hydraulic systems to keep operating temperatures within a range suggested by the manufacturer.

- The most common heat exchangers used in hydraulic systems include air-cooled radiators, shell-and-tube units, and electric powered immersion-type units. Finned tubes, brazed plate, and immersed pipes and tubes are also used to cool or warm system fluid.

- Radiators consist of a series of small fluid carrying lines that pass through a series of metal fins. As system fluid passes through the tubes, heat is exchanged through the tube to the metal fins that pass it on to the air moving over the fins.

- Shell-and-tube exchangers consist of a bundle of tubes enclosed in a metal shell. Cooling or heating water passes through the tubes exchanging heat with the hydraulic fluid moving through the space between the shell and the bundled tubes.

- Heat exchangers may be needed in a hydraulic system if reservoir oil temperatures during continuous operation are outside the recommended temperature range of 110–150°F. Seal failures, system fluid oxidation, rapid varnish buildup, and elevated fluid temperatures indicate the potential need for a heat exchanger.

Internet Resources

The following are some useful resources available on the Internet. Enter a company or organization name into a search engine to access its website. Explore the various areas of the sites to discover useful fluid power resources.

API Heat Transfer, Inc. Details of one type of shell-and-tube heat exchanger. Provides a link to download the related brochure that provides specifications and calculations.

Donaldson Company, Inc. Technical reference guide covering a wide variety of filter information, ranging from types of contamination to locating filters in systems.

Automotive Aftermarket Suppliers Association (AASA). Frequently asked questions (FAQs) about heavy-duty hydraulic filters. Several responses provide links to additional resources.

Hydraulics & Pneumatics Magazine. Provides an introduction to heat exchangers in hydraulic systems, including basic thermodynamics, heat transfer mechanisms, and cooling methods. Information is provided on exchanger designs and application considerations.

Parker Hannifin Corporation. Provides extensive information about contamination, filter media, filter selection, and filter circuits. Hydraulic fluid analysis is also described, including suggestions for collecting samples.

Chapter Review

Answer the following questions using information in this chapter.

1. The fluid-conditioning system of hydraulic-powered equipment provides fluid that is clean and maintains acceptable operating _____.
   A. levels of wear
   B. temperatures
   C. viscosity
   D. lubrication levels

2. Contaminants in a hydraulic system can _____.
   A. reduce the operating efficiency
   B. increase in operating costs
   C. increase the rate of wear of component parts
   D. All of the above.

3. _____ percent or more of the prime mover horsepower is converted into heat that raises the fluid temperature in a hydraulic system.
   A. 10
   B. 20
   C. 30
   D. 40

4. _____ is a viscous, often gel-like residue that forms when water and other contaminants interact with hydraulic fluid.
   A. Varnish
   B. Viscosity
   C. Sludge
   D. Immersion exchanger

5. *True or False?* Dirt particles reduce the ability of hydraulic fluid to provide lubrication.

6. Reservoirs contain _____ that provide a long route for the fluid to travel between the return line and pump inlet line, allowing time for dirt particles to settle.

7. *True or False?* The National Fluid Power Association defines a strainer as a coarse filter.

8. A(n) _____-type filter is often considered a coarse filter that removes only the largest of the contaminants that have entered the hydraulic fluid.

9. To trap contaminants, _____-type filters use a mass of porous material that provides many different flow routes.

10. For all hydraulic systems, the minimum filtration level recommended by the National Fluid Power Association is _____ microns.
    A. 10
    B. 15
    C. 20
    D. 25

11. When a filter becomes clogged with contaminants, the _____ valve opens to allow system fluid to flow past the filter.

12. Which of the following is a common location for a filter in a hydraulic system?
    A. Pump outlet line.
    B. System working lines between the pump and actuators.
    C. Inside the actuator.
    D. All of the above.

13. The _____ filtration routing method requires the use of a filter bypass valve to prevent the development of excessive pressure drop across a clogged filter.

14. Lower-than-recommended system temperatures _____ fluid viscosity.

15. A heat _____ in a hydraulic system adds or subtracts heat from system fluid to keep the fluid temperature within a desired range.
    A. radiator
    B. conductor
    C. bypass
    D. exchanger

16. _____ heat exchangers are constructed of a bundle of tubes enclosed in a metal shell.
    A. Shell-and-tube
    B. Brazed-plate
    C. Immersion
    D. Alternating-line

17. During operation of the shell-and-tube heat exchanger, _____ flows through the tubes, while the hydraulic fluid flows around the tubes.

18. *True or False?* The number of times water flows through a fluid power heat exchanger controls the amount of heat transfer.

19. *True or False?* Multiple-pass shell-and-tube heat exchangers are made possible by dividers located in the interior of the shell.

20. Most manufacturers of hydraulic equipment indicate 110°F to _____°F as an acceptable reservoir operating temperature.
   A. 120
   B. 130
   C. 140
   D. 150

Apply and Analyze

1. What aspects of reservoir design are crucial for contaminant removal? How do these design aspects or features assist in the removal of contaminants?

2. Filters can be placed in three general locations in a hydraulic circuit to remove contaminants.
   A. What are the advantages and disadvantages of each general location?
   B. How would the possible sources of contamination affect your choice of filter location?
   C. How does your choice of filter location affect the type of filter you might install?

3. If you were designing a hydraulic circuit for a system, what factors would you consider in choosing proportional, full-flow, or off-line filtration? Explain your answer.

Research and Development

1. Visit a local construction or industrial site where hydraulic equipment is operated. If possible, interview the equipment operator about typical operating procedures. Identify possible hydraulic fluid contaminants, the sources of these contaminants, and propose measures that could be taken to minimize contamination.

2. Research one design of hydraulic fluid filter currently available on the market. Prepare a 1–2 page fact sheet that provides information about materials, method of filtration, contaminants removed, basic operating conditions, and typical applications to share with your classmates.

3. Find and photograph as many examples of heat exchangers in your home and community as you can. Group these examples into general categories and provide a brief overview of the basic principles of operation of each general design type.

4. Design, construct, and test a simple heat exchanger that could be used to heat or cool fluid in a separate reservoir. Compare and contrast your heat exchanger design with those currently in use in hydraulic systems.

Applying Hydraulic Power

Typical Circuits and Systems

To a general observer, hydraulic systems often appear to be large, complex, and confusing because of the number of valves, lines, and actuators. However, basic systems are often used as subsystems in complex equipment. Large systems typically use smaller subsystems or circuits to control pressure and flow rate, time the motion of actuators, or enhance operator safety. Being aware of the design and operation of these smaller elements can often make a large system easier to understand. This chapter provides examples of basic circuits. Each example has been selected to provide insight into how various components can be combined to produce desired pressure, flow rate, sequence of motion, or other specialized feature.

Learning Objectives

After completing this chapter, you will be able to:

- Name common, basic subsystems that compose complex hydraulic systems and describe their function.
- Compare the design and operation of basic pressure control circuits.
- Contrast the operating characteristics of meter-in, meter-out, and bleed-off flow control circuits.
- Compare the design and operation of several typical flow control circuits.

- Compare the design and operation of basic motion control circuits.
- Describe the design and operation of rapid-advance-to-work circuits.
- Describe various hydraulic circuits designed to provide for operator safety and the protection of system components.

Key Terms

back-pressure check valve
bleed-off circuit
deceleration circuit
deceleration valve
decompression valve
high-low pump circuit
maximum operating pressure
meter-in circuit

meter-out circuit
multiple maximum system
   operating pressures
overrunning load condition
rapid-advance-to-work circuit
reduced-pressure section
regenerative-cylinder-
   advance circuit

remote location
sequencing
synchronization
system cycle time
two-hand-operation circuit
vent connection

13.1 Basic Circuits and Their Use in System Analysis and Design

Understanding the operation of the hydraulic system of a complex piece of equipment can be a major challenge for even experienced hydraulic personnel. These systems typically involve a group of components that closely work together to produce a desired performance. Being familiar with the design and operation of a variety of basic control circuits can be very helpful when working with any hydraulic system. This is true in troubleshooting problems in an existing system, upgrading a system on existing equipment, or designing circuits for new pieces of equipment.

The majority of basic hydraulic systems can be divided into four function groups:

- Pressure control.
- Flow control.
- Motion control.
- Miscellaneous functions.

A typical hydraulic system includes segments from all four groups. These segments are merged to produce the operating system. The complexity of the individual segments depends on the performance required of the total system.

The following sections provide schematics of several hydraulic circuit variations that can be used to produce certain types of system operation. The descriptions of the circuits explain how each segment operates. These circuit segments are only a few of the many designs that may be used in operating systems.

13.2 Design and Operation of Basic Pressure-Related Circuits

The output force of a linear actuator and the output torque of a hydraulic motor are determined by system pressure. Controlling the output force and torque of these actuators in an operating hydraulic system is, therefore, closely associated with the ability to regulate and control system pressure. Chapter 10, *Controlling the System*, covers the basic structure and operation of pressure control valves. This section illustrates several basic circuits that provide pressure control under a variety of situations.

13.2.1 Limiting Maximum Pressure in a System

The circuit shown in **Figure 13-1** is a basic hydraulic circuit. The relief valve in this circuit is concerned only with limiting the *maximum operating pressure* of the system. The maximum force generated by the actuator is determined by multiplying the pressure setting of the relief valve by the effective operating surface of the actuator. When the actuator is moving, system pressure is determined by the external actuator load and any internal resistance to fluid movement. When the load stalls the actuator or the actuator reaches the end of its travel, the pressure increases to the maximum setting of the relief valve with all pump output returned to the reservoir through that valve.

Figure 13-2 shows the same circuit with the two-position, four-way directional control valve replaced by a three-position, four-way valve with a tandem center. This modification allows the same circuit

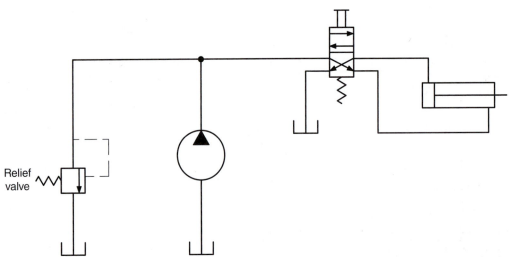

Figure 13-1. Pressure control in a basic hydraulic circuit can consist of a simple, direct-operated relief valve to limit maximum system pressure.

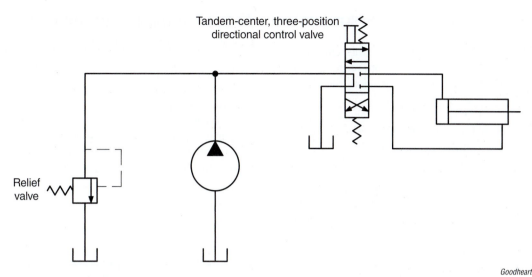

Tandem-center, three-position
directional control valve

Relief
valve

Goodheart-Willcox Publisher

Figure 13-2. The addition of a three-position directional control valve to a basic hydraulic circuit can add greater flexibility to pressure control. The tandem center allows increased system efficiency by reducing system pressure to near zero when the directional control valve is centered.

performance during actuator extension and retraction. However, energy is saved by unloading the pump to the reservoir when the directional control valve is shifted to the center position.

13.2.2 Remote Control of System Pressure

In some situations, it is desirable to control the maximum pressure setting of the system relief valve from a *remote location*. The circuits shown in this section use the venting feature of a compound relief valve to achieve this type of circuit operation.

In **Figure 13-3**, the *vent connection* of the relief valve is piped to the reservoir through a solenoid-operated shut-off valve. When the shut-off valve is closed, the system relief operates normally. This allows the pressure to climb to the selected maximum pressure setting before opening to protect the system against over pressurization. However, when the solenoid is activated, the shut-off valve opens to vent the relief valve control chamber. This allows the relief

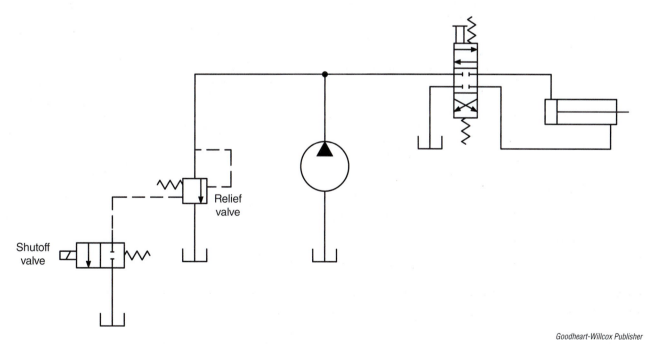

Relief
valve

Shutoff
valve

Goodheart-Willcox Publisher

Figure 13-3. The pressure in a hydraulic circuit may be controlled from a remote location by using the vent feature of a compound relief valve. Opening the solenoid-operated shut-off valve drops system pressure to near zero.

valve to open, which, in turn, drops system pressure to near 0 psi.

Figure 13-4 shows a variation of this circuit that allows the remote adjustment of the operating pressure of the system relief valve. In this variation, the shut-off valve is replaced by a low-flow-capacity, direct-operated relief valve. This remotely located relief valve controls the maximum setting of the system relief valve. Adjusting the remote relief valve allows maximum system operating pressure to be quickly and easily modified.

13.2.3 Multiple Operating Pressures in a System

Hydraulic circuits often provide *multiple maximum system operating pressures*. In some systems, varying the pressure is the only practical way to produce the varying force needed to complete the work. In other systems, the varying pressure settings may be related to reducing energy consumption during extended *system cycle times*. **Figure 13-5** illustrates a simple circuit that can provide

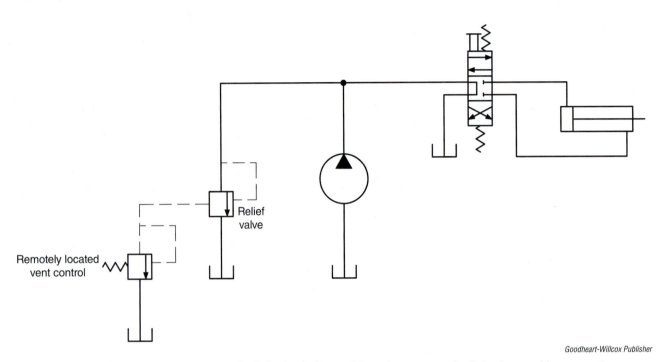

Goodheart-Willcox Publisher

Figure 13-4. Placing a small, direct-operated relief valve in the vent line of a compound relief valve provides a remote means of adjusting the operating pressure of the system.

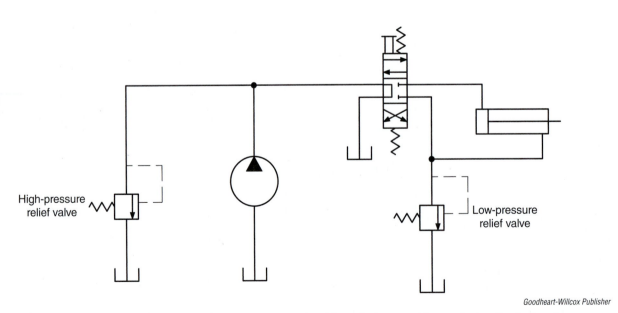

Goodheart-Willcox Publisher

Figure 13-5. Different maximum operating pressures may be obtained for cylinder extension and retraction by locating a second, lower-pressure relief valve in the working line between the directional control valve and the cylinder.

one maximum operating pressure for extension and a different one for retraction. This system uses two relief valves with full system flow capacity. The valve located on the pump side of the directional control valve must be adjusted for the higher pressure.

Figure 13-6 illustrates a circuit that can provide three different maximum system operating pressures. The primary relief valve, which is set for the highest maximum system pressure, is a vented, compound relief valve. This valve must have flow capacity adequate to handle the rated output flow of the pump. The two remaining relief valves can be direct operated and only large enough to accommodate the fluid discharged through the vent of the primary valve. The fourth valve shown in this circuit is a three-position, four-way directional control valve with a closed center. In this example, the directional control valve is spring centered and solenoid shifted.

When the solenoid-controlled directional control valve is not activated, the primary relief valve controls maximum system pressure. This is because the vent line of the valve is blocked by the closed center of the directional control valve. When the directional control valve is shifted in either direction, the vent line is connected to one of the small, direct-operated relief valves. Fluid from the

control chamber of the primary valve then flows through the smaller direct-operated valve. This sets a reduced operating pressure for the system. Shifting the directional control valve to connect the second direct-operated relief valve produces the third pressure level.

13.2.4 Reduced Pressure in a System

Some hydraulic circuits contain sections requiring a maximum operating pressure that is lower than the setting of the primary relief valve. This *reduced-pressure section* requires a normally open valve that closes to restrict fluid flow into the section once the reduced pressure is reached. In order to function, these pressure-reducing valves must be externally drained with the pilot sensing pressure from the downstream side of the valve.

Figure 13-7 shows the placement of a pressure-reducing valve to control the maximum pressure of the clamp cylinder in a clamp-and-machining sequence. Once the pressure at the clamp cylinder reaches the setting of the pressure-reducing valve, the valve allows only enough hydraulic fluid to pass to maintain the reduced pressure.

Figure 13-8 illustrates a circuit that needs only a minimal amount of pressure for an operation. One of

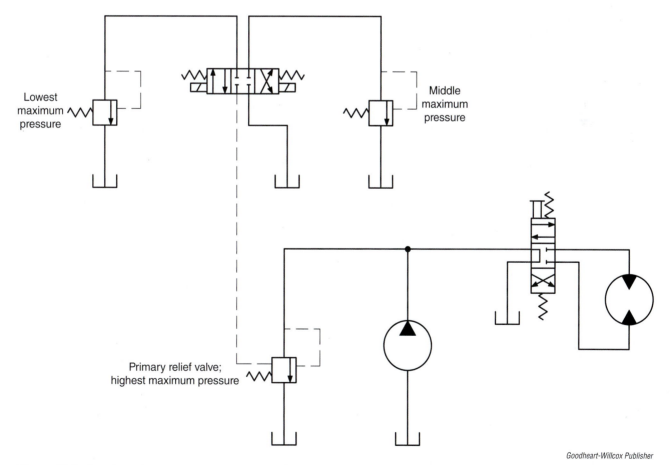

Goodheart-Willcox Publisher

Figure 13-6. A variety of system pressures may be obtained by connecting multiple, small, direct-operated relief valves to the vent of the primary system relief valve through a three-position directional control valve.

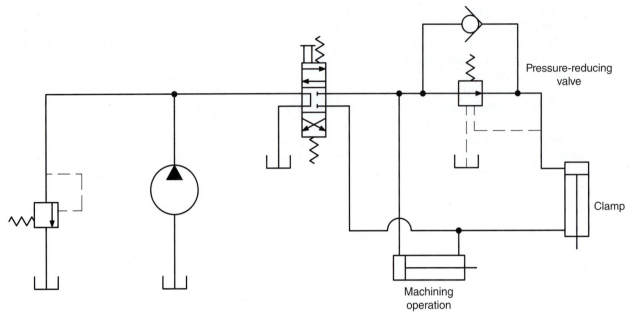

Goodheart-Willcox Publisher

Figure 13-7. A pressure-reducing valve may be used to operate a portion of a circuit at a pressure lower than the setting of the system relief valve. Pressure-reducing valves must be externally drained to function.

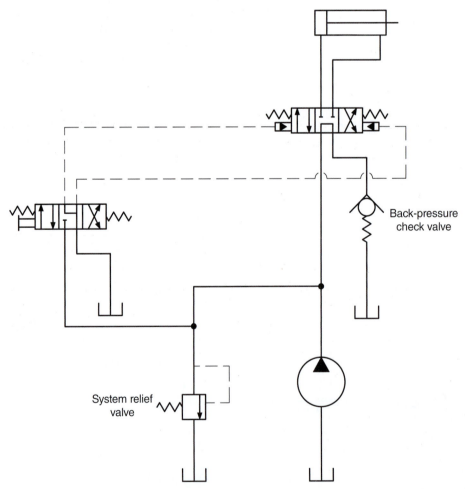

Goodheart-Willcox Publisher

Figure 13-8. A back-pressure check valve can be placed in the return line of a system to maintain a minimum pressure when an open- or tandem-center directional control valve is used. This allows the use of pilot operation to shift directional control valves.

the primary purposes of open- and tandem-center, three-position directional control valves is to reduce system pressure to a minimum when the valves are in the center position. This characteristic provides energy-efficient operation when the system is idle. However, this low pressure can cause problems if the system needs pilot pressure for the operation of directional control valves and other components. A spring-loaded, *back-pressure check valve* is placed in the return line between the primary directional control valve and the reservoir. This valve provides the restriction needed to produce the relatively low pressure for pilot operation.

13.3 Design and Operation of Basic Flow-Related Circuits

The movement of actuators is the result of fluid flow. Controlling the movement and speed of hydraulic actuators is, therefore, closely associated to the ability to regulate system fluid flow. Chapter 10 discusses the basic structure and operation of fluid flow control valves. This section illustrates several basic circuits that control fluid flow to obtain the desired operating speed and performance of actuators.

13.3.1 Basic Flow Control Methods

Three basic methods are used to control fluid flow in hydraulic systems:

- Meter-in circuits.
- Meter-out circuits.
- Bleed-off circuits.

Each of these circuits can be used with linear or rotary actuators. Each circuit has advantages and disadvantages, depending on the application and actuator type. The three circuit variations are illustrated in **Figures 13-9** through **13-11**. These figures show the placement of the flow control valves to control the extension speed of the actuators. It must be kept in mind that these same valves can also be rearranged to control the retraction speed of the actuators. For purposes of the following discussion, the pump in each circuit is rated at 10 gpm, the system relief valve is set at 500 psi, and the flow control valve is set at 5 gpm.

Meter-in circuits

A *meter-in circuit* directly controls fluid flow into an actuator. **Figure 13-9** shows a schematic of a basic meter-in flow control circuit that controls actuator extension speed. In this circuit, the flow control valve (*Valve D*) is placed between the directional control valve (*Valve B*) and inlet port of the actuator. When the directional control is shifted to extend the actuator, a portion of the pump output is metered into the blind end of the cylinder through the orifice in the flow control. The extension speed of the cylinder is controlled by adjusting the orifice size, which changes the fluid flow rate.

The pump in the circuit is producing a flow of 10 gpm. The flow control valve is set to pass 5 gpm to the actuator to produce the desired extension speed. The remainder of the fluid is returned to the reservoir through the system relief valve (*Valve A*), which is set at 500 psi. This means that when the circuit is activated, the pressure at point 1 is always 500 psi, regardless of the load on the actuator.

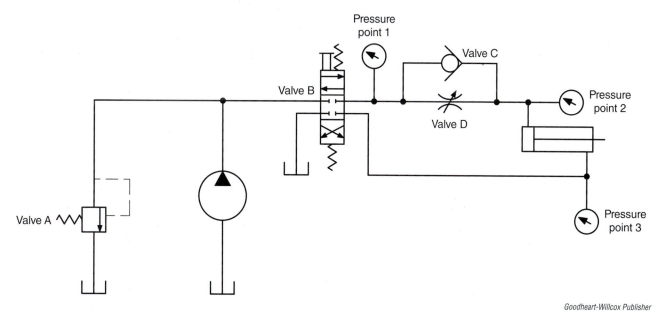

Goodheart-Willcox Publisher

Figure 13-9. Meter-in flow control circuits directly control fluid flow into an actuator. These circuits should be used only in applications where the workload maintains a positive load on the actuator.

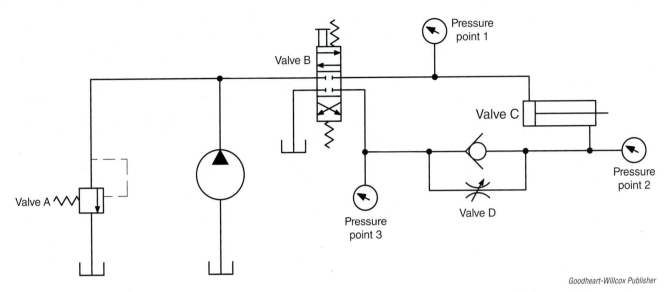

Figure 13-10. Meter-out flow control circuits control the rate of fluid flow out of an actuator. These circuits provide accurate speed control under either positive- or overrunning-load conditions.

The fluid that enters the actuator acts against the piston of the cylinder. This extends the cylinder rod against the resistance of the load. The pressure at *Pressure point 2* indicates the resistance of the load and any flow resistance in the lines leading to the reservoir. The pressure at *Pressure point 3* is only the resistance to flow of the fluid returned to the reservoir from the blind end of the cylinder.

The actuator stalls when the load increases the pressure at *Pressure point 2* to the setting of the relief valve. When this occurs, all of the fluid from the pump is returned to the reservoir through the relief valve.

Meter-in flow control circuits should be used only in applications that maintain a positive load on the actuator. Applications in which the actuator is pulled, rather than resisting extension, create a vacuum in the line between the flow control valve orifice and the actuator. As a result, the pressure drop across the flow control valve orifice increases beyond the conditions under which the valve flow rate was set. This actually increases the flow rate supplied to the actuator application. The increased actuator speed that results is often referred to as an *overrunning load condition*.

Meter-out circuits

A *meter-out circuit* controls the rate of fluid flow out of an actuator. **Figure 13-10** shows a schematic of a basic meter-out flow control circuit that controls actuator extension speed. In this circuit, the flow control valve is placed between the outlet port of the actuator and the directional control valve. When the directional control valve (*Valve B*) is shifted to extend the actuator, the pump output is directed to the blind end of the cylinder. This fluid acts on the cylinder piston to extend the cylinder rod. The fluid in the rod end of the cylinder is

pushed out of the cylinder into the line leading to the flow control valve (*Valve D*). The extension speed of the cylinder is controlled by adjusting the orifice size of that valve.

The pump in this circuit is producing a flow of 10 gpm with the system relief valve (*Valve A*) set at 500 psi. The desired extension speed is obtained by setting the flow control valve to allow 5 gpm to exit from the rod end of the cylinder. This meter-out circuit differs from the meter-in circuit described above in the fact that fluid is metered *out* of the rod end of the cylinder. The pump supplies fluid to the larger volume of the cylinder blind end. This means that slightly less than 5 gpm is returned to the reservoir through the system relief valve.

A key concept in the operation of meter-out circuits is the pressure generated in each end of the cylinder. When there is no load on the cylinder, *Pressure point 1* is 500 psi, while *Pressure point 2* is higher because of the difference in effective area between the two sides of the piston. The amount of pressure increase is proportional to the difference in effective areas. The force generated by the pressure at *Pressure point 1* multiplied by the area of the blind side of the piston must equal the force generated by the pressure at *Pressure point 2* multiplied by the area of the rod side of the piston.

When the actuator encounters a load, the forces on the two sides of the cylinder piston must remain equal. The fluid that enters the actuator acts against the piston. This extends the cylinder rod against the resistance of both the load and the force generated by the pressure at *Pressure point 2* multiplied by the effective area of the rod end of the piston. The higher the resistance of the load, the lower the pressure at *Pressure point 2*. The cylinder extends because of the pressure drop across the orifice of the flow control valve. This pressure drop exists because the pressure at *Pressure point 3* is low due to the direct reservoir connection through the directional control valve.

The actuator stalls when the load balances the force generated by the pressure at *Pressure point 1* multiplied by the area of the blind side of the piston. At that point, all of the fluid from the pump is returned to the reservoir through the relief valve. The pressure at *Pressure point 1* is the relief valve setting of 500 psi and the pressure at *Pressure points 2* and *3* is 0 psi.

In this circuit, the pump operates against the maximum relief valve pressure setting, regardless of how light or heavy the load is. When an overrunning load is encountered, the pressure at *Pressure point 2* increases to balance the forces acting on the two sides of the cylinder piston. With the forces balanced, the circuit maintains a uniform operating speed.

Bleed-off circuits

A *bleed-off circuit* sends a measured fluid flow directly back to the reservoir with the remaining pump output used to maintain actuator speed. **Figure 13-11** shows a schematic of a basic bleed-off circuit. In this circuit, the flow control valve (*Valve C*) is piped via a tee connection to the working line between the directional control valve (*Valve B*) and the inlet port of the actuator. The line connected to the outlet port of the flow control valve is routed directly to the reservoir. When the directional control is shifted to extend the actuator, the pump output is directed to both the inlet port of the flow control valve and the blind end of the cylinder. Fluid entering the blind end of the cylinder acts on the cylinder piston to extend the cylinder rod. Fluid entering the inlet

port of the flow control valve passes through the valve orifice and moves directly to the reservoir. The extension speed of the cylinder is controlled by adjusting the orifice size of the flow control valve.

The pump is producing a flow of 10 gpm with the system relief valve (*Valve A*) set at 500 psi. The desired extension speed is obtained by setting the flow control valve to allow a maximum of 5 gpm to exit directly to the reservoir. This circuit allows the system to operate at a pressure only high enough to move the load. This is unlike the meter-in and meter-out circuits, which continuously operate at the setting of the system relief valve. As a result, the system is more energy efficient, but not as accurate. The loss of accuracy is due to the metered flow going to the reservoir, rather than to the actuator. Any pump slippage and flow losses in the lines and directional control valve cause slight variations in the flow entering the actuator.

When there is no load on the actuator, the pressure at *Pressure point 1* is only as high as the pressure needed to overcome resistance of fluid flow through the system lines and the internal friction in the cylinder. If the flow control valve is properly selected, it will accurately bleed 5 gpm to the reservoir, while the remaining 5 gpm from the pump extends the cylinder rod at the selected rate.

The actuator stalls when the load increases the pressure at *Pressure point 1* to 500 psi, which is the system relief valve setting. At that point, 5 gpm of the fluid from the pump is returned to the reservoir through the relief valve. The other 5 gpm is returned through the flow control valve. The pressure at *Pressure point 2* is never higher than the resistance to fluid flow as it returns to the reservoir.

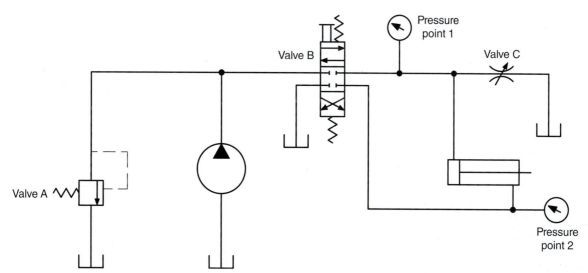

Figure 13-11. Bleed-off flow control circuits send measured fluid flow directly back to the reservoir with the remaining pump output used to maintain actuator speed. The circuit is energy efficient, but not usually considered an accurate control of actuator speed under varying load conditions.

Key characteristics of flow control methods

Three operating characteristics should be considered when determining which of the basic flow control methods to use in a hydraulic circuit. These include:

- Accuracy of flow control.
- Operating efficiency.
- Ability to control varying loads.

The meter-in and meter-out circuits provide the most accurate flow control as they directly meter fluid into or out of the actuator. The bleed-off circuit is less accurate since it meters fluid to the reservoir with the remainder of pump output used for control of actuator speed. This remaining flow can vary because of variations in pump delivery and system leakage.

The best operating efficiency is achieved by the bleed-off flow control circuit. It allows the system power unit to operate at a pressure just high enough to move the load. In contrast, the meter-in and meter-out flow control circuits continuously operate at the system relief valve pressure. This increases prime mover energy consumption and produces considerable heat. Excess heat can require increased costs to cool system fluid or replace it on a shorter maintenance interval.

Meter-in and bleed-off flow control circuits are only effective with opposing loads that resist the movement of the actuator. The meter-out flow control circuit functions equally well for either positive loads that resist movement or overrunning loads that tend to pull the load.

13.3.2 Rapid-Advance-to-Work Circuits

Many hydraulic-powered systems involve equipment that requires considerable movement of machine members before the primary operation begins. This application requires a *rapid-advance-to-work circuit*. One or more of the following features is often included in the circuit. These features include the ability to:

- Provide large volumes of fluid to quickly fill large-bore cylinders.
- Provide high-pressure operation after an extension phase performed at low pressure.
- Rapidly decelerate a moving machine member to prevent shock pressures in the system.

Each of these features requires a circuit that controls system flow in a unique way to provide the needed fluid volume while controlling initial equipment costs and operating costs.

High-low pump circuit

Figure 13-12 shows a *high-low pump circuit*. This circuit can economically provide high-volume flow during low-pressure-demand periods. It can also instantly switch to low-volume output when high-pressure operation is required. The circuit contains two pumps, which may be a dual-pump arrangement within a single case. The remaining portion of the circuit includes a low-pressure unloading valve, high-pressure relief valve, and check valve.

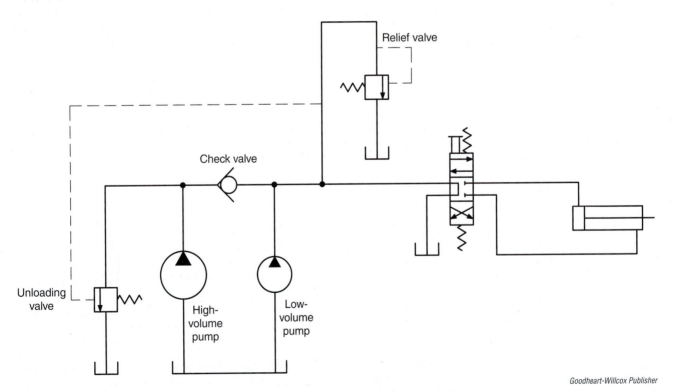

Goodheart-Willcox Publisher

Figure 13-12. A high-low pump circuit design can efficiently provide two different fluid flow rates.

When the system is started, the directional control valve is in the center position. This allows the output of both pumps to be returned directly to the reservoir. At this point, both the unloading and relief valves are closed. Shifting the directional control valve to extend the cylinder directs fluid from both pumps to the blind end of the cylinder. The pressure in the system is just sufficient to overcome the resistance of the cylinder as it extends without an applied load. Pressure in the system lines is well below the setting of both the unloading and relief valves. Under these conditions, all of the fluid from the large pump passes through the check valve on its way to the cylinder.

As a load is applied to the cylinder, pressure in the system lines begins to rise in an attempt to overcome the resistance of the load. When the line pressure reaches the setting of the unloading valve, that valve begins to dump fluid to the reservoir. As system pressure increases with an increasing load, the unloading valve opens further to allow more fluid to be dumped to the reservoir. At some point during this pressure increase, the unloading valve is dumping all of the fluid from the high-volume pump. The fluid from the low-volume pump also attempts to flow to the unloading valve. However, this reverse fluid flow closes the check valve. This restricts the output of the small pump to the side of the circuit with the high-pressure relief valve. The low-volume pump continues to supply fluid to the cylinder until load resistance increases the pressure to the setting of the high-pressure relief valve. At that point, fluid from the low-volume pump is returned to the reservoir, while the cylinder is held in position by the high pressure.

This circuit has the ability to provide a high fluid-flow rate at low pressure for rapid movement of an actuator and high-pressure, low-volume fluid to perform the work. These characteristics match two of the three desirable features for rapid-advance-to-work circuits discussed earlier. This circuit is generally considered energy efficient as both the high-volume, low-pressure and the low-volume, high-pressure features provide maximum performance for the energy consumed.

Regenerative-cylinder-advance circuit

Figure 13-13 shows a *regenerative-cylinder-advance circuit*. This circuit produces a rapid actuator advance until the load is encountered. This is followed by a slower extension speed to perform the required work. Rapid cylinder extension results when the fluid in the rod end of the cylinder is directed to the blind end to increase the amount of flow extending the cylinder. Fluid movement from the rod end to the blind end is achieved by using the force difference generated by system pressure acting on the blind and rod ends of the cylinder piston.

When the directional control valve is shifted to position *1* to extend the cylinder, the blind end of the cylinder is pressurized. The line leading directly from the pump to the rod end of the cylinder is also pressurized up to the check valve in the line. The force generated on the blind end of the piston attempts to extend the cylinder. Movement of the piston pressurizes the fluid in the rod end of the cylinder. The fluid is forced into the line leading to the sequence valve and check valve. The difference in area between the blind side

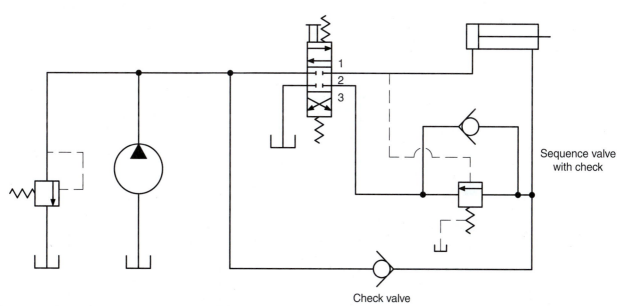

Goodheart-Willcox Publisher

Figure 13-13. A regenerative-cylinder-advance circuit provides a means of obtaining rapid cylinder advance under no-load conditions, followed by a reduced flow rate when a load is encountered.

and rod side of the piston results in a greater force generated on the blind side of the piston. As a result, the cylinder extends. The fluid in the rod end of the cylinder flows through the return line, passing through the check valve, and joins pump output at the junction just before the directional control valve. The sequence valve and its associated check valve remain closed.

When a load is encountered, the pressure increases in the blind end of the cylinder and the working line between the directional control valve and the actuator. The external pilot line of the sequence valve senses system pressure in this area. The increased pilot line pressure causes the sequence valve to open. Once open, the sequence valve allows fluid from the rod end of the cylinder to return directly to the reservoir. Also, the check valve in the line that had been returning fluid from the rod end to the cylinder input line closes. As a result, the flow rate entering the blind end of the cylinder is reduced. In turn, the extension speed of the cylinder is reduced, while the cylinder is allowed to apply maximum force to the load.

Cylinder retraction occurs when the directional control valve is switched to position 3. Fluid from the pump is directed to the rod end of the cylinder through the check valve used in conjunction with the sequence valve. Fluid from the blind end of the cylinder is returned directly to the reservoir.

There are several important characteristics of regenerative-cylinder-advance circuits. The circuit provides a rapid extension of the cylinder by combining pump output and the return flow from the cylinder blind end. This rapid extension is used to quickly approach the workpiece. Once this position is reached, the cylinder slows to an extension speed determined by the volume of the blind end of the cylinder and

pump output flow. During retraction, the cylinder rod moves at a speed controlled by the output flow of the system pump and the volume of the rod end of the cylinder. During the rapid-extension phase, the effective force of the cylinder is limited to the pressure of the system multiplied by the cross-sectional area of the cylinder rod.

Deceleration circuit

Figure 13-14 shows a third type of rapid-advance-to-work circuit. This circuit systematically decelerates the actuator as it approaches a work position. A *deceleration circuit* is especially helpful in situations where a considerable mass must be accelerated and then slowed as it approaches the selected work position. The circuit shown in the figure involves meter-out flow control with the addition of a mechanically operated *deceleration valve* built into a bypass line around the flow control valve. The deceleration valve is basically an on/off valve opened and closed by a ramp attached to the cylinder rod.

During the rapid extension phase, the deceleration valve is open to allow the cylinder to extend at maximum speed. As the desired machine position is approached, the ramp begins to shift the deceleration valve to block fluid flow. The cylinder slows at a controlled rate until the valve is completely closed. At that point, cylinder extension speed is controlled by the size of the orifice in the meter-out flow control valve. The rate of deceleration is controlled by both the slope of the ramp and the shape of the internal spool of the deceleration valve.

When the cylinder is retracted, fluid is routed through the check valve to bypass both the flow control and deceleration valves. This provides maximum retraction speed without the deceleration control.

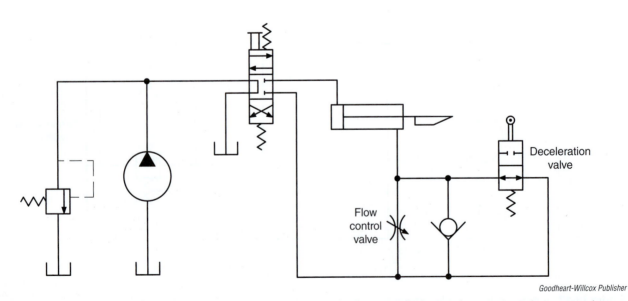

Figure 13-14. A deceleration feature in a rapid-advance-to-work circuit allows a cylinder to advance at maximum speed, decelerate at a controlled rate, and then slowly move to complete a job.

13.4 Design and Operation of Basic Motion-Related Circuits

Directional control valves are used to start, stop, and create varied movement in hydraulic systems. A number of these actions are discussed in Chapter 10, *Controlling the System*. However, motion can be controlled using a number of other valves or basic principles. Two types of motion control that are usually considered outside of normal directional control valve function are actuator sequencing and synchronization. This section describes several designs used to provide these actions using components other than valves typically associated with directional control.

13.4.1 Sequencing of Actuators

Hydraulic system *sequencing* can be defined as the control of actuators to provide a specific order of movement. An example is a circuit in which a cylinder extends followed by a second cylinder. After work has been completed by the system, the second cylinder retracts followed by the first cylinder. This type of motion control can be achieved in hydraulic systems using pressure sensing, mechanical operation of valves, or electrical means. These operating methods may be used by themselves or in combination to produce the desired movement. Pressure sensing is used in the following example circuits.

Figure 13-15 shows a circuit that uses spring-loaded check valves to produce controlled back pressure in the working line between the directional control valve and the actuators. This results in a sequence of *Cylinder 1* extending followed by *Cylinder 2*. During retraction, the same order of movement occurs.

Shifting the directional control valve to position *1* directs fluid into the line that goes to the blind end of each cylinder. The spring-loaded, back-pressure check valves in the line going to *Cylinder 2* block flow to that cylinder. This forces *Cylinder 1* to extend. When *Cylinder 1* encounters resistance, the pressure in the system increases. This forces one check valve to open. The resulting flow through the check valve forces the extension of *Cylinder 2*.

Shifting the directional control valve to position *3* directs fluid to the rod end of each cylinder. The spring-loaded check valves in the return line of *Cylinder 2* block the retraction of that cylinder. *Cylinder 1*, therefore, retracts first. When *Cylinder 1* is fully retracted, pressure in the working lines increases and one of the spring-loaded check valves opens. This allows *Cylinder 2* to retract.

This is a relatively simple, low-cost sequencing system. However, to work effectively, the tension of the springs in the check valves must be carefully matched to the application. The tension must allow a force adequate to move *Cylinder 1* to be generated. This force must overcome both the internal friction of the cylinder and the external load.

Figure 13-16 shows a circuit that uses pressure control valves manufactured specifically to function as sequence valves. These normally closed valves have built-in check valves to allow reverse flow. This facilitates cylinder retraction. External drains from the control chambers of these valves allow drainage of fluid even when the working lines on either side of the valve are pressurized. This circuit produces a sequence of *Cylinder 1* extending first followed by *Cylinder 2*. During retraction, *Cylinder 2* retracts first followed by *Cylinder 1*.

Shifting the directional control valve to position *1* to extend the cylinders directs fluid into the line that goes to the blind end of *Cylinder 1* and to the inlet port of

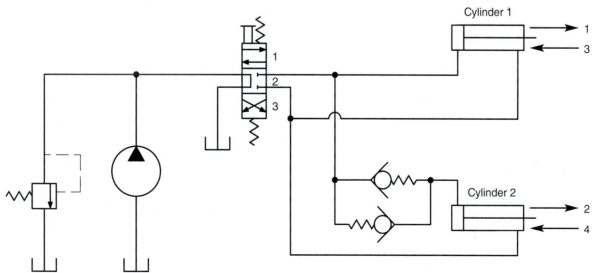

Figure 13-15. Basic actuator sequencing can be achieved by using simple, spring-loaded check valves in a selected working line to control fluid flow.

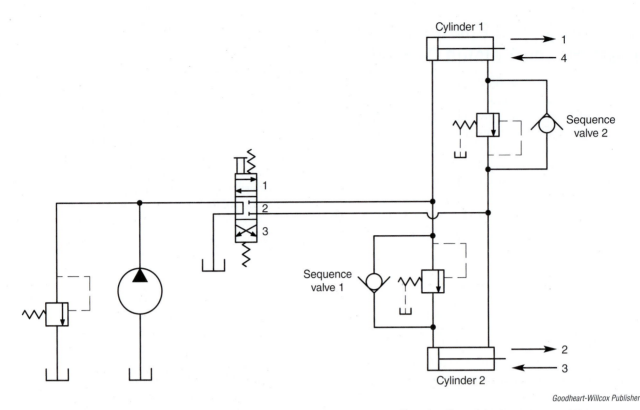

Goodheart-Willcox Publisher

Figure 13-16. This circuit uses control valves to sense pressure in the working lines leading to the actuators and then accurately controls the operating sequence of the circuit actuators.

Sequence valve 1. The normally closed sequence valve and the built-in check valve block fluid flow into *Cylinder 2*. All fluid is diverted to *Cylinder 1*. As *Cylinder 1* extends, fluid from the rod end of the cylinder is returned to the reservoir through the built-in check valve in *Sequence valve 2* via the direction control valve.

When *Cylinder 1* encounters resistance during extension, the pressure in the line increases in an attempt to overcome it. When the pressure reaches the setting of *Sequence valve 1*, that valve opens to allow fluid to pass to *Cylinder 2*. As a result, *Cylinder 2* extends. Fluid from the rod end of *Cylinder 2* is returned directly to the reservoir through the directional control valve.

Shifting the directional control valve to position *3* to retract the cylinders directs fluid to the rod end of *Cylinder 2* and the inlet port of *Sequence valve 2*. The normally closed sequence valve and the built-in check valve block fluid flow into the blind end of *Cylinder 1*. All fluid is diverted to *Cylinder 2*. As *Cylinder 2* retracts, fluid from the blind end of the cylinder is returned to the reservoir through the check valve built into *Sequence valve 1* via the directional control valve.

When *Cylinder 2* encounters resistance during retraction, the pressure in the line begins to increase. When the pressure reaches the setting of *Sequence valve 2*, the valve opens to allow fluid to pass to *Cylinder 1*. As a result, *Cylinder 1* retracts. Fluid from the blind end of *Cylinder 1* is returned directly to the reservoir through the directional control valve.

This circuit is commonly used to illustrate the use of sequence valves. The use of sequence valves such as these, rather than the back pressure check valves shown in the previous circuit, results in a more easily adjusted and accurate system. The level of performance of circuits such as this depends on the valves used. Sequence valves are available in both direct- and pilot-operated designs, producing considerable variation in performance. Pilot-operated valves produce greater accuracy and repeatability. This results in a more efficient system.

13.4.2 Synchronization of Linear Actuators

Synchronization is when two or more actuators move in unison. Synchronizing the movement of two cylinders in a hydraulic circuit is a fairly common design problem. When first confronted with synchronization, it is often considered a minor problem that can be solved by the use of simple flow control valves or a mechanical connection between the cylinder rods. In actual practice, it often becomes a challenge due to uneven loading between the actuators, internal leakage, and friction variations. Even if the cylinders are equal in size, slight variations that occur during each stroke can grow into major problems in systems that cycle many times in each hour of operation.

Figure 13-17 shows a flow control arrangement used to synchronize the extension stroke of two cylinders.

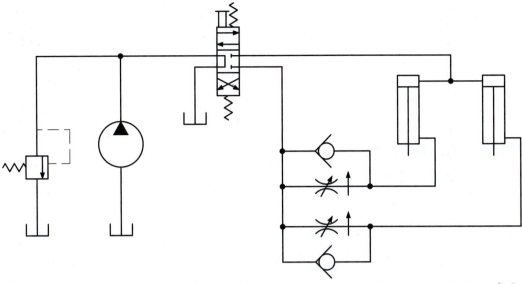

Goodheart-Willcox Publisher

Figure 13-17. Obtaining accurate synchronization of actuators requires a consistent rate of fluid flow into the inlet of each actuator. One method involves using meter-out flow control circuits containing pressure-compensated flow control valves.

The circuit has a pressure-compensated flow control valve placed in a meter-out flow configuration for each cylinder. The flow control valves can be fine-tuned to obtain identical extension times with each cylinder. The check valves allow fluid to bypass the flow control valves during cylinder retraction. The cylinders are returned to identical starting positions for the beginning of each extension stroke. This assists in maintaining synchronization.

> **NOTE**
>
> The circuit shown in **Figure 13-17** allows cylinders of different sizes to be used even when identical extension rates are required.

An effective flow-dividing device can be created by mechanically connecting the output shafts of two hydraulic motors. **Figure 13-18** shows this device applied to a circuit that requires cylinder synchronization. The mechanical leakage forces the motors to turn at the same speed. This produces equal fluid discharge from each motor. The equal discharge rates produce equal extension rates in the two cylinders. The circuit shown also synchronizes the retraction of the cylinders. The motors meter an equal rate of fluid out of the blind end of each cylinder. If synchronization is not needed on cylinder retraction, insert check valves in lines around the motors. This bypasses the metering effect of the motors.

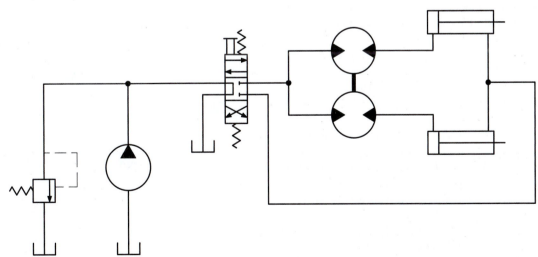

Goodheart-Willcox Publisher

Figure 13-18. Hydraulic motors with mechanically connected output shafts can be used to produce two equal flows. Output flow from the motors ensures synchronized movement of the two cylinders.

The performance of these systems must be monitored to ensure the circuit is maintaining the required extension rate. Internal leakage, unexpected load variations, and fluid viscosity changes can cause variations in those systems that depend on fluid flow rates to produce synchronization of movement.

13.5 Design and Operation of Circuits with other Functions

The previous sections present a small selection of pressure, flow, and motion control methods. There is a large number of additional circuits. For example, additional examples include accumulator, locking, motor-control, pressure-intensification, pump-control, and safety-related circuits. The following three sections describe additional circuits representing important concepts of system operation. These selected circuits represent only a small percentage of the methods that can be used to fill a specific design need in hydraulic powered equipment.

13.5.1 Hydraulic Motor Control

In hydraulic motor circuits, system pressure establishes torque output and flow rate controls speed. In addition to these two basic factors, the design of a hydraulic motor circuit can be used to produce other operating characteristics. For example, it may be desirable to allow the motor to freewheel while coasting to a stop. On the other hand, it may be necessary to brake the motor and associated load to a controlled stop. If these factors are not considered, high shock pressures and physical component damage may occur.

Figure 13-19 shows a basic hydraulic motor circuit. It allows the motor to be powered by the prime mover, but the machine operator can allow the motor to freewheel or be braked to a stop by the system relief valve. When the directional control valve is in the center position (2), the majority of pump output is returned to the reservoir through the open center. The hydraulic motor can freewheel by taking fluid from the motor-inlet side and directing any fluid coming from the outlet side to the reservoir.

When the directional control valve is shifted to position 1, the motor rotates. Fluid from the motor outlet is returned directly to the reservoir. When the load stalls the motor, system pressure increases to the relief valve setting. Pump output is then returned to the reservoir through the relief valve.

Any time during operation, the system operator can allow the motor to freewheel by shifting the directional control valve to position 2. If the inertia of the load causes the motor to continue to rotate, the motor can be braked by shifting the directional control valve

to position 3. This forces the motor to pump fluid over the relief valve. The result is a braking action that quickly slows and then stops motor rotation.

Figure 13-20 shows an advanced hydraulic motor circuit that provides more flexible control than the previous circuit. Two pilot-operated pressure control valves are used to provide control. They serve as a relief valve and a counterbalance valve. The directional control valve controls the vent chambers of the pressure control valves.

During system startup, when the directional control valve is in position 2, both the relief valve and counterbalance valve are vented. This allows all pump output to return to the reservoir at near-zero pressure and the hydraulic motor to freewheel. Shifting the directional control valve to position 3 blocks the vent of the relief valve and opens the vent of the counterbalance valve. The power unit drives the hydraulic motor at pressures up to the relief valve setting. The motor fluid discharge is returned to the reservoir at near-zero flow resistance through the open counterbalance valve.

The hydraulic motor continues to turn the load until the directional control valve is moved to either position 1 or 2. Moving the valve to position 2 vents both the relief and counterbalance valves. This allows the inertia of the load to continue to turn the motor (freewheel). Shifting the directional control valve into position 1 opens the vent of the relief valve and closes the vent of the counterbalance valve. The relief valve then opens. This allows pump output to return to the reservoir. The activated counterbalance valve creates a back pressure on the motor, braking it to a stop.

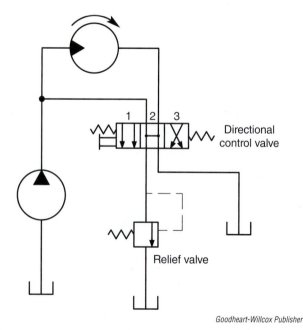

Figure 13-19. The inertia of the load on a hydraulic motor can present special problems. This relatively simple circuit can allow a motor to be powered, freewheel, or braked to a stop.

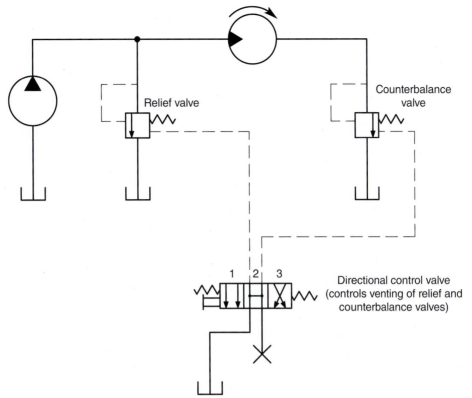

Figure 13-20. The venting capability of the pressure control valves in this circuit provides remote control of the valves and accurate pressure setting for both motor power output and braking action.

Once the motor stops, the closed counterbalance valve holds the motor and load in a fixed position.

13.5.2 System Decompression

It is generally assumed that hydraulic fluids compress so little as not to be considered important when designing, operating, or maintaining hydraulic systems. This thinking may be appropriate when working with systems containing only a small volume of fluid. However, in major systems containing a large volume of fluid, the energy of compressed fluid must be considered as well as the energy stored by machine members under tension. The sudden release of either pressure can cause unexpected flow or pressure surges in a system. Both of these factors are safety issues and issues concerning service life of the hydraulic system or machine.

Figure 13-21 shows a simple circuit used with a forming press that has a large-diameter cylinder containing a large volume of hydraulic fluid. The circuit includes two check valves. The first is a staged, pilot-operated check valve that acts as a *decompression valve*. This valve first decompresses the fluid in the system and then acts as the primary route to return fluid to the reservoir as the platen quickly lowers. The second check valve is a simple check valve that acts to prevent the reverse flow of fluid through the working line leading from the directional control valve to the blind end of the cylinder.

When the directional control valve is in position 2, the tandem center returns the output of the pump to the reservoir at near-zero pressure. The cylinder is locked in place by the directional control and check valves. Shifting the directional control valve to position 1 directs hydraulic fluid to the blind end of the cylinder. This opens the simple check valve in the cylinder inlet line and closes the pilot-operated check valve leading to the reservoir. When the press closes, the pressure in the system increases to the setting of the relief valve. This maintains a constant force on the platen.

When the pressure portion of the cycle is completed, the directional control valve can be returned to position 2 to reduce the load on the system power unit. This occurs because pump output is directed to the reservoir. At this point, the weight of the platen pressurizes the fluid in the blind end of the cylinder to keep both check valves closed.

When the press platen needs to be lowered, the directional control valve is shifted to position 3. This connects the piloted-check valve to the pump. As a result, pressure in the pilot line quickly increases to the point where the pilot mechanism opens a small orifice in the check valve. This small opening decompresses

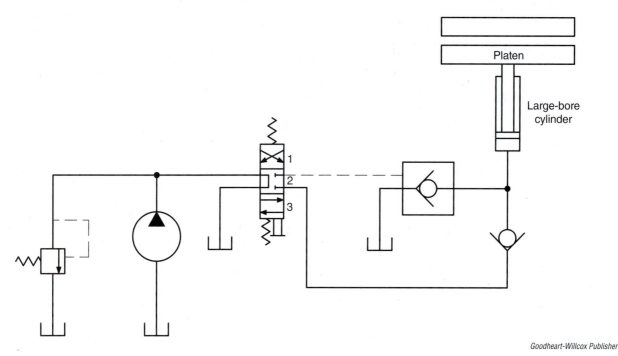

Platen

Large-bore
cylinder

1

2

3

Figure 13-21. The decompression of circuits having large fluid capacities and static loads is sometimes necessary to reduce shock pressure. The process may also be used to reduce unanticipated machine member movement, which is a safety issue.

the fluid in the cylinder. The pressure in the blind end of the cylinder drops, allowing the pilot mechanism to completely open the check valve. Gravity acting on the heavy platen forces the fluid in the cylinder to the reservoir and rapidly retracts the cylinder.

13.5.3 Safety-Related Circuits

Safe operation should always be a concern for equipment designers, manufacturers, and operators. The hydraulic field is not an exception to this concern. High pressures, immense forces, high flow rates, and high rates of movement are common elements in hydraulic systems. A variety of safety circuits are used to ensure:

- Equipment can be operated without injury.
- Equipment cannot unexpectedly operate.
- Costly equipment is protected against excessive pressure and operating rates.

The *two-hand-operation circuit* is common on presses and other equipment. This circuit requires both hands of the machine operator be kept clear of dangerous areas. The circuit as described here operates using hydraulic principles. However, versions of this circuit may use pneumatics or electricity.

The circuit shown in **Figure 13-22** uses two manually shifted, spring-offset, two-position, four-way directional control valves to shift the system directional control valve. The system directional control valve is a

pilot-operated, spring-centered, three-position valve with a closed center. When the power unit is not operating, the internal elements of the control valves should all be in position 2, as shown in the circuit schematic. This positioning is established by spring offsets in the two manual valves and spring centering in the system directional control valve. The cylinder is locked into position by the closed center of the system directional control valve.

Starting the power unit pressurizes the pilot lines. This causes the system directional control valve to shift into position 3. As a result, the cylinder fully retracts. Pressing and holding *Valves A* and *B* in position 1 redirects pilot pressure to shift the system directional control valve into position 1. This causes the cylinder to fully extend and hold that position until *Valves A* and *B* are released. When these valves are released, the pilot pressure shifts the system directional control valve to position 3 to fully retract the cylinder.

No other combination of *Valve A* and *Valve B* operation will fully extend and retract the cylinder or allow the cycle to be repeated. Releasing either *Valve A* or *Valve B* during cylinder extension causes the centering springs in the system directional control valve to shift the valve into position 2. This stops cylinder movement. Tying down either *Valve A* or *Valve B* to free one hand will not allow a new cylinder cycle to begin. As a result, the operator must have *both* hands on *Valves A* and *B* for the cylinder to operate.

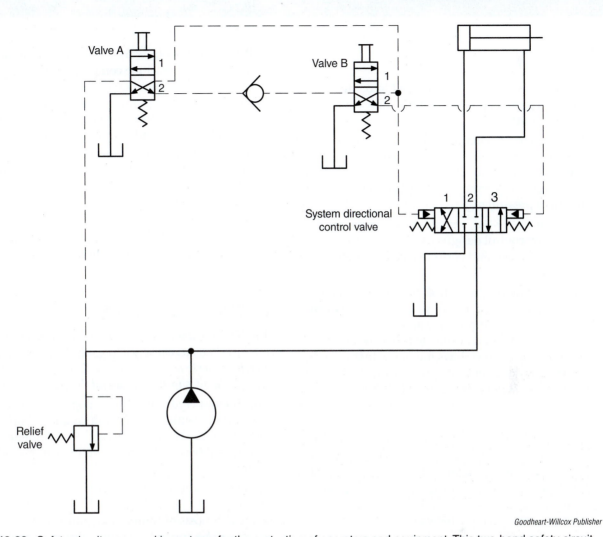

Figure 13-22. Safety circuits are used in systems for the protection of operators and equipment. This two-hand safety circuit will not function unless *Valves A* and *B* are activated at the same time.

Summary

- Hydraulic systems can be divided into four function groups: pressure control, flow control, and motion control.

- Pressure control circuits limit the maximum operating pressure of a hydraulic system. Variations of this function include circuits that can control pressure from a remote location or select multiple pressures in a single circuit.

- Meter-in, meter-out, and bleed-off circuits are three basic flow control designs used to produce desired operating speed and performance of actuators in the system.

- The meter-in design controls fluid flow into the actuator by placing a flow control valve between the directional control valve and the actuator inlet. Only the fluid needed to move the actuator at the desired speed flows into the actuator while remaining pump output flows back to the reservoir. The prime mover driving the pump is always operating against the maximum pressure setting of the system relief valve.

- The meter-out design controls fluid flow out of the actuator by placing a flow control valve between the outlet of the actuator and the directional control valve. The prime mover driving the system pump is always operating against the maximum pressure setting of the system relief valve.

- The bleed-off design places a flow control valve in a line teed off the working line that connects the directional control valve and the inlet port of the actuator. The outlet of the flow control valve diverts a metered portion of flow to the reservoir, while the remaining pump output operates the actuator. The prime mover driving the system pump operates against a pressure only high enough to move the workload encountered by the actuator.

- The meter-in and meter-out flow control circuits provide the most accurate actuator speeds as they both meter fluid flow delivered directly to the actuator.

- The bleed-off flow control circuit is less accurate as the flow control valve in the circuit meters flow back to the reservoir while the remaining pump output establishes actuator speed. This remaining flow will vary because of pump efficiency and system leakage.

- The bleed-off flow control circuit is the most energy efficient flow control design.

- Rapid-advance-to-work circuits allow considerable movement of machine components before the primary work operation begins. These circuits include high-low pump circuits to provide large volumes of fluid, regenerative circuits that recirculate fluid, and deceleration circuits.

- In hydraulic motor circuits, system fluid flow rate establishes motor speed. System pressure establishes torque output.

- Safety-related circuits address machine operator safety, prevent unexpected system movement, and protect expensive systems and costly equipment from excessive operating pressures.

Internet Resources

The following are some useful resources available on the Internet. Enter a company or organization name into a search engine to access its website. Explore the various areas of the sites to discover useful fluid power resources.

Continental Hydraulics. A catalog of one series of vane pumps with extensive information about pump specifications. The material can serve to identify factors that need to be considered when designing a hydraulic system.

Eaton Corporation. This guide covers many aspects of a hydraulic system, ranging from basic formulas to the diagnosis of component failure. Many topics can be applied to system design as well as troubleshooting.

Hydraulics & Pneumatics Magazine. Provides links to different types of hydraulic circuits. Selecting a link displays sample schematics for the type of circuit, accompanied by a brief explanation of circuit purpose and operation.

Chapter Review

Answer the following questions using information in this chapter.

1. The output force of a hydraulic actuator and output torque of a hydraulic motor is determined by system _____.
   A. pressure
   B. bleed-off
   C. unloading
   D. All of the above.

2. The _____ connection of a compound relief valve can be used to adjust maximum system pressure quickly and easily.
   A. deceleration
   B. sequencing
   C. safety
   D. venting

3. Pressure-reducing valves must be _____.
   A. decompressed before use
   B. motor-controlled
   C. pneumatically operated
   D. externally drained

4. A maximum pressure lower than the relief valve setting may be obtained by using a(n) _____ valve in a circuit.

5. In a system with an open-center directional control valve, the pressure required to shift pilot-operated valves may be obtained by inserting a(n) _____ valve in the return line to the reservoir.
   A. remote control
   B. shut-off
   C. check
   D. decompression

6. *True or False?* A meter-in circuit directly controls fluid flow into the directional control valve.

7. *True or False?* In the meter-out flow control circuit, the flow control valve is placed between the outlet port of the actuator and the directional control valve.

8. *True or False?* In a meter-out flow control circuit controlling the extension of a single-acting cylinder, the pressure in the rod-end chamber of the cylinder may be higher than the maximum pressure setting of the relief valve.

9. What are the three characteristics of basic flow control circuits that need to be considered when selecting a design for use in a hydraulic system?
   A. Accuracy of flow control, operating efficiency, and control of varying loads.
   B. Required pressure, flow requirements, and system efficiency.
   C. Pump flow rate, actuator size, and extension speed.
   D. Cylinder bore, safety needs, and speed control.

10. Which flow control systems are considered to provide the most accurate flow control?
    A. Check valve and high-low.
    B. Meter-in and meter-out.
    C. Bleed-off and meter-in.
    D. Meter-out and high-low.

11. *True or False?* Rapid-advance-to-work circuits often provide high-pressure operation after an extension performed at low pressure.

12. *True or False?* A high-low pump circuit is designed to provide high volume flow during low-pressure demand periods.

13. The control of actuators to provide a specific order of movement is called _____.
    A. speed
    B. decompression
    C. freewheeling
    D. sequencing

14. In hydraulic motor circuits, system pressure establishes motor _____.
    A. torque
    B. sequencing
    C. speed
    D. direction

15. *True or False?* Decompression valves lower the level of pressure in a hydraulic unit to a selected level in order to complete service tasks.

16. Hydraulic safety circuits are designed to _____.
    A. reduce injury when operating fluid power circuits
    B. prevent unexpected circuit operation
    C. prevent operation of fluid power equipment at excessive pressure and operating rates
    D. All of the above.

Apply and Analyze

1. Design and draw a circuit that allows two cylinders to be operated at different pressures at the same time.
   A. Label the components used.
   B. Describe the pressures throughout the circuit as the cylinders are extended and retracted.
   C. Design and draw a second circuit that performs the same operation.
   D. Which design would you choose to implement? Why?
   E. Modify one of your circuit designs to allow one of the cylinders to retract at a lower pressure than required for extension.

2. Describe the operation of the circuit shown in **Figure 13-8** as the manually operated directional control valve moves through its three positions.

3. Design and draw a circuit that allows a motor to be operated at two different speeds depending on direction of rotation.
   A. Label the components used.
   B. Describe the fluid pressure throughout the circuit as the motor is operated in both directions.
   C. Design and draw a second circuit that performs the same operation. How does the operation of the circuit compare to your first circuit?
   D. Which design would you choose to implement? Why?
   E. How might your circuit design be modified if your application involves a heavy rotating load?

4. Modify one of the cylinder sequencing circuits shown in **Figure 13-15** or **Figure 13-16** or design your own circuit so that the following occurs during circuit operation:
   A. *Cylinder 1* extends to clamp, *Cylinder 2* extends to drill, *Cylinder 2* retracts drill, and *Cylinder 1* retracts clamp.
   B. Cylinder speeds are faster during movement with no load applied.
   C. As the load increases, cylinder speed must be reduced so that load is applied slowly.

Research and Development

1. Find an example of hydraulic equipment that uses actuator sequencing in its operation. Create a poster that shows the hydraulic circuit, identifies the components used, and lists the steps that occur during operation.

2. Examine the hydraulic circuits used in some type of mobile hydraulic equipment. Identify the components, explain the role they play in the operation of the circuit and discuss how the choice of components affect the operation of the equipment. Present your results.

3. Investigate the safety concerns associated with hydraulic power press brake machines and the safeguards that have been installed to protect operators. Prepare a report that focuses on those safeguards that are part of the hydraulic circuit of the machine.

4. Design, construct, and test a hydraulic device that grasps and raises a small block and places it on a platform. Demonstrate your device to your classmates.

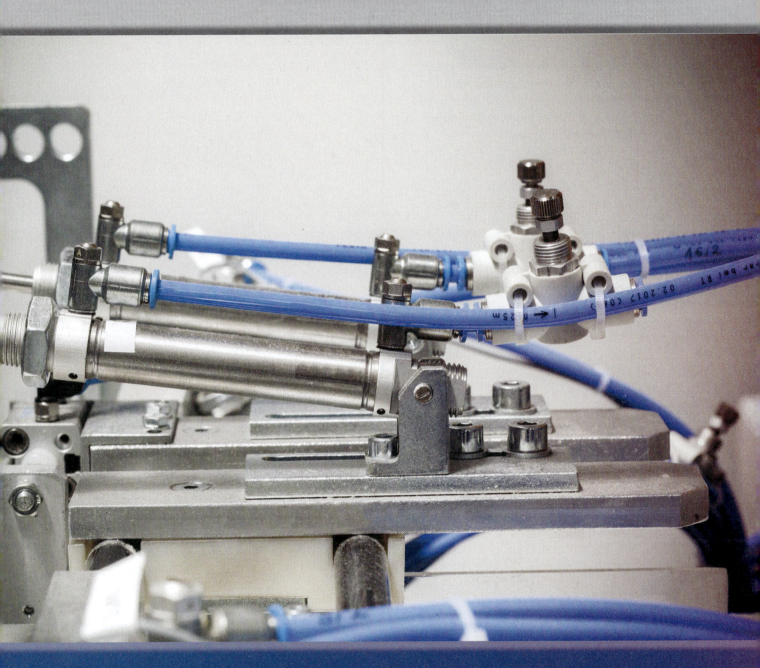

14 Compressed Air

The Energy Transmitting Medium

In pneumatic systems, energy is transmitted using a gas. This contrasts with the liquid used in hydraulic systems, electron movement in electrical circuits, and belts and gears used in mechanical systems. To a casual observer, gas may appear to be a simple and inexpensive approach to energy transmission. However, this is not the case in actual practice. Free air from the atmosphere must be conditioned before it can be used in pneumatic systems. This chapter deals with the composition of free air and the nature of the compressed air needed in a pneumatic system.

Learning Objectives

After completing this chapter, you will be able to:

- Describe the characteristics of free air.
- Identify the problems associated with the use of unconditioned air in a pneumatic system.
- Compare the composition of compressed air in a pneumatic system with the characteristics of free air.
- Explain the terms used to describe characteristics of compressed air.
- Compare the scales used to measure the pressure of free and compressed air.
- Compare the isothermal, adiabatic, and actual modes of air compression.
- Apply the principles of the general gas law to the compression and expansion of the air used in a pneumatic system.

Key Terms

adiabatic compression
adiabatic expansion
atmosphere
dew point
dry air
free air

ionosphere
isothermal compression
isothermal expansion
lubricant
mesosphere
ozone layer

relative humidity
saturation point
stratosphere
troposphere
water vapor

14.1 Basic Source of System Air

The source of the air used in a pneumatic system is the *atmosphere* that blankets earth. This blanket is a layer of gas approximately 620 miles deep. This layer of air sustains life as we know it, supporting activities we take for granted. These include acting as our personal source of oxygen as we breathe, the source of weather that distributes moisture to our food crops, our protection from solar radiation, and insulation from the frigid temperatures of space. In addition, the atmosphere supplies the oxygen that supports combustion, which is our primary source of heat in many activities.

14.1.1 Structure of the Atmosphere

The atmosphere is made up of distinct layers separated by transition zones, **Figure 14-1**. The layer in which we live is called the *troposphere*. It extends from earth's surface to a maximum of 10 miles above the surface. The *stratosphere* extends from the troposphere to approximately 30 miles above earth's surface. The *ozone layer* that filters ultraviolet energy is located in the upper limits of this layer. The troposphere and stratosphere contain almost all of the gases that we normally consider *air*. The *mesosphere* begins above the stratosphere

and goes to approximately 50 miles above earth's surface. The uppermost layers of the atmosphere contain the *ionosphere*, which may extend to the edge of space about 600 miles from the earth's surface.

From this general description it can be seen that, in theory, we have an enormous quantity of air available for use in pneumatic systems. However, in actual practice, careful conditioning is required to make atmospheric air usable in pneumatic systems.

14.1.2 Composition of Atmospheric Air

Many definitions indicate the composition of atmospheric air only by the proportion of the individual gases that make up the mixture. This may be critical for users who need one or more of the individual gases for use in a chemical process. However, when used as the energy-transmitting medium in pneumatic systems, additional factors must also be considered if the system is to operate effectively.

When only the gases are considered, air is made up of approximately 78.09% nitrogen, 20.95% oxygen and <1% other gases, **Figure 14-2**. These other gases include argon, ozone, carbon dioxide, and a number of additional gases in very small quantities. Other components include water vapor and entrapped dirt. The gas composition is relatively consistent, however, the other components vary depending on location.

Some water is always present in atmospheric air in a gaseous state. The amount of *water vapor* varies from very little to the point of saturation. The *saturation point* is reached when the maximum amount of water vapor is held for a given temperature of air. This basic concept is an important factor. Water vapor forms liquid water that accumulates in pneumatic system lines and other components as atmospheric air is compressed and then cooled in the system.

Dust and dirt may become suspended in atmospheric air and cause severe problems in a pneumatic system. The amount of dust and dirt depends on the location.

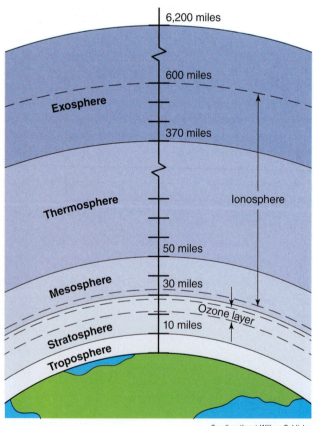

Goodheartheart-Willcox Publisher

Figure 14-1. The atmosphere of earth is made up of a series of layers. We live in the lowest level, called the troposphere.

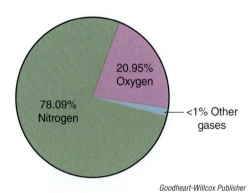

Goodheart-Willcox Publisher

Figure 14-2. The major portion of the atmosphere is made up of nitrogen and oxygen gases. The remaining portion is other gases, water vapor, and entrapped dust and dirt.

Dust and dirt in a pneumatic system can be directly associated with internal wear of the compressor and other system components.

The weight and pressure of atmospheric air must also be considered. In daily activities, these factors are usually not a conscious concern. However, the weight of an atmosphere that is 600 miles deep is a real factor. It produces an atmospheric air pressure of 14.7 pounds per square inch at sea level (14.7 psia or 0 psig). See **Figure 14-3**. Also, at 68°F, one cubic foot of air weighs approximately 0.0752 pounds.

14.2 Pneumatic System Compressed Air

Atmospheric air, typically referred to as *free air*, needs to be conditioned before distribution and use in a pneumatic system. Conditioning involves four factors

to produce air that results in an efficiently operating system with minimal equipment service:

- Removal of entrapped dirt.
- Removal of water vapor.
- Removal of heat.
- Incorporating oil for component lubrication.

The location of the pneumatic system has a great deal of influence on the amount of dirt in free air. This dirt must be removed before the free air enters the compressor intake. The free air in clean environments needs minimal filtration. In contrast, a compressor at a dusty construction site needs several levels of filtration to prevent excessive wear of the compressor and air tools, **Figure 14-4**. Details of the methods used to remove dirt from intake air are found in Chapter 16, *Conditioning and Distribution of Compressed Air*.

Water that accumulates in a pneumatic system can be detrimental to system components and must be

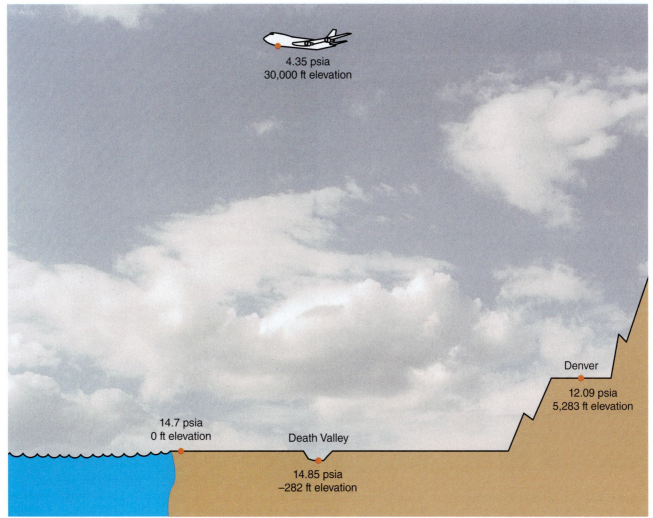

4.35 psia
30,000 ft elevation

Denver

12.09 psia
5,283 ft elevation

14.7 psia
0 ft elevation

Death Valley

14.85 psia
−282 ft elevation

Figure 14-3. Atmospheric pressure varies depending on the vertical location at which the measurement is taken. The higher the location, the lower the pressure reading.

Figure 14-4. The amount of entrapped dirt in free air varies, depending on local conditions. A—The air at construction sites often needs additional filtration, but free air in clean environments needs little filtration. B—Quarrying operations are very dusty.

removed, **Figure 14-5**. It can also affect the manufactured product if used for cleaning or in any other step of the process. Designers and operators of pneumatic systems need to be aware of air's ability to hold water vapor. They also need to understand the relationship of humidity and temperature to the formation of water in a system.

The amount of water carried in air largely depends on the temperature of the air. Hot air can hold more water than cool air. When air cannot hold additional water at a given temperature, it is saturated.

Relative humidity is the amount of water vapor actually carried by a given volume of air. This is a percentage showing the relationship between the actual amount of water in the air and the maximum amount the air can hold at the given temperature. For example, a relative humidity of 100% indicates that the air is saturated with water vapor. This point is called the ***dew point***. Beyond this point, any additional water vapor is released as a liquid, **Figure 14-6**.

Figure 14-5. Compressed air containing an excessive amount of water vapor results in liquid water forming in distribution lines. This water may be collected by the use of water legs placed in strategic locations in the distribution lines.

Peerapong Poongkorn/Shutterstock.com

Figure 14-6. Liquid water forms when the temperature of the air containing water vapor drops to the dew point. At this point, water forms in pneumatic system lines and components just the same as the water droplets that form on grass.

An appreciation of humidity and temperature is critical when developing a system to produce and deliver dry, compressed air to a workstation. Free air typically carries some water vapor. During air compression and distribution, temperature changes can cause this vapor to be released in the system as a liquid. A well-designed pneumatic system removes a sufficient quantity of water vapor to ensure the delivery of dry air to the workstation. *Dry air* contains water vapor, but should have a relative humidity sufficiently low to prevent formation of liquid water at the ambient temperature of the workstation.

In a typical pneumatic installation, the compressed air is also the carrier of *lubricant* for components. Although oil-free systems are required in areas such as food processing, a typical, industrial pneumatic system provides a means to add a lubricant to the air delivered to components. The lubricant is atomized and transported with the air through the working lines. The lubricant is typically introduced at the workstation that is located at the beginning of the working circuit, **Figure 14-7.** The air transported through the distribution lines between the compressor unit and the workstations typically does not contain lubricants. Details of the components used to add lubricants to system air are found in Chapter 16, *Conditioning and Distribution of Compressed Air*.

14.3 Compression and Expansion of Air

Working with compressed air introduces a variety of concepts different from those experienced with the

Goodheart-Willcox Publisher

Figure 14-7. Lubricants are added to compressed air at the workstation located just before the working circuit.

liquids used in hydraulic systems. A review of the sections in Chapter 4, *Basic Physical Principles*, relating to the physical principles of fluid power systems may be helpful. This will reinforce the similarities and differences between concepts basic to the two systems. The remaining sections of this chapter expand on these concepts. This approach provides a practical understanding of air and the way it reacts to changes in temperature, pressure, and volume in operating pneumatic systems.

During pneumatic system operation, system air is constantly changing as it is compressed, transmitted, and used. The system users must be conscious of these changes to better appreciate the function and importance of system and component design features.

When air is compressed, the volume of the air decreases and the pressure increases. Compressing the air also produces heat that increases the temperature of the air. If it were possible to remove all of this added heat, the air temperature would remain constant. This constant-temperature process is referred to as *isothermal compression*. Under these compression conditions, the process follows Boyle's law. That formula can be used to easily calculate final system pressure.

In contrast, if all of the heat from compression is retained in the air, the pressure is increased above the pressure produced by the volume change alone. This process is known as *adiabatic compression*.

These principles also relate to expansion of air. In *isothermal expansion*, it is necessary to add heat to the air to maintain a constant temperature as the volume increases. In *adiabatic expansion*, all heat needs to be retained in the air with none transferred to the metal of the compressor unit.

In actual practice, the compression and expansion characteristics are somewhere between the isothermal and adiabatic conditions. However, formulas do exist that take these factors into account for the design of pneumatic equipment.

Figure 14-8 shows graphs indicating volume and pressure differences for isothermal, adiabatic, and actual compression. The area between the isothermal and adiabatic curves represents the volume and pressure differences when working with identical initial quantities of air. The area between the actual and adiabatic curves represents the differences caused by heat lost to the atmosphere in typical equipment operation, rather than the ideal calculated adiabatic condition.

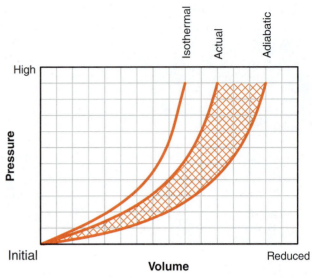

Figure 14-8. The plots represent the volume and pressure variations resulting from the compression of an equal volume of air following isothermal, adiabatic, and actual compression modes.

14.4 Reaction of Air to Temperature, Pressure, and Volume

The concepts involved with isothermal, adiabatic, and actual compression and expansion of air begin to identify the complexities of handling air. However, these complexities may be more easily understood if two basic principles are remembered:

- A change in either air temperature, pressure, or volume results in a change in at least one, and possibly both, of the other factors.

- The relationships expressed are based on the absolute scales of temperature and pressure.

In an operating system, the variations that occur can be complex. They can be a challenge when analyzing the operation or performance of a pneumatic system. However, it must be remembered that these changes adhere to the general gas law:

$$\frac{P_1 \times V_1}{T_1} = \frac{P_2 \times V_2}{T_2}$$

In addition to the pressure, volume, and temperature of the air, many outside variables make direct application of the general gas law impractical in an operating system. Several sources, including publications of component manufacturers and data handbooks for compressed air and gas, provide engineering data that can be applied to compensate for additional variables. The resulting calculations are improved estimates of how a component or system design will respond in an operating situation.

The following sections discuss several aspects of how air in an operating pneumatic system responds to the general gas law and the other variable factors. The discussion does not involve calculations, but is intended to illustrate the complex relationship between changes in air pressure, volume, and temperature and the influence of other system factors.

14.4.1 Pressure Variations

According to the general gas law, any change in the pressure of air will result in a temperature or volume change. For example, when shifting the system directional control valve, air flows from the receiver to the cylinder, **Figure 14-9**. The air entering the cylinder increases the pressure in the confined space of the cylinder chamber. The increase in cylinder air pressure is due to air from the system receiver entering the fixed volume of the stalled cylinder and the slight increase in temperature resulting from the heat of compression.

When the force generated by the air pressure acting on the piston surface is adequate to extend the cylinder, the chamber volume begins to increase. As a result, the cylinder rod moves. Air from the receiver continues to flow into the cylinder chamber at a pressure adequate to move the load. If the load remains constant, the pressure and temperature of the air in the cylinder chamber remain constant.

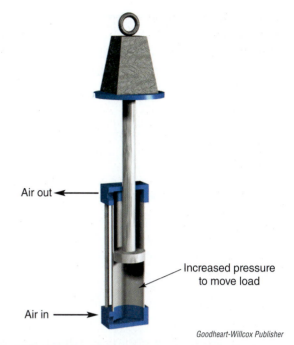

Figure 14-9. To extend the cylinder under load, both air pressure and volume must be increased. Air temperature also changes in the process due to rapid pressure changes.

14.4.2 Volume Variations

Changes in the volume of air result in a pressure or temperature change. An example is the reduction of volume in a cylinder during the compression stroke of a piston-type compressor, **Figure 14-10**. The compression stroke increases air pressure by both the reduction of volume in the cylinder chamber and the heat of compression that increases air temperature. The smaller the volume left in the cylinder at the end of the compression stroke, the higher the pressure produced and the higher the air temperature.

EXAMPLE 14-1

Reaction of Air to Temperature Changes at Constant Volume

An air tank has an initial temperature of 60°F with a gauge pressure of 1,000 psig. On a sunny day, the gas temperature may rise to 100°F. At that temperature, what pressure should the gauge read?

To complete this calculation, pressure must be calculated in atmospheric pressure (psia) rather than in gauge pressure (psig) and temperature must be converted to the Rankine scale.

$$P_1 \text{ (psia)} = P_1 \text{ (psig)} + 14.7 \text{ psia}$$
$$= 1,000 \text{ psig (initial gauge pressure)} + 14.7 \text{ psia}$$
$$= 1,014.7 \text{ psia}$$

$$T_1 \text{ (°R)} = T_1 \text{ (°F)} + 459.67$$
$$= 60°F + 459.67$$
$$= 519.67°R$$

$$T_2 \text{ (°R)} = T_2 \text{ (°F)} + 459.67$$
$$= 100°F + 459.67$$
$$= 559.67°R$$

The general gas law will be used to solve this problem.

$$\frac{P_1 \times V_1}{T_1} = \frac{P_2 \times V_2}{T_2}$$

Because air tank volume is constant ($V_1 = V_2$), volume can be removed from the equation:

$$\frac{P_1}{T_1} = \frac{P_2}{T_2}$$

Solving for final pressure:

$$P_2 = \frac{T_2 \times P_1}{T_1}$$

The converted values can now be inserted into the general gas law equation to solve for final pressure:

$$P_2 \text{ (psia)} = \frac{T_2 \times P_1}{T_1}$$
$$= \frac{559.67°R \times 1,014.7 \text{ psia}}{519.67°R}$$
$$= 1,092.8 \text{ psia}$$

$$P_2 \text{ (psig)} = 1,092.8 \text{ psia} - 14.7 \text{ psia}$$
$$= 1,078.1 \text{ psig}$$

Figure 14-10. Reducing the volume of air during the compression stroke in a piston-type compressor increases the pressure of the air. The rapid air compression also produces heat that contributes to an additional increase in pressure of the air.

EXAMPLE 14-2

Reaction of Air to Temperature Changes at Constant Pressure

A 2″ diameter cylinder in a pneumatic circuit is extended 10″ to raise and hold a load of 100 lb. Air temperatures during the initial cylinder extension were 100°F, but quickly drop to 50°F. What happens to the position of the load assuming no additional air is added to the cylinder?

To complete this calculation, temperature must be converted to the Rankine scale.

$$T_1 \, (°R) = T_1 \, (°F) + 459.67$$
$$= 100°F + 459.67$$
$$= 559.67°R$$

$$T_2 \, (°R) = T_2 \, (°F) + 459.67$$
$$= 50°F + 459.67$$
$$= 509.67°R$$

The general gas law will be used to solve this problem.

$$\frac{P_1 \times V_1}{T_1} = \frac{P_2 \times V_2}{T_2}$$

Because the load is constant ($P_1 = P_2$), pressure can be removed from the equation:

$$\frac{V_1}{T_1} = \frac{V_2}{T_2}$$

Solving for final volume:

$$V_2 = \frac{T_2 \times V_1}{T_1}$$

Volume is equal to the area of the cylinder multiplied by the length of the extension:

$$V_1 = \text{Area}_1 \, (in^2) \times \text{Extension Length}_1 \, (in)$$
$$= A_1 \times L_1$$

$$V_2 = \text{Area}_2 \, (in^2) \times \text{Extension Length}_2 \, (in)$$
$$= A_2 \times L_2$$

Substituting volume in the general gas law:

$$A_2 \times L_2 = \frac{T_2 \times (A_1 \times L_1)}{T_1}$$

Because the area of the cylinder is fixed ($A_1 = A_2$), area can be removed from the equation:

$$L_2 = \frac{T_2 \times L_1}{T_1}$$

The converted values can now be inserted into the general gas law equation to solve for final pressure:

$$L_2 = \frac{509.67°R \times 10″}{559.67°R}$$
$$= 9.1″$$

14.4.3 Temperature Variations

An increase or decrease in the temperature of air results in a pressure or volume change. For example, the pressure in the fixed space of the system receiver changes as the temperature of the air changes, **Figure 14-11**. As heated air coming from the compressor enters the receiver, the existing air is warmed by both the heat of the incoming air and the heat generated by the heat of compression as the air existing in the receiver is further compressed. Both of these factors increase the pressure in the fixed-volume receiver.

Temperatures in a variety of situations can cause changes in the operation of a pneumatic system. Examples include compressor intake air temperature, compressor room temperature, and the temperature of the air surrounding the system distribution lines. The temperatures existing in each of these situations can produce variations in system performance.

EXAMPLE 14-3

Reaction of Air to Volume Changes at Constant Temperature

During the compression stroke of a piston-type compressor, the volume of the air is reduced to 1/8 of its original volume. If the air enters the compressor at atmospheric pressure (14.7 psia) and compression happens isothermally, what is the final gauge pressure of the air leaving the compressor?

$$\frac{P_1 \times V_1}{T_1} = \frac{P_2 \times V_2}{T_2}$$

If the compression happens isothermally ($T_1 = T_2$), temperature can be removed from the equation:

$$P_1 \times V_1 = P_2 \times V_2$$

Solving for final pressure:

$$P_2 = \frac{P_1 \times V_1}{V_2}$$

Substituting in $V_2 = 1/8 \times V_1$ leaves:

$$P_2 = \frac{P_1 \times V_1}{1/8 \times V_1}$$
$$= P_1 \times 8$$

To complete this calculation, pressure must be calculated in atmospheric pressure (psia) rather than in gauge pressure (psig).

$$P_1 \, (psia) = 14.7 \, psia$$
$$P_2 \, (psia) = 14.7 \, psia \times 8$$
$$= 117.6 \, psia$$

$$P_2 \, (psig) = 117.6 \, psia - 14.7$$
$$= 102.9 \, psig$$

Heat transfers
from the receiver
to the air in
the room.

Goodheart-Willcox Publisher

Figure 14-11. The rate of heat loss to the atmosphere through the external walls of the receiver influences the final pressure of the air in the unit.

Summary

- The source of air for use in pneumatic systems is the atmosphere. Atmospheric air is 78.09% nitrogen, 20.95% oxygen, and less than 1% other gases. The atmosphere also contains variable amounts of water vapor and entrapped dirt.

- Air pressure can be measured in atmospheric pressure (psia) or gauge pressure (psig).

- Atmospheric air must be conditioned before use in a pneumatic system. Conditioning of compressed air involves incorporation of lubricants and the removal of entrapped dirt, water vapor, and heat.

- The amount of water vapor air can hold is dependent on the temperature of the air. At higher temperatures, more water vapor may be retained.

- Relative humidity expresses the percentage of water in the air compared to the maximum amount that can be held at the specified temperature.

- System compressed air will contain water vapor but should have a relative humidity sufficiently low to prevent the formation of liquid water at the ambient temperature of the pneumatic workstation.

- Dew point is the temperature at which water vapor in the saturated air begins to be released in liquid form.

- The general gas law expresses the changes in temperature, pressure, and volume that occur when air is compressed.

- Two different theoretical compression models are used to describe air compression. In the isothermal compression model, all the heat generated by compression is removed, resulting in a constant temperature. In the adiabatic compression model, all heat is retained resulting in increased temperature and pressure of the air. Actual compression results are somewhere between the theoretical isothermal and adiabatic compression models.

Internet Resources

The following are some useful resources available on the Internet. Enter a company or organization name into a search engine to access its website. Explore the various areas of the sites to discover useful fluid power resources.

Georgia State University—Dept. of Physics and Astronomy: HyperPhysics. Explains relative humidity, dew point, and saturation vapor pressure. Calculation methods are shown for each of these factors.

Norgren. Provides basic engineering information related to determining water content in compressed air systems and establishing pressure drop in pneumatic circuits.

Wikipedia: The Free Encyclopedia—Gas Laws. Provides a brief description of gas laws with related mathematical equations.

Wikipedia: The Free Encyclopedia—Adiabatic Process. Provides a basic definition of the adiabatic process. Several examples are given with links to other sites that provide more detailed mathematical explanations.

Wikipedia: The Free Encyclopedia—Isothermal Process. Provides a basic definition of the isothermal process. Several examples are given with links to other sites that provide more detailed mathematical explanations.

Chapter Review

Answer the following questions using information in this chapter.

1. The layer of atmosphere in which we live is called the _____.
   A. ozone layer
   B. mesophere
   C. adiabatic
   D. troposphere

2. Air is made up of approximately _____ nitrogen.
   A. 78.09%
   B. 14.2%
   C. 1%
   D. 5%

3. *True or False?* An increase or decrease in the temperature of air results in a pressure or volume change.

4. The amount of water vapor present in atmospheric air varies from very little to the point of _____.

5. Standard atmospheric air pressure is _____ pounds per square inch at sea level.
   A. 78.09
   B. 14.7
   C. 0.0752
   D. 12.09

6. The amount of water vapor actually carried by a volume of air compared to the maximum amount it can carry at the specified temperature is _____.
    A. entrapped water vapor
    B. dew point
    C. relative humidity
    D. vapor level

7. *True or False?* Lubricants typically are introduced into a pneumatic system just before the working circuit.

8. The process that assumes all heat is retained in the air during compression is known as _____ compression.
    A. isothermal
    B. actual
    C. adiabatic
    D. None of the above.

9. *True or False?* A change in either air temperature, pressure, or volume results in a change in at least one, and possibly both, of the other factors.

10. According to the general gas law, what happens when the pressure of air is changed?
    A. The volume increases or decreases.
    B. The temperature may change.
    C. Both A and B.
    D. None of the above.

Apply and Analyze

1. A cylinder in a pneumatic circuit extends under adiabatic conditions to lift a load. Predict the system behavior over time if the cylinder remains extended under constant load.

2. Predict what will happen in the following circumstances:
    A. Air in a closed tank is heated.
    B. Air in an open tank is heated. The tank is then sealed and cooled.
    C. Air is quickly compressed in a cylinder.
    D. Air is slowly added to a tank until a needed pressure is reached. The tank then sits for several days.

3. A 3" diameter pneumatic piston is used to lift a load of 100 lb.
    A. What air pressure is required to accomplish this task?
    B. What volume of air at this pressure is required to extend the piston 10"?
    C. What volume of air at atmospheric pressure is needed to perform this operation?
    D. Assume the piston is extended 10" while supporting this 100 lb load. If the temperature drops from 80°F to 65°F while the piston is extended, how far will the piston retract due to the contraction of the air?
    E. If the temperature instead remains constant at 80°F, but the load increases to 150 lb, how far will the piston retract? Assume the piston is initially extended to 10".
    F. What are the implications of your calculations for the operation of pneumatic systems?

Research and Development

1. Find several photographic examples online or in your community of pneumatic equipment used in different environments. For each environment, identify the possible ramifications of using unconditioned air to operate the equipment.

2. Design and conduct an experiment to test the reaction of air to changes in pressure, temperature, and volume. Share your results with your classmates.

3. Investigate how seasonal variations in the weather conditions in your area might affect the operation of pneumatic equipment used outside.

15 Source of Pneumatic Power
Compressed-Air Unit and Compressor

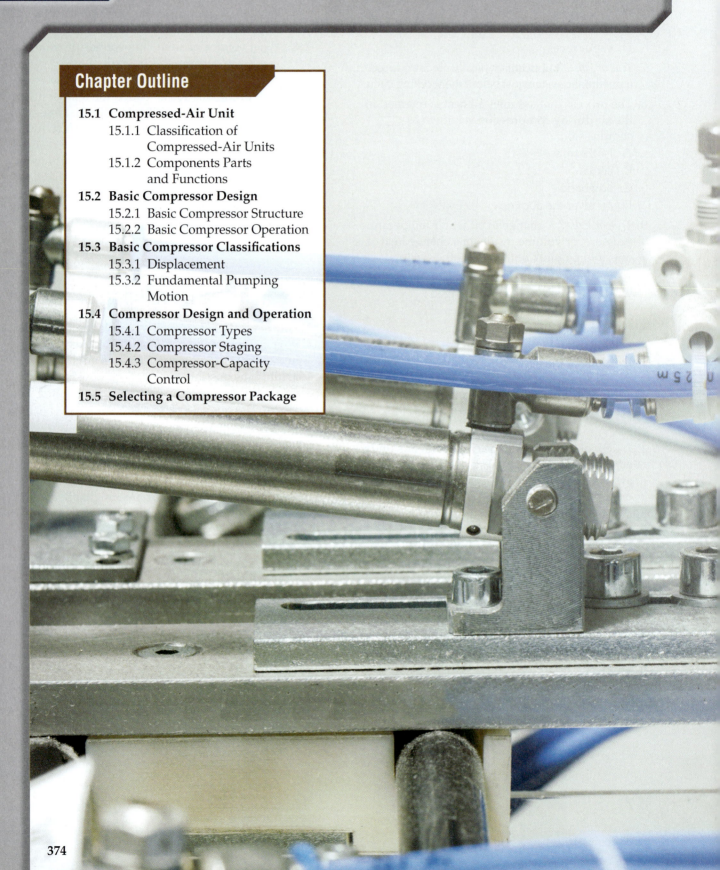

The compressed-air unit is the source of power for the remainder of the pneumatic system. These units range in size from small, portable compressors to huge units that provide compressed air to large manufacturing facilities. The units may be relatively simple, consisting of a prime mover, compressor, receiver, and simple pressure switch. In contrast, major installations not only include larger versions of these components, but additional components designed to provide a higher level of conditioning for cleaner and drier compressed air. This chapter introduces each of the components and then discusses compressor design in more detail to provide an understanding of compressor operation and application.

Learning Objectives

After completing this chapter, you will be able to:

- Describe the function of a compressed-air unit.
- Name and explain the function of each of the components in a compressed-air unit.
- Identify the basic designs used in air compressor construction.
- Compare the operating characteristics of positive- and nonpositive-displacement air compressors.
- Compare the operating characteristics of rotary and reciprocating air compressors.

- Describe the general construction characteristics of the various compressor types.
- Explain the operation of the various systems used to control the maximum air pressure available from the compressed-air unit.
- Identify the factors that must be considered to estimate the required output of a compressor to meet the air demands of a pneumatic system.
- Interpret performance data supplied by a compressor manufacturer.

Key Terms

air filter
capacity-limiting system
central air supply
compressed-air unit
compressor
compressor-capacity control
cooler
coupling
double-acting compressor
dryer

dynamic compressor
lobe-type compressor
nonpositive-displacement
   compressor
portable unit
positive-displacement
   compressor
prime mover
pumping chamber
pumping motion

receiver
reciprocating compressor
rotary compressor
rotary screw compressor
rotary sliding-vane
   compressor
single-acting compressor
slippage
staging

15.1 Compressed-Air Unit

The *compressed-air unit* produces pressurized air for the pneumatic system. The design and appearance of this unit varies, depending on the use of the air and the size of the system it must operate. The unit may be a very simple compressor supplying a fraction of a cubic foot per minute for an artist's airbrush or portable air tool, **Figure 15-1**. On the other hand, it may be a large, stationary installation in an industrial plant, **Figure 15-2**. Industrial units may produce several thousand cubic feet per minute of conditioned air for the operation of a number of system workstations.

15.1.1 Classification of Compressed-Air Units

A compressed-air unit is classified as a portable unit or central air supply. A *portable unit* is one that can be relatively easily moved from one location to another. Portable units are moved to the work location and provide compressed air to remote work sites, **Figure 15-3**. A *central air supply* has a permanently mounted compressor and air distribution lines. Central air supplies provide air to fixed workstations in a permanent installation.

Christina Richards/Shutterstock.com

Figure 15-1. A small, portable air compressor is commonly used at home building sites to operate nailing equipment or other tools.

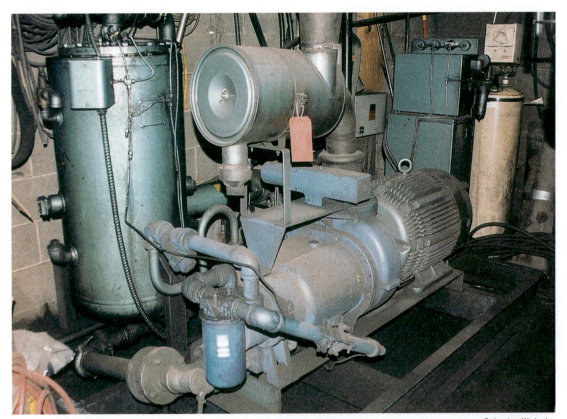

Badger Iron Works, Inc.

Figure 15-2. Industrial plants use central air stations that supply compressed air to numerous workstations.

Figure 15-3. Portable air compressors mounted on trucks or trailers are commonly used at highway and other construction sites.

Physical size may be a factor in selecting the classification of unit to use, but certainly cannot be used as the primary deciding feature. Physical size could be considered a major difference if the small, portable compressors that have become a popular consumer tool were the only portable compressor units, **Figure 15-4**. However, portable compressors used in the construction, mining, and quarrying fields often exceed the size of central air supplies in small manufacturing plants.

Even though the appearance of these units is different, they do include a number of elements that serve identical functions. Units under both classifications:

- Provide the prime mover energy needed to compress the air.
- Control air temperature, retained water vapor, and entrapped dirt.
- Provide a means of controlling the maximum air pressure in the system.
- Store the conditioned air until needed at the workstation.

The components in the compressor unit that provide these functions are described in the next section.

15.1.2 Components Parts and Functions

The compressed-air unit serves as the overall source of energy for a pneumatic system. A *prime mover* drives

Figure 15-4. Portable units allow the compressor to be operated where the work is being performed.

the compressed-air unit. Typically, the primer mover is dedicated to the operation of the compressor. The prime mover in a central air supply is most often an electric motor, although steam and turbine engines may be used in very large applications. Small, portable units are typically powered by electric motors. Internal combustion engines are used to power portable units for use in remote locations, such as new building sites and road construction, where electricity is not available. The energy that drives the prime mover comes from electric current for a motor or fuel for an internal combustion engine.

The prime mover is connected to the input shaft of the compressor by a mechanical *coupling*. Several different coupling designs are used, including direct, mechanical couplers, and belt drives. See **Figure 15-5**. The design used in a specific compressed-air unit is based on rated speed, load, and allowable shaft misalignment between the prime mover and compressor.

The *compressor* is driven by the prime mover and generates compressed air. A variety of compressor designs are available for use in pneumatic systems. The design selected for an application depends on physical size restrictions, required system airflow rates, operating pressure, and a variety of other requirements. All of the compressor designs increase the pressure of the air from atmospheric pressure to the desired operating pressure. This is done by moving additional atmospheric air into the fixed volume of the components that comprise the compressed-air unit.

Another major component of the compressed-air unit is the *receiver*. This component serves a number of purposes. First, it is the sealed, fixed-volume chamber that holds the compressed air until needed by the pneumatic system. When the system needs air for the operation of a workstation circuit, that air is drawn from the receiver. The receiver also assists in cooling the air and removing water vapor to ensure the distribution of dry air to the system.

Another vital part of the compressed-air unit is a *capacity-limiting system*. This system is used to control the maximum air pressure produced by the unit. The maximum pressure is somewhat higher than the pressure needed to operate the actuators in the system. This higher pressure ensures efficient air distribution in the system. The capacity-limiting function is commonly performed by an electrically operated pressure switch that automatically starts and stops the prime mover at preselected minimum and maximum pressures. In systems where the prime mover is an internal combustion engine, the pressure switch opens and closes a valve that vents compressor output when the preselected maximum system pressure is reached. The compressed-air unit also includes a mechanically operated safety valve. If the electrically operated valve should fail, the safety valve opens to vent pressure before a dangerous pressure is reached.

The compressed-air unit includes a number of additional components, depending on the size and complexity of the unit. A critical unit is an *air filter* to remove dirt from the atmospheric air. *Coolers* and

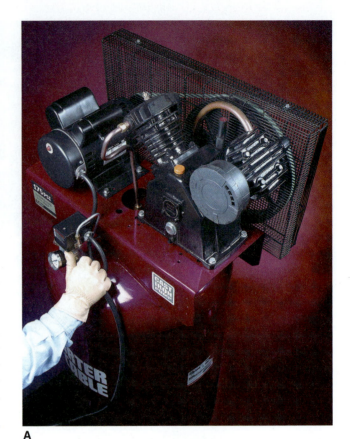

A

B

DeVilbiss Air Power Company

Figure 15-5. Belts (A) or direct couplers (B) are used to transfer energy from the system prime mover to the compressor.

dryers are also used to ensure the delivery of clean, cool, dry air to the distribution lines and workstations of the system.

Typically, compressor manufacturers are willing to provide all of the necessary components and electrical controls as a package. This provides a compressed-air unit that meets the requirements of an installation without the trouble of assembling the system on site.

15.2 Basic Compressor Design

A number of compressor designs are used to convert the mechanical power output of the prime mover into high-pressure air for use in a pneumatic system. The capacity of these units varies significantly, as does the general appearance and structure. However, basic operating concepts are similar. This section examines the conceptual design and operation of a compressor to illustrate how air compression is accomplished.

15.2.1 Basic Compressor Structure

A pneumatic system compressor should be thought of as a specialized air pump. It is simply a pump designed to deliver air at an elevated pressure. The following explanation uses a basic reciprocating compressor design to illustrate both compressor structure and operation. The design has been used for years in the pneumatic field and is still common today.

The structure of a basic reciprocating compressor can be compared to an internal combustion engine minus the fuel and ignition systems. The structure of such a compressor is shown in **Figure 15-6**. The space above the pistons form *pumping chambers* that vary in size as the crankshaft rotates to move the pistons up and down.

The cylinder head of the compressor contains ports. Each port is fitted with a valve. The inlet ports are open to the atmosphere through the filtering system. The inlet valves are open as the pistons move down in the cylinder and closed as the pistons move up. The outlet ports are connected to the inlet port of the system receiver. The outlet valves are open when the pistons are moving up in the cylinder and closed as the pistons move down.

15.2.2 Basic Compressor Operation

The following sections describe the basic operation of a compressor. Refer to **Figure 15-7**. There are three basic phases of operation:

- Air intake.
- Air compression.
- Air discharge.

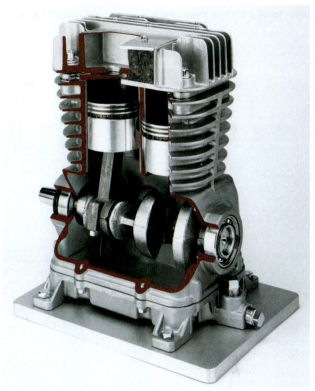

DeVilbiss Air Power Company

Figure 15-6. A reciprocating compressor is similar in appearance and structure to an internal combustion engine.

Air intake

When the rotating crankshaft pulls the piston down, the pumping-chamber volume increases. This increased volume lowers the air pressure in the cylinder. As a result, atmospheric pressure forces air into the pumping chamber through the open inlet valve. Ideally, the air pressure in the pumping chamber will be as close to atmospheric pressure as possible during the intake phase. This maximizes the amount of air moved through the compressor. When the continued rotation of the crankshaft begins to move the piston upward, the inlet valve closes. The volume of the pumping chamber also begins to decrease.

Air compression

As the crankshaft continues to rotate, the piston is forced upward in the cylinder. This decreases the volume of the pumping chamber. The decrease in volume causes the pressure of the air in the chamber to increase. This increase in pressure indicates two important items:

- The amount of actual decrease in chamber volume.
- The amount of energy from the prime mover that has been converted to potential energy, as represented by the higher air pressure.

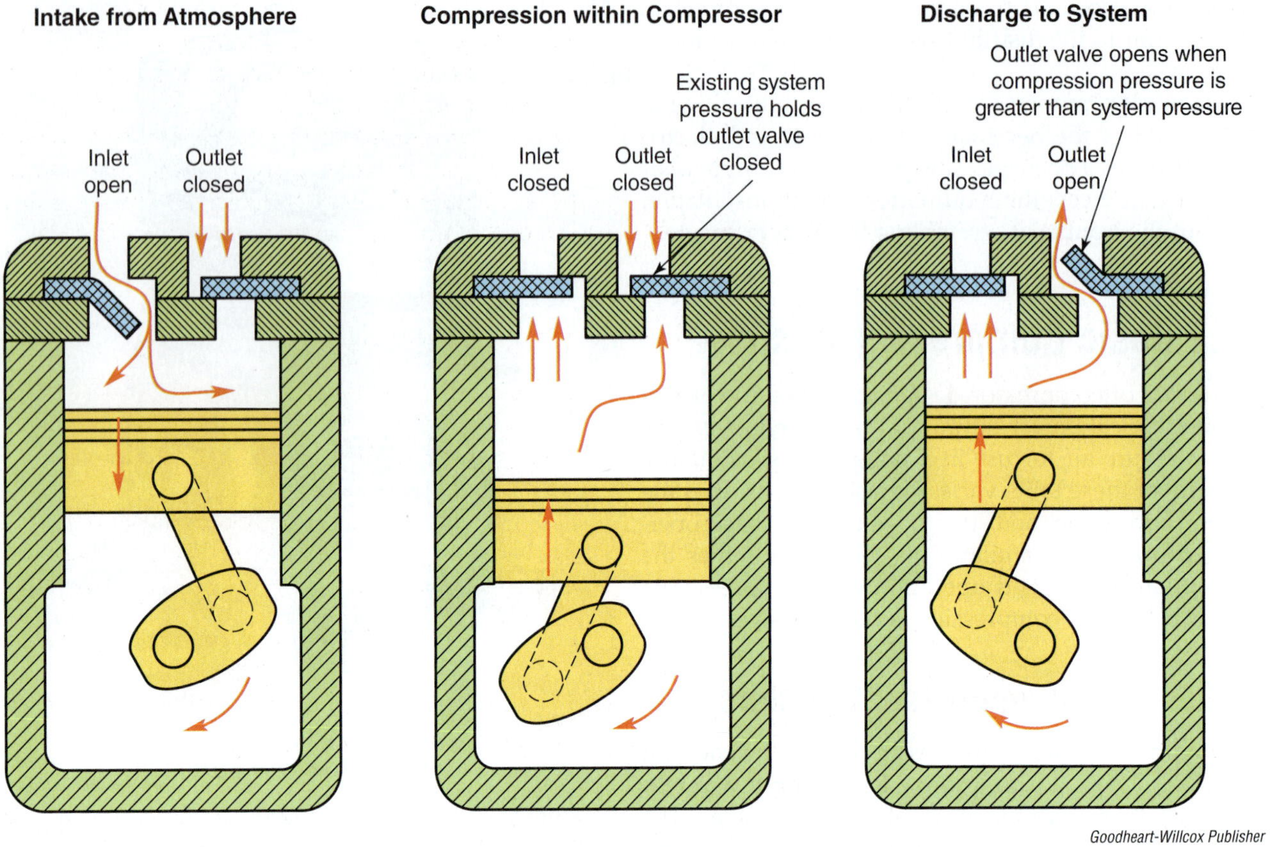

| Intake from Atmosphere | Compression within Compressor | Discharge to System |

Goodheart-Willcox Publisher

Figure 15-7. The point at which air flows from a compressor depends on the existing pressure in the system.

Air discharge

The pressurized air in the pumping chamber only moves into the system receiver when the pressure in the chamber is higher than the pressure in the receiver. When the receiver and system pressures are low, the transfer to the receiver is early in the compression stroke. When the system pressure is close to the maximum pressure setting of the compressed-air unit, the transfer is late in the compression stroke. The response of a compressor to the air demands of a pneumatic system is determined by:

- Size of the pumping chamber of the compressor.
- Operating speed of the compressor.
- Pressure and volume of the air needed to operate the circuits of the system workstations.

15.3 Basic Compressor Classifications

All compressors used in pneumatic systems are classified using two methods. These classifications involve the design, structure, and operation of the compressors relating to:

- Displacement.
- Fundamental pumping motion.

15.3.1 Displacement

The displacement of a compressor is the volume of air displaced per unit of time. It is usually stated in cubic feet per minute (cfm). The method used to calculate displacement depends on the compressor design. Compressors are classified as either having positive displacement or nonpositive displacement.

Positive displacement

Positive-displacement compressors confine air in a chamber that can be mechanically reduced in volume to increase the pressure. When the air chamber volume is decreasing, air is transferred to the receiver of the compressed-air unit after the chamber pressure exceeds the air pressure in the system. Reciprocating, sliding vane, and rotary screw compressors are considered positive-displacement designs, **Figure 15-8.**

Nonpositive displacement

Nonpositive-displacement compressors compress air by using impellers or vanes rotating at high speed. This movement increases both air velocity and pressure. In the past, these designs were often called *blowers* because of their relatively low pressure output. Today, they are typically referred to as *dynamic compressors*.

Figure 15-8. The reciprocating compressor design is a common example of a positive-displacement compressor.

Both centrifugal- and axial-flow dynamic compressors are considered nonpositive-displacement units.

15.3.2 Fundamental Pumping Motion

The *pumping motion* classification relates to the basic concepts used to increase the air pressure and move air through the compressor. Compressors are rated either as reciprocating or rotary under this classification.

In the reciprocating design, a cylinder fitted with a piston produces a variable-volume chamber. A *reciprocating compressor* produces an up-and-down movement of the piston to bring in atmospheric air, compress the air, and then move it to the receiver of the compressed-air unit. The process follows the basic compressor operation described earlier. However, a number of refinements are used to increase compressor efficiency, including multiple cylinders and single- or double-acting piston designs.

In the rotary design, a rotating component inside of the compressor produces the variation needed to compress the air. One of a number of different variations is used in a *rotary compressor* to bring in the atmospheric air, complete compression, and then move the air to the system receiver. In each case, the process is continuous, rather than the stop-and-go performance of a reciprocating compressor. Sliding-vane, screw, and rotary-liquid-piston compressors are examples of rotary, positive-displacement designs. In each of these designs, a moving chamber that varies in size provides a more continuous compression than a single-stage reciprocating unit.

Nonpositive-displacement, dynamic compressors also operate with a continuous, rotary motion. The high-speed rotation of both centrifugal and axial compressors causes airflow through the units. The cross-sectional area of the discharge of these compressors is much smaller than the intake, resulting in an increased pressure. These compressors generally use multiple stages that produce the desired pressure increase.

15.4 Compressor Design and Operation

System designers have a variety of air compressors to consider when developing a new pneumatic system. The compressor should be selected only after a careful analysis matches it to the performance needs of the system. Factors to consider include:

- Compressor capacity.
- Expected length of service.
- Environmental conditions.
- Initial unit and installation costs.
- Ongoing operating costs.

This section examines the basic designs and features of both positive- and nonpositive-displacement compressors commonly found in pneumatic installations.

15.4.1 Compressor Types

The basic principles involved in the operation of compressors are similar in the fact atmospheric air must be drawn into the compressor and then reduced in volume to increase the pressure to a desired level. However, there are some differences between designs. The following sections describe the various compressor designs that may be encountered in the field.

Reciprocating-piston compressors

Reciprocating-piston compressors, or simply reciprocating compressors, are very common in pneumatic systems. They are available as both single- and double-acting units. At the low end of performance, they may be units designed for the home user. These small units usually have a single cylinder with a displacement that may be as low as 1 cubic foot per minute. System pressure may range from 40–120 psi. At the high end of performance, multiple-cylinder industrial units with prime movers of several thousand horsepower produce extremely high operating pressures. See **Figure 15-9.**

Single-Acting Compressors. *Single-acting compressors* are, from a technical standpoint, the simplest of the compressors. They use cylinders, pistons, connecting

Atlas Copco

Figure 15-9. Reciprocating compressors in large, industrial, central air systems may have multiple cylinders and deliver large volumes of air at very high pressures.

rods, and crankshafts of the same type found in internal combustion engines. See **Figure 15-10**. Air is compressed in the cylinder chamber, which is on the top of the piston. The piston bottom is exposed to the crankcase, which serves to lubricate the crankshaft bearings and other bearing surfaces. Most units use pressure-operated reed valves to control airflow. Some models use cam-operated valves synchronized with the position of the piston and crankshaft.

During compressor operation, atmospheric air enters the cylinder through the inlet valve. This results from the reduced pressure in the cylinder produced by the increasing volume of the cylinder chamber as the piston moves downward. When the piston moves upward, the chamber volume is reduced. This increases the air pressure. When the pressure becomes higher than the air pressure in the compressor discharge line, the outlet valve opens. This allows the high-pressure air to move toward the system receiver. With this design, there is intake of atmospheric air each time the piston moves down and a discharge of higher-pressure air each time the piston moves upward.

Single-acting compressors are available as single- or multiple-cylinder units. Multiple-cylinder units may have from two to eight cylinders in an inline, opposed, V, W, or semiradial arrangement. The W arrangement is typically used with three-cylinder compressors. The semiradial arrangement has a single crankshaft throw with a master connecting rod.

Both air- and water-based cooling systems are used to remove the heat of compression from single-acting compressors. Air-cooled units have cooling fins on the outside of the cylinder to transfer heat to the atmosphere. The cooling air is moved across the fins by a fan that is often an integral part of the compressor

flywheel. The flywheel helps dampen the natural irregular power needs of reciprocating machines. Large units that operate in dirty or extremely hot environments may require a water cooling system. This involves a cooling jacket around the compressor cylinders and an external radiator to eliminate the excessive heat. Water circulated through the cooling jacket picks up heat and then releases it to the atmosphere via the radiator.

Double-Acting Compressors. Double-acting, reciprocating compressors are normally considered heavy-duty units designed for continuous service. *Double-acting compressors* have separate cylinder compression chambers at each end of a piston. **Figure 15-11** shows the internal structure of a single-cylinder, double-acting unit. The piston is powered by a fixed piston rod that extends through a stuffing box located between the piston and a crosshead. The end of the piston rod outside of *Cylinder compression chamber 2* is attached to the crosshead that slides in a cylinder bore with walls parallel to the walls of the compression chambers. The crosshead is attached to the compressor crankshaft with a connecting rod. This mechanical arrangement allows the energy of the prime mover to be transferred through the crankshaft and connecting rod to the crosshead. The reciprocating motion of the crosshead allows the piston to accurately travel within the cylinder, while the stuffing box packing around the piston rod provides a positive seal for that compression chamber.

A double-acting compressor produces a maximum volume of compressed air. During each crankshaft revolution, the compressor experiences two intake strokes and two compression strokes. For example, as the piston extends during the first 180° of rotation, there is compression in *Cylinder compression chamber 2*

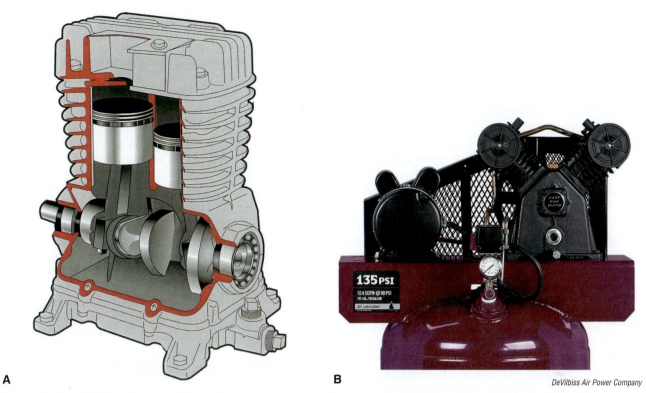

A   B

DeVilbiss Air Power Company

Figure 15-10. Multiple-cylinder, reciprocating compressors may have an inline, V-type, or other cylinder arrangement. A—Inline. B—V-type.

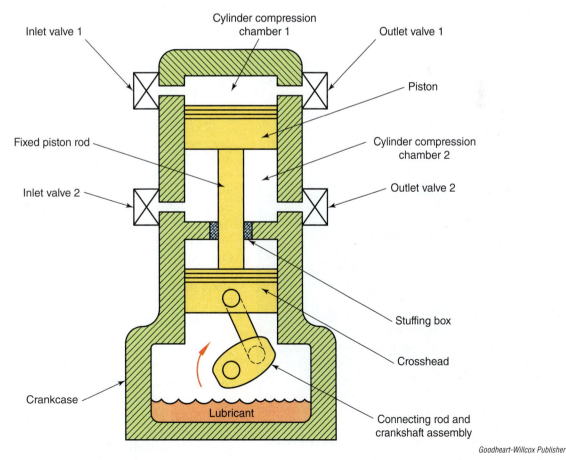

Goodheart-Willcox Publisher

Figure 15-11. The double-acting compressor design allows air intake and air compression to simultaneously occur.

and intake in *Cylinder compression chamber 1*. During the second 180° of rotation, there is intake in *Cylinder compression chamber 2* and compression in *Cylinder compression chamber 1*.

Cooling and lubrication are critical to the performance and service life of these units. They often operate on a continuous basis. However, the exact cooling system used heavily depends on the application. Water cooling is recommended in most situations. Both cylinders and cylinder heads are fitted with water jackets. Splash lubrication, pressure-feed lubrication, or a combination of both is used with these compressors. Pressure-feed lubrication is typically considered the best.

Very large and complex single-cylinder, double-acting compressors have been built. They also exist as multiple-cylinder units. The pistons may be arranged in the same manner as previously described for single-acting compressors.

Rotary sliding-vane compressors

Rotary sliding-vane compressors use rotary motion directly to produce an almost pulsation-free compressing action. They are positive-displacement units. The compressor basically consists of a housing forming a compression chamber with inlet and outlet ports and a slotted rotor. The rotor contains movable vanes that use centrifugal force to maintain a seal with the walls of the compression chamber. See **Figure 15-12**. The rotor and vanes form the compression mechanism of the unit.

The slotted rotor is directly attached to the input shaft of the compressor, which is turned by the system prime mover. The rotor is offset in the compressor housing. This results in a series of varying-size compression chambers between the rotor, vanes, and walls of the housing. Intake and compression occur as the

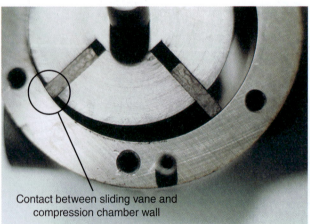

Contact between sliding vane and compression chamber wall

Goodheart-Willcox Publisher

Figure 15-12. Centrifugal force keeps the vanes of the sliding-vane compressor in constant contact with the walls of the compression chamber.

size of these chambers increases and then decreases as the rotor turns. As the chambers are increasing in size, they are exposed to the inlet port. This allows atmospheric air to enter. As the rotor continues to turn, the chambers begin to reduce in size. When the chambers are exposed to the outlet port, the pressurized air is allowed to move into the system.

The vanes of these units are subjected to friction resulting from centrifugal force pressing the vane tip against the compression chamber wall. This pressure and the scraping action against the chamber wall results in considerable vane wear. To reduce this wear and increase sealing, oil is often injected into the compression chamber. Oil separators are placed in the compressor discharge line to remove any liquid oil that remains in the compressed air.

Rotary, sliding-vane compressors are typically water cooled. The cylindrical compression chamber housing and the housing heads contain water jackets. The oil injected to improve sealing and reduce wear also helps remove some of the heat of compression.

Rotary screw compressors

Rotary screw compressors use intermeshing, helical screws to form chambers that linearly move air through the compressor, **Figure 15-13**. The design provides a continuous, positive displacement of air to deliver a nonpulsating flow at all operating speeds. This group of compressor designs has become very popular in larger industrial applications because of:

- Lower initial cost for high-volume outputs.
- Low maintenance.
- Availability of sophisticated control systems.

These compressors may be dry or oil flooded. The dry version uses timing gears to turn the screw elements. These units are constructed with an extremely small clearance between the screw surfaces. This requires little or no lubrication. In the oil-flooded design, the screws are driven by direct screw contact. As a result, the timing gear mechanism is not required. In both versions, when the screws are turned, a volume of atmospheric air is trapped in the open spaces between the screw threads. This trapped air is pushed along the axis of the compressor to the discharge port leading to the system receiver.

The oil-flooded compressor design uses an oil separator to remove the oil from the compressed air. The separated oil is reused after it is cooled and filtered. An oil separator is normally not needed with the dry compressor as the small screw clearance eliminates the need for lubricating oil. Both designs require a cooling system to remove the heat of compression. The oil in the oil-flooded version acts as a coolant to remove some heat, as well as lubricate and seal.

Atlas Copco

Figure 15-13. Rotary screw compressors have intermeshing, helical screws. The screws form sealed chambers during rotation to transport a volume of air from the atmosphere to the receiver and distribution lines.

Dynamic compressors

Dynamic compressors are typically used to compress air or other gases in large applications such as petroleum refineries, chemical plants, and steel mills. In recent years, the pressure capability of the design has increased from under 40 psi to several hundred psi. Also, these designs are most often available as high-flow-output units producing thousands of cubic feet per minute.

Dynamic compressors use the rotary design and may be of centrifugal or axial design. The designs vary considerably in physical appearance, but achieve compression using basically the same theory. Kinetic energy of moving air is converted into pressure by slowing the airspeed with a fixed diffuser or stationary blades. In the centrifugal compressor, the airflow is radial from the shaft of the unit. Compression is primarily due to changes in centrifugal forces. Airflow in the axial compressor is parallel to the shaft. Compression is caused by the action of alternate rows of rotating and fixed blades.

The basic internal structure of a centrifugal-flow compressor is shown in **Figure 15-14**. The compressor consists of an inlet nozzle, impeller, and volute (spiral) collector. The impeller consists of a circular disk fitted with blades that taper from a deep, center intake section to a thin discharge section on the exterior edge. See **Figure 15-15**. The impeller is surrounded by a housing that forms the volute collector. The impeller is normally directly driven by the prime mover. As

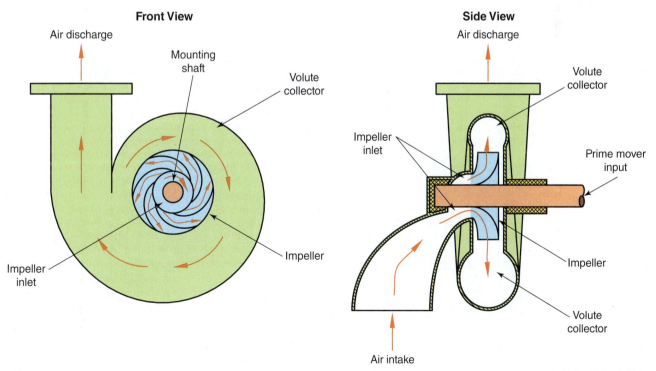

Goodheart-Willcox Publisher

Figure 15-14. The impeller is the only moving part in a centrifugal compressor.

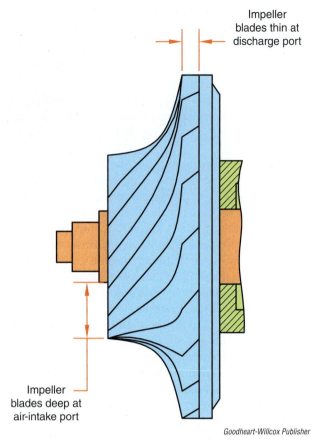

Impeller blades thin at discharge port

Impeller blades deep at air-intake port

Goodheart-Willcox Publisher

Figure 15-15. An impeller turning at high speed moves air through a centrifugal compressor. Energy from the prime mover is converted into kinetic energy in the rapidly moving air.

the impeller rotates at a high speed, centrifugal force moves atmospheric air into the inlet nozzle, through the impeller, and on to the collector. Energy from the prime mover is converted into kinetic energy as the airspeed rapidly increases through the impeller. This kinetic energy is then converted into pressure as the airspeed sharply decreases as air moves into the collector and on to the system distribution lines.

Figure 15-16 shows the internal structure of a multistage, axial-flow compressor. The unit is actually a series of basic compressors functioning together to compress the air. Each basic compressor stage consists of a row of moving rotor blades and a row of fixed stator blades. See **Figure 15-17**. Each of these basic units moves the air parallel to the compressor shaft. The length of the rotor and stator blades decreases from the first-stage compressor to the last stage. This results in a reduced volume of air transferred from each of the basic compressors. See **Figure 15-18**.

In axial compressors, pressure is created in both the rotor and stator sections of the unit. As the prime mover turns the rotor blades, energy is transferred to the air as its speed increases. When the moving air impacts the fixed stator blades, it is slowed. This converts the kinetic energy into pressure. In the second stage, the rotating rotor blades of the second basic compressor again increase the airspeed. This increasing speed transfers additional energy from the prime mover to the moving air. Impact with the fixed stator blades of the second compressor again slows the air.

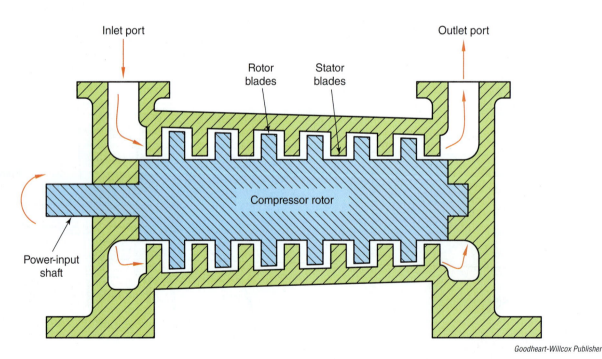

Inlet port

Outlet port

Rotor blades

Stator blades

Compressor rotor

Power-input shaft

Goodheart-Willcox Publisher

Figure 15-16. Most axial flow compressors are really several staged, basic compressor units working together to increase air pressure.

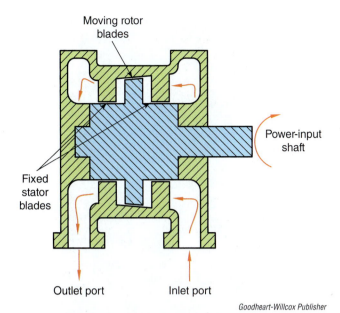

Figure 15-17. Representation of an axial flow, dynamic compressor with a single set of rotor blades. The rotor moves air through the stator blades to produce an increase in air pressure.

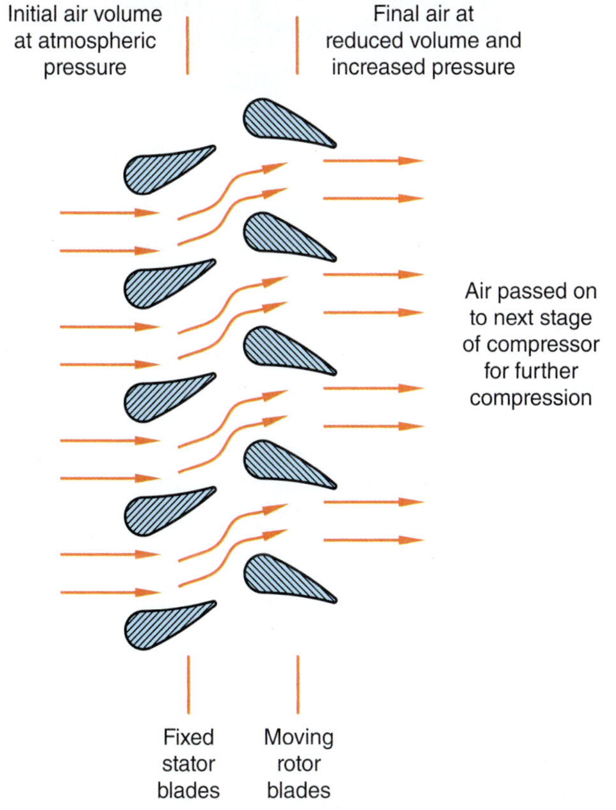

This converts additional kinetic energy into pressure. This process repeats through each of the basic compressor units until the last unit discharges the air into the system distribution lines at the final pressure.

Dynamic compressors do not require internal lubrication for either sealing or friction control. Compressor shaft bearings are outside of the compression chamber in these units. The bearings can be externally lubricated without contributing to air contamination. Therefore, the units can provide large volumes of oil-free air without the use of an oil separator.

Lobe-type compressors

Most models of *lobe-type compressors* only develop air pressures ranging from 10–20 psi. Therefore, they are often referred to as *blowers*. The compressors have two impellers containing two or three lobes each. See **Figure 15-19**. The impellers operate in an elongated chamber located in the compressor body. Inlet and discharge ports are also built into this chamber. The impellers rotate in opposite directions and are synchronized by timing gears. A constant, minute clearance is maintained between the surfaces of the impeller lobes. There is also a minute clearance between the tips of the impeller lobes and the surface of the elongated chamber.

In this design, there is no reduction of air volume in the compressor as the impellers turn. The air is simply swept from the inlet port to the discharge port. Compression is achieved when the transported volume of air is added to the volume of air already trapped in the receiver or distribution lines. Atmospheric air is trapped on both the right and left sides of the inlet port as the impellers turn. With further rotation of the impellers, both of these air volumes are added to

Figure 15-18. The speed of the air increases as it passes through the rows of fixed-position stator blades and rotating rotor blades. The increased airspeed increases the kinetic energy in the air, which results in increased air pressure.

Atlas Copco

Figure 15-19. The internal structure of lobe-type compressors allows a volume of atmospheric air to be swept through the unit and added to the air already confined in the receiver and distribution lines.

the pressurized system air when they are transported to the discharge port. While discharge is occurring, additional atmospheric air enters the inlet port. This fills the next chambers that will sweep air through the compressor.

The minute clearances involved between the impeller lobes and the walls allow air leakage between the discharge and intake port areas. This leakage is called *slippage* and reduces the efficiency of these compressors. Because of the lost operating efficiency caused by this slippage, compressors of this type need to be operated at speeds close to, but not exceeding, the maximum recommended by the manufacturer. Some design variations use oil or other liquid injection to improve sealing and help cool the air. When this is done, a separator is required to remove the liquids to ensure the delivery of clean, dry air to the system.

15.4.2 Compressor Staging

It is more effective to compress air to high pressures in multiple stages, rather than in a single step. This is related to a number of factors:

- Discharge temperature of the air.
- Compression-ratio limitations of the compressor.
- Balancing of loads on compressor bearings.
- Power requirements of the prime mover needed for the installation.

Staging involves a number of compressor units to compress air to the desired pressure by increasing pressure in small increments.

Figure 15-20 illustrates a two-stage, reciprocating-piston compressor. The compressor contains two cylinder and piston assemblies connected to a crankshaft containing one throw (journal) for each cylinder. Each cylinder contains inlet and outlet valves. The inlet of the large-displacement, first-stage cylinder is open to the atmosphere. The outlet of that cylinder is connected to the inlet of the small-displacement, second-stage cylinder by a transfer line. This line transports air between the compressor stages, but also acts as a cooler to remove heat from the air. The outlet of the second-stage cylinder is connected to the system receiver and distribution lines. The throws on the compressor crankshaft not only move the pistons, but synchronize their movement. This allows the pistons of the two stages to smoothly move air through the unit.

During operation, when the first-stage piston is moving downward, atmospheric pressure pushes air into the cylinder through the open inlet valve. The inlet valve closes as the piston begins to move upward. This upward piston movement reduces the volume of the trapped air, thus increasing its pressure. The outlet valve opens as the piston moves upward to allow the pressurized air to move into the transfer line.

DeVilbiss Air Power Company

Figure 15-20. This two-stage, reciprocating compressor contains two cylinders connected in series. Other staged designs use individual compressors arranged in series to produce similar results.

The downward movement of the second-stage piston is timed to occur as the first-stage piston is moving upward. The pressurized air from the first-stage cylinder fills the second-stage cylinder through the open second-stage inlet valve. The second-stage piston then moves upward, reducing cylinder volume. This further increases the air pressure. The high-pressure, second-stage air is discharged to the receiver and system distribution lines through the outlet valve in the second-stage cylinder.

The heat of compression results in discharge air temperature higher than the ambient atmospheric air that enters the compressor inlet. However, multistage compressor designs produce compressed air that is discharged at a temperature below that produced by a comparable-size, single-stage compressor unit. This results from the removal of heat in the transfer line as the first-stage compressed air moves to the inlet of the second-stage compressor. As a result, multistage compressors do not require as much energy to operate as single-stage units.

Staging can be used with most of the compressor designs discussed in this chapter. Some of the designs easily allow several basic, single-acting compressors to be combined into a single body. In other designs, the single-stage units are externally combined in series to achieve the desired compression.

15.4.3 Compressor-Capacity Control

The volume of compressed air produced by a compressor must closely match the air consumption of the pneumatic system. The compressor section of a system running at a constant speed produces a relatively consistent volume of air. However, the air demanded by the many system workstations fluctuates, depending on the performed tasks. Therefore, a means must be provided to match these air quantities to produce an efficient system. The matching of compressor-air output and system-air demand is typically accomplished by providing some type of *compressor-capacity control*.

A number of methods have been used in pneumatic systems for compressor-capacity control. These range from a simple bypass valve that bleeds excess system compressed air to the atmosphere to complex prime mover speed control on dynamic compressors that varies compressor output to match system demand. The continued development of computer-based controls has also influenced the type of systems used on compressors. The following sections briefly discuss several of the most common capacity controls.

Bypass control

Bypass compressor-capacity control is a very simple system that does not actually control the capacity of the system compressor. With this system, air is continuously delivered to the system receiver and distribution lines at the maximum flow rate of the compressor. The system maintains a constant pressure by allowing air to exhaust through a relief-type valve. This prevents the development of pressure higher than that desired for the operation of the system.

When demand is below the delivery rate of the compressor, system pressure increases to the setting of the bypass valve. The valve then opens to bleed pressurized air to the atmosphere. The system operating pressure is held at the bypass pressure setting.

This type of control is not considered desirable. The prime mover must continuously operate under full system pressure. This results in a poor overall operating efficiency.

Start-stop control

An automatic start-stop compressor-capacity control method is commonly used with smaller, reciprocating compressors having electric motors. This system uses a pressure switch that controls operation of the prime mover. See **Figure 15-21**. The motor drives the compressor until a preset maximum pressure is reached in the receiver and distribution lines. At that point, the pressure switch opens and the compressor unit stops. See **Figure 15-22**. As air is used, the pressure drops until the switch closes at a preset minimum pressure and the motor restarts. The compressor again supplies air to the pneumatic system until the maximum pressure setting is reached. Then, the cycle repeats. The setting of the minimum pressure switch must be higher than the desired minimum operating pressure of the system actuators.

The start-stop system operates well for pneumatic systems with intermittent air consumption. However, in systems with continuous air use, the constant starting and stopping can lead to heat buildup. This can reduce the service life of the motor.

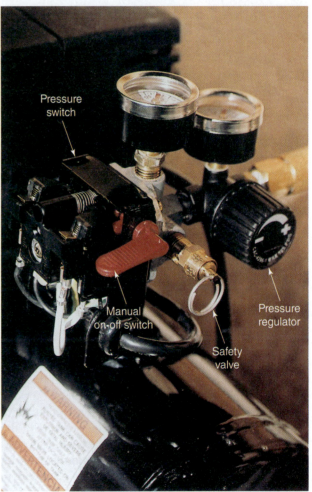

Goodheart-Willcox Publisher

Figure 15-21. The start-stop compressor-capacity control method makes use of a pressure switch to control the operation of the prime mover.

A

B

Goodheart-Willcox Publisher

Figure 15-22. A—With start-stop compressor-capacity control, the start pressure is slightly above the workstation operating pressure. B—The stop pressure setting is the selected maximum line pressure.

Inlet valve unloading control

Another common method of compressor-capacity control used with reciprocating compressors is inlet valve unloading. This method operates the system prime mover at a constant speed while controlling compressor output by manipulation of the inlet valve. When compressed air is needed, the inlet valve operates normally. This allows air to be drawn into the compressor, compressed, and discharged to the receiver and distribution lines. When the maximum pressure of the pneumatic system is reached, the inlet valve is held open. This allows all of the air drawn into the cylinder on the intake stroke to be returned to the atmosphere during the compression stroke. The mechanism that unloads the inlet valve is operated by sensing air pressure in the receiver.

Other capacity control methods

Varying compressor speed is another method that can be used to control compressor output. This method is used with both reciprocating and dynamic compressor designs. It is primarily used on larger, industrial installations. In these systems, the speed of the prime mover is governed by sensors monitoring pressure in the air storage and distribution lines. A speed-control unit converts pressure variations into signals used to modify prime mover speed. The result is a compressor that runs at varying speeds. The compressor only produces the amount of air needed to maintain the pressure selected for the system.

Capacity control can also be achieved by varying the size of the compressor inlet. The system restricts the volume of air that can enter the compressor, which is operated at a constant speed. Compressed air output varies directly with the size of the compressor inlet opening. This method is primarily used with dynamic compressors.

15.5 Selecting a Compressor Package

The variety of compressor designs, load variations in a pneumatic system, the variety of auxiliary equipment available, and the demands of future growth of the system make the selection of a pneumatic system compressor difficult. All of these factors need to be considered when designing an air compressor package. When developing specifications for and installing a pneumatic compressor package, take the following steps.

1. Establish needed capacity for the pneumatic system.

   This step requires a listing of all tools and equipment to be operated in the system. These data should include air consumption, pressure requirements, and percentage of operating time per unit.

2. Estimate increases in system demand.

   Analyze production data to identify trends that can be used to estimate future increases in system demand.

3. Select the compressor type.

   This selection is based on system demand, considering the required volume and pressure of the output air and the efficiency of the compressor. Reciprocating compressors are generally considered most effective for typical shop pressure and flow requirements, although screw compressors are becoming more common.

4. Select the type of system prime mover.

   Factors that need to be considered are power requirements, mobility of the unit, and availability of an energy source for the prime mover. Electric motors, reciprocating internal combustion engines, and turbines are all used to drive compressors. Electric motors are very commonly used to drive compressors. Motors are found on small portable units and very large stationary units. Internal combustion engines are used as the power source where electricity is not readily available. Gas and steam turbines are primarily used in large, stationary installations.

5. Determine the regulation method to control the volume of air delivery.

   Compressor-capacity control methods provide this regulation. Smaller, electric motor-driven units often use start-stop control. Portable units driven by an internal combustion engine use inlet valve unloading. The compressor and prime mover types influence control selection.

6. Select auxiliary components of the compressor package.

   The compressor package typically includes a number of components in addition to the prime mover and compressor. These generally include a receiver for air storage, safety valve, and pressure gauges. Coolers, separators, driers, and a distribution manifold may also be included. Selection factors relating to these components are determined by the application of the overall pneumatic system. Details concerning the selection of these additional components are discussed in Chapter 16, *Conditioning and Distribution of Compressed Air*.

7. Identify compressor package instrumentation.

   The instruments on a compressor package may be basic, electrical devices; complex electronic controls; or computer controls. Instrumentation for a simple, bypass valve capacity control is a mechanical, relief-type valve and an on-off switch. However, large-capacity, dynamic-compressor packages may use computer control to maintain consistent flow rates at a constant pressure. The design, operation, and maintenance of these systems are beyond the scope of this text.

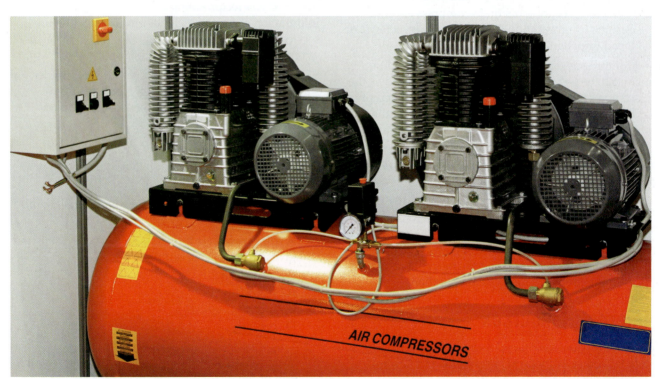

Baloncici/Shutterstock

Dual compressors are sometimes used in installations where high volume air consumption is limited. The end result may be reduced operating and service costs.

Summary

- Compressed-air units produce pressurized air for the pneumatic system.
- A compressed-air unit consists of a prime mover, a compressor, and components that condition and store the pressurized air used by the system workstations.
- Compressed-air units can be portable or permanently mounted to provide air to fixed workstations from a central supply.
- The basic operation of any compressor design used in a pneumatic system includes three phases: air intake, air compression, and air discharge.
- Compressors are classified by displacement (positive-displacement or nonpositive-displacement designs) and fundamental pumping motion (rotary or reciprocating designs).
- Positive-displacement compressor designs mechanically reduce a compression chamber in size to compress air. Nonpositive-displacement compressors use impellers or vanes rotating at high speed to increase air velocity and pressure.
- Reciprocating-piston compressors use a cylinder and a reciprocating piston to compress air. Rotary designs use continuously rotating vanes, screws, or lobed impellers to move and compress the air.
- Compressor-capacity control systems are used to match the volume of compressed air produced by the compressor to the volume of compressed air consumed by the working pneumatic system. The closer the compressor air output matches system consumption, the more cost effective the system operation.
- Compressor-capacity control systems include bypass, start-stop, inlet valve unloading, speed, and inlet size controls.
- Determining the required output of a compressor can be accomplished by identifying the actuators used, the volume of compressed air needed for the operation of each item, and the percentage of time each functions during system operation.

Internet Resources

The following are some useful resources available on the Internet. Enter a company or organization name into a search engine to access its website. Explore the various areas of the sites to discover useful fluid power resources.

Atlas Copco. Enter "screw compressors" in the product search box to review product information on the various rotary screw compressors manufactured by the company. Includes stationary and portable units.

Dresser-Rand. Describes a centrifugal compressor designed for a process application in the oil, gas, or chemical industries.

FS-Elliott Co., LLC. Describes a line of centrifugal compressors. Illustrated with a general description of the operation of a two-stage unit including intercoolers.

Hydraulics & Pneumatics Magazine. Review two articles: *Air Compressors–Part 1* and *Air Compressors–Part 2*. These articles cover the styles of compressors available in the pneumatic field, emphasis on rotary compressors. Also includes information on power and efficiency, lubricated or lubrication-free compressors, cooling methods, and capacity control.

Ingersoll Rand Company. Provides information on reciprocating air compressors manufactured by the company. The brochure linked on the page refers to single- and two-stage air compressors.

Jenny Products, Inc. Guide to selecting an air compressor for a small business operation or home shop. Covers factors such as the need for staging, type of prime mover, and specific questions that should be answered before purchasing a compressor.

Saylor-Beall Manufacturing Company. Manufacturer that produces reciprocating and screw compressor units. Information contained includes product brochures and service manuals with illustrated, informative materials covering general questions and answers.

SENCO Brands, Inc. Product manual for a single-stage, reciprocating compressor. Applications for these compressors range from the home shop to industrial situations.

Chapter Review

Answer the following questions using information in this chapter.

1. A compressed air unit classified as a portable unit gives a(n) _____ air supply.
   A. industrial
   B. prime mover
   C. receiver
   D. central

2. *True or False?* Double-acting reciprocating compressors have compression chambers at each end of the pistons.

3. *True or False?* The crankshaft is the only moving part in a centrifugal compressor.

4. *True or False?* Displacement of a compressor is the volume of air displaced per revolution of the unit.

5. *True or False?* Single-acting compressors use cylinders, pistons, connecting rods, and crankshafts similar to the type found in internal combustion engines.

6. An automatic compressor capacity control method is commonly used with a smaller compressor using _____ as the prime mover.
   A. steam power
   B. electric motors
   C. pneumatics
   D. gasoline engines

7. *True or False?* The housing around the centrifugal compressor forms a volute collector that directs the airflow through the compressor.

8. Lobe-type compressors typically only develop air pressures ranging from 10 to _____ psi.
   A. 20
   B. 35
   C. 60
   D. 100

9. Typically, today's nonpositive-displacement compressors compress air using _____.
   A. impellers
   B. cylinders
   C. chambers
   D. pistons

10. *True or False?* An impeller turning at relatively low speed moves air through a centrifugal air compressor.

11. In a two-stage, reciprocating-piston compressor, the outlet port of the first compression chamber is connected to the _____ port of a second compression chamber.

12. Using a number of compressor units to increase pressure in small increments is called _____.
    A. start-stop
    B. capacity control
    C. bypass
    D. staging

13. Compressor-air output and system-air demand are matched by using some type of _____ control system.

14. A common method of compressor-capacity control used with reciprocating compressors is inlet valve _____.

15. *True or False?* When selecting a pneumatic system compressor, only one or two factors need to be considered.

Apply and Analyze

1. Which type of compressor would you choose for each of the situations listed below? Why? Identify specific aspects of compressor design or operation that allow the compressor to meet the stated needs.
   A. A steady, continuous supply of air.
   B. Low volumes of high-pressure air.
   C. Very large volumes of oil-free pressurized air.
   D. Minimal vibration of the compressor.
   E. Large volumes of air, but space for installation is limited.

2. For each of the following compressor types, where does energy loss occur during operation?
   A. Reciprocating-piston compressor.
   B. Rotary sliding-vane compressor.
   C. Centrifugal compressor.

3. You are assisting a company that is installing new compressors and updating their pneumatic system. The company is interested in minimizing not only initial investment costs, but also costs of operation.
   A. What specific advice would you give them in their choice of compressors?
   B. What information would you need to determine the best method(s) of compressor-capacity control for improving the efficiency of the system during operation? Explain your answer.

Research and Development

1. Lubrication is critical for the performance and service life of many air compressors. Investigate the different methods of lubrication used for the main types of positive-displacement compressor. Prepare a chart that illustrates the benefits and drawbacks of each lubrication method.

2. Using air compressor manufacturers' product listings found online, find and compare two compressors which could be used for similar applications, but which have different basic designs, such as rotary and reciprocating. Develop a fact sheet to be used to help a potential customer choose between the two compressors.

3. Investigate the historical development of air compressors. Produce a time line that shows design changes as well as the use of new materials or machining methods to improve performance.

4. Research an application that uses dynamic compressors to pressurize air. Write a report that describes the application, discusses the design and capacity of the compressor, and explains why this type of compressor is used.

The energy-transmission medium for pneumatic systems, air, is readily available as atmospheric air that blankets earth. This huge volume of air is available for our use without apparent cost. This is often reinforced by the old saying, "free as the air we breathe." While air may be free for the taking, the statement is not accurate when referring to the compressed air used in a pneumatic system. Pneumatic-system air must be conditioned, compressed, stored, and then distributed to a workstation. Each of these processes requires equipment and energy input, resulting in operating costs that can be substantial. This chapter discusses each of the phases required to convert ambient, atmospheric air into the conditioned air delivered to an actuator at a workstation.

Learning Objectives

After completing this chapter, you will be able to:

- Compare the various methods used to remove dirt from ambient air entering the compressor and in the final filtering of air distributed to the workstations.
- Identify the benefits of controlling the temperature of compressed air and the methods used to remove excess heat.
- Describe the problems caused by excess moisture in the compressed air of a pneumatic system.
- Identify the source of moisture in a pneumatic system and various methods used to remove liquid water from a system.
- Describe the functions of the pneumatic system receiver and identify construction features of typical designs.

- Explain the factors that must be considered when establishing the size and location of a receiver for a pneumatic system.
- Compare the design and operation of pneumatic pressure-regulator valves.
- Compare the various types of pipe, tubing, and hose used in pneumatic systems.
- Identify and explain the factors that should be used when selecting a conductor for use in a pneumatic system.
- Explain the design and construction of the various air-distribution systems.
- Describe the purpose, construction, and operation of the components used for the final preparation of compressed air at a pneumatic system workstation.

Key Terms

aftercooler
air-distribution system
air-drying equipment
auxiliary-air receiver
balanced-poppet valve
  regulator
centralized grid
decentralized grid
desiccant
diaphragm-chamber
  regulator

direct-operated regulator
draincock
drop line
dry filter
flexible conductor
flexible hose
FRL unit
humidity
intake-line filter
intercooler
load factor

loop system
lubricator
moisture separator
oil-bath filter
oil-wetted filter
pilot-operated regulator
pressure regulator
relieving-type regulator
rigid conductor
time factor
water trap

16.1 Conditioning and Storing Pneumatic System Air

The quality of atmospheric air greatly varies from one location to another. Also, air quality can rapidly change at a given location based on weather conditions and surrounding work activities, **Figure 16-1**. However, for maximum operating efficiency and service life, it is essential to operate a pneumatic system using compressed air that is consistently clean, free from moisture, and at a relatively uniform temperature. Producing compressed air with these qualities on a continuous basis can be done cost effectively with a correctly designed and maintained compressor installation and air-distribution system. See **Figure 16-2**.

The first contributing factor in providing quality, conditioned air to a pneumatic system is the quality of the atmospheric air at the system air intake. Quality can be enhanced by positioning the compressor intake

DeVilbiss Air Power Company

Figure 16-1. Atmospheric air varies by location, weather condition, and worksite activities. The pneumatic system must be provided with clean, dry air.

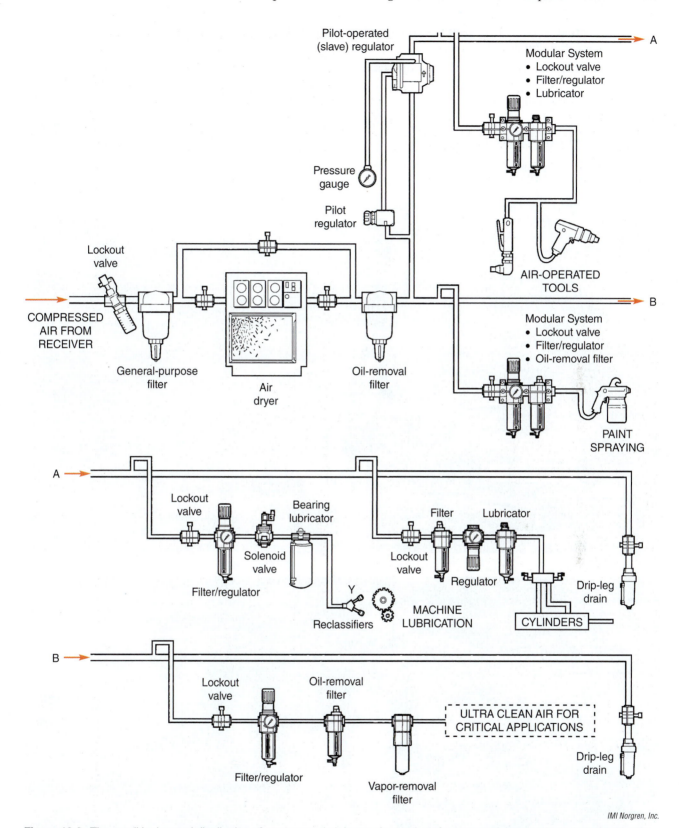

Pilot-operated (slave) regulator

Modular System
• Lockout valve
• Filter/regulator
• Lubricator

A

Pressure gauge

Pilot regulator

AIR-OPERATED TOOLS

Lockout valve

COMPRESSED AIR FROM RECEIVER

General-purpose filter

Air dryer

Oil-removal filter

B

Modular System
• Lockout valve
• Filter/regulator
• Oil-removal filter

PAINT SPRAYING

A

Lockout valve

Filter/regulator

Solenoid valve

Bearing lubricator

Reclassifiers

MACHINE LUBRICATION

Filter

Lockout valve

Regulator

Lubricator

CYLINDERS

Drip-leg drain

B

Lockout valve

Filter/regulator

Oil-removal filter

Vapor-removal filter

ULTRA CLEAN AIR FOR CRITICAL APPLICATIONS

Drip-leg drain

IMI Norgren, Inc.

Figure 16-2. The conditioning and distribution of compressed air is a major portion of a pneumatic system in an industrial facility.

to provide air that is as clean, dry, and as cool as possible for the location. Intakes should be:

- Located away from dust-producing conditions.
- Shielded from the weather to eliminate the entrance of rainwater.
- Protected from direct exposure to sunlight. Direct sunlight can significantly increase the intake-air temperature.

16.1.1 Dirt Removal

Atmospheric air contains dirt, even in the best of conditions. However, in situations such as road or building construction, the level of airborne dirt may be extremely high, **Figure 16-3**. Preventing dirt from entering the pneumatic system is essential to the service life and performance of the total system. Dirt allowed to enter a system produces unnecessary component wear. It also leads to the

A *Atlas Copco*

B *DeVilbiss Air Power Company*

Figure 16-3. Dust and moisture in atmospheric air must be removed to protect pneumatic components from excessive wear (A) and system-powered processes from contamination (B).

Figure 16-4. An air filter in the compressor intake line is a first step in the production of clean compressed air.

accumulation of dirt in lines and sensing orifices of control valves. These conditions can quickly reduce the efficiency of compressors and workstation circuits.

The *intake-line filter* removes dust and dirt from atmospheric air before the air is allowed into the compressor inlet, **Figure 16-4**. The type of filter depends on atmospheric conditions, the size and type of compressor, and the final use of the compressed air. Many different filter designs are used with pneumatic systems. However, filters can generally be placed in one of three categories:

- Dry.
- Oil wetted.
- Oil bath.

Dry filters are available in a large number of configurations using paper, plastic, cloth, ceramic, or metal elements. The filter element remains dry as it traps dirt.

Oil-wetted filters use metal-packed, woven wire or other material held in a metal frame. A light oil coating is used on this material to trap the dust and dirt entering the filter.

Oil-bath filters have a filter medium that is constantly bathed in oil. The oil captures the dirt and transports it to a sump where it settles and is removed during periodic service. This style of filter was popular in older compressor systems and is still used in some applications today.

Filters require regular servicing, regardless of their design. A filter that is clogged with dirt reduces system efficiency by limiting the volume of air intake. Limiting intake volume reduces the output volume of the compressor and can cause overheating of the unit. A regular schedule should be developed for service of the intake filter based on the environment surrounding the pneumatic system and the duration of compressor operation.

16.1.2 Controlling Air Temperature

The temperature of both the intake air and compressed air are critical considerations when working with a pneumatic system. According to the general gas law, any change in the temperature of a gas is directly reflected in gas pressure or volume change. Therefore, maintaining a consistent air temperature throughout a system would be a real advantage. Ideally, the temperature of intake air should be low, the temperature of the air should not increase during compression, and the compressed air distributed to workstations should be near the air temperature where the workstation is located, **Figure 16-5**.

Air temperature relates to both the energy consumed by the prime mover and the water vapor brought into the system with the atmospheric air. Compressing air produces heat. This increases the temperature and pressure of the air in the confined space of the pneumatic system. This warmed, compressed air retains the water vapor present in the atmospheric air brought into the compressor.

Two conditions result when this warm, humid, high-pressure air cools as it moves through the system distribution lines. The first of these conditions is a reduction in the pressure of the air as it cools. The amount of pressure reduction is based on the general gas law. The second condition relates to the dew point of the compressed air. As the temperature of the compressed air drops to the dew point, liquid water forms in system components and distribution lines. This liquid water can damage internal components and cause operating problems when a mixture of air and water is discharged at the distribution system outlets.

	Pneumatic System Function		
	Air Intake	**Air Compression**	**Air Distribution and Use**
System Location	System atmospheric air intake	Compressor station	Distribution lines and workstation
Ideal Air Temperature	Lowest ambient temperature available	Minimal air temperature increase	Near-ambient air temperature

Figure 16-5. Ideally, air temperature should change little as it is processed from the system inlet to use at the workstation.

Air can be cooled either before, during, or after compression. In industrial pneumatic systems, control of the air temperature before compression is typically restricted to locating the compressor intake in the coolest atmospheric air available. Cooling of the air during compression involves the general cooling system used for the protection of the compressor. In multiple-stage compressor designs, a separate *intercooler* unit is used to cool the air between stages. An *aftercooler* is a separate unit used to cool the air after compression is complete, **Figure 16-6**.

In industrial pneumatic systems, intercoolers and aftercoolers are typically surface coolers using air or water as the cooling medium. Refrigeration equipment may also be used in some situations. When air is used for cooling, the system may be as simple as external fins on a tube or pipe exposed to air circulated by the compressor cooling fan. Double-pipe, shell-and-tube coolers are used in more complex air systems or in water-cooled applications.

16.1.3 Controlling Moisture in Compressed Air

Atmospheric air contains moisture that is essential to the operation of our environment. This moisture, which is in vapor form, is commonly referred to as *humidity*. Generally, humidity goes unnoticed until the moisture content becomes excessive, which causes us to become physically uncomfortable. However, the water vapor in air is an important and constant factor that must be considered when compressing air for use in a pneumatic system. Water in vapor form does no harm in a pneumatic system. The problem develops when this vapor condenses into liquid water. When working with a pneumatic system, it is important to understand the terminology and relationships involving humidity, vapor condensation, and compressed air.

Water vapor and compressed air

The amount of water vapor that can be retained in air depends on the temperature and volume of the air. When air at a given temperature is holding the maximum amount of water vapor, it has reached a level known as the saturation point. Reducing the temperature of this saturated air results in the formation of liquid water. The temperature at which this water vapor begins to condense to form liquid water is known as dew point.

The importance of this principle to the operation of a pneumatic system can be illustrated by using an example of free air at 70°F. At this temperature, air can hold 1.14 pounds of water vapor per 1,000 cubic feet at the saturation point. This is equal to approximately 31.5 cubic inches of liquid water, which is slightly more than 1 pint. Compressing this saturated air to system operating pressures considerably reduces the volume of the air. However, the ability of air to hold water vapor is dependent on the temperature and pressure of the air. Increasing temperature increases the ability of

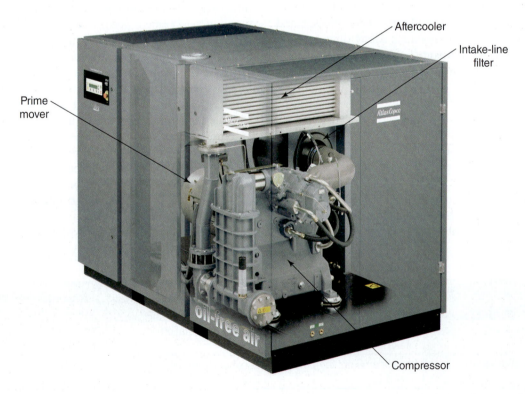

Atlas Copco

Figure 16-6. An air-cooled aftercooler is often used to reduce the temperature of the compressed air as it moves from the compressor outlet to the receiver.

air to hold vapor. Increasing pressure raises the dew point temperature of the air, essentially decreasing the ability to hold water vapor. During the compression process in a pneumatic system, as air pressure increases, the air temperature may also increase depending on how much heat is retained in the system. The change in water vapor content with compression can be difficult to predict, but typically, as the air cools as it moves through the system, the saturation point or dew point will be reached and liquid water will form. The total volume of liquid water that forms depends on the volume of air that moves through the system, the initial water vapor content of the air, and the changes in pressure and temperature that occur during system operation.

The figures in this example are relatively small. If these figures are multiplied by the large volume of air that passes through even a small pneumatic system, you can see the significant quantity of liquid water that can be produced in a system. It must be kept in mind that whenever the temperature of compressed air drops below the dew point, liquid water is produced within the system components and lines.

Water removal

Water in a compressed-air system can cause a number of serious problems:

- Washing away of lubricants in system tools.
- Increased component wear and maintenance.
- Inconsistent operation of system control valves and actuators.
- Freezing of water that accumulates in a distribution line.
- Lowering of the finished quality of processes directly using the air (such as painting).

These problems warrant the lowering of the water vapor level in the system air as early as possible during air compression and distribution.

Reducing moisture problems should begin with the location of the atmospheric air intake for the compressor. A protected location that provides the driest-available air to the compressor helps reduce problems even before the air is compressed. The next step is an intercooler located between the stages of a multistage compressor. This component cools the compressed air to increase compressor efficiency, but, on larger units, may also be used to remove some moisture. Typically, the aftercooler is a more critical component for moisture control. This unit lowers the air temperature below the dew point. Part of the water vapor condenses in the unit where it can be removed. A *moisture separator* is often teamed with the aftercooler to increase the effectiveness of the moisture removal process. The liquid water removed by these separators is often drained by automatic traps. See **Figure 16-7**.

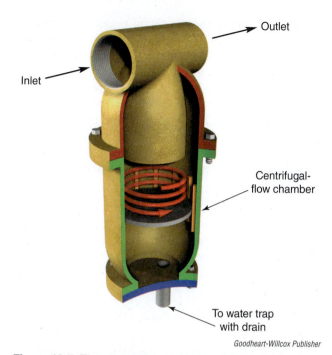

Goodheart-Willcox Publisher

Figure 16-7. The air entering a moisture separator forms a centrifugal flow that forces liquid and dirt to the outer wall of the unit. These materials collect in a trap for easy removal.

Other locations for moisture removal in a compressed-air system are the receiver and the system distribution lines. Both of these areas allow further air cooling. Liquid water is formed in these areas whenever the temperature of the compressed air drops below the dew point. Those aspects of moisture control are discussed in those specific sections of this chapter.

Moisture can also be removed from the compressed air using *air-drying equipment* especially designed to remove water vapor. These units are typically located between the receiver and the main distribution line or at specific locations in a system where dry air is essential. These driers use one of three methods:

- Chemical *desiccant* to absorb moisture.
- Refrigeration unit to condense water vapor.
- Membranes made from a material that allows water vapor to pass directly from the compressed air to the atmosphere.

A variety of designs and capacities are available in these units.

16.1.4 Compressed Air Storage

Compressed air in a pneumatic system is stored in a specialized tank called a receiver. This tank not only stores compressed air for periods of heavy demand that exceed compressor capacity, but also serves a number of other functions. One of these functions is dampening air pulsations resulting from compressor operation. These pulsations result from the varying air output inherent in the operating cycle of reciprocating compressors and some other designs. The receiver also acts to remove additional water vapor from the air as the air temperature continues to drop after passing through the system aftercooler. In addition, the stored volume of air in the receiver reduces the frequency of compressor cycling in capacity-control systems using start-stop operation or inlet valve unloading.

Receiver design and construction

A receiver is typically cylindrical with domed ends, although other designs exist. The domed, cylindrical configuration provides an easily constructed unit that can withstand the pressures used in consumer and industrial systems. Standardizing organizations provide construction specifications and pressure ratings that are often listed as part of consumer and industrial safety codes.

> **CAUTION**
>
> Care should be taken to be certain codes are followed. A large volume of highly compressed air can present serious safety issues.

The receiver is constructed with inlet and outlet ports, fittings to allow the connection of various associated components, and, in some designs, a hole to facilitate cleaning. An internal baffle may also be incorporated to ensure liquefied water, pipe scale, and rust particles that enter from the connecting compressor line fall out of the airstream. Receivers are also fitted with feet for vertical or horizontal installation. In smaller units, they may include fittings to allow for the easy attachment of compressors and prime movers.

Accessory components often directly connected to the receiver are an electrical pressure switch for operating the prime mover, pressure gauges, liquefied-water drain valve, and maximum-pressure safety valve for the system. The size and complexity of the pneumatic system has a great deal to do with the included accessories and if they are located on the receiver or in some other location in the system.

Sizing the receiver

An adequately sized receiver must be provided if a pneumatic system is to operate efficiently. Formulas are available that can be used to provide a reasonable estimate of the needed size of receiver for a system. The variables involved in the operation of a pneumatic system make the strict application of a receiver-sizing formula questionable. It is often suggested that a factor based on operating experience be used as a part of a recommended mathematical formula. For example, one major manufacturer suggests a factor of 1.5 to 3.0; this is based on experience alone.

A receiver-sizing formula suggested by the Compressed Air and Gas Institute is:

$$V = \frac{T \times C \times P_A}{P_1 - P_2}$$

where:

V = Capacity of the receiver (ft³)

T = Time to reduce receiver pressure from P_1 to P_2 (min)

C = Air requirement (ft³/min)

P_A = Atmospheric air pressure (psia)

P_1 = Initial receiver air pressure (psig)

P_2 = Final receiver air pressure (psig)

This formula assumes a constant air temperature in the receiver and a standard air pressure of 14.7 psia. The formula also assumes that no additional air is being supplied to the system. In other words, the compressor is stopped or unloaded using a start-stop or inlet valve capacity control system. If air is being added to the system, that additional volume needs to be subtracted from variable C of the formula.

EXAMPLE 16-1

Sizing the Receiver

An air receiver must supply air to a pneumatic system for 10 minutes while the compressor is unloaded. The system is consuming air at 10 cubic feet per minute (cfm) during this time and the receiver air pressure drops from 100 psig to 50 psig during this time. Temperature is constant and the atmospheric air pressure is 14.7 psia.

$$V = \frac{T \times C \times P_A}{P_1 - P_2}$$

Substituting in:

$$T = 10 \text{ min}$$
$$C = 10 \text{ cfm}$$
$$P_A = 14.7 \text{ psia}$$
$$P_1 = 100 \text{ psig}$$
$$P_2 = 50 \text{ psig}$$

$$V = \frac{10 \text{ min} \times 10 \text{ cfm} \times 14.7 \text{ psia}}{100 \text{ psig} - 50 \text{ psig}}$$

$$= \frac{1,470 \text{ ft}^3}{50}$$

$$= 29.4 \text{ ft}^3$$

16.2 Air-Distribution System

The purpose of the *air-distribution system* is to carry high-pressure, conditioned air from the system receiver to workstations. Delivery must be done with minimal pressure drop in the lines to ensure satisfactory operation of the workstation tools and minimize operating costs. A variety of designs are used to ensure the transported air remains pressurized, is delivered at the required volume, and is free from moisture and other contaminants. A distribution system may be as simple as a flexible hose leading from a portable compressor to a portable hand tool. On the other hand, it may be as complex as an elaborate grid delivering air in a large production plant.

16.2.1 Distribution System Designs

Distribution systems may be grouped into four general categories:

- Fixed-piping, centralized-grid system.
- Fixed-piping, decentralized-grid system.
- Fixed-piping, loop system.
- Flexible hoses used with portable compressors.

The design that is selected depends on the size of the installation and the level of demand for compressed air.

The *centralized grid* design distributes air from one central location to all workstations in a facility. This design uses one or more compressors at a single compressor station with air piped to each workstation in the facility via fixed piping. The single compressor station is connected to a large central line with small feeder lines branching out to individual workstations. This design is sometimes called a *dead end* or *tree* system. **Figure 16-8** illustrates a simple, centralized grid that still exists in many industrial facilities. The illustration shows feeder lines that balance distribution capability on both sides of the central line. This distribution design allows the use of larger compressor equipment while providing easier access for maintenance because of the lower number of compressor stations.

In theory, the centralized grid allows flexibility for rearranging workstations since the grid can be accessed at any point. These systems can be effective if original planning of the facility provided adequate capacity for the increased air consumption. However, the single, main supply line can create pressure drops on the end branches if the air demand of a workstation close to the compressor station substantially increases.

The *decentralized grid* design uses individual compressors in several locations to provide compressed air for one or, at the most, a small number of workstations. This design allows each of the decentralized grids to function as independent systems. **Figure 16-9** shows a decentralized grid system containing three independent compressor installations. This design is sometimes referred to as a *unit* system.

The air from each independently operating compressor powers a single, large piece of equipment or a small grid serving a group of workstations in a manufacturing cell. This design is considered more flexible than the centralized grid, although the capacity of each unit must be adequate to absorb changes caused by relocating or resizing equipment. In some installations, the individual, decentralized systems are connected together. This is an attempt to equalize pressure to improve the effectiveness of each of the unit sections.

The *loop system* distributes air through a main line that forms a continuous loop, **Figure 16-10**. The loop-shaped main distribution line provides maximum airflow with a minimum of flow resistance. The design may use a single compressor location or multiple compressors that are dispersed around the loop.

If the system uses a single compressor, the air is fed into the loop at one location. A workstation located

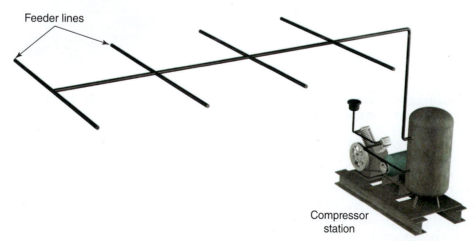

Feeder lines

Compressor station

Goodheart-Willcox Publisher

Figure 16-8. A centralized grid is a common design for an air-distribution system. An actual system is usually not as symmetrical as shown in this illustration.

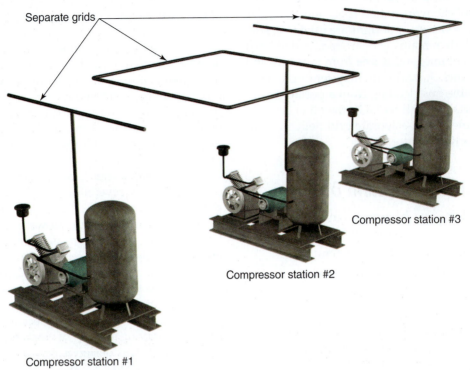

Separate grids

Compressor station #3

Compressor station #2

Compressor station #1

Figure 16-9. A decentralized grid uses a number of compressor stations each with a smaller, independent distribution grid.

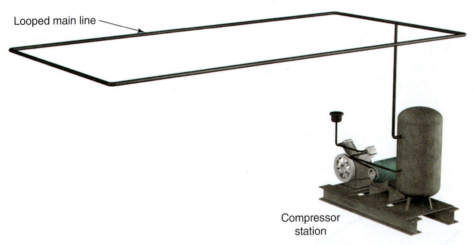

Looped main line

Compressor
station

Figure 16-10. A loop distribution system provides a main distribution line that forms a continuous loop. This arrangement allows the distributed air to travel to an operating workstation following two possible routes rather than a single line.

on the loop can receive air from either leg of the loop. This dual-line concept significantly reduces flow resistance, providing compressed air to any workstation on the loop with minimal pressure loss. Several compressors can be located at points around the loop system to provide increased airflow while maintaining minimal pressure drop caused by higher airflow rates.

Fixed air-distribution systems are often unique to a building. However, the objective of any design is a system that can effectively deliver compressed air as needed by a workstation. In addition, a well-designed, fixed system needs to have the capability to handle a reasonable increased load or equipment relocation without major modifications.

DeVilbiss Air Power Company

Figure 16-11. Distribution lines for air delivered by portable compressors are typically hose. Care must be taken to provide a system with minimal pressure drop by minimizing hose length and the number of fittings and eliminating bends or kinks.

Distribution of air from portable compressors typically involves *flexible hoses*, **Figure 16-11**. At some construction sites where work will last for extended periods, pipe main lines may be installed. Hoses, like the pipes of fixed systems, must be sized to provide air with minimal pressure drop. Care must be taken to protect the hose from cuts and abrasions. The lines should be kept straight and free from excessive kinks, although this is difficult in a work situation. Pressure drops in a twisted hose are as real as any pressure drop caused by excessive fittings in a fixed distribution system.

16.2.2 Sizing of Distribution System Lines

The size of the pipe used in the distribution system should be based on the air needed by system actuators. However, determining this volume of air becomes a problem as consumption varies considerably during the operation of most systems. This variation is based

on the fact that a distribution system serves multiple workstations. In addition, each of these workstations contains actuators that intermittently operate under varying loads. These multiple variables make application of a simple formula for pipe sizing impractical. The solution recommended by many experienced pneumatic system staff is to design new systems based on information from similar, proven, existing systems.

This process begins by calculating the total air consumption of the actuators planned for the new system. Individual consumption data are available from manufacturer literature or engineering data handbooks. System air consumption can be determined by totaling the air consumption for each actuator to be included in the system. This figure will be greater than the actual volume of air used by the operating system, as few of the system actuators operate on a continuous basis or at full load, **Figure 16-12**. In most systems, the actual amount of air needed is considerably less than the combined consumption figure.

Next, multiply the estimated air consumption by two additional factors. First, multiply by a *time factor* that is the percentage of time actuators function during

DeVilbiss Air Power Company

Figure 16-12. Estimating air consumption for use in sizing distribution lines for a new system is difficult, as actuator operating times and load factors vary. This is emphasized by the many variables at typical workstations.

system operation. A fairly accurate time factor can be established by averaging the percentages of operating times of actuators based on time and motion studies using similar operations in existing equipment. Then, multiply by a *load factor* that is an estimated percentage of maximum actuator load delivered during operation. The load factor is more difficult to establish and requires the assistance of experienced pneumatic system staff.

Once the final estimated air consumption has been determined, the pipe sizes can be selected for the distribution system. Selection is based on the anticipated airflow rates and pressure drop caused by the lines and fittings in the system.

16.2.3 Other Distribution System Considerations

A number of additional factors may affect the distribution of air in a pneumatic system. Two common factors encountered are the presence of liquid water in the distribution lines and excessive pressure drop caused by high, short-term air usage in sections of the system.

The formation of liquid water is the result of the continued cooling of air in the lines. As previously discussed, the components in the compressor station remove a large portion of the water vapor that enters the system with the atmospheric air. However, some water vapor still remains in the compressed air. Liquid water forms if the temperature of the compressed air in the lines drops below the dew point. This situation may become a serious problem when distribution lines are exposed to low temperatures, such as during cooler seasons of the year.

The liquid water that forms in the distribution lines can be removed by designing the lines with a slope. A pitch (slope) of 1″ per 10′ of line allows the water to accumulate at selected low points in the system. *Drop lines* from these low points lead to water traps where the water is removed, either manually or with an automatically operated device. See **Figure 16-13**.

Another design feature that greatly reduces the entrance of liquid water into actuators is how the workstation is connected to the distribution system. The drop lines leading to the workstation should be attached to the top side of the distribution line. See **Figure 16-14**. The drop line should also include a *water trap* or *drain trap* that is lower than the workstation outlet. This allows the accumulation and removal of any liquid water that forms in the drop line.

If the system includes a high-demand piece of equipment, the air pressure in that section of the distribution system may drop excessively. If this equipment only occasionally operates, it may not be necessary to increase the size of the total system. The solution may be an *auxiliary-air receiver* to supply extra compressed

air. The additional receiver is located close to the high-demand equipment. It supplies the extra air needed to prevent an excessive drop in distribution-line pressure. This arrangement is cost effective as it can be used to reduce construction cost related to the compressor and distribution line. It can also accommodate unexpected resizing or relocation of existing workstations. Auxiliary receivers may also be used in loop-type distribution systems to maintain a more uniform pressure as air demand changes in sections of the line.

16.3 Final Preparation of Air at the Workstation

Compressed air is often used directly in process applications. These applications may be involved in chemical, food, petroleum, or metal processing. In many of these situations, the air must be cleaner than typical, available at multiple pressure levels, or contain lubricants to prolong the life of system components. These situations are common in manufacturing applications involving pneumatic-powered hand tools or complex machines powered and controlled by pneumatic actuators.

This final preparation is usually provided at each workstation. Three individually constructed components assembled as a combined filter/pressure regulator/lubricator (FRL) unit are normally used for this purpose. **Figure 16-15** illustrates a typical *FRL unit* of the type used at individual workstations. The FRL unit removes any remaining unwanted particles in the compressed air, provides a means to readily adjust workstation air pressure, and provides lubricant to prolong the life of workstation components.

16.3.1 Air Filtration at the Workstation

A filter is the first component of the FRL unit. It is designed to remove any remaining unwanted particles in the compressed air. These particles include:

- Airborne dirt.
- Rust and scale that have broken free from the walls of the distribution lines.
- Liquid water that has condensed in the distribution lines.
- Atomized oil particles from the compressor.

A variety of designs are used by filter manufacturers to remove these contaminants. The following is a brief description of a filter designed for an FRL unit.

A two-stage filtering arrangement is commonly used in an FRL unit. The design involves both centrifugal force and porous materials to separate and trap unwanted particles. The porous material may be paper, metal, or ceramics. Air entering the intake port of the

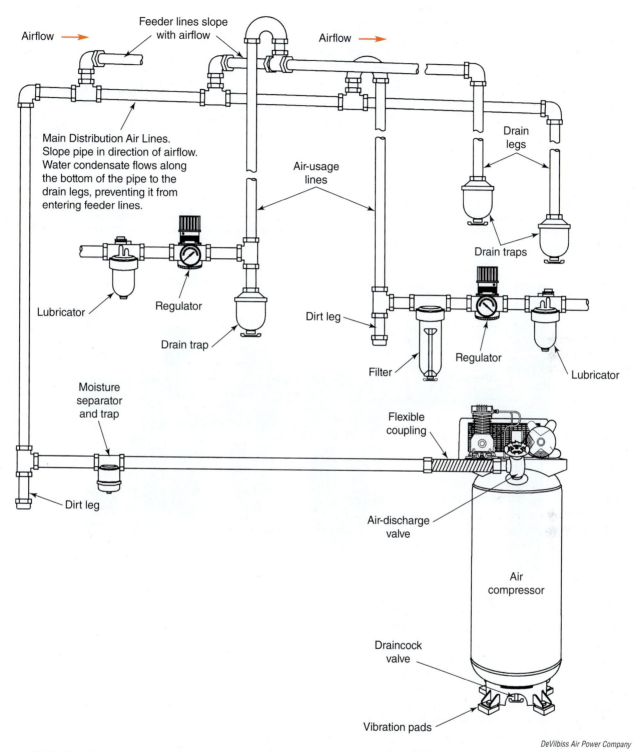

Airflow →

Feeder lines slope with airflow

Airflow →

Main Distribution Air Lines. Slope pipe in direction of airflow. Water condensate flows along the bottom of the pipe to the drain legs, preventing it from entering feeder lines.

Air-usage lines

Drain legs

Lubricator

Regulator

Drain trap

Dirt leg

Filter

Regulator

Drain traps

Lubricator

Moisture separator and trap

Flexible coupling

Dirt leg

Air-discharge valve

Air compressor

Draincock valve

Vibration pads

DeVilbiss Air Power Company

Figure 16-13. The distribution lines in a permanent installation should be sloped to allow drainage of liquid water that has condensed in the lines. Dirt legs provide a point for the accumulation and removal of this water.

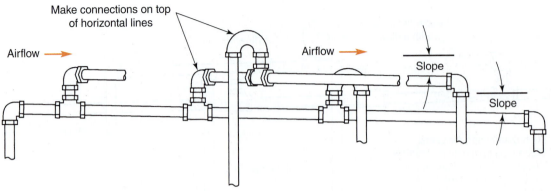

DeVilbiss Air Power Company

Figure 16-14. Workstation drop lines should be attached to the top side of distribution lines to prevent liquid water from draining into the workstation. This arrangement also reduces the entrance of rust and scale that has formed in the main distribution lines.

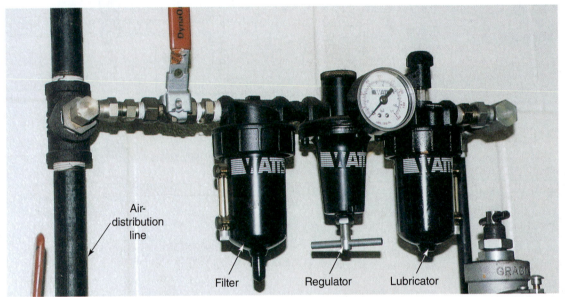

Used with permission of CNH America LLC

Figure 16-15. Final air preparation at the workstation is provided by a unit that consists of an air filter, pressure regulator, and air lubricator. This unit is known as an FRL unit.

filtering unit is swirled. This causes any water, oil droplets, or solid particles to be thrown against the walls of the filter case. These materials then settle by gravity to the bottom of the filter bowl where they can be physically removed during system service. The air then passes through a series of small holes in the porous material. The porous material traps particles, which are removed during filter replacement or cleaning. **Figure 16-16** shows the internal structure of a typical filter.

The water, oil, and dirt that accumulate at the bottom of the filter bowl need to be periodically removed. The filter bowl is often equipped with a *draincock* to facilitate this service, **Figure 16-17**. Opening the valve allows system pressure to force the liquid and dirt out of the bowl. Automatic draincocks are available. These valves automatically open when the liquid and dirt

reach a predetermined level. The valve resets after system air pressure forces the accumulated liquid and dirt out of the bowl.

Factors that must be considered when selecting a filter unit for a workstation include:

- Conductor size on the filter body.
- Pressure rating of the filter body.
- Particle size (in microns) of the contaminants the filter will remove.
- Airflow rating of the filter.
- Pressure drop across the unit at the airflow rate required by the system.
- Need for special features such as pressure gauges and automatic drains.

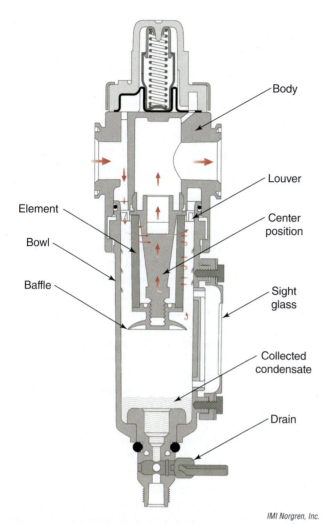

IMI Norgren, Inc.

Figure 16-16. Many air filters used in the final preparation of air at the workstation use both centrifugal force and filter elements to remove dirt and water from the air distributed by the system.

Information and specification sheets available from the manufacturer supply these details for the various filter units it produces.

16.3.2 Pressure Regulation at the Workstation

The *pressure regulator* is the second of the components in the FRL unit. The regulator maintains a uniform air pressure at the workstation. This promotes consistent actuator operation for tools and circuits powered by the station.

A regulator is necessary as the air pressure in the distribution system is higher than that normally used for workstation operation. Also, the air pressure in the distribution line varies because of the characteristics inherent to compressor-capacity control systems. Regulators maintain the desired output pressure for

Goodheart-Willcox Publisher

Figure 16-17. Accumulated dirt and water need to be regularly removed from the filter bowl. A manual draincock or automatically operated valve is found on most filter units.

the station as long as the pressure in the distribution system does not drop below that selected pressure.

Several different regulator designs and features are available. These designs include:

- Direct operated.
- Basic, diaphragm chamber.
- Relieving.
- Balanced-poppet valve.
- Pilot operated.

A number of variations and combinations of these designs are available. The type of regulator selected for a particular system depends on the accuracy and control needed in the system.

Direct-operated regulators

Figure 16-18 shows the internal structure of a *direct-operated regulator*. This is the most basic of the units. The unit consists of a cast body with inlet and outlet ports, spring-loaded poppet valve, flexible diaphragm, transfer pin (located between the poppet valve and the diaphragm), control spring, and threaded screw (for adjusting the tension of the control spring).

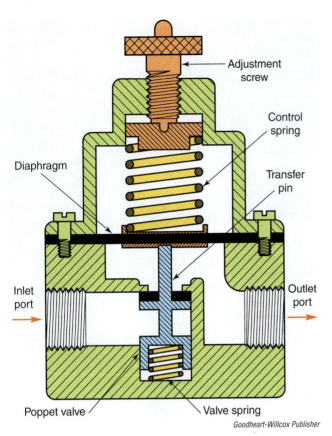

Figure 16-18. The diaphragm of a direct-operated pressure regulator is directly exposed to the pressurized air in the outlet port.

When the regulator is installed, the inlet port is connected to a line from the air distribution system. The outlet is connected to the workstation line containing the pneumatic-powered actuators. The maximum operating pressure of the regulator is set by tightening the adjustment screw to compress the control spring. This applies force to the atmospheric side of the diaphragm. Application of this force moves the transfer pin, which opens the poppet valve and compresses the valve spring. System air is then allowed to move through the valve to the outlet port and on to the actuator.

When the actuator encounters resistance, the pressure of the incoming air begins to increase. This produces force in the actuator that attempts to move the actuator load. At the same time, the pressure in the outlet port of the regulator increases. This produces a force on the system side of the diaphragm, which begins to balance the force applied by the control spring.

When the pressure reaches the setting of the regulator pressure, the force generated on the surface of the diaphragm overcomes the control spring force. This allows movement of the diaphragm. When the diaphragm moves, it no longer applies force to the poppet valve through the transfer pin. The valve spring can then adjust the valve. The adjusted valve position allows only sufficient air to pass to maintain the desired working pressure.

If the air-powered equipment does not need air, the pressure climbs to the regulator setting and the valve completely closes. If there is a need for air, the regulator adjusts until only sufficient air is allowed through the valve to compensate for the required flow.

This regulator design provides pressure control but directly exposes the diaphragm to system airflow. This exposure subjects the diaphragm to any foreign materials in the air, which may shorten the life of the component. Direct exposure also affects regulator accuracy and sensitivity.

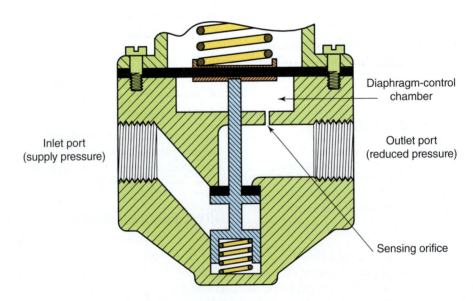

Figure 16-19. The diaphragm of a basic, diaphragm-chamber regulator is separated from the direct airflow through the valve. The pressure in the outlet port is transmitted to the diaphragm through a pressure-sensing orifice in the valve body.

Basic, diaphragm-chamber regulators

Figure 16-19 shows the internal structure of a *diaphragm-chamber regulator*. This regulator incorporates a separate diaphragm chamber to provide more accurate control of system pressure. The diaphragm-control chamber separates the regulator side of the diaphragm from direct contact with air in the outlet port. This control chamber senses the pressure in the outlet port through a small-diameter sensing orifice. The other construction features of this unit are very similar to the direct-operated regulator. Setting the desired system pressure for this regulator design is much the same as for the direct-acting unit.

Air from the outlet port must pass through the sensing orifice to keep the pressure in the diaphragm-control chamber balanced with the pressure in the outlet port. When the pressure in the outlet port begins to increase, additional air is metered through the sensing orifice to keep the pressure balanced. This increased pressure applies greater force on the diaphragm to overcome the force produced by the control spring. The resulting diaphragm movement allows the poppet valve to adjust position in the same way as in the direct-operated design.

The area of the diaphragm in a diaphragm-chamber regulator is typically much larger than in an equivalent direct-operated unit. The larger diaphragm produces more accurate regulator performance.

Relieving-type regulators

In the direct-operated and diaphragm-chamber regulators, external forces acting on the outlet side of the regulator may increase pressure above the regulator setting. Situations that can cause the pressure to exceed the regulator setting are:

- Heating of the system from an external source, such as exposure to direct sunlight.
- Unexpected load increases on actuators that move internal elements, turning the actuator into a pump.
- Leakage of the regulator valve, allowing system pressure to increase to the pressure level in the distribution line.

Figure 16-20 shows the internal structure of a *relieving-type regulator*. The relieving feature eliminates this pressure increase.

Overall, the construction of a relieving-type regulator is very similar to the basic, diaphragm-chamber design. The primary difference is a venting orifice located in the center of the diaphragm. The end of the transfer pin keeps the orifice closed while outlet port pressure remains below the maximum setting of the regulator. However, the orifice allows the diaphragm-control chamber to be vented to the atmosphere if the pressure in the outlet port exceeds the pressure setting of the regulator.

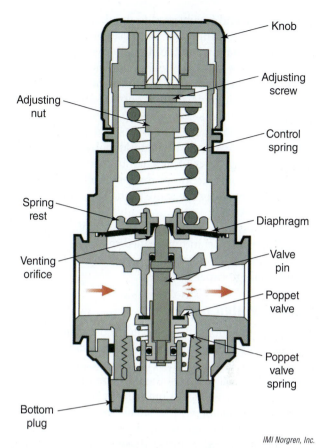

IMI Norgren, Inc.

Figure 16-20. A relieving-type pressure regulator uses a venting orifice located in the center of the diaphragm to prevent excessive pressure buildup in the regulator outlet port.

The venting orifice functions only when the maximum desired operating pressure has been reached in the workstation lines and the regulator poppet valve is completely closed. This means the control spring and the diaphragm force are balanced. Any increase in outlet port pressure then generates additional force on the control-chamber side of the diaphragm and further compresses the control spring. The resulting diaphragm movement opens the venting orifice. Air is allowed to escape to the atmosphere, preventing any pressure increase above the maximum regulator setting.

This regulator design acts much like a small relief valve to prevent the system from exceeding the pressure setting of the regulator. However, the design should *not* be used as the primary relief valve in a compressed air system.

Balanced-poppet valve regulators

The direct-operated, diaphragm-chamber, and relieving-type regulators use unbalanced poppet valves. Basically, this means that the air pressures on the ends of the poppet are different. Typically, the end of the poppet facing the inlet port is subjected to atmospheric pressure.

The opposite end is subjected to varying, higher system pressures. The difference in these pressures causes operating variations as the regulator compensates for load variations in the workstation lines.

Figure 16-21 shows a *balanced-poppet valve regulator*. Both ends of the poppet are exposed to approximately the same pressure. This produces balanced forces acting on the valve. The valve is positioned only by the force generated by the pressure in the diaphragm-control chamber and the poppet spring. The construction of the remainder of this type of valve is primarily the same as the basic, diaphragm-chamber regulator.

This design results in reduced fluctuations in the regulated pressure and improved response during the normal range of operation. The feature is most often found on large-capacity units.

Pilot-operated regulators

In some installations, the regulator must be located in a remote location to maintain maximum system efficiency. These remote locations often have limited access, which makes it difficult to adjust the pressure of the regulator. In these situations, a pilot mechanism is used to control a primary regulator with a small, secondary regulator located at an easily accessed location. **Figure 16-22** shows the structure of a *pilot-operated regulator*.

In a pilot-operated regulator, the diaphragm-control spring and adjustment screw are replaced by a sealed, pilot-air chamber. This chamber is connected to the secondary regulator located at the operator workstation via a small-diameter air line. See **Figure 16-23**.

The setting of the secondary regulator controls the air pressure in the pilot-air chamber of the primary regulator.

During operation, the air pressure in the pilot-air chamber generates a force on the regulator diaphragm. This force is in place of the force produced by the control spring of the basic regulator design. The secondary regulator provides pilot pressure adjustment, which replaces the adjustment screw of the basic regulator. Basically, the pressurized air in the pilot-air chamber acts as an adjustable air spring that establishes the pressure setting of the pilot-operated regulator. The construction and operation of the remainder of the regulator are the same as the basic, diaphragm-chamber regulator.

Regulator selection

Factors that must be considered when selecting a regulator for a system are:

- Choice of a direct-operated; basic, diaphragm-chamber; relieving-type; balanced-poppet valve; or pilot-operated style, depending on system size and performance demands.
- Pressure range of the unit, which considers the pressure in the input distribution line compared to the desired output pressure at the workstation.
- Airflow range of the unit.
- Conductor connection size.

Manufacturer information and specification sheets supply these types of detail for the various models produced.

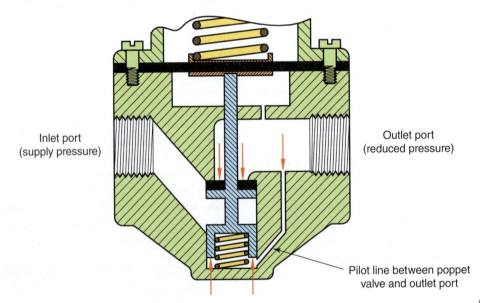

Inlet port (supply pressure)

Outlet port (reduced pressure)

Pilot line between poppet valve and outlet port

Figure 16-21. A balanced-poppet valve regulator uses a design that exposes both ends of the poppet to equal pressure. This increases performance accuracy.

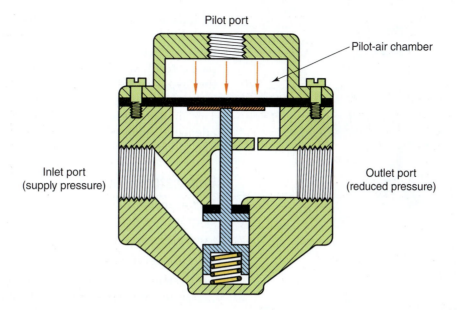

Goodheart-Willcox Publisher

Figure 16-22. In a pilot-operated regulator, the control spring and adjustment screw are replaced with a sealed air chamber. Varying the air pressure in this chamber allows the regulator to be adjusted to the desired operating pressure.

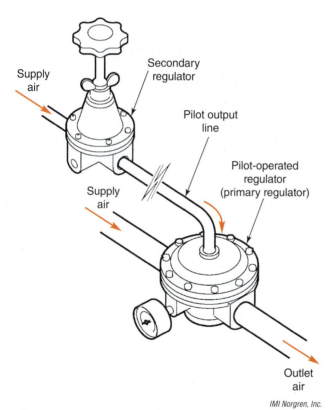

IMI Norgren, Inc.

Figure 16-23. The small, remotely located regulator is used to adjust the main system regulator by controlling pilot chamber pressure.

16.3.3 Air Line Lubrication at the Workstation

A lubricator is the third of the components in the FRL unit. A *lubricator* provides the oil needed to improve the performance and service life of pneumatic system valves, actuators, and air-powered tools. A lubricator atomizes a measured amount of oil and mixes it with the compressed air used at system workstations. The lubricator must accurately meter the oil to provide component lubrication. Providing inadequate lubrication can result in rapid component wear. Adding excessive amounts of oil is wasteful and can contaminate work areas around component exhausts.

Lubricator design and operation

Figure 16-24 shows the internal structure of one type of lubricator. The major external parts are a cast body containing inlet and outlet ports and a clear glass or plastic reservoir. Some designs have a metal reservoir. The internal elements are:

- Air passageway leading from the inlet port to the reservoir. This passageway contains a venturi.
- Oil-siphon tube connecting the reservoir and the venturi in the air passageway.
- Adjustable oil-drip orifice.
- Passageway leading from the reservoir to the outlet port.
- Spring-loaded bypass valve that directly connects the inlet and outlet ports.

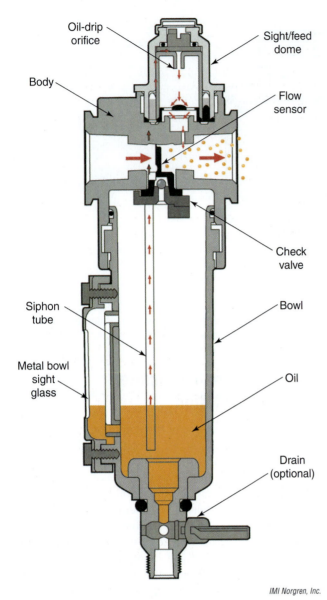

Oil-drip orifice

Sight/feed dome

Body

Flow sensor

Check valve

Siphon tube

Bowl

Metal bowl sight glass

Oil

Drain (optional)

IMI Norgren, Inc.

Figure 16-24. A lubricator atomizes a measured amount of oil and mixes it with the compressed air used at system workstations.

During operation, system air passes through the lubricator inlet port and passageway leading to the reservoir. The venturi in the passageway creates a reduced pressure. This causes oil to move through the siphon tube and enter the airstream through the oil-drip orifice. The rapidly moving air breaks up the oil drops into a mist that passes with the system air into the reservoir airspace. The airspace reduces turbulence in the air/oil mist. This allows larger droplets of oil to return to the oil supply. The oil mist then moves to the outlet port and on to the workstation air supply line. The amount of oil added to the air is controlled by adjusting the size of the oil-drip orifice.

During periods of low air use, all of the system air passes through the venturi of the lubricator. During periods of high air use, the higher pressure drop produced by the venturi causes the spring-loaded bypass valve to open. This allows a portion of the system air to pass directly into the supply line. The unlubricated air that bypasses the venturi is mixed with lubricated air before being passed out of the lubricator outlet port. This air contains less oil than normal, but does provide adequate component lubrication during brief periods of high-air-volume operation.

Lubricator selection

Factors that must be considered when selecting a lubricator for a system are:

- Conductor connection size.
- Oil reservoir capacity.
- System pressure drop created by the volume of air that must pass through the unit to operate the components of the workstation.

Manufacturer information and specification sheets supply these types of detail for the various models produced.

16.4 Distribution System Conductors and Fittings

Moving compressed air from the compressor to the point of use at the workstation is a critical part of any pneumatic system. Any reduction in pressure between those points represents a loss of energy, which cannot be recovered. Previous sections of this chapter discuss general styles of distribution systems and their layout. This section is devoted to information about the pipe, hose, and fittings that form these systems.

The volume of products and variety of available designs can make the expansion of an existing system or the development of a new system confusing. The following materials provide general guidelines and suggestions for the selection of appropriate conductors and fittings required for installation.

16.4.1 Rigid Conductors

Pipe is the most common *rigid conductor* used for major compressed-air-distribution systems. Steel, copper, and ridged-plastic tubing may be found in smaller systems and on air-powered machines. Some of these materials may not be recommended for all installations, but they may provide adequate service in some installations. However, the conductor must be correctly rated for pressure and protected from external abrasives and activities that could crush the lines.

When constructing or modifying a distribution system, pressure losses caused by system lines should receive careful consideration. The sizing of conductors can be done only after the volume of air and the location of the workstations have been established. The objective is to hold pressure losses between the receiver at the compressor station and the workstation to less than 10% of the receiver pressure. Both the conductors and the associated fittings must be considered, **Figure 16-25**.

Tables are available in engineering handbooks that provide data on the air pressure loss in rigid conductors transferring a specific volume of compressed air expressed in cubic feet per minute (cfm). These pressure-loss figures are typically given as the loss per 100 feet of conductor of a specific nominal size and schedule rating. Tables are also available indicating the pressure losses caused by the common fittings used in distribution lines. The pressure loss caused by these fittings is given in feet of comparable-sized straight conductor.

Determining the correct size of conductor for various sections of the distribution system requires the development of a map of the air needed, in cfm, at the junction points of the main distribution lines and each workstation drop line. This map can be constructed by starting at the point that is the greatest distance from the compressor station receiver and adding the air needed by each station. These figures must be increased by 10% to cover air leakage in the system. An estimated air volume should be added to these figures to cover the air needed for future system expansion. The resulting flow rates can then be used to select a pipe size that will deliver air at a pressure drop lower than the recommended level.

Adequately sized distribution lines are critical to the operation of a pneumatic system. The cost of the conductors is only a small part of the total installation cost. Therefore, when an estimated flow rate falls between conductor sizes, the *larger* size should be installed. The larger conductor will increase the performance of the current system and provide increased capacity for future system expansion.

16.4.2 Flexible Conductors

Hoses and semirigid plastic tubing are examples of *flexible conductors* used extensively in pneumatic systems, **Figure 16-26**. Hose applications include temporary distribution lines for portable-compressor setups, connections between movable machine members, and connections between the FRL unit of a workstation and air-powered tools. Semirigid plastic tubing is often used in small equipment as the primary conductor between components. It is also used as sensing and control lines in many other compressed-air applications.

Hose is constructed with a minimum of three layers of materials:

- Inner tube providing a smooth surface to minimize airflow resistance.
- Middle section containing woven reinforcing materials to withstand system pressure.
- Outer layer to protect the hose from damage resulting from handling and the abrasive and corrosive materials encountered in the work environment.

IMI Norgren, Inc.

Figure 16-25. The fittings used in distribution lines produce a pressure loss. Pressure loss for each fitting must be included as part of the total loss in the system.

Atlas Copco

Figure 16-26. A wide variety of hose and semirigid tubing is available for use in pneumatic systems.

Manufacturers provide a selection of hose types to meet various system working conditions. Many of these are built to meet standards established by industry standardizing groups.

Ideally, hoses should be no longer or larger in diameter than necessary to supply the compressed air needed at a work location, **Figure 16-27**. This ensures the delivery of air with minimal pressure drop. It also ensures minimal hose and installation costs. However, in work conditions that are often less than ideal, this may not be possible. These points should be considered when working with hose to minimize installation costs and pressure drops:

- Hoses should be no longer than necessary.
- Hoses should be no larger in diameter than necessary.
- A minimum number of fittings and couplings should be used.
- Lay out lines to eliminate kinks and reduce the number of bends.

Goodheart-Willcox Publisher

Figure 16-27. Hoses should be no longer or larger in diameter than necessary to supply the compressed air needed at a work location.

Summary

- Maximum pneumatic system operating efficiency is achieved when system compressed air is consistently clean, free from moisture, and at a relatively uniform temperature.

- Dry, oil-wetted, and oil-bath filters on the intake line of pneumatic compressors remove dirt and dust from atmospheric air.

- Because changes in temperature affect air pressure and volume, it is important to monitor the temperature of the intake air and the compressed air moving through the system. In addition, temperature influences the ability of air to retain water vapor.

- Water vapor in air is referred to as humidity. When the temperature of the compressed air drops to the dew point, the vapor condenses into liquid water.

- Liquid water in a pneumatic system can wash away lubricants, increase component wear, cause inconsistent system operation, and lower the finished quality of products using the air directly in the manufacturing process.

- Locating the air intake of the compressor in a protected area; cooling the air below the dew point to condense, collect, and remove liquid water in system components; and installing specific air dryers between the system receiver and distribution lines are all methods to reduce moisture in the pneumatic system.

- The receiver is typically a metal cylindrical tank with domed ends, inlet and outlet ports, and an internal baffle that ensures that any particles or liquid water fall out of the airstream.

- The receiver stores compressed air for use in the pneumatic system, dampens system pressure pulsations, removes water vapor, and may serve as the mount for the system prime mover and compressor.

- The size of receiver needed for a pneumatic system depends on the cubic feet of free atmospheric air needed per minute and the desired cycle time, as well as the atmospheric, initial receiver, and final receiver air pressures.

- The pressure regulator maintains a consistent pressure for use by workstation tools and circuit actuators. The unit is necessary as the air pressure in the distribution line fluctuates because of varying air demands and the inherent characteristics of compressor capacity control systems.

- Pressure regulators range from direct-operated regulators to pilot-operated units that can be controlled from a remote location.

- Pressure regulators commonly use a flexible diaphragm to sense outlet-line pressure and provide the balancing force needed to control airflow through a poppet valve. A diaphragm-control chamber may be used to separate the regulator side of the diaphragm from direct contact with outlet port airflow. This arrangement dampens the reaction of the diaphragm, providing more sensitive and efficient workstation pressure control.

- Lubricators meter oil into pressurized system air at the workstation to provide lubrication for system valves, actuators, and air-powered tools.

- Conductors moving compressed air through a distribution system can be grouped into two general classes: rigid and flexible. Pipe is the most common rigid conductor, and hose is the most common flexible product.

- Conductors must be sized and assembled properly to produce system lines that transport compressed air from the compressor to the system actuator with a minimum of pressure drop. Tables are available in engineering handbooks and other publications that provide data on air pressure loss in standard rigid and flexible conductor sizes using various flow rates.

- Hose selection, application, and maintenance is critical in both fixed and portable systems to assure air distribution with a minimum loss of pressure.

Internet Resources

The following are some useful resources available on the Internet. Enter a company or organization name into a search engine to access its website. Explore the various areas of the sites to discover useful fluid power resources.

Condit Company. The web page provides information on the importance of air filtration for pneumatic systems, and includes a number of links to discussion of other system functions. The product brochure provides detailed information on one series of pneumatic filters.

eCompressedair. Discusses the piping system involved in a pneumatic system, including illustrations with details concerning the total air distribution system. A number of links are provided to related topics, such as the compressor and air-quality control.

CompAir (Gardner Denver, Inc.). FAQs regarding air treatment in compressed air systems.

ABAC Air Compressors. Describes the function and operation of the aftercooler, separator, and automatic condensate drain for large compressor units.

R.P. Adams, Subsidiary of Service Filtration Corporation. Provides a brief explanation of how cyclone separators remove condensed liquid and other contaminants from the primary airstream of a pneumatic system distribution line.

Wilkerson Corp., Pneumatic Division. Provides extensive information on pneumatic system filtration, pressure control, and lubrication. Includes especially helpful information concerning workstation FRL units.

Chapter Review

Answer the following questions using information in this chapter.

1. *True or False?* Conditioning and distribution of compressed air is the major portion of tasks in an industrial pneumatic system.

2. *True or False?* The first factor contributing to the quality of air for a pneumatic system is the quality of atmospheric air surrounding the system.

3. The three filter categories used in pneumatic systems are dry, oil bath, and _____.
   A. paper
   B. pneumatic
   C. oil wetted
   D. intake line

4. *True or False?* According to general gas law, any change in the temperature of a gas is only reflected in gas pressure.

5. The temperature at which water vapor in the air begins to condense to form liquid water is known as the _____.
   A. dew point
   B. poppet point
   C. saturation point
   D. balance point

6. The _____ distribution system distributes air through a main line that forms a continuous circle.
   A. centralized
   B. drop
   C. FRL
   D. loop

7. A pitch (slope) of _____ per 10′ of distribution lines allows water to accumulate and be drained.
   A. 1/2″
   B. 1″
   C. 5″
   D. 8″

8. *True or False?* Workstation drop lines should be attached to the top side of the distribution lines to prevent liquid water from draining into the workstation.

9. *True or False?* The diaphragm of a direct-operated pressure regulator is directly exposed to the pressurized air in the outlet port.

10. In a pilot-operated pressure regulator, the control spring and adjustment screw are replaced by a _____.
    A. sealed, pilot-air chamber
    B. poppet spring
    C. solenoid-operated, two-way valve
    D. herringbone gear

11. *True or False?* A lubricator atomizes a measured amount of oil and mixes it with compressed air during the intake stroke of the compressor.

12. The most common rigid conductor used for major compressed-air systems is _____.

13. *True or False?* Tables are available in engineering handbooks that provide only data on air pressure loss based on ten feet of straight conductor.

14. *True or False?* Flexible conductors constructed as pneumatic lines should contain a minimum of three layers of materials.

15. *True or False?* The middle sections of flexible pneumatic conductors are typically woven, reinforcing layers designed to withstand system abrasive handling.

16. *True or False?* Ideally, hoses should be no larger in diameter than necessary to supply the needed volume of air to operate the equipment.

For questions 17–20, match the pressure regulator designs listed below with the feature associated with that design.
   A. Venting orifice.
   B. Sealed, pilot-air chamber.
   C. Control spring.
   D. Poppet spring.
   E. Diaphragm-control chamber.

17. Direct operated.
18. Basic, diaphragm chamber.
19. Relieving.
20. Balanced-poppet valve.
21. Pilot operated.

Apply and Analyze

1. On a hot and humid day, ambient air entering a compressor may be as warm as 90°F with a dew point as high as 70°F.
   A. What are the possible impacts of this air on the operation of your pneumatic circuit?
   B. What methods or equipment could be used to reduce these impacts?
   C. Which of these methods or equipment would you implement first? Why?

2. How do the design and functional roles of a typical pneumatic receiver compare to those of a typical hydraulic reservoir?

3. Calculate the capacity of a receiver required to supply air to a pneumatic system while the compressor is stopped. The system consumes 25 cfm of free air for 5 minutes between 100 psig and 80 psig.

4. Additional pneumatic actuators are added to a pneumatic circuit in which an adequately sized receiver is already installed. Explain the possible effects of this change on the operation of other actuators in the circuit.

5. You have been asked to design the air distribution system for a small company that uses pneumatics in its manufacturing process. What questions would you ask before you propose a system? How will the answers to each of these questions affect your design?

Research and Development

1. Investigate one basic type of filter (dry, oil wetted, or oil bath) used to remove dust and other contaminants from air used in pneumatic circuits. Determine the variety of configurations and materials available, the basic operation of the filter, service requirements, and typical application for which this type of filter is best suited. Share your results in a written report.

2. Visit an industrial site where pneumatic equipment is operated. If possible, interview the equipment operator or site manager about the design and layout of the pneumatic system. Identify the equipment used to condition and distribute air and determine how the choice of components and layout affect system operation.

3. The final preparation of air at the workstation is typically provided by an FRL unit. These units vary widely depending on the requirements of the application. Making the wrong choice can be costly. Prepare a guidebook for correctly choosing one or more of the following components of an FRL unit:
   - Filter.
   - Pressure regulator.
   - Lubricator.

4. Design and conduct an experiment to investigate the impact of conductor size and rate of airflow on pressure drops in a pneumatic system. Share your results in a written report.

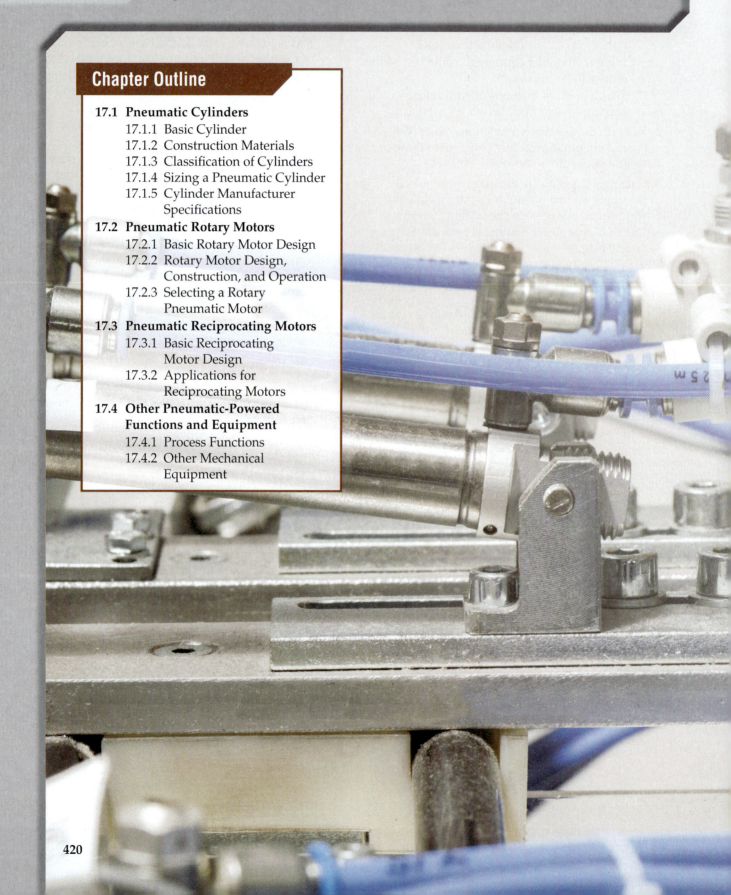

17 Work Performers of Pneumatic Systems

Cylinders, Motors, and Other Devices

$\mathbf{P}$neumatic systems must convert the potential energy contained in compressed air to force and movement. Without effective energy conversion, the storage, transmission, and control advantages of pneumatic systems cannot be realized. Actuators are used in pneumatic systems to produce the linear and rotary motion basic to the operation of industrial equipment. The actuators are designed to effectively produce the required force, torque, and speed of movement needed in an application by using the characteristics of compressed air. Actuators include basic cylinders and motors, as well as specialized devices.

Learning Objectives

After completing this chapter, you will be able to:

- Describe the construction features of basic pneumatic linear and rotary actuators.
- Compare the design and operation of pneumatic cylinders.
- Compare the design and operation of pneumatic motors.
- Explain the performance characteristics used to rate the operation of pneumatic motors.

- Describe the basic design and operation of specialized pneumatic tools commonly used in consumer and industrial applications.
- Size pneumatic cylinders and motors to meet the force and speed requirements of a basic work application.
- Interpret manufacturer specifications for basic pneumatic cylinders, motors, and power tools.

Key Terms

absolute pressure ratio
air consumption
air nozzle
blowgun
chipping hammer
corrosion resistance
double-acting cylinder
drying
force
gripper
impact wrench

linear actuator
material agitation
material transfer
metal rolling
nail driver
nutsetter
paving breaker
piston motor
pneumatic reciprocating
    motor
rammer

resilient material
riveting hammer
rock drill
rotary actuator
scaling hammer
single-acting cylinder
spraying
tamper
turbine motor
vane motor

17.1 Pneumatic Cylinders

A pneumatic cylinder is the system component that converts the energy in compressed air to linear force and movement, **Figure 17-1**. Cylinders are often called *linear actuators*. The force generated by the actuator is determined by the area of a cylinder piston or diaphragm and the air pressure in the system. The rate of movement is determined by the volume of the actuator chamber per inch of stroke and the amount of air allowed to enter the actuator. The maximum distance that a linear actuator can move is determined by the physical length of the unit.

Controlling air pressure and flow rates can produce a wide range of actuator force outputs and operating speeds. However, the compressibility of air affects the output of both force and speed. Operation of a pneumatic actuator is a practical example of the general gas law.

17.1.1 Basic Cylinder

A basic pneumatic cylinder involves a piston and piston rod assembly operating within a close-fitting, cylindrical tube called the barrel. One end of the barrel is closed. An opening in the opposite end of the barrel allows the piston rod to extend outside of the component. Seals are used to prevent leakage to the exterior of the cylinder. Seals also prevent leakage around the piston that separates the barrel into two sealed chambers. Ports located near the ends of the cylinder allow air to be forced into and vented from each of the sealed chambers. **Figure 17-2** illustrates the construction of such a basic cylinder.

Parker Hannifin

Figure 17-1. Cylinders are used to generate force to produce linear movement in a variety of industrial equipment.

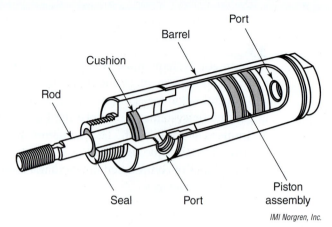

IMI Norgren, Inc.

Figure 17-2. A pneumatic cylinder uses relatively simple construction to convert the potential energy of compressed air into linear force and movement.

The basic structure of pneumatic cylinders is very similar to that of hydraulic cylinders. Chapter 9, *Actuators*, examines the features and structure of hydraulic cylinders in detail. That chapter should be reviewed for details of cylinders. Features are similar for hydraulic and pneumatic cylinders. The remainder of this chapter is devoted to features and operating concepts primarily limited to cylinders used in pneumatic applications.

17.1.2 Construction Materials

Cylinders must be designed to withstand harsh operating conditions. Both hydraulic and pneumatic cylinders are subjected to corrosion and abrasion that can produce rapid wear. However, in pneumatic applications, water vapor produces corrosive conditions that require construction materials different than those needed in hydraulic applications. Another condition that is different between the two systems is operating pressure. Pneumatic applications typically operate at lower pressures than hydraulic systems. Therefore, pneumatic actuators do not need the same structural strength that must be built into hydraulic actuators.

Increased *corrosion resistance* is provided by:

- Use of resistant materials for the construction of component parts.
- Plating of parts subjected to high corrosion.

Brass is a corrosion-resistant material that has been used to construct component parts. Chromium-plated steel is commonly used for piston rods and the bore surface of cylinder barrels. Stainless steel and anodized aluminum are also used to provide corrosion resistance.

Aluminum parts, including machined heads and caps and extruded barrels, are used on some models of pneumatic cylinders to reduce the weight. These materials easily provide the strength required for the low forces encountered in most pneumatic applications.

In most pneumatic cylinders, *resilient materials* are used for piston seals. The style of the piston seal ranges from standard, commercially available designs to special forms designed especially for a given product. Resilient materials are also used between the cylinder rod and cylinder head and between the barrel and cylinder head and cap, **Figure 17-3**.

Some manufacturers produce a product line of nonlubricated pneumatic cylinders. These operate without adding oil to the compressed air of the system. Nonlubricated cylinders are used in applications requiring oil-free air, such as in the food industry. Lubrication is provided by coating the surfaces of the cylinder bore, piston, and bearing surfaces with lubricant that adheres, but is not scraped off by the seals.

Pneumatic, linear actuators are also produced with a diaphragm instead of a piston. The diaphragm is attached to the rod, which extends from the body the same as in a conventional cylinder. This actuator design typically has a short stroke and is single acting with a spring return.

17.1.3 Classification of Cylinders

A number of pneumatic cylinder designs are available. These designs can be classified in much the same way as hydraulic cylinders, as described in Chapter 9, *Actuators*. Some variations exist that must be considered when selecting a cylinder for use with compressed air. However, like hydraulic cylinders, pneumatic cylinders may be classified based on basic operating principles and construction type.

Operating principles

Pneumatic systems use both single- and double-acting cylinders. The operating principles of these designs are basically the same as for hydraulic cylinders. *Single-acting cylinders* can generate force only in one direction. The force required to return the piston and rod to their original position is provided by an internal spring or an external weight. See **Figure 17-4**. *Double-acting cylinders* can generate force during both extension and retraction of the piston and rod assembly. As in hydraulic cylinders, there is a difference in the effective area of each side of the piston. This is due to the cross-sectional area of the piston rod. As a result, extension and retraction strokes have different force outputs.

Construction types

The head and cap are attached to the cylinder barrel in pneumatic cylinders using methods similar to those used for hydraulic cylinders. Many cylinders use the external tie-rod design to secure these parts. The catalog of one manufacturer lists pneumatic tie-rod cylinders with bores ranging from 3/4" to 14". The manufacturer indicates these cylinders can be produced with strokes of "any practical length."

Other cylinders use threads or snap rings to attach the cylinder head and cap to the barrel. These methods result in a compact actuator with a relatively smooth surface that provides easy cleaning. Ease of cleaning is important in many applications. Also, these cylinders

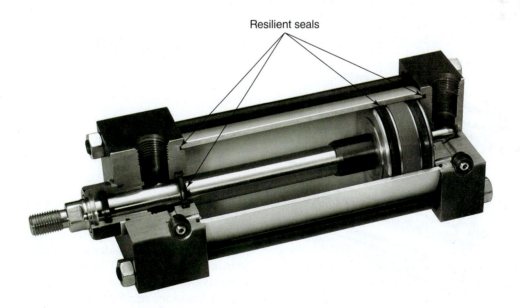

Resilient seals

IMI Norgren, Inc.

Figure 17-3. Resilient materials are typically used in pneumatic cylinders to ensure a positive seal between the bore of the cylinder and the piston.

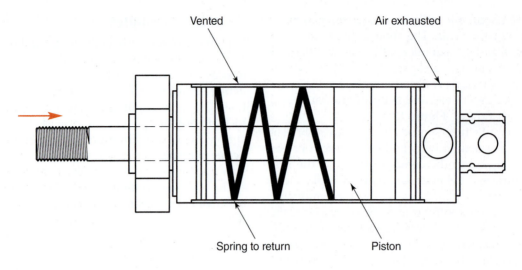

IMI Norgren, Inc.

Figure 17-4. A spring-return cylinder uses an internal spring to return the cylinder to the initial starting position after the work phase is completed.

allow easy disassembly for service. Bores of typically 1/2″ or less are available. The maximum stroke of these smaller-diameter units is often limited to 6″.

The head and cap can also be permanently attached to the barrel. In hydraulic cylinders, this is often done by welding because of the high system pressures the cylinder must withstand. The lower operating pressure of pneumatic systems allows the process to be done using a *metal rolling* technique. See **Figure 17-5**. The internal components of these cylinders cannot be serviced. The cylinders are considered throwaway parts. However, improved materials and seal designs allow these cylinders to provide service for extended periods before replacement is required.

Parker Hannifin

Figure 17-5. The barrel of a pneumatic cylinder can be attached to the end caps using metal rolling techniques. This method may be used because of the relatively low operating pressure of pneumatic systems.

17.1.4 Sizing a Pneumatic Cylinder

When selecting a cylinder for use in a pneumatic system, two factors must be considered to be certain the actuator will provide the desired performance:

- Required force output.
- Air consumption needed to provide the desired cycle-operating rate.

Force output is associated with the pressure available at the workstation. Cycle operation is directly related to the ability of the air-distribution system and, ultimately, the compressor to supply the volume of air needed to operate the actuator at the selected rate.

Required force output

Force is equal to pressure times the area affected:

$$F = P \times A$$

where:

$$F = \text{Force}$$
$$P = \text{Pressure}$$
$$A = \text{Area}$$

In pneumatic cylinders, as in hydraulic cylinders, force is directly related to the effective area of the cylinder piston. Refer to Chapter 9, *Actuators*, for details on calculating the cylinder size needed to produce the desired force output. The maximum operating pressure available to a pneumatic cylinder is controlled by the pressure regulator located at the workstation. However, the maximum pressure of the air in the distribution lines is normally higher than the workstation pressure. This can provide some flexibility when designing circuits and determining the required cylinder diameter.

EXAMPLE 17-1

Required Force Output of a Pneumatic Cylinder

Determine the force generated by a double-acting pneumatic cylinder during extension and retraction. The cylinder has a piston diameter of 1.5″ and a rod diameter of 0.5″. The system pressure is 40 psi.

$$\text{Force (lb)} = \text{Pressure (psi)} \times \text{Area (in}^2)$$

$$\text{Piston Area} = \pi \times \text{Radius}^2$$

$$= \pi \times \left(\frac{1.5''}{2}\right)^2$$

$$= 1.77 \text{ in}^2$$

$$\text{Rod Area} = \pi \times \text{Radius}^2$$

$$= \pi \times \left(\frac{0.5''}{2}\right)^2$$

$$= 0.196 \text{ in}^2$$

$$\text{Extension Force (lb)} = \text{System Pressure (psi)}$$
$$\times \text{Piston Area (in}^2)$$
$$= 40 \text{ psi} \times 1.77 \text{ in}^2$$
$$= 70.8 \text{ lb}$$

$$\text{Retraction Force (lb)} = \text{System Pressure (psi)}$$
$$\times (\text{Piston Area (in}^2)$$
$$- \text{Rod Area (in}^2))$$
$$= 40 \text{ psi} \times 1.57 \text{ in}^2$$
$$= 62.8 \text{ lb}$$

Determining air consumption

The *air consumption* of a cylinder in a functioning circuit should be determined in cubic feet of free atmospheric air per minute. This is how the capacities of a compressor and the air-distribution system are rated. Therefore, it is easier to check the effect of the circuit containing the actuator on the operation of the compressed-air system. If the air consumption of the cycling actuator approaches or exceeds the capacity of the compressor or distribution system, the actuator circuit will not operate effectively.

The air consumption of a cylinder in an operating circuit can be determined using this formula:

$$\text{Air Consumption} = V \times P_R \times N$$

where:

V = Volume of compressed air per cycle

$$V = \frac{(A_B \times S) + (A_R \times S)}{1,728}$$

where:

A_B = Cross-sectional area in square inches of the blind end of the cylinder piston

A_R = Cross-sectional area in square inches of the rod end of the cylinder piston

S = Length of cylinder stroke in inches

1,728 = Cubic inches per cubic foot

P_R = Absolute pressure ratio

$$= \frac{\text{Operating psig} + 14.7}{14.7}$$

N = Number of cylinder cycles per minute

The *absolute pressure ratio* (P_R in the formula) adjusts the volume of free air needed to produce the volume of air consumed at system pressure. The above formula is based on Boyle's law, which assumes the temperature of the air is constant throughout the process. In practice, temperatures will vary. Two additional factors can cause inaccuracies when using this formula:

- Air leaks in the circuit.
- Variations in the volume of the cylinder air chambers resulting from design factors such as cushions.

However, the calculation will produce the best estimate available. Individuals experienced in the design of pneumatic circuits often add a correction factor to the formula to compensate for various cylinder features, circuit designs, and load conditions.

EXAMPLE 17-2

Determining Air Consumption

A double-acting pneumatic cylinder has a piston blind area of 4 in², a rod end area of 3 in², and a stroke of 10″. If the cylinder operates 75 psi and completes 20 cycles per minute, how many cubic feet of free air are consumed in 10 minutes?

$$\begin{array}{c}\text{Total Free Air} \\ \text{Consumed (ft}^3)\end{array} = \begin{array}{c}\text{Air Consumption (cfm)} \\ \times \text{Time (min)}\end{array}$$

$$\text{Air Consumption (cfm)} = V \times P_R \times N$$

$$V \text{ (ft}^3) = \frac{(A_B \times S) + (A_R \times S)}{1,728}$$

$$= \frac{(4 \text{ in}^2 \times 10'') + (3 \text{ in}^2 \times 10'')}{1,728}$$

$$= \frac{40 \text{ in}^3 + 30 \text{ in}^3}{1,728}$$

$$= 0.04 \text{ ft}^3$$

$$P_R = \frac{75 \text{ psig} + 14.7 \text{ psia}}{14.7 \text{ psia}}$$

$$= 6.1$$

Substituting these values in the air consumption equation:

$$\text{Air Consumption (cfm)} = 0.04 \text{ ft}^3 \times 6.1 \times 20$$
$$= 4.88 \text{ cfm}$$

$$\text{Total Free Air Consumed (ft}^3) = 4.88 \text{ cfm} \times 10 \text{ min}$$
$$= 48.8 \text{ ft}^3$$

17.1.5 Cylinder Manufacturer Specifications

Manufacturers provide considerable detail concerning the various cylinders they produce. This includes information on:

- Construction materials.
- Available sizes.
- Pressure ratings.
- Specifications regarding cylinder features.

These details include data and illustrations covering factors such as:

- Duty service rating.
- Construction type.
- Pressure rating.
- Bore size.
- Piston rod diameter.
- Stroke length.
- Mounting style.
- Cushions.
- Rod-end configuration.
- Port fitting size.

Review the cylinder section of Chapter 9, *Actuators*, for details that apply to both pneumatic and hydraulic cylinders.

17.2 Pneumatic Rotary Motors

A pneumatic motor is the system component that converts the energy in compressed air into torque and rotary movement. These actuators are often called air motors or *rotary actuators*. See **Figure 17-6**. The torque generated by the motor is determined by the area of the driving mechanism and the air pressure in the system. The speed is determined by the volume of the internal motor chambers per revolution and the amount of air allowed to enter the component.

Controlling air pressure and flow rates can produce a wide range of torque outputs and operating speeds. However, the compressibility of air often affects output and speed.

17.2.1 Basic Rotary Motor Design

Pneumatic rotary motors use many of the basic design concepts found in hydraulic pumps, hydraulic motors, and pneumatic compressors. The motor uses these concepts to convert the potential energy of compressed air into torque and rotation. Several different designs are used. However, each design must contain internal chambers that can vary in size to allow air to move through the motor. The design must also contain surfaces that air pressure can act against to produce the desired output shaft rotation and torque.

A vane design may be used to explain the construction and operation of a basic pneumatic motor. The motor is constructed using three component groups:

- Housing.
- Rotary internal parts.
- Power output shaft.

The housing contains inlet and outlet ports, an internal chamber to hold the rotating motor parts, bearings to support the power output shaft, and fittings to plumb and mount the motor. The rotating internal parts include a rotor containing slots fitted with sliding vanes. This rotating assembly is mounted

Atlas Copco

Figure 17-6. Air motors are used in pneumatic systems to produce torque and rotary movement.

on the output shaft that delivers the power produced by the motor.

The rotating internal parts of the motor are offset in the internal chamber of the housing. This allows the chambers formed by the vanes and the surfaces of the rotor and housing to change in size as the rotor turns. When compressed air is directed to the inlet port, it flows into the smallest-volume chamber. As air pressure increases in the chamber, pressure is applied to the exposed surface of the vanes. The force produced causes the rotor and vane assembly to turn, producing the torque output of the motor. When the rotor has turned to the point that the chamber size begins to decrease in volume, the chamber is connected to the output port. This allows the air to exhaust to the atmosphere. This

process continues as each motor chamber is exposed to system air pressure. See **Figure 17-7**.

The speed of an air motor is determined by the displacement of the motor chambers and the volume of air allowed to pass through the unit. Motor torque depends on the internal structure of the motor and the air pressure available to the motor circuit.

17.2.2 Rotary Motor Design, Construction, and Operation

Several different designs of air motors are used in pneumatic systems. These designs are used in applications ranging from small, portable hand tools to high-horsepower industrial applications. However, each

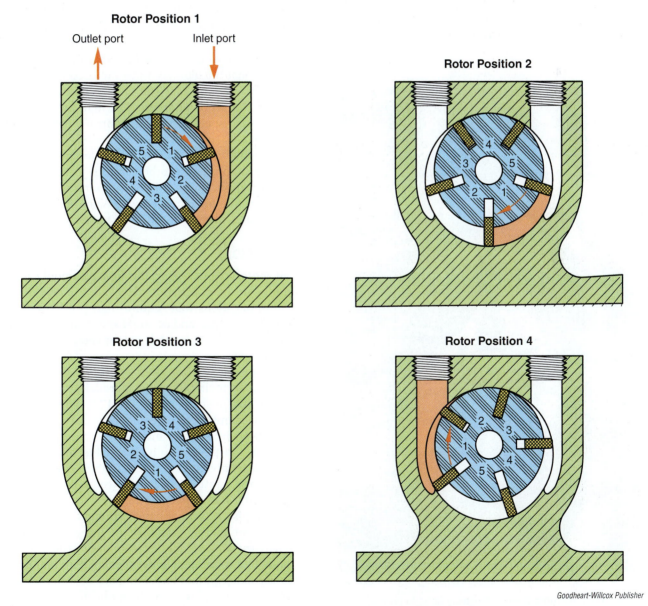

Goodheart-Willcox Publisher

Figure 17-7. The sequence of events in the operation of a vane motor begins with the exposure of the expanding air chambers to high-pressure air and ends with the exhaust of the air to the atmosphere.

design depends on compressed air as the source of air volume and pressure needed for operation.

Air motors are available in vane, piston, gerotor, turbine, and other designs. Generally, these units are similar to the hydraulic motors discussed in Chapter 9, *Actuators*. This section briefly reviews each design with primary emphasis placed on variations related to pneumatic applications. Designs that are only used in pneumatic applications are discussed in detail.

Vane motors

Vane motors are the most common of the rotary air motor designs. They are used in a variety of portable tool applications, but are also found in stationary machines. Portable applications include screwdrivers, grinders, drills, impact wrenches, and many other tools, **Figure 17-8**. Stationary applications can include serving as the prime mover of a complex machine or as the driver for a single machine function, such as mixing or drilling.

Atlas Copco

Figure 17-8. Air motors are extensively used to produce rotary movement in both portable hand tools and stationary equipment.

The construction and operation of a vane motor is discussed earlier and shown in **Figure 17-7**. Manufacturers typically produce vane motors with three or more vanes. Increasing the number of vanes reduces internal air leakage. This results in a motor with better low-speed performance. However, increasing the number of vanes also increases the manufacturing costs and internal friction. Internal friction decreases overall motor efficiency.

The seal between the tips of the vanes and the walls of the motor body is an important factor in vane motors. The seal must be maintained over the full range of speeds and loads for the motor. If this seal is not maintained, the motor will be difficult to start under load. Also, efficiency is lost at operating speeds. Springs or air pressure are often used to maintain the seal. To pressurize the underside of the vane, a passageway leads from the inlet port to the chamber formed by the rotor slot and vane. Once the motor is operating, centrifugal force also helps create a seal.

Vane motors are available in a wide range of sizes. Tiny, fractional-horsepower units are used in portable tools. Large motors used in stationary applications may produce 50 horsepower or more. Designs are available with operating speeds from less than 100 rpm to over 20,000 rpm.

Both nonlubricated and lubricated designs are available. Lubrication needs to be carefully considered in high-speed applications. It is also important if a motor will remain idle for long periods. When the motor is idle, lubricants may drain from the vane-tip area.

Piston motors

Piston motors were one of the first commonly used designs of air rotary actuators. They are still very desirable in applications that require high power output, high starting torque, and accurate speed control. These motors are available in both axial-piston and radial-piston designs.

Axial-piston motors have cylinders and pistons arranged with their centerlines parallel to the centerline of the output shaft. See **Figure 17-9**. A swash plate or other mechanism converts the reciprocating motion of the pistons into rotary motion of the output shaft. Positioning the pistons inline with the output shaft provides a relatively compact actuator. This may be important if the ratio of output power to physical size is important for an application.

Radial-piston motors have cylinders and pistons perpendicular to the centerline of the output shaft. Designs vary considerably. Some manufacturers use a design with a rotating cylinder block. Other manufacturers use a design that has stationary cylinders. High starting torque and a smooth application of power makes radial-piston motors desirable for applications where heavy starting loads are encountered.

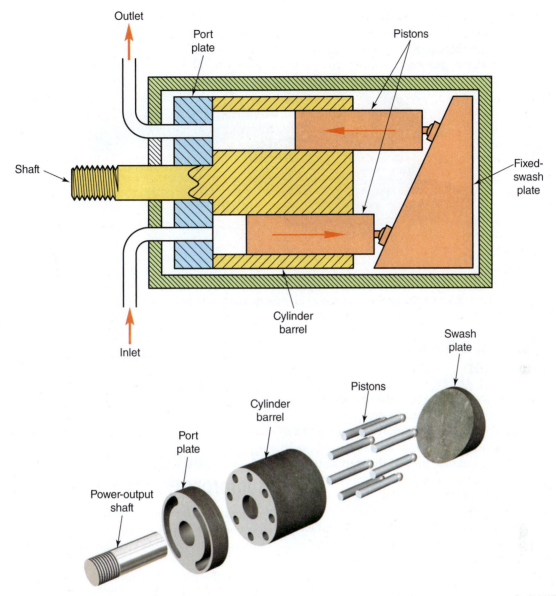

Figure 17-9. The basic internal structure of an axial-piston pneumatic motor.

Goodheart-Willcox Publisher

Piston motors are available in a variety of sizes, from less than 1 horsepower to over 50 horsepower. Axial designs are found in a wide variety of applications, from hand tools to stationary machines. Radial designs are typically used in applications requiring lower maximum speeds. The radial design is especially suitable for use in situations that require high starting torques.

Piston motors may be nonlubricated or lubricated. If lubrication is required, oil is supplied as a mist from a standard lubricator located in the air-distribution system. A separate oil sump and pump located in the motor crankcase may be used to lubricate the crankshaft and valve mechanisms of larger, radial-piston motors.

Turbine motors

Turbine motors use high-speed air directed onto a turbine wheel to produce rotary motion, **Figure 17-10**. High airspeed is produced when compressed air is passed through a nozzle. The nozzle converts the potential energy of the static, high-pressure air into kinetic energy in high-speed air. When the moving air strikes the surface of the turbine, it slows and transfers energy to the turbine. As a result, the turbine moves and rotates the output shaft.

Turbine motors are limited in their application because of their high output speeds. Output speeds can exceed 100,000 rpm. Their most common application is for small, high-speed grinding operations, especially in tool and die work.

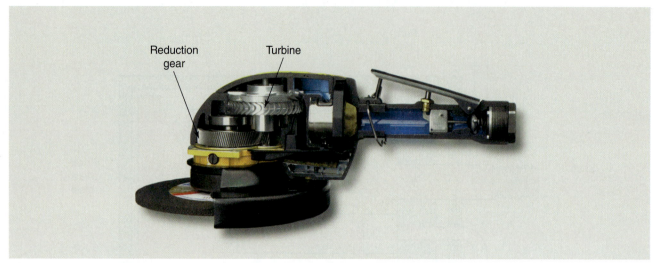

Atlas Copco

Figure 17-10. The internal design of turbine motors produces high-velocity air movement that turns the turbine and output shaft at a very high speed. In some designs, a gear reduction may be used to reduce the output speed.

17.2.3 Selecting a Rotary Pneumatic Motor

Manufacturers supply considerable detail concerning the performance of the motors they produce. Tables are usually available that provide details on maximum speed, power output, torque, and air consumption. Many companies also provide graphs that can be used to identify horsepower, torque, and air consumption over a wide range of operating speeds.

It is easy to change the speed and power of a pneumatic motor by controlling system air pressure and airflow rate. However, several factors need to be kept in mind when selecting a motor for an application:

- The consumption of air increases as motor operating speed or air pressure increases.
- Higher torque output is required to start a load than for maintaining the load at operating speed.
- The torque output of a motor varies with the size of the load.
- The operating speed of a pneumatic motor changes as the load on the motor increases or decreases.
- Pneumatic motors can be held in the stalled condition without causing damage to the motor or system.

Selecting a pneumatic motor for a given situation requires careful analysis of the loads that will be encountered. Once this has been completed, examine the resulting data using the five general factors listed above. Then, make adjustments to minimize the possibility of oversizing or undersizing the motor. A motor can be selected using this information and data sheets from manufacturers. See **Figure 17-11**.

One manufacturer suggests that the selected motor provide the needed power using only 2/3 of the workstation pressure. This safety factor allows the additional air pressure to be used for starting loads and unexpected overloads.

17.3 Pneumatic Reciprocating Motors

Pneumatic reciprocating motors produce linear motion that completes tasks with either percussive or nonpercussive techniques to transfer energy. Percussive devices use multiple impacts of the cylinder piston to produce the desired results, **Figure 17-12**. Nonpercussive devices use repeated, reciprocating movement to complete a required task. Reciprocating motors are designed to continuously operate. This provides multiple cycles for each minute of tool operation.

17.3.1 Basic Reciprocating Motor Design

A percussive reciprocating motor is a cylinder closed at one end. The opposite end is fitted with the base of a working tool or a separate component called an anvil. The anvil is used on some large motors to increase energy transfer in heavy-load situations. The cylinder is divided into two air chambers by a free-floating piston. Air entering or exiting the two cylinder chambers is controlled by a control valve or ports.

When the chamber on the closed end of the cylinder is connected to the compressed-air-distribution line, the incoming air moves the piston toward the tool base or anvil. Air on the tool side of the piston is

Model	Free Speed (rpm)	Maximum Power (hp)	Torque at Maximum Power (in lb)	Length (in)	Weight (lb)
JRD 4500	4,500	.75	19.5	6.375	2.25
JRD 2500	2,500	.75	24.6	6.375	2.25
JRD 2000	2,000	.75	37.9	6.375	2.25
JRD 850	850	.50	89.7	8.125	3.50
JRD 500	500	.50	152.8	8.125	3.50
JRD 350	350	.50	223.5	8.125	3.50
JRD 100	100	.25	364.3	9.626	4.25
JRD 50	50	.25	364.3	9.625	4.25

Goodheart-Willcox Publisher

Figure 17-11. Information provided by the manufacturer needs to be carefully examined when selecting a pneumatic motor for a specific application.

Atlas Copco

Figure 17-12. A jackhammer is a very common application of a reciprocating air motor.

vented to the atmosphere. When the rapidly moving piston reaches the end of its travel, it strikes the tool base or anvil, thus transferring energy to the tool.

Next, the control valve or ports redirect the airflow to return the piston to its original position. When the piston reaches the closed end, the air is redirected to again power the piston toward the tool base or anvil. This process is continually repeated to produce the desired results. The impact force of the piston is determined by the system air pressure and area of the piston.

A nonpercussive reciprocating motor has a rod attached to the cylinder piston. The external end of the rod is attached to the component that requires reciprocating movement for operation. As in the percussive design, a valve or ports in the cylinder provide the control needed to produce a continuous, reciprocating action.

17.3.2 Applications for Reciprocating Motors

A wide variety of industries make use of reciprocating motors, including construction, mining, and manufacturing. These applications range from handheld tools to large units mounted on heavy support equipment. The following descriptions provide basic information on several common applications.

The *paving breaker* is the reciprocating motor application that is probably most visible to the general public. Often called a jackhammer, it is used at many street maintenance and building demolition sites to break up concrete or asphalt. These units can also be fitted with a spade blade to cut through hard-packed soil. Paving breakers are available in several sizes. Size is usually designated by the physical weight of the unit. For example, one manufacturer produces models weighing 30–70 pounds. When operating at 90 psi, these units consume approximately 40–70 cubic feet per minute of compressed air. These models produce 1,500–1,800 impacts per minute at their rated pressure and air consumption. Large versions of these units are often mounted on tractor, skid steer, or crawler units to provide larger capacity and greater mobility.

Scaling, chipping, and riveting hammers are other examples of reciprocating motor application. *Scaling hammers* are small hand tools used to remove flux and spatter from welds in fabricated products. Various-shaped chisels are used with the tool to fit the configuration of the material around the weld. *Chipping hammers* perform tasks as diverse as cleaning foundry castings or roughing the surface of concrete to ensure the attachment of surface coatings, **Figure 17-13**. These tools are usually larger than scaling hammers. *Riveting hammers* are designed to form the head of the rivets used in the

construction of airplanes and other equipment. These units typically provide greater control over the speed of operation and impact force to ensure better forming of the rivet head. Riveting tools are available in a number of sizes. Maximum strokes range from under 1" to over 6". Impact rates range from around 1,000 to over 2,500 per minute.

Tampers and *rammers* are used to compact sand in foundry molds and dirt at construction sites, **Figure 17-14**. These tools vary slightly in design from the previously described tools. The reciprocating piston, piston rod, and tool are directly attached to form a single unit. This allows piston movement to withdraw the tool, as well as produce impact during extension. These tools are available as relatively small hand tools for ramming sand in small foundry molds. For compacting dirt at building sites, tampers may be several feet in length.

Atlas Copco

Figure 17-13. Scaling and chipping hammers powered by pneumatic reciprocating motors are commonly used in foundries to clean metal castings.

Badger Iron Works, Inc.

Figure 17-14. Pneumatic reciprocating motors power rammers that are used to compact sand in foundry molds.

17.4 Other Pneumatic-Powered Functions and Equipment

Compressed air is used in a wide variety of processes and equipment. Many of these are not visible to the general public. Over a period of years, equipment has been designed or modified to reduce costs, allow smaller operations, and even allow consumers to use techniques formerly restricted to major manufacturers. The following sections discuss several processes or equipment designs contributing to our lifestyles that depend on compressed air systems for operation.

17.4.1 Process Functions

Compressed-air systems power processes without supplying force to traditional linear or rotary actuators. These processes include spraying, drying, material agitation, material transfer, and others. In many of these situations, air is seldom considered a primary factor. In reality, however, air serves as a critical part of the process.

Spraying

Painting is perhaps the most recognized spraying operation. However, painting is only one of many situations in which materials are applied by *spraying*. The application of agricultural chemicals for weed and insect control is also an example of spraying, **Figure 17-15**. Spraying heavy-texture materials in building construction is another example. Although many airless techniques have been developed for the spray application of materials, air still plays an important role in many spraying processes. Quality clean and dry air is critical when compressed air is used to power many of these application systems.

Drying

Air is often used for *drying* purposes. These situations range from drying cars at the local car wash to the concentration of materials in chemical producing industries. The movement of air over a surface or through a liquid removes unwanted liquid. This produces a drier solid or more concentrated liquid.

Material agitation

A number of processes require continuous *material agitation* to prevent the settling of solids or to maintain a uniform distribution of materials in a mixture. In some situations, compressed air can be directly used to continuously agitate the materials. A factor that must be considered when thinking about this type of application is the fact that air can cause oxidation. This may interfere with the primary process. In other applications, such as sewage treatment, agitation with

USDA

Figure 17-15. Applying insecticide or herbicide is an example of spraying. When compressed air is used for spraying, it is a process function, rather than a producer of force and movement.

compressed air not only keeps the material mixed, but also supplies oxygen to the bacteria converting the waste materials, **Figure 17-16**.

Material transfer

Compressed air is often used to expedite the movement of liquid, powder, or granular materials. Various techniques of *material transfer* are used in industries as diverse as food processors and cement production. Compressed air is used for the unloading and loading of railroad cars and trucks. Materials are often transferred to processing areas and then returned to storage using air-powered equipment.

17.4.2 Other Mechanical Equipment

A wide variety of pneumatic tools and equipment have been developed to serve various industrial situations. Many of these have been refined and modified to fit many situations beyond their original application. The following sections discuss a number of air-powered items that are commonly found in industrial or consumer applications.

Nozzles

Air nozzles convert the potential energy of static, compressed air into kinetic energy of rapidly moving air.

They are commonly used with pneumatic systems. Most people who have worked with even the simplest of compressed air systems have probably used a nozzle as a cleaning device. Nozzles may look simple, but the theory of their design can be complex. They can be used with turbines to produce rotor motion or ejectors to produce vacuum or mix fluids.

Nozzles in their simplest forms are either the convergent or convergent-divergent designs shown in **Figure 17-17**. Varying the length of the curved portion of these designs provides a means to control the velocity and pressure of the air as it passes through the nozzle. On the other hand, in a simple ejector, a vacuum is formed at the nozzle. This vacuum draws in and mixes a second fluid with the air as it moves through a diffuser section at the outlet of the device.

Figure 17-18 shows simple *blowguns* used in many shops for cleaning. They use the principles of a nozzle in their operation. These may be simple devices, but, from a safety standpoint, they need to be treated with a great deal of respect. Eye protection is needed due to flying chips and particles from cleaning operations. Also, the air pressure may be high enough to inject air through the skin. This can occur if the nozzle is brought into direct contact with the skin. The blowgun shown in **Figure 17-19** eliminates this risk by including a simple bleed-off just before the nozzle discharge. This bleed-off prevents the buildup of high-air pressure if the nozzle is blocked.

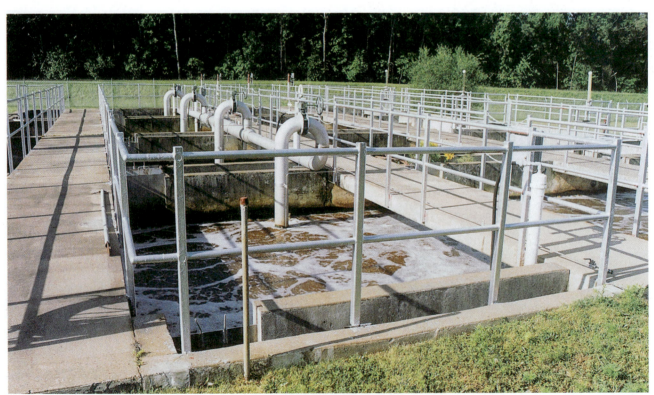

Goodheart-Willcox Publisher

Figure 17-16. The agitation of sewage in a waste-treatment facility is often done with compressed air.

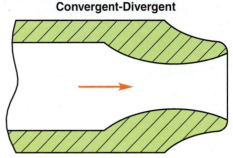

Convergent   **Convergent-Divergent**

Figure 17-17. Nozzles convert the potential energy of static, compressed air into kinetic energy in rapidly moving air.

CAUTION

Never point a blowgun at anyone. Blowguns are capable of causing serious injuries, especially to the eyes and ears.

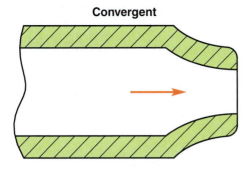

Figure 17-18. The simple blowgun, based on the principles of a nozzle, is used to clean dust and chips from the surfaces of machined parts.

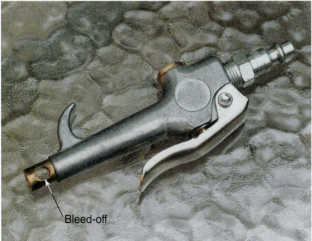

Bleed-off

Figure 17-19. This blowgun design incorporates a vent or bleed-off near the nozzle tip that serves as a safety device.

Impact wrenches

Manufacturing and service operations often involve the tightening or removal of nuts or bolts. *Impact wrenches* can make this process quicker and more efficient, **Figures 17-20** and **17-21**. Compressed air provides rapid acceleration of the impact mechanism. It also provides good control of torque and rotational speed.

These power tools produce rotary motion to run down the nut or bolt until resistance is met. Once resistance is met, a series of rapid, rotary blows tighten the nut or bolt to a preset torque. Reversing the mechanism allows the wrench to deliver a series of blows to loosen the nut or bolt and then rapidly turn the nut or bolt for quick removal. These tools are available in models

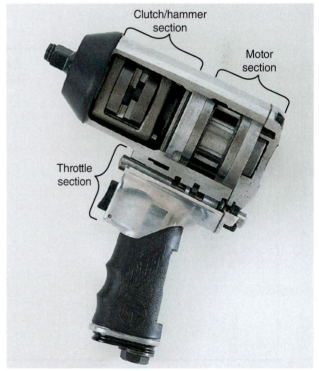

Clutch/hammer section

Motor section

Throttle section

Figure 17-20. The internal mechanism of an impact wrench produces a series of rapid blows that apply torque to a threaded fastener for either installation or removal.

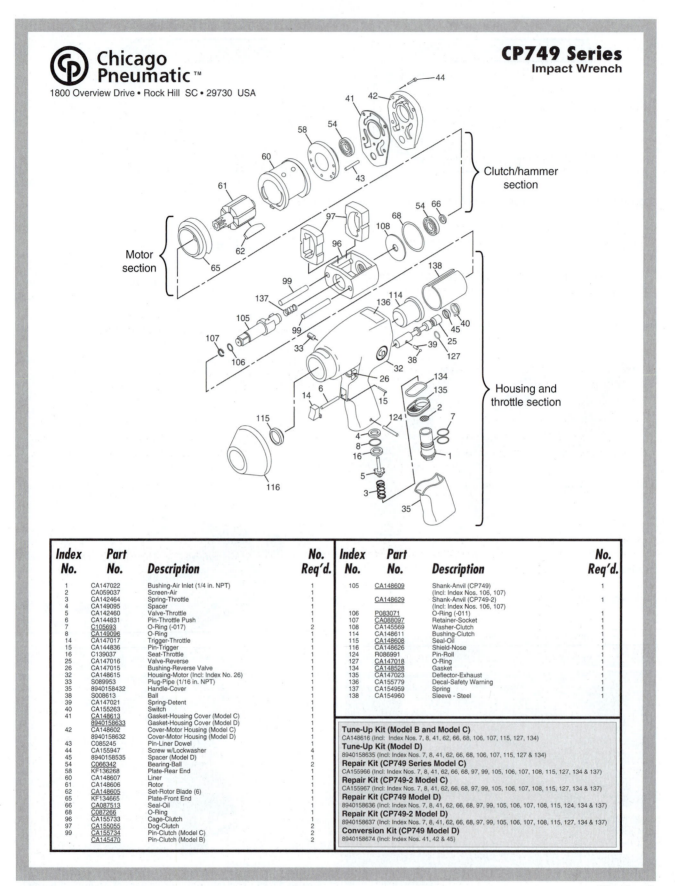

Chicago Pneumatic™
1800 Overview Drive • Rock Hill SC • 29730 USA

CP749 Series
Impact Wrench

Clutch/hammer section

Motor section

Housing and throttle section

Index No.	Part No.	Description	No. Req'd.
1	CA147022	Bushing-Air Inlet (1/4 in. NPT)	1
2	CA059037	Screen-Air	1
3	CA142464	Spring-Throttle	1
4	CA149095	Spacer	1
5	CA142460	Valve-Throttle	1
6	CA144831	Pin-Throttle Push	1
7	C105693	O-Ring (-017)	2
8	CA149096	O-Ring	1
14	CA147017	Trigger-Throttle	1
15	CA144836	Pin-Trigger	1
16	C139037	Seat-Throttle	1
25	CA147016	Valve-Reverse	1
26	CA147015	Bushing-Reverse Valve	1
32	CA148615	Housing-Motor (Incl: Index No. 26)	1
33	S089953	Plug-Pipe (1/16 in. NPT)	1
35	8940158432	Handle-Cover	1
38	S008613	Ball	1
39	CA147021	Spring-Detent	1
40	CA155263	Switch	1
41	CA148613	Gasket-Housing Cover (Model C)	1
	8940158633	Gasket-Housing Cover (Model D)	1
42	CA148602	Cover-Motor Housing (Model C)	1
	8940158632	Cover-Motor Housing (Model D)	1
43	C085245	Pin-Liner Dowel	1
44	CA155947	Screw w/Lockwasher	4
45	8940158535	Spacer (Model D)	1
54	C066342	Bearing-Ball	2
58	KF136268	Plate-Rear End	1
60	CA148607	Liner	1
61	CA148606	Rotor	1
62	CA148605	Set-Rotor Blade (6)	1
65	KF134665	Plate-Front End	1
66	CA087513	Seal-Oil	1
68	C087266	O-Ring	1
96	CA155733	Cage-Clutch	1
97	CA155055	Dog-Clutch	2
99	CA155734	Pin-Clutch (Model C)	2
	CA145470	Pin-Clutch (Model B)	2

Index No.	Part No.	Description	No. Req'd.
105	CA148609	Shank-Anvil (CP749) (Incl: Index Nos. 106, 107)	1
	CA148629	Shank-Anvil (CP749-2) (Incl: Index Nos. 106, 107)	1
106	P083071	O-Ring (-011)	1
107	CA088097	Retainer-Socket	1
108	CA155569	Washer-Clutch	1
114	CA148611	Bushing-Clutch	1
115	CA148608	Seal-Oil	1
116	CA148626	Shield-Nose	1
124	R086991	Pin-Roll	1
127	CA147018	O-Ring	1
134	CA148528	Gasket	1
135	CA147023	Deflector-Exhaust	1
136	CA155779	Decal-Safety Warning	1
137	CA154959	Spring	1
138	CA154960	Sleeve - Steel	1

Tune-Up Kit (Model B and Model C)
CA148616 (Incl: Index Nos. 7, 8, 41, 62, 66, 68, 106, 107, 115, 127, 134)
Tune-Up Kit (Model D)
8940158635 (Incl: Index Nos. 7, 8, 41, 62, 66, 68, 106, 107, 115, 127 & 134)
Repair Kit (CP749 Series Model C)
CA155966 (Incl: Index Nos. 7, 8, 41, 62, 66, 68, 97, 99, 105, 106, 107, 108, 115, 127, 134 & 137)
Repair Kit (CP749-2 Model C)
CA155967 (Incl: Index Nos. 7, 8, 41, 62, 66, 68, 97, 99, 105, 106, 107, 108, 115, 127, 134 & 137)
Repair Kit (CP749 Model D)
8940158636 (Incl: Index Nos. 7, 8, 41, 62, 66, 68, 97, 99, 105, 106, 107, 108, 115, 124, 134 & 137)
Repair Kit (CP749-2 Model D)
8940158637 (Incl: Index Nos. 7, 8, 41, 62, 66, 68, 97, 99, 105, 106, 107, 108, 115, 127, 134 & 137)
Conversion Kit (CP749 Model D)
8940158674 (Incl: Index Nos. 41, 42 & 45)

Chicago Pneumatic

Figure 17-21. An exploded view of an impact wrench and its associated parts list.

ranging in sizes suitable for use with 1/4" diameter bolts to large bolts 10" or more in diameter requiring torque of several thousand foot-pounds.

Multiple-spindle versions of these devices called *nutsetters* are used on many industrial assembly lines. See **Figure 17-22**. They can be used to set 25 or more fasteners at one time using fastener-feed devices. This significantly reduces assembly times. Some nutsetters use spindle stall, rather than impact control, to establish the maximum torque applied to the fastener.

Nail drivers

Nail drivers are used to drive fasteners in building construction, furniture assembly, and other applications, **Figure 17-23**. These devices have become very popular with several different designs manufactured. They are used by small contractors, major home developers, and manufacturing operations. Today, they can also be found in the home shop. Models are available to drive fasteners such as staples, small brads, large-head shingle nails, and nails 2" or more in length.

The drivers contain a cylinder, piston, piston rod that impacts the fastener, return air chamber, and valve to trigger the driving action. In addition, a mechanism to hold and deliver the fasteners and a safety mechanism are incorporated into the driver, **Figure 17-24**. The safety mechanism prevents the driving action from being triggered unless the tip of the driver is in firm contact with the material to be fastened. This mechanism is

Goodheart-Willcox Publisher

Figure 17-23. Pneumatic nail drivers are used in building construction to drive nails and other fasteners.

essential to reduce the possibility of fasteners becoming projectiles, which can cause serious injury.

When the driver is connected to the compressed-air line, the piston is in the retracted position. The piston rod is aligned with a fastener ready to be driven. Pressing the tip of the driver against the material to be fastened releases the safety catch. At that point, pulling the trigger of the driver releases compressed air into

Atlas Copco

Figure 17-22. Pneumatic nutsetters apply multiple fasteners at one time and are commonly used on assembly lines.

the cylinder above the piston. This causes rapid piston movement. The rapidly moving piston and piston rod impact the fastener, driving it into place. The movement of the piston also compresses the air on the rod side of the piston. This develops pressure in the return-air chamber.

When the trigger is released, the compressed-air supply is blocked and the high-pressure air used to drive the piston is released to the atmosphere. The pressure in the return-air chamber is then used to retract the piston and piston rod to their starting position. When the driver is raised from the material being

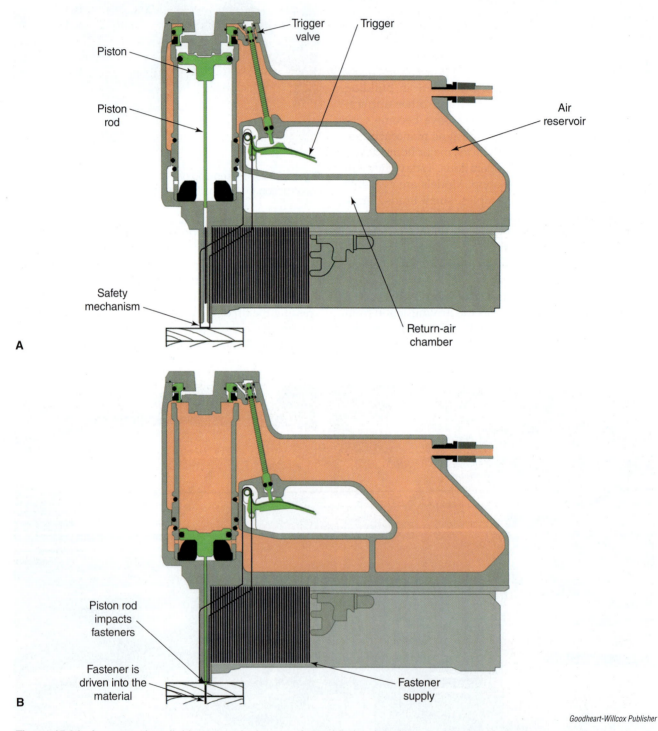

Figure 17-24. A pneumatic nail driver uses the impact of a rapidly moving piston to drive brads, staples, and nails. A—Idle position. B—The tool is on the material surface so the safety is released. When the trigger is pulled, the fastener is driven into the material.

fastened, the safety mechanism is reset and the driver is ready for driving another fastener at the next location.

Rock drills

Pneumatic *rock drills* are often used to make rock cuts in quarries, mines, and highway construction. The result of this drilling is often visible along highways constructed in hilly and mountainous areas. The sheer rock faces may show a series of bore marks spaced about every 2 1/2 feet. These bore holes allow placement of explosive charges that break the rock between the holes to produce a relatively smooth face. The rock and soil in front of this line of bore holes are then removed to expose the wall. The drilling and removal of rock in quarries and mines use similar methods, **Figure 17-25**.

A number of different methods are used to drill holes in rock. Pneumatic rotary drilling equipment is typically used for these operations. The compressed air is also directed to the bottom of the hole to cool the drill bit. The air also blows rock chips and other waste material to the surface. Some techniques use an air-powered, percussive tool at the end of the drill to break up the rock and increase the cutting speed, **Figure 17-26**. This down-the-hole drill produces a high blow rate using the principles of a pneumatic reciprocating motor. It is highly effective in drilling through rock. The units are available in a number of sizes.

Atlas Copco

Figure 17-25. Compressed air powers drilling equipment commonly used to bore holes in rock in road construction, rock quarrying, and mining operations.

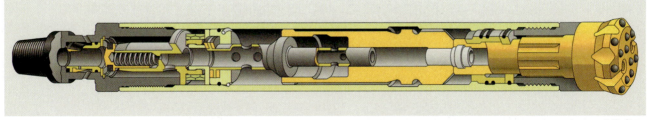

Atlas Copco

Figure 17-26. Pneumatic down-the-hole rock drills are used in the well-drilling, mining, quarrying, and construction industries.

Rock drills are available as hand tools. In large drilling operations, however, they are major pieces of equipment mounted on a truck, tractor, crawler, or even barge. Included in this mounted equipment are a prime mover, power-transmission system, mast or tower to support the drill, hoisting equipment to lift the drill, dust-control equipment, and equipment-leveling devices.

Grippers

The use of robots in manufacturing has added the gripper to the list of actuators available in the fluid power area. A *gripper* is used by a robot to place and remove parts in a machining, assembly, or other operation. Many of these units are pneumatic powered, although models are available using hydraulic actuation.

Figure 17-27 shows two models of grippers. Some grippers allow adjustment of the opening and closing of the jaw and the force applied to the part being handled. Robot programming is responsible for locating and timing the opening and closing of the gripper. However, the structure of the gripper must provide the accuracy required for consistent jaw clearance and gripping force.

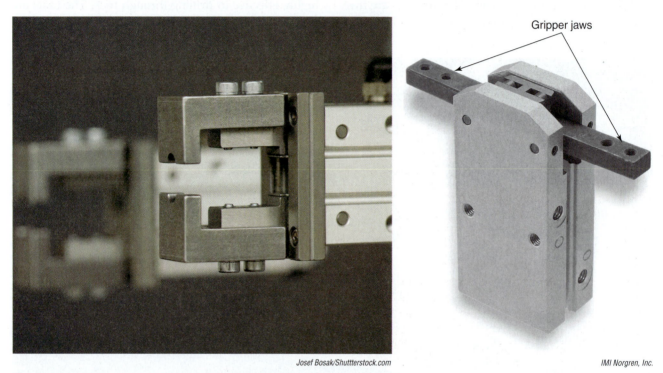

Gripper jaws

Josef Bosak/Shuttterstock.com

IMI Norgren, Inc.

Figure 17-27. Grippers are used by industrial robots to pick up and place materials.

Summary

- Pneumatic cylinders, motors, and a variety of other actuators convert the potential energy of compressed air into force and movement.

- Cylinders provide straight-line movement and force for use in mechanically operated equipment.

- Lower pneumatic system operating pressures allow light materials and metal rolling to be used in cylinder construction. Water vapor present in compressed air requires the use of corrosive-resistant materials and coatings for many component parts.

- Cylinder force output and absolute air consumption needed to produce the desired system performance must be considered when selecting the appropriate size for use in a pneumatic circuit.

- System air pressure and the effective area of the cylinder piston determine the force output of a cylinder.

- Manufacturers provide a variety of specific information about the cylinders they produce, including construction materials, sizes available, pressure ratings, and detailed information on specific features, such as cushions and mountings.

- Pneumatic rotary motors convert the potential energy of compressed air into torque and rotational movement that is used to power mechanical devices. The torque produced is dependent on the air pressure available for system operation and the internal structure of the motor. The operating speed is determined by the internal displacement of the motor per revolution and the volume of compressed air passing through the motor.

- Pneumatic rotary motors are available using vane, piston, geroter, turbine, and a number of other specialized designs.

- Pneumatic rotary motors range in size from fractional horsepower units used in handheld power tools to motors producing over 50 horsepower used in stationary installations.

- Most rotary air motor manufacturers publish tables and graphs that provide details about motor horsepower, torque, and air consumption over a wide range of operating speeds.

- Reciprocating pneumatic motors use percussive or nonpercussive techniques to transfer energy from compressed air to a workpiece. Percussion tools provide multiple physical impacts to overcome resistance. Nonpercussion devices generally repeat a cycle to provide repeated linear motion that operates a machine.

- Reciprocating motors are used to power scaling, chipping, riveting, tamping, and ramming tools found in the foundry, construction, and general metal fabrication industries.

Internet Resources

The following are some useful resources available on the Internet. Enter a company or organization name into a search engine to access its website. Explore the various areas of the sites to discover useful fluid power resources.

Atlas Copco. Provides links to the wide selection of products offered by this company. Review information on several of the air-powered tools available.

Bimba Manufacturing Company. Lists library resources. Technical Tips include listings of several factors users must understand and regularly follow to maximize cylinder performance and operating life.

Apex Tool Group. Provides information on the construction and performance of vane, axial-piston, and radial-piston pneumatic motors. Includes basic air motor selection procedures and formulas for one brand of motor.

Chapter Review

Answer the following questions using information in this chapter.

1. The force produced by a pneumatic cylinder is dependent on _____.
   A. system pressure
   B. total external area of the piston
   C. system pressure times cylinder rod area
   D. system pressure times the area of the blind side of the cylinder piston

2. *True or False?* Nonlubricated pneumatic cylinders are used in applications requiring oil-free air, such as the food industry.

3. The lower operating pressure of pneumatic systems allows the head and cap of the cylinder to be attached using a _____ technique.
   A. welding
   B. threading
   C. metal rolling
   D. torque

4. *True or False?* The air-consumption rate of a pneumatic cylinder during operation is expressed in cubic feet of free air per minute.

5. What primary type of information is usually provided by cylinder manufacturers in their catalogs and data sheets?
   A. Construction materials.
   B. Pressure ratings.
   C. Specifications on cylinder features.
   D. All of the above.

6. What determines the speed of a pneumatic motor?
   A. Air displacement and airflow.
   B. Torque and air speed.
   C. Force output and torque.
   D. All of the above.

7. Name two techniques used to ensure the seal between vanes and walls of the internal chamber of a vane air motor.
   A. Centrifugal force and air pressure.
   B. Vane motor and swash plate.
   C. Centrifugal action and air pressure.
   D. Speed of the swash and port plates.

8. *True or False?* Piston air motors are available in either radial-piston or balanced-vane designs.

9. Some _____ air motor designs can operate at speeds exceeding 100,000 rpm.
   A. piston
   B. vane
   C. turbine
   D. gerotor

10. An air motor should be sized to operate on _____ of the workstation pressure to allow for the extra power needed for starting loads and unexpected overloads.
    A. 1/4
    B. 1/3
    C. 1/2
    D. 2/3

11. A nonpercussive reciprocating motor has a(n) _____ attached to the cylinder piston.

12. *True or False?* A riveting hammer makes use of nonreciprocating motors in its design.

13. *True or False?* A nutsetter is used in an assembly line to run down and torque multiple fasteners in a single operation.

14. Rock drills are available as _____ tools or large drilling operations equipment.

15. *True or False?* Robots often make use of pneumatic grippers to pick up and place materials.

Apply and Analyze

1. Compare and contrast the components, construction materials, and operation of pneumatic cylinders with hydraulic cylinders. What factors account for the similarities and differences?

2. List three applications in which a single-acting pneumatic cylinder is typically used. For each application, explain why this type of cylinder is the best choice.

3. You must choose a single-acting cylinder to move a 100 lb load 10″. Maximum system pressure is 50 psi, but pressure at the cylinder could be reduced to 20 psi if an additional pressure regulator is installed in the circuit.
   A. What cylinder piston diameters would provide the required force for 50 psi or 20 psi?
   B. For both pressures and cylinders, how much free air in cubic feet per minute is required if the cylinder must operate at 50 cycles per minute?
   C. Which cylinder size would you choose to install? Why?

4. A double-acting cylinder has a piston blind side area of 12.5 in^2, a rod side area of 10.7 in^2, and a stroke of 12″. The cylinder completes 30 cycles per minute at pressures of 50 psi.
   A. What is the force generated on the blind side and rod side of the piston?
   B. How much standard free air in cfm is required to operate the cylinder for 60 minutes?
   C. If the retraction of the piston requires the generation of less force, how much would air consumption be reduced if the air pressure delivered during this portion of the cycle were delivered at 20 psi?

5. A simple pneumatic circuit consists of a loaded vane motor driven by air supplied by a compressor.
   A. What changes in system operation will occur if the load on the motor is doubled?
   B. What specific changes could you make to your system to increase the speed of the pneumatic motor's output shaft? How would these changes affect overall system operation?
   C. If you replace the vane motor with one that has twice as many blades, how might your system operation be affected?

Research and Development

1. Applications in which oil-free air is used require special pneumatic actuators. Research a manufacturer's line of nonlubricated pneumatic cylinders or motors to determine the construction methods and materials used to allow these actuators to operate effectively without additional lubrication. Prepare a report that compares these methods and materials with those used in a basic pneumatic cylinder or motor.

2. Cushioning in a cylinder is used to slow the movement of the piston as it nears the end of its stroke, reducing the stress on the cylinder components as well as reducing system noise and vibration. Investigate the various cushioning designs used in pneumatic cylinders. Prepare a pamphlet that explains cushioning designs and how to choose appropriate cushioning for your application and system.

3. Choosing the right pneumatic motor for an application can be difficult. Prepare a fact sheet that explains the design and operating characteristics of one type of motor, the technical specifications typically provided by manufacturers, and the primary factors that should be considered when making your choice.

4. Compare and contrast two different models of a tool that uses a pneumatic reciprocating motor or rotary motor to operate, such as a jackhammer, impact wrench, or nail driver. Prepare a report that addresses how components and design affect tool capacity, operation, safety, and cost.

5. Research air nozzle designs currently available. Focus on the application for which the nozzle is designed, the nozzle's performance and operating conditions, and any possible safety concerns or regulations involved in its use.

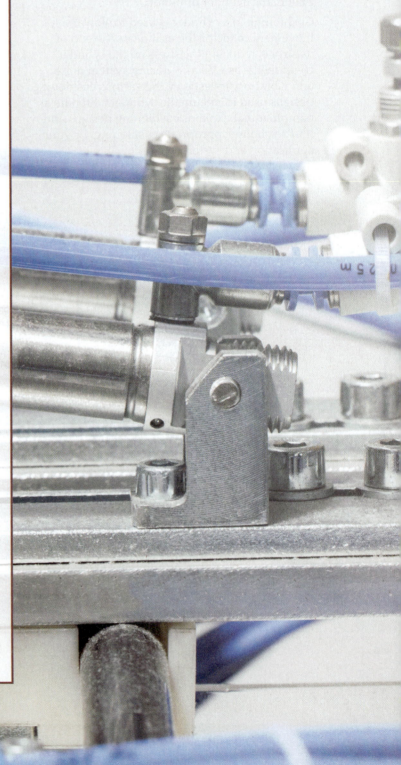

18 Controlling a Pneumatic System

Pressure, Direction, and Flow

The operation of a pneumatic system requires the control of the compressed air supplied by the system compressor. This control is provided by a number of specialized components located between the compressor and the workstation actuator. These components provide control of the compressed air in the distribution lines, at the system workstation, and in various actuator and specialized tool circuits. A variety of valve configurations designed for pneumatic use are available to provide the required system control. Understanding the structure and operation of these various valves will allow you to design, install, and maintain pneumatic circuits that obtain maximum performance from a system. This chapter provides information on basic valve design and operation in each of the control areas.

Learning Objectives

After completing this chapter, you will be able to:

- Explain the function of the three general types of control valves used in pneumatic systems.
- Describe the methods used to control air pressure in a pneumatic system.
- Describe the design and operation of pneumatic system directional control valves.
- Name and compare the various types of construction used to ensure sealing of the internal passages found in pneumatic directional control valves.

- List and compare the methods used to position control spools in pneumatic directional control valves.
- Describe the structure and operation of flow control valves used to control actuator speed in pneumatic circuits.
- Explain how flow control valves are used in pneumatic circuits to control actuator speed.

Key Terms

bypass control system
check valve
compressor-capacity control
diaphragm
five-port directional control valve
fixed orifice
four-way directional control valve
meter-in circuit

meter-out circuit
muffler
needle valve
packed-bore design
packed-spool design
piston
poppet valve
pressure booster
quick-exhaust valve
safety valve

shutoff valve
shuttle valve
sliding plate
spool
three-way directional control valve
timing-volume reservoir
valve body

18.1 Primary Control Functions in a Pneumatic Circuit

Control valves allow pneumatic circuits to produce the type of motion and level of force required to operate a machine. The valves control air pressure, the direction of airflow, and the rate of airflow in selected parts of a circuit. These specialized valves are needed to provide the conditions to produce the desired machine operation.

Air pressure in a pneumatic circuit is controlled to provide a pressure required to produce actuator force. Pressure regulation is also used to limit maximum system pressure. This is done to prevent excessive force that can damage circuit components, machine parts, or the part being produced.

The direction of compressed airflow in pneumatic system conductors is controlled by directional control valves. Controlling the direction of airflow provides a means to control the direction of actuator movement. These valves are used to extend and retract cylinders and change the rotation of pneumatic motors.

Controlling the rate of airflow through system lines controls the operating speed of actuators. The flow rate of air in a pneumatic circuit depends on conductor size, the size of the orifice, and the pressure drop across the orifice. Changing the cross-sectional area of an orifice adjusts the airflow rate through the valve.

Knowing the function and operation of each of the control components is basic to understanding the operation of pneumatic circuits. This chapter provides a foundation in the design and operation of each valve type.

18.2 Basic Control Valve Design and Structure

The external appearance and internal structure of pneumatic valves vary considerably. However, many of the valves incorporate similar component parts and operating principles. A variety of these factors are discussed in this section. Basic valve structure is introduced and the function of components is discussed.

18.2.1 Valve Body Construction and Function

The *valve body* is the primary structural component of most fluid power valves, **Figure 18-1**. In high-pressure hydraulic systems, this can be a fairly heavy component. However, the relatively low operating pressure of pneumatic systems allows the use of lightweight, compact materials for valve body construction. Heavy castings may be found on some pneumatic valves designed

A

B   Goodheart-Willcox Publisher

Figure 18-1. The body of a pneumatic valve may be a metal casting or it may be machined from a section of standard metal bar stock. A—Casting. B—Bar stock.

for locations where the valve is exposed to conditions that may cause damage. However, in most situations, the weight of the body of a pneumatic valve is determined more by the structural strength needed to hold valve parts and allow secure mounting of the valve.

The valve body may be made by casting metal. This results in a unique exterior shape for the valve. However, more and more valves are being produced from standard round or square metal bar stock. The stock is cut to the required length and machined to produce the required bores and passageways for the pistons, control springs, orifices, and movement of air. The valve body also includes inlet, outlet, and control ports.

18.2.2 Valve Control Elements

All pneumatic control valves operate by regulating system air movement. These valves have elements designed to allow, direct, meter, or stop the flow of compressed air. In pneumatic valves, these elements typically take the form of a fixed orifice, needle valve, spool, piston, diaphragm, poppet valve, or sliding plate. The elements are used as a single unit or in various combinations to obtain desired valve performance.

A *fixed orifice* is a precision hole that controls airflow through passageways. See **Figure 18-2**. Varying the pressure drop across the orifice causes the flow rate of the air passing through the passageway to increase or decrease.

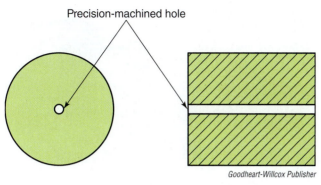

Figure 18-2. A fixed-size orifice is basically a precision-machined hole in a valve component.

A *needle valve* is a similar flow control device. It consists of a fixed orifice fitted with a tapered needle. See **Figure 18-3**. The needle is used to vary the cross-sectional area of the orifice. Reducing the area of the orifice reduces the flow rate. Increasing the area increases the flow rate. Both fixed orifices and needle valves are used either as basic flow control devices or as a part of complex control valves.

A *spool* is a cylindrical component that slides in a bore machined into the valve body. The spool contains grooves and lands. In combination with passageways in the body, these provide various routes for air to flow through the valve. Positioning the spool in various locations within the bore connects the passageways to system lines to obtain a desired circuit action. In any position, an airtight seal must exist between the spool and bore to prevent internal air leakage. Spools are commonly used in both pneumatic and hydraulic directional control valves. **Figure 18-4** shows a typical spool with the primary parts labeled.

A *piston* is used in pneumatic valves to operate other valve parts. They are cylindrical and operate in a machined bore. Piston chambers located on either end of the piston are connected to passageways that sense pressure. Pressure may be sensed in a section of the valve or an external part of the circuit. The sensed

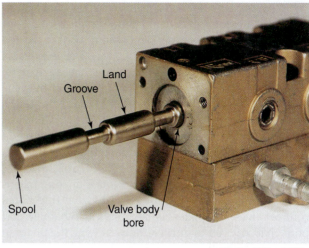

Goodheart-Willcox Publisher

Figure 18-4. A spool is a cylindrical control element with machined clearances that serve as passageways to direct the flow of the compressed air through the valve.

pressures acting on the ends of the piston generate forces that move the piston. The resulting piston movement operates other valve components that produce the desired air pressure or rate of airflow in the valve or circuit. Often, a biasing spring is used in conjunction with the piston. This limits piston movement until selected pressures are reached in the valve chambers.

A *diaphragm* is often used in pneumatic valves as a control element, **Figure 18-5**. These are made from a relatively thin, flexible material. They are often used in place of pistons to sense pressure and operate other control elements. Control chambers can be located on one or both sides of the diaphragm. A biasing spring can be used to limit diaphragm movement until selected pressures have been reached. Diaphragms function well as control elements in the relatively low operating pressure of a pneumatic system.

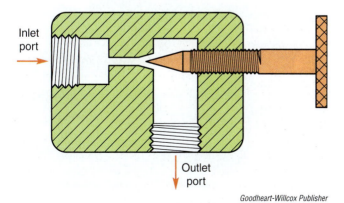

Goodheart-Willcox Publisher

Figure 18-3. A needle valve consists of an adjustable, tapered needle positioned in the center of a fixed-size orifice.

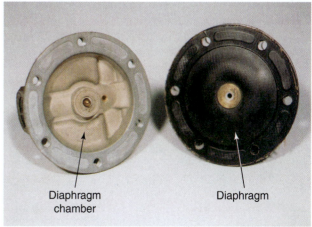

Goodheart-Willcox Publisher

Figure 18-5. A flexible diaphragm is commonly used as a control element in a pneumatic pressure regulator.

A *poppet valve* is used as a control element in a number of pneumatic directional control applications. Poppet valve assemblies consist of a seating surface located in the valve body and a valve face located on a movable poppet. In pneumatic valve applications, a resilient material is generally used on the poppet face, rather than the traditional, machined-metal face and seat. This design variation provides a positive seal at a reduced production cost. An advantage of the poppet-valve design is a large-diameter poppet can allow maximum flow through a valve with minimal movement of the control element.

A precision-machined *sliding plate* is another means used to control airflow through pneumatic directional control valves. The plates contain a series of channels and ports that are used to direct airflow. Moving the plates in relation to each other connects the channels and ports in various combinations to produce different airflow patterns. Some designs use rotary movement of the plates to change the airflow combinations, **Figure 18-6**. Other designs use linear travel to operate the valve.

18.2.3 Positioning Control Valve Internal Elements

Pneumatic pressure, direction, and flow control valves are operated by varying the position of their various control elements. Forces both internal and external to the valve are used to position the control elements.

Internal force refers to forces generated by air pressure in the internal passageways of the valve or by internal valve components. Springs and air pressure are the typical sources of internal force. Springs are used to position piston, spool, and diaphragm control elements, **Figure 18-7**. They may also be used to resist the movement of these elements until a selected pressure is reached. Air pressure sensed in valve passageways or control chambers is also used to move elements.

Spring

Goodheart-Willcox Publisher

Figure 18-7. Internal springs are commonly used to establish the normal position of the spool in pneumatic directional control valves.

External force refers to forces applied by sources outside of the valve. Manually applied force, pilot air pressure, and electromagnets are external forces commonly used to position the internal control elements of valves. Manually applied force may be as simple as an operator shifting the handle of a directional control valve. Air pressure is often sensed at a remote location in the circuit and transmitted by lines to a control valve. Electromagnets operate solenoids, **Figure 18-8**. Solenoids are commonly used in both simple control circuits and complex, computer-controlled systems.

Additional details relating to operation of control elements can be found in Chapter 10, *Controlling the System*. The initial sections of that chapter discuss in more detail the basic features of control elements. Many of the features of hydraulic and pneumatic components

Goodheart-Willcox Publisher

Figure 18-6. Precision-machined sliding plates are often used as the control element in rotary-style pneumatic directional control valves.

Goodheart-Willcox Publisher

Figure 18-8. Solenoids are used to shift the position of the control spools in many pneumatic directional control valves.

are similar. However, it must always be remembered that air is compressible and the liquid of hydraulic systems is, for all practical purposes, not compressible.

18.3 Pressure Control Methods and Devices

Several methods are used to control air pressure in a pneumatic system. The method used depends on several factors. These factors include the section of the system being controlled, required accuracy of control, effect on efficient system operation, and safety of both system components and machine operators. Three system sections can be identified that use different pressure control methods:

- System distribution lines.
- Workstations providing compressed air to various system circuits.
- Portions of circuits that require pressures above or below the level set by the workstation controls.

18.3.1 System Distribution Line Pressure Control

The pressure in the distribution lines of a pneumatic system can be controlled by two basic methods. The first method, called a bypass control system, bleeds excess air from the compressor to the atmosphere to maintain a constant, maximum pressure in the distribution line. The second method controls the capacity of the compressor to maintain distribution line pressures from slightly above the required pressure to a predetermined maximum pressure, **Figure 18-9**.

Goodheart-Willcox Publisher

Figure 18-9. A pressure switch operates the start-stop compressor-capacity control system commonly used with small to medium compressors.

In a *bypass control system*, the compressor continuously operates at a constant speed. When the desired pressure is reached in the distribution line, a relief-type valve discharges excess compressor output to the atmosphere. The system is generally considered inefficient, except in specific situations.

Compressor-capacity control is the most common means of controlling distribution line pressure. It is accomplished by using control over compressor starting-stopping, inlet valve unloading, compressor speed, or, in the case of dynamic compressors, the size of the inlet opening. Each of these methods is discussed in Chapter 15, *Source of Pneumatic Power*. That section provides the details of compressor-capacity control and explains how distribution line pressures are maintained.

Several designs are used when the compressor capacity method is used to control distribution line pressure. Each of these designs allows the compressor to deliver only sufficient compressed air to maintain the desired distribution line pressure. Air pressure varies in the distribution lines when these designs are used. However, it will not drop below the level needed for effective operation of the circuits of the overall pneumatic system. These methods are considered more efficient for most pneumatic installations.

18.3.2 Workstation Pressure Control

A second pressure-control point is at system workstations. The pressure regulator of the filter/pressure regulator/lubricator (FRL) unit further controls air pressure. This additional pressure control is necessary for two reasons. First, the distribution line air pressure is higher than needed for the operation of working circuits. Second, the air pressure in the distribution lines varies considerably because of the inherent characteristics of the compressor-capacity control system. Therefore, pressure regulators located at workstations are necessary to provide air at a consistent pressure that is below distribution line pressure, **Figure 18-10**.

Air pressure regulators maintain a desired output pressure for a workstation as long as the distribution system pressure does not drop below the selected pressure. Several different regulator designs and features are available for use in pneumatic systems. The regulator selected for a system depends on the accuracy and type of control needed by the system. Designs include direct operated; basic, diaphragm chamber; relieving; balanced poppet; and pilot operated. Many manufacturers produce air line regulators, **Figure 18-11**. Some designs include a combination of design features.

Each of these regulator features is discussed in Chapter 16, *Conditioning and Distribution of Compressed Air*. That section provides the details of design, construction, and operation of air pressure regulators.

IMI Norgren, Inc.

Figure 18-10. Workstation pressure control is provided by a pressure regulator that is often part of a filter/pressure regulator/lubricator (FRL) unit.

Goodheart-Willcox Publisher

Figure 18-11. Properly sized pressure regulators maintain a uniform workstation operating pressure even though pressure in the system distribution line varies.

18.3.3 Pressure Control within System Circuits

Pneumatic circuits may need higher or lower air pressure than that supplied by the workstation pressure regulator. Lower pressures can be easily provided by the placement of additional regulators. These additional regulators easily reduce pressure in the subcircuit to the desired level. **Figure 18-12** shows a circuit in which a low pressure is available to operate an air motor. The higher setting of the workstation regulator controls the higher pressure available to the remainder of the circuit.

When the required pressure is higher than that available from the system distribution line, it may be obtained by using a *pressure booster*. Boosters, often called *pressure intensifiers*, may operate using air-to-air or air-to-oil designs. Air-to-oil designs are primarily used when very high pressures are needed. Both types are limited to situations in which only a small volume of high-pressure air is needed for circuit operation.

Figure 18-13 shows a circuit containing an air-to-air booster. The unit is basically a reciprocating pump with two cylinder chambers. The piston in the larger cylinder chamber is powered by air coming from the system distribution line. The force produced by this air

pressure drives the second, smaller-diameter piston. This compresses air in the smaller cylinder and moves it into the circuit subsection needing high-pressure air. Valves in the booster cause the pistons to reciprocate. This continuing movement produces a small volume of air at a pressure higher than that of the air distribution lines of the pneumatic system. The operation of an air-to-oil booster is similar.

18.3.4 Limiting Maximum System Pressure to a Safe Level

Pneumatic systems must be fitted with a positive-acting *safety valve* to limit the system to a safe maximum pressure, **Figure 18-14**. This valve is required to prevent the development of dangerous pressure levels if compressor capacity controls and workstation pressure regulators fail. Although these valves are often called relief valves, the unit operates differently from a hydraulic system relief valve. A pneumatic safety valve is designed to fully open as soon as the pressure setting is reached. A hydraulic relief valve gradually opens until it is fully open at the maximum system pressure setting. This difference is required because of the different reaction of a compressed gas and a noncompressible liquid.

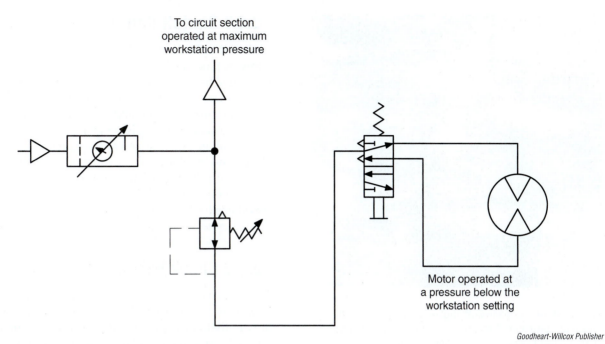

Figure 18-12. A pressure lower than that selected as the workstation pressure may be easily provided by the addition of a second pressure regulator.

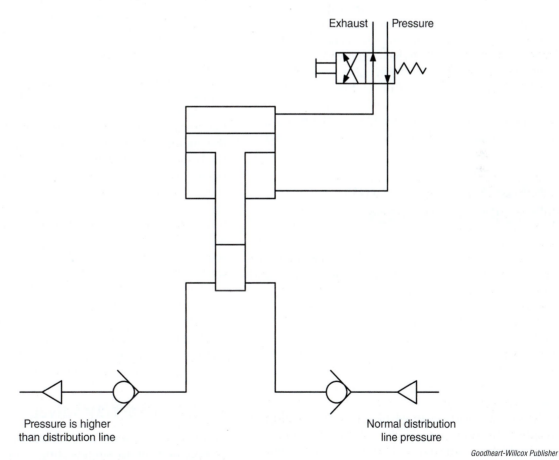

Figure 18-13. An air-to-air booster needs to be used if a portion of a workstation circuit must operate at a pressure greater than that available from the system distribution line.

Goodheart-Willcox Publisher

Figure 18-14. A positive-acting safety valve limits the maximum pressure that can be developed in the system even if other pressure-control devices fail.

Only safety valves that have the capacity to pass the full-load, compressed-air output of the compressor should be installed in a system. Safety valves are normally installed between the compressor and the first point in the air distribution line where airflow can be shut off. The pressure setting of the safety valve depends on the operating pressure of the system. A setting 25% above the maximum operating pressure is often suggested for systems operating between 50 and 125 psi. The percentage varies at operating pressures above or below this range.

18.4 Directional Control Methods and Devices

Directional control in pneumatic systems consists of channeling compressed air to cylinders, air motors, or other system work devices. This is done to start and stop their operation or produce a desired direction of movement. The valves that provide this control can be separated into four general classifications:

- Shutoff valves.
- Check valves.
- Three-way valves.
- Four-way valves.

These classifications are based on the routing of air through passageways in the valve. This section discusses these classifications as they relate to pneumatic valves. Additional details are discussed in Chapter 10, *Controlling the System*. That section should be reviewed for details about directional valve design, construction, and operation.

18.4.1 Shutoff Valves

A *shutoff valve*, often called a *two-way valve*, is used to allow or block airflow in pneumatic distribution lines. The valve has a single internal passageway connecting two external ports. Shutoff valves are commonly used to isolate circuit components to allow system service.

This classification includes globe, gate, ball, spool, and needle valves. All of these valves are relatively simple in design and involve basic construction techniques. Except for needle valves, these designs function best when used as shutoff devices operating only in a fully open or fully closed position. The needle valve design can be effectively used as either a shutoff or metering valve.

18.4.2 Check Valves

The primary purpose of a *check valve* is to allow free flow of air through a line in only one direction. Reverse flow in that line is prevented. Check valves are used in a variety of situations. For example, check valves are used in compressor discharge lines to prevent reverse airflow through the compressor when the unit is not operating. Another common application of a check valve is to allow the free reverse flow of air around flow control valves. This application produces a metered airflow for controlled cylinder extension and unrestricted airflow for rapid retraction. See **Figure 18-15**.

A simple check valve consists of a valve body with a passageway connecting inlet and outlet ports. The passageway contains a chamber in which a poppet or ball element can move and a machined surface around the inlet line opening. This surface acts as the seat for the ball or poppet element. Air entering the inlet port passes unrestricted through the passageway, around the poppet or ball, and through the outlet port. When air movement is reversed, the poppet or ball element is forced against the seat to block flow. Many designs use a lightweight spring to hold the poppet or ball against the seat. This ensures effective valve operation in any installed position. See **Figure 18-16**.

Check valves are available with a number of design variations. In addition to lightweight springs, many pneumatic check valves have a resilient seal material on the poppet or seat surface to ensure a positive seal. Check valves can also be designed to provide restricted reverse flow, rather than completely blocking airflow.

18.4.3 Three-Way Valves

Three-way directional control valves are used to direct compressed air to single-acting cylinders for extension. When the valve is shifted to allow cylinder retraction, the air returning from the cylinder is allowed to directly pass to the atmosphere through the valve exhaust port. See **Figure 18-17**.

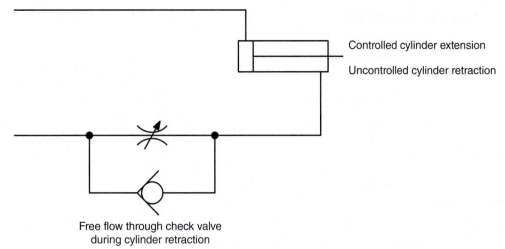

Controlled cylinder extension

Uncontrolled cylinder retraction

Free flow through check valve
during cylinder retraction

Goodheart-Willcox Publisher

Figure 18-15. Check valves are often used to allow free reverse airflow around flow control valves.

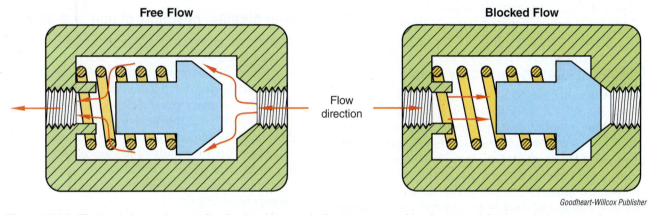

Free Flow

Blocked Flow

Flow
direction

Goodheart-Willcox Publisher

Figure 18-16. The basic internal parts of a check valve are a ball or poppet, machined seat, and, in some designs, a spring.

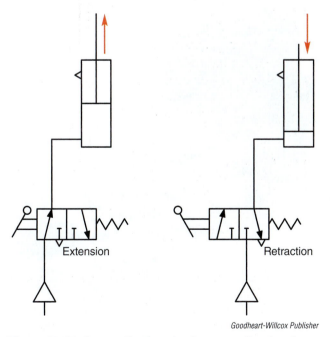

Extension

Retraction

Goodheart-Willcox Publisher

Figure 18-17. One application of a three-way directional control valve is to operate a single-acting cylinder.

Three-way valves consist of a valve body with passageways connecting a pressure port, actuator port, exhaust port, and sliding spool. The sliding spool operates in a bore machined into the valve body. It can be shifted to connect the ports in various combinations. One of these combinations powers the extension of single-acting cylinders. Another combination exhausts air from the cylinder to allow retraction.

18.4.4 Four-Way Valves

Four-way directional control valves are used to direct the extension and retraction of double-acting cylinders, select the direction of air motor rotation, or operate other workstation circuit configurations requiring alternate compressed air delivery. A basic four-way valve has a minimum of four external port connections. These include a port connected to the pressurized-air supply, inlet and outlet ports connected to the actuator, and an exhaust port leading directly to the atmosphere.

Basic operation of four-way valves

Shifting a four-way valve in one direction connects the air supply to one port of the actuator. This allows pressurized air to move the actuator in one direction. In this shifted position, the valve also connects the second port of the actuator to the exhaust port. Air is returned to the atmosphere through the exhaust port. Shifting the valve in the opposite direction connects the air supply and exhaust ports to the opposite ports of the actuator. These connections pressurize the opposite side of the actuator, causing it to move in the opposite direction. Air returned from the actuator is again exhausted to the atmosphere through the exhaust port.

Four-way valves are also available using a design that has five ports. A *five-port directional control valve* provides the same basic operation as the four-port design described above. The primary difference is the fact that a separate exhaust port is provided for each valve position. The symbols shown in **Figure 18-18** illustrate the airflow paths of four-way directional control valves with the four-port and five-port designs.

Internal sealing in four-way valves

The control element in a four-way, pneumatic directional control valve is generally a sliding plate or spool. Movement of these elements connects internal passageways in the valve body, thus routing airflow to produce the desired actuator movement. In order to efficiently operate using these control elements, the valves must route the compressed air through the valve without leakage around internal parts and between the internal passageways. Both machined surfaces and resilient materials are used in these valves to provide this tight seal.

Manufacturers use a number of seal variations to provide an airtight seal. Seals made of resilient materials are most common in pneumatic valves. This is primarily due to the fact that the close tolerance required

to prevent air loss with precision-fit parts is both expensive to create and difficult to maintain. Even so, some pneumatic directional control valves use a precision-fit spool in combination with the lubricating oil added to the system as the only sealing method. However, two resilient seal designs are more commonly used in pneumatic valves to provide the seal between the spool and bore.

The first of the resilient seal designs is sometimes called a *packed-spool design*. It uses resilient seals mounted on the spool lands. This provides the positive seal needed between the spool and the machined metal bore located in the valve body. See **Figure 18-19**.

The second of the resilient seal designs is called the *packed-bore design*. It uses resilient seals placed in grooves machined into the surface of the bore in which the spool operates. See **Figure 18-20**. This provides the positive seal.

Control element positions of four-way valves

The above description of four-way directional valves is based on a two-position valve. The first of these positions activates the actuator in one direction. Shifting the valve to the second position reverses the movement. However, many pneumatic circuits more effectively function when a third valve position is provided to allow additional control. In three-position valves, the third position is typically called the center position, **Figure 18-21**. The center position can be used simply to stop actuator movement or provide some additional operating characteristic.

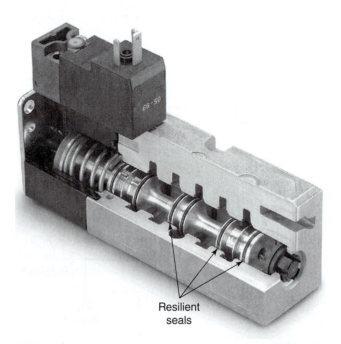

IMI Norgren, Inc.

Figure 18-19. The packed-spool design of directional control valves has resilient seals on the lands of the valve spool.

Four-Port Configuration

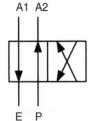

A1 A2

E   P

Five-Port Configuration

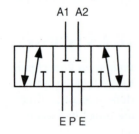

A1 A2

E P E

P = pressure
E = exhaust
A1 = actuator connection 1
A2 = actuator connection 2

Goodheart-Willcox Publisher

Figure 18-18. Pneumatic four-way directional control valves are available having either four or five port connections.

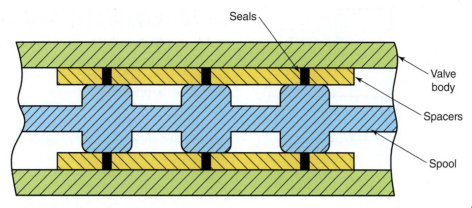

Seals

Valve body

Spacers

Spool

Goodheart-Willcox Publisher

Figure 18-20. The packed-bore design of directional control valves has resilient seals on the surface of the valve bore.

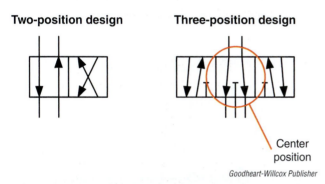

Two-position design

Three-position design

Center position

Goodheart-Willcox Publisher

Figure 18-21. The center position of a three-position, four-way directional control valve provides additional circuit operating options.

Center position configurations

There is a variety of center position configurations. Three of these can be considered basic designs that are readily available and commonly used in pneumatic equipment. These three are:

- Blocked center.
- Open center.
- Pressure center.

Figures 18-22, **18-23**, and **18-24** show these center designs as used in a five-port, three-position directional control valve operating a double-acting cylinder.

The blocked-center configuration blocks the pressure port supplying compressed air from the system distribution line. See **Figure 18-22**. Both lines leading to the circuit actuator are also blocked. This design is often called a *closed center* as the movement of air is blocked at all port connections. No compressed air can enter the circuit, and actuator movement is restricted. The amount of restriction is based on the compressibility of the air in the lines connecting the actuator and the directional control valve.

The open-center configuration is often called the *exhausted-center design*. In this configuration, the pressure port is blocked. Both ports connected to the actuator are connected to exhaust. See **Figure 18-23**. This allows

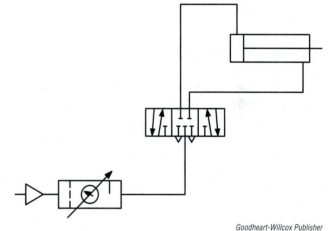

Goodheart-Willcox Publisher

Figure 18-22. The blocked-center configuration blocks all air movement through a three-position, four-way directional control valve.

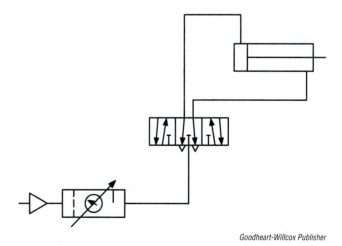

Goodheart-Willcox Publisher

Figure 18-23. The open-center configuration of a three-position, four-way directional control valve blocks the pressure port and opens the actuator ports to exhaust.

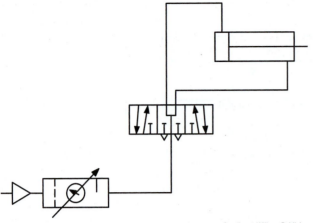

Goodheart-Willcox Publisher

Figure 18-24. The pressure-center configuration of a three-position, four-way directional control valve connects the pressure port to both actuator ports.

the actuator to float, depending on the external forces applied to it.

The pressure-center design connects the pressure port of the valve to both lines leading to the actuator. See **Figure 18-24**. When the directional control valve is in the center position, equal air pressure is applied to both sides of the cylinder piston. Because of the difference of force generated on the rod and blind ends of the piston, the actuator will rapidly extend. This regenerative operating concept produces a faster extension speed, but reduces the usable force. However, the pressure-center design can also be used in other ways, such as supplying pressurized air to more than one actuator or subcircuit.

18.5 Flow Control Methods and Devices

Controlling the flow of compressed air in a pneumatic system controls the operating speed of cylinders, air motors, and other actuators. Control is achieved by using flow control valves that contain an orifice. The orifice meters the amount of air that can enter a circuit. Because of the compressibility of air, accurate speed control is difficult in pneumatic systems. The accuracy of a pneumatic flow control valve is influenced by system load changes that cause:

- Variations in system pressure.
- Changes in internal system friction, which cause load variations.
- Increases or decreases in the temperature of the compressed air, which cause variations in system pressure.

18.5.1 Types of Flow Control Valves

Fixed-size orifices and needle valves are commonly used to control actuator speeds in pneumatic systems. However, the flow control valve most commonly found in a pneumatic circuit is a component that combines a needle valve and check valve in the same valve body. This *combination valve* uses the needle to accurately adjust the rate of airflow in one direction, while the check valve allows unrestricted return flow. **Figure 18-25** illustrates the internal structure of a flow control valve that combines these two features.

It is not recommended that shutoff devices, such as ball, globe, and gate valves, be used as metering devices in pneumatic systems. The internal structure and method of valve adjustment of these valve designs do not allow consistently accurate metering of the airflow.

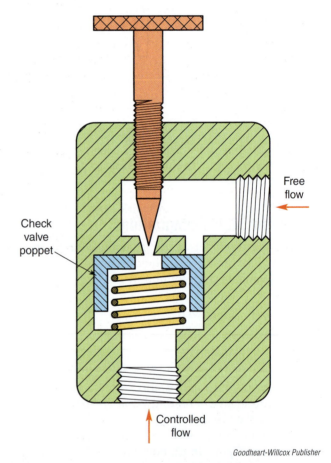

Check valve poppet

Free flow

Controlled flow

Goodheart-Willcox Publisher

Figure 18-25. Pneumatic system flow control valves are often needle valves combined with a check valve that allows free flow in the reverse direction.

18.5.2 Placing Flow Control Valves in Circuits

Flow control valves can be placed to create either meter-in or meter-out circuits. In a *meter-in circuit*, the valve is placed in the inlet line of an actuator. In a *meter-out circuit*, the valve is placed in the outlet line. See **Figure 18-26**. However, a meter-out circuit is the method preferred by system designers.

Meter-out flow control is preferred because of the compressibility of air. Backpressure that develops in the line between the directional control valve and actuator in a meter-out circuit results in stable line pressure and actuator movement. If a meter-in circuit is used, the air in this line will compress and expand during varying actuator loads. These changes will result in poor actuator speed control.

In pneumatic meter-out flow control circuits, the flow control valve can be placed between the actuator and the directional control valve or on the exhaust ports of the directional control valve. When a four-port, four-way directional control valve is used, the flow control valve must be placed between the actuator and the directional control valve. In this arrangement, the flow control valve should include an integral check valve to allow unrestricted return flow when the direction of the actuator is reversed. When a five-port, four-way directional control valve is used, place the flow control valve in the lines leading from the direction control valve exhaust ports. A check

valve is not needed in this case as airflow through the valve is only toward the exhaust opening. **Figure 18-27** shows schematics of circuits using four- and five-port directional control valves.

18.6 Special-Purpose Control Valves and Other Devices

Several special-purpose valves and devices used in pneumatic circuits provide increased control and efficiency. A number of these are exclusively produced by a single company, while many others are available from most manufacturers. The following sections discuss several of the special-purpose valves and devices often found in pneumatic circuits.

18.6.1 Quick-Exhaust Valves

A *quick-exhaust valve* is used to increase the rate at which air is exhausted from a cylinder. Increasing the exhaust rate allows the cylinder to extend or retract as rapidly as possible. **Figure 18-28** shows a circuit that includes a four-way directional control valve, double-acting cylinder, and quick-exhaust valve. The quick-exhaust valve is located in the line between the cylinder port on the blind end and the directional control valve.

When the directional control valve is shifted to retract the cylinder, the pressure and movement of

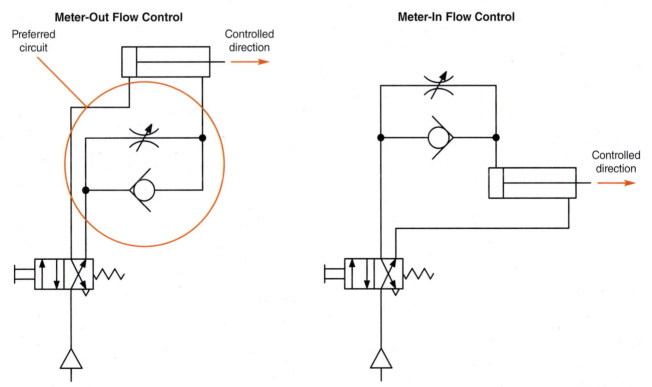

Goodheart-Willcox Publisher

Figure 18-26. The preferred flow-control circuit for pneumatic systems is the meter-out design.

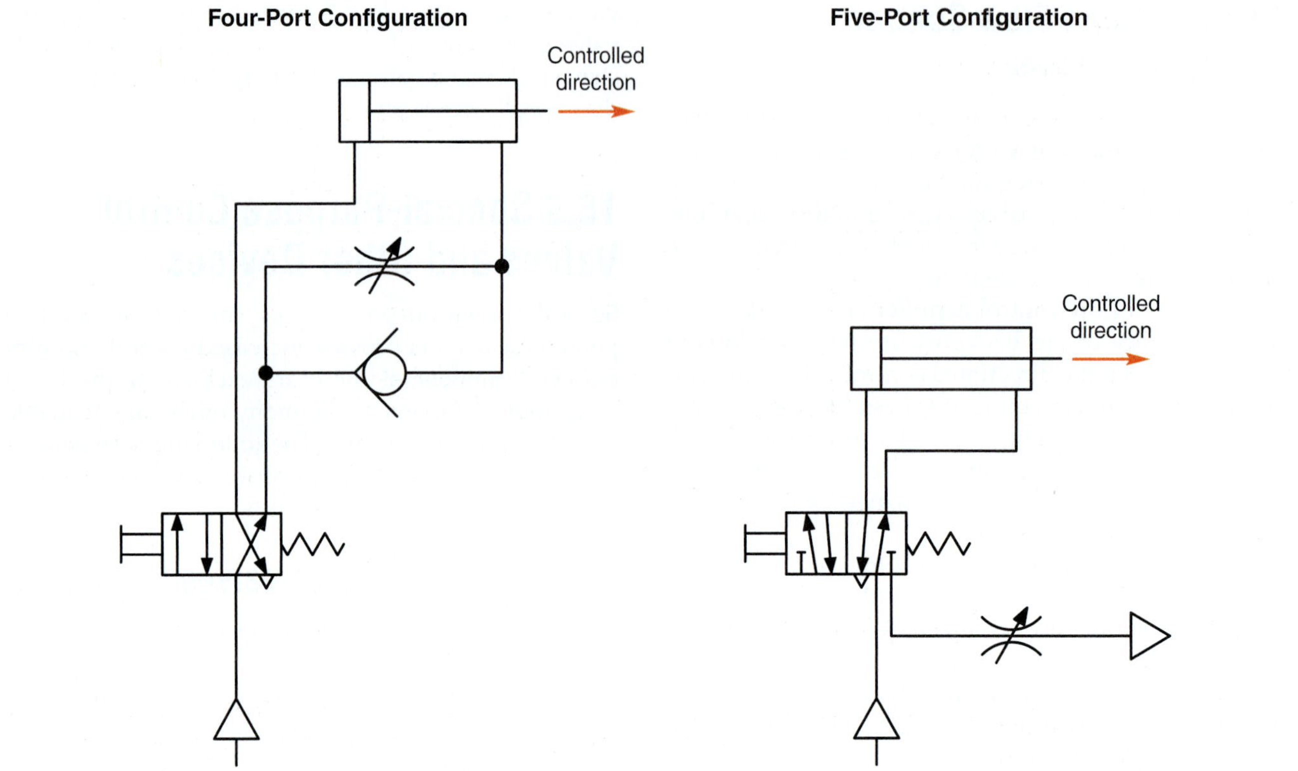

Four-Port Configuration

Five-Port Configuration

Controlled
direction

Controlled
direction

Goodheart-Willcox Publisher

Figure 18-27. When using a four-port directional control valve, place the flow control valve between the directional control valve and the actuator. The preferred location of the flow control valve when using a five-port directional control valve is on the exhaust port of the directional control valve.

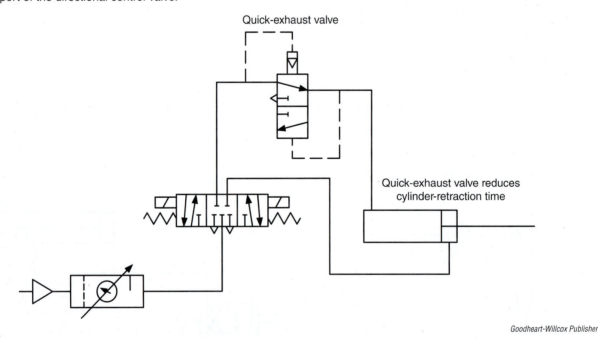

Quick-exhaust valve

Quick-exhaust valve reduces
cylinder-retraction time

Goodheart-Willcox Publisher

Figure 18-28. Quick-exhaust valves allow the rapid release of large quantities of exhaust air from a cylinder.

the air coming from the blind end opens the quick-exhaust valve. This air then quickly escapes through the valve to the atmosphere. As a result, backpressure in the line is reduced to below the level required to force the air through the four-way valve. Due to this lower resistance to air movement, the cylinder retracts at a higher rate of speed. When the directional control valve is shifted to extend the cylinder, the quick-exhaust valve shifts and directs air to the cylinder to produce a normal extension speed.

Cylinder Extension

From directional
control valve

Cylinder Retraction

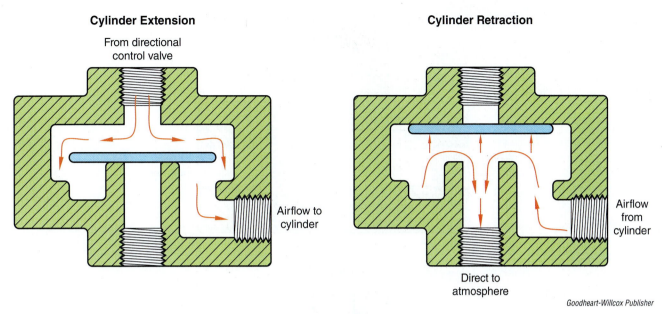

Airflow to
cylinder

Airflow
from
cylinder

Direct to
atmosphere

Goodheart-Willcox Publisher

Figure 18-29. A common model of quick-exhaust valve uses a floating, flexible disk to block and open ports.

Figure 18-29 shows the internal structure of a typical quick-exhaust valve. The valve body contains three ports. The first port is connected to an inlet line coming from the circuit directional control valve. The second port is connected to a line leading to one cylinder port. The third port leads directly to the atmosphere. A flexible disk freely floats in a cavity connected to the three ports. When the inlet port is pressurized, the disk moves to cover the exhaust port. This directs compressed air to the cylinder port. When air is being forced out of the cylinder, the disk moves to cover the inlet-line port. This allows air from the cylinder line to quickly exhaust to the atmosphere. Very low air pressure and airflow are required to shift the disk.

18.6.2 Shuttle Valves

Shuttle valves are used to automatically select the higher of two air pressures and connect that source to the outlet port of the valve. The valve is basically a quick-exhaust valve used in a different application. The inlet and exhaust ports of the valve are connected to two different pressure sources. The third port is connected to the actuator.

Activating only one of the pressure sources shifts the shuttle valve and connects that pressure source to the actuator. If both pressure sources are active at the same time, the higher pressure source automatically shifts the shuttle to connect the higher-pressure line to the actuator. If the relationship of the pressures changes during operation, the shuttle valve shifts to keep the higher pressure connected to the actuator. See **Figure 18-30**.

18.6.3 Timing-Volume Reservoirs

Timing-volume reservoirs are air chambers used to time the shifting of pilot-controlled pneumatic valves.

The component is usually a tube fitted with end caps containing ports for connection to the pilot-control portion of a circuit. The end caps also contain a means for mounting the device. One manufacturer has reservoirs available in sizes ranging from 5 to 225 cubic inches.

One port of the reservoir is connected to the pilot actuator of a control valve. The other port is connected to a pressure source. Air is bled into the reservoir through a needle valve. The rate of airflow and the volume of the reservoir determine the time it takes to increase the pressure in the reservoir to a level sufficient to shift the piloted valve.

When the control valve shifts, the air in the reservoir is exhausted. The reservoir can then repeat the function during the next cycle of the circuit. **Figure 18-31** shows a schematic of a circuit using a timing-volume reservoir to time the shifting of a control valve.

18.6.4 Exhaust Mufflers

Mufflers are commonly used to reduce to an acceptable level the noise of air exhausting from pneumatic control valves and tools. In addition, these devices prevent dirt and other foreign materials from entering the system. Mufflers are fitted to the exhaust ports. Muffler designs range from compact, sintered-bronze units to designs that use sheet metal outer shells filled with plastic or metal materials. See **Figure 18-32**. Sizes are readily available for 1/8″ to 1 1/2″ diameter ports with pipe threads.

In clean mufflers, the resistance to airflow is relatively low. However, as a muffler is used, resistance to airflow can increase to excessive levels. This increase is due to system lubricating oil, condensed water, and rust and scale particles that pass through the muffler with the exhausted air. These materials slowly build up

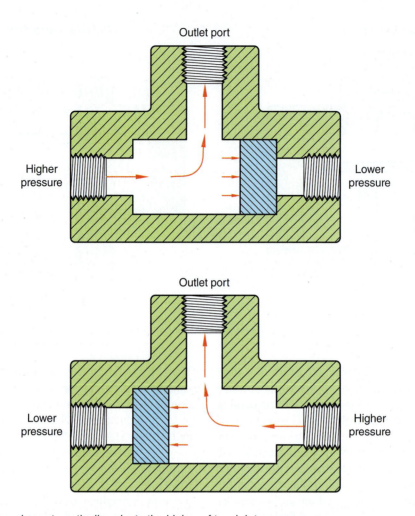

Outlet port

Higher pressure

Lower pressure

Outlet port

Lower pressure

Higher pressure

Figure 18-30. A shuttle valve automatically selects the higher of two inlet pressure sources.

in the muffler and eventually restrict airflow. Mufflers should be periodically inspected. If there appears to be an excessive buildup of dirt or oil, the muffler should be cleaned or replaced. The design of most mufflers allows them to be easily removed and cleaned.

18.7 Selecting and Sizing Pneumatic Components

Manufacturers of pneumatic components provide considerable detail about the units they produce. This information is available from both company catalogs and websites. In many situations, these sources provide the detail required to select appropriate features and correctly size units for a circuit.

18.7.1 Information Available Concerning Component Features

Catalog and website information includes a general description of the components; the operating temperature,

flow, and pressure ranges; and schematic symbols for variations. In addition, line drawings of the units provide physical dimensions of the exterior, as well as port and mounting-hole sizes.

Pressure drop and flow rate data are also displayed in these descriptions, if that information is important for selection. The illustrations shown in **Figure 18-33** are examples of part of the information provided in catalog and data sheets.

18.7.2 Determining Required Valve Size

Selecting the size of a control valve for a pneumatic system is a relatively difficult task. This is due to the many variables involved in system operation. However, selection of a properly sized control valve for an application is critical for effective operation of a system. Proper sizing is also important because selecting a valve that has a larger capacity than necessary increases both construction and operating costs for the system. It also consumes extra physical space, which can be important in some situations.

Several approaches described by different manufacturers can be used to determine a valve capacity

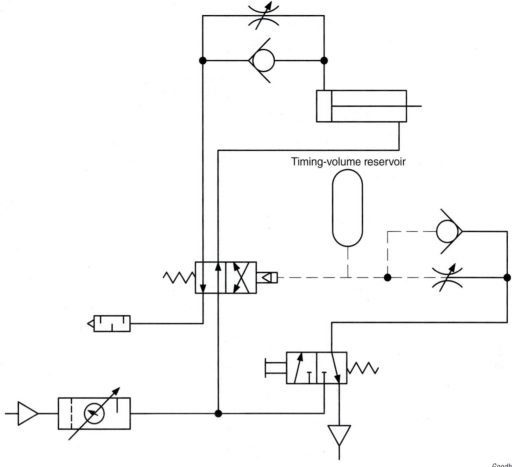

Timing-volume reservoir

Goodheart-Willcox Publisher

Figure 18-31. Timing-volume reservoirs allow control of the sequence of a circuit using pilot-activated directional control valves.

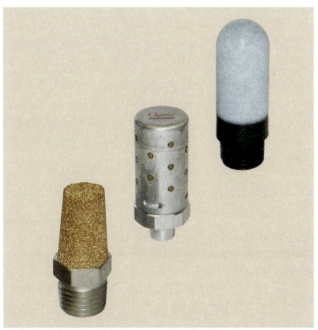

IMI Norgren, Inc.

Figure 18-32. A variety of mufflers are used on the exhaust ports of components to reduce the noise of component operation.

appropriate for an application. Most of these approaches begin with a relatively simple formula. The data required for this calculation (for a circuit containing a cylinder) typically include:

- Cylinder piston area.
- Cylinder stroke.
- System compression.
- System pressure drop.
- Cylinder stroke time.

The approaches to this calculation are similar in design, but use variations of formulas and data from tables provided by the manufacturer. Due to this, specific calculations are beyond the scope of this text. However, it will be beneficial if you select a manufacturer and study its approach to valve sizing. Printed brochures and Internet-based calculators are often provided by companies.

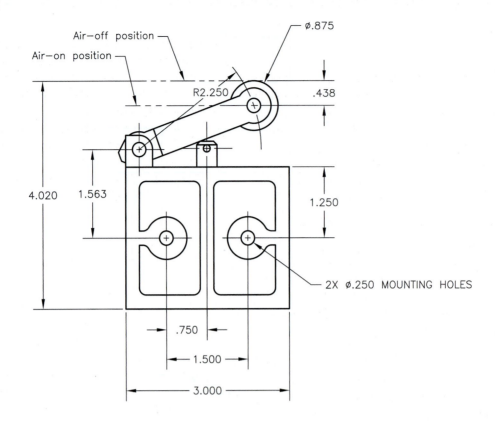

A

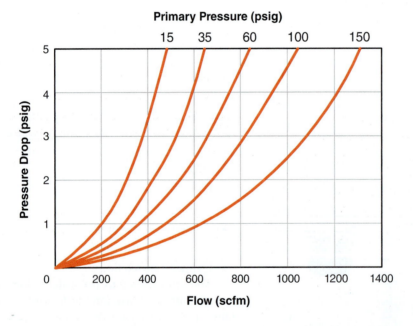

B

Figure 18-33. Manufacturer catalogs provide considerable technical information about components, including operating details, physical dimensions, pressure ratings, and airflow rates (often provided in graph form). A—Dimensions of critical features for a switch. B—Pressure drop and flow for a filter.

Summary

- Pneumatic valves control air pressure, flow direction, and airflow rate to produce the level of force and the type of motion needed to operate a machine.

- Pressure control valves limit maximum system pressure and regulate the force or torque an actuator generates. Directional control valves control flow direction in system lines to establish the direction of actuator movement. Flow control valves regulate the airflow rate through system lines to control operating speed of the actuator.

- The body of pneumatic system valves may be formed from a metal casting or machined from standard bar stock material. Machined precision bores and passageways hold component parts and allow airflow that controls valve operation.

- Internal control elements of valves allow, direct, meter, or stop the flow of compressed air. These elements typically include fixed orifices, needles, spools, pistons, diaphragms, poppets, or sliding plates.

- Internal control elements are positioned by internal springs, internal air pressure, external pilot pressure, electromagnets, or manually applied force.

- Pressure control valves limit maximum system pressure and regulate the force or torque an actuator generates.

- Directional control valves control the airflow direction in system lines to establish the direction of actuator movement. Directional control valves can be classified as shutoff, check, three-way, or four-way valves.

- Resilient materials are used to seal the space between the bore and the spool in most pneumatic directional control valves. The sealing materials may be attached to the valve spool lands or to the bore in the valve body. Valves using these seal locations are often referred to as packed-spool and packed-bore valves.

- Flow control valves regulate the airflow rate through system lines to control operating speed of the actuator.

- Fixed orifices and needle valves are the most commonly used methods of controlling the rate of pressurized airflow in pneumatic circuits.

- Because of the compressibility of air, a meter-out design is the preferred method of flow control in pneumatic circuits. In a meter-out design, the flow control valve is placed on the outlet line of the actuator.

- Special purpose valves and devices, such as quick-exhaust valves, shuttle valves, timing reservoirs, and exhaust mufflers, provide increased control and improved efficiency of pneumatic circuits.

- Manufacturer catalogs typically contain pressure ranges, flow rates, physical dimensions, and graphs illustrating performance under various conditions, and may also provide formulas to determine the size of control valves needed in a specific pneumatic circuit.

Internet Resources

The following are some useful resources available on the Internet. Enter a company or organization name into a search engine to access its website. Explore the various areas of the sites to discover useful fluid power resources.

Bimba Manufacturing Company. This page allows you to select product catalogs for all the products this company offers, including actuators. The actuator product catalogs provide details regarding construction, specifications (such as pressure and flow ratings), and available options.

ControlAir Inc. Regulators produced by the company are listed, with links to information that includes photographs and line drawings. Details include specifications, operation, installation, and maintenance.

Numatics, Incorporated. Valve catalogs for pneumatic directional control valves. Include detailed technical and operating information using photographs, cutaway graphics, and symbols. Product ordering information introduces one variation of a system often used to structure component identification numbers.

Power Aire, Inc. A listing of one line of fittings for rapid assembly of pneumatic components using plastic tubing. Similar fittings also are available from other manufacturers.

Sensata Technologies, Inc. Brochure describes industrial sensing technologies. The information covers sensors and switches produced by one manufacturer, and includes pressure, temperature, airflow, humidity, and other conditions related to pneumatic systems. Introduces basic sensing concepts.

United Electric Controls Co. These brochures provide overviews of a typical model of pressure switch, including general specifications and information on installation and adjustment.

Chapter Review

Answer the following questions using information in this chapter.

1. The three factors control valves regulate to produce a desired performance in a pneumatic circuit are _____.
   A. pressure, direction, and flow rate
   B. check, needle, and shuttle
   C. pressure, exhaust, and direct control
   D. timing, pressure drop, and four-way

2. The valve control element that consists of a precision hole used to control the rate of airflow is called a(n) _____.
   A. inlet port
   B. land
   C. fixed orifice
   D. outlet port

3. A _____ valve varies the cross-sectional area of the orifice, reducing the flow rate.
   A. needle
   B. gate
   C. ball
   D. globe

4. *True or False?* In a bypass control system, the compressed air used to operate an actuator is typically returned to the atmosphere after the circuit cycle is completed.

5. A _____ valve fully opens as soon as the maximum allowable system operating pressure is reached.
   A. relief
   B. safety
   C. shutoff
   D. meter-out

6. *True or False?* A shutoff valve is often called a two-way valve.

7. *True or False?* The only function of a four-way directional control valve is to direct the extension and retraction of double-acting cylinders.

8. *True or False?* The open-center configuration of a three-position, four-way pneumatic directional control valve blocks the pressure port and opens both actuator ports to exhaust.

9. *True or False?* Because of the compressibility of air, accurate speed control is difficult to obtain in pneumatic systems.

10. Pneumatic system designers prefer the meter-out flow control design because of _____.
    A. component size
    B. airflow rate
    C. air compressibility
    D. safety

11. The purpose of a _____ valve is to increase the rate at which air can be exhausted from a pneumatic cylinder.
    A. shuttle
    B. three-way
    C. quick-exhaust
    D. None of the above.

12. *True or False?* Shuttle valves are used to automatically select the higher of three or more air pressures.

13. Timing-volume reservoirs are air chambers used to time the shifting of _____ pneumatic valves.
    A. pilot-controlled
    B. quick-exhaust
    C. manually operated
    D. All of the above.

14. *True or False?* Mufflers are located on the intake line of pneumatic circuits to reduce the noise level of system operation.

15. Which of the following basic data information items would be the most helpful in a pneumatic component catalog description?
    A. General operation, pressure range, and flow rating.
    B. Cost, pressure rating, and operating formulas.
    C. Pressure rating, port sizes, and mounting-hole sizes.
    D. Cylinder size, pressure rating, and schematic symbols.

Apply and Analyze

1. A small production plant uses a centralized grid system to distribute compressed air to multiple workstations. At each workstation, parts are clamped and drilled. These two tasks can be performed at different air pressures. What specific methods and equipment would you use to control the air pressure in each section of the pneumatic system? Justify your choices.

2. Design and draw a circuit that allows you to control both the extension and retraction rates of a cylinder.
   A. Label the type of directional and flow control valves used in your circuit. Explain your choices.
   B. Describe the airflow through the circuit during its operation.
   C. If you had to be able to stop the cylinder mid-stroke, what changes would you make to your circuit? Why?
3. Examine the circuit in **Figure 18-12**.
   A. Describe the sequence of events that occur as the directional control valve is shifted from one position to the other.
   B. If the motor only rotates in one direction, what changes could you make to the circuit? Explain.
   C. Redraw the circuit to allow the operator to control the speed of the motor rotation in the clockwise direction. What changes did you make? Why?
4. Describe the sequence of events that occur as the manually operated directional control valve in the circuit shown in **Figure 18-31** is shifted from one position to the other.
5. What modifications would you make to the circuit shown in **Figure 18-31** if both cylinder extension and retraction rates were required to be controlled?

Research and Development

1. Make a poster that compares and contrasts the design, construction materials, and operation of poppet- and spool-type directional control valves used in pneumatic systems.
2. Investigate one type of special purpose valve. Prepare a presentation that describes the design, operation, and applications in which the valve could be used.
3. Create a video that shows how center-position configurations affect the operating characteristics of a circuit.
4. Additive manufacturing, a common method of 3-D printing, could have major effects on the design and construction of pneumatic valves. Investigate how additive manufacturing is currently used in pneumatic valve production, and examine the possible impact of this process on future production.

19 Applying Pneumatic Power
Typical Circuits and Systems

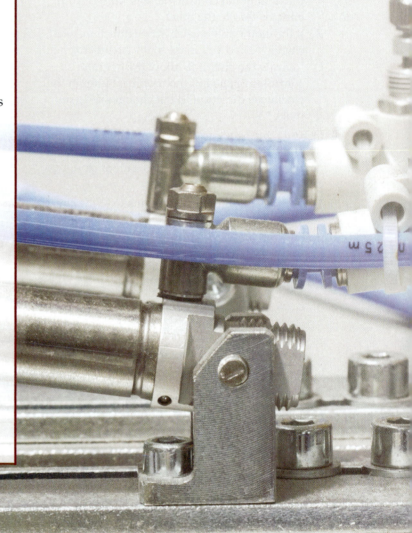

The ability to read and interpret a schematic diagram of a complex pneumatic system begins with familiarity of the operation of basic circuit segments. Basic segments are often found in a variety of pneumatic systems. Even though a circuit or system may look complex, a careful analysis of the valves, lines, and actuators should reveal a number of component groups used to control air pressure, airflow rate, and direction of airflow. This chapter presents examples of basic circuits. These examples were selected to illustrate how various, basic components can be combined to produce desired actuator force, operating speed, operating motion, or other function.

Learning Objectives

After completing this chapter, you will be able to:

- Describe basic circuits that are frequently used to assemble complex pneumatic systems.
- Identify the placement and explain the function of workstation air-preparation components.
- Compare the design and operation of basic pressure-control circuits used in pneumatic systems.
- Explain the design and operation of control circuits used to regulate actuator operating speeds in pneumatic systems.
- Describe the operation of circuits using four- and five-port pneumatic four-way directional control valves.
- Explain the design and operation of basic motion-control circuits for pneumatic cylinders.
- Explain the design and operation of pneumatic quick-exhaust valves.
- Describe a two-handed safety circuit used to provide operator safety.

Key Terms

automatic reciprocation
automatic return
booster
composite symbol

logic function
memory circuit
multiple-pressure circuit
remote pressure adjustment

time-delay circuit
trio unit
two-hand safety circuit

19.1 Basic Circuits and System Analysis and Design

The assembly of valves, lines, and actuators in a pneumatic system can become confusing. This is especially true for individuals beginning work in the field. Designing, installing, and maintaining the pneumatic system of a machine can be a challenge even to experienced fluid power personnel. However, this challenge can be considerably reduced if the systems are considered groups of basic components and subcircuits working together to produce a final desired action. Being familiar with basic component groups and circuits can be very helpful when working on pneumatic systems.

The majority of the basic pneumatic component groups and circuits can be divided into five function groups:

- Workstation air preparation.
- Pressure control.
- Flow control.
- Motion control.
- Other miscellaneous functions.

A typical pneumatic circuit includes segments from several of these groups merged together. The basic components and circuits selected for the final version of any circuit depend on the functions and level of sophistication demanded by the machine powered or controlled by the circuit.

The following sections illustrate several pneumatic circuit segments that can be used to produce various types of system operation. Each of these circuit segments is briefly explained in the text and shown in an accompanying figure. It must be emphasized that the segments shown here are only a few common examples of the many possibilities.

19.2 Workstation Air-Preparation Components

The compressor and system distribution lines leading to various workstations are designed to deliver clean, high-pressure air. This air must be further conditioned to:

- Reduce the pressure to the level needed for circuit operation.
- Remove remaining contaminants.
- Add a lubricant, if needed.

Chapter 16, *Conditioning and Distribution of Compressed Air*, discusses these individual processes in detail. This section illustrates examples of the placement of the basic conditioning components in operating circuits to assure clean, lubricated air at the required pressure level.

The circuit shown in **Figure 19-1** shows the placement of the air filter, pressure regulator, and lubricator in a basic pneumatic circuit. These components are usually assembled into a close-fitting group often considered a single component. This FRL assembly may be referred to as the *trio unit* in some literature. The component enclosure symbol surrounding the filter, regulator, and lubricator symbols in this circuit indicates the close relationship of the components.

In an operating circuit, the conditioning units are located between the system distribution line and the circuit. Incoming air from the distribution line is first filtered to remove dirt and condensed water. It is then routed to the regulator, which reduces the air pressure to the level needed to operate the circuit. The air then passes through the lubricator where atomized oil is added to the airstream. This provides lubrication for the system components. The clean, pressure-controlled, lubricated air is then distributed throughout the workstation circuit.

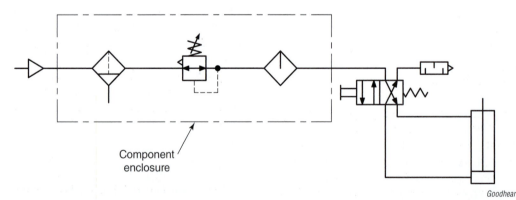

Component enclosure

Figure 19-1. The final preparation of air in a pneumatic system occurs at the workstation. The air is filtered, the pressure is adjusted to the requirements of the working circuit, and needed lubricants are added.

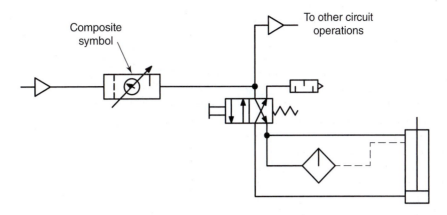

Composite symbol

To other circuit operations

Goodheart-Willcox Publisher

Figure 19-2. This circuit shows one method that may be used to add lubricant to remote or difficult-to-lubricate actuators. The three air-preparation components are indicated by a standardized composite symbol.

Figure 19-2 shows a circuit designed to provide additional lubrication to a specific system component. Because of a remote location or internal design characteristic, some components may not receive adequate lubrication from the FRL unit lubricator. The design shown in the figure has an additional lubricator located in a line parallel to the normal supply line for the cylinder. The parallel line must be correctly sized to assure appropriate airflow through the secondary lubricator.

The diagram shown in **Figure 19-2** uses the *composite symbol* to designate the FRL unit. This symbol is typically used instead of three separate symbols in an enclosure. The symbol is much less complex, but indicates the three conditioning elements. This symbol is used in previous chapters. It will be used in the remainder of the circuits shown in this chapter unless individual filters, pressure regulators, or lubricators are needed for the operation of the circuit.

19.3 Controlling System Air Pressure

The pressure of the air delivered to a pneumatic cylinder or motor determines the force and torque output of the units. The maximum air pressure in an operating circuit is controlled by the regulator that is part of the final air-preparation components at the workstation. However, it is also important to have the capability to control pressure beyond limiting the maximum pressure. Any portion or phase of a circuit operating below the maximum system pressure can reduce the volume of atmospheric air that must be compressed. Reducing the volume of high-pressure air needed increases system efficiency and reduces the operating costs of the pneumatic system. This section illustrates several basic

pressure control circuits that can be used to control circuit pressure. These circuits either:

- Control maximum circuit pressure.
- Provide multiple levels of pressure.
- Allow remote adjustment of pressure level.

19.3.1 Controlling Maximum Circuit Pressure

Figure 19-3 shows a circuit that provides different maximum pressure levels to operate three different circuit functions. One pressure level is provided for cylinder extension. A second pressure is set at a level only high enough to retract the cylinder. The third pressure is based on the level required to shift the pilot-operated directional control valve.

The highest pressure of this circuit is controlled by the setting of the workstation pressure regulator. When the circuit is pressurized, the normal position of *Valve C* causes pilot pressure to shift *Valve D*. This retracts the cylinder. The setting of the pressure regulator identified as *Valve A* controls the maximum air pressure available for retraction. When *Valve C* is manually shifted, it directs pilot pressure to shift *Valve D* to extend the cylinder using maximum circuit pressure. The pressure regulator identified as *Valve B* controls the maximum pressure available to operate the pilot portion of the circuit.

19.3.2 Providing Multiple Pressures

The circuit shown in **Figure 19-4** is a basic *multiple-pressure circuit*. It provides high pressure to extend a cylinder and a reduced pressure for retraction. Another feature of this design is the use of a three-way directional control valve to provide both extension

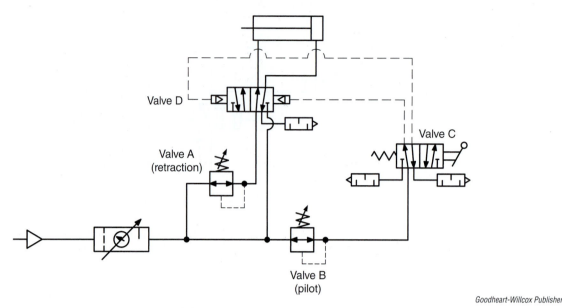

Figure 19-3. Several operating pressures may be obtained within a single pneumatic circuit by the installation of additional pressure regulators.

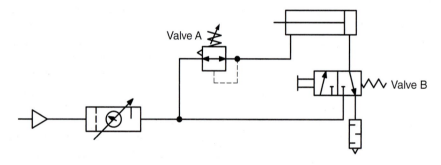

Figure 19-4. The difference in effective piston area and a relieving-type pressure regulator allow this cylinder to be operated using different extension and retraction pressures.

and retraction of a double-acting cylinder. Three-way directional control valves are typically lower in cost than similar four-way directional control valves.

The key to the operation of this circuit is the control valve identified as *Valve A*. This valve must be a relieving-type pressure regulator. It acts as a relief valve when the downstream pressure exceeds the pressure setting of the valve. *Valve B* is a three-way directional control valve that controls the extension of the cylinder. Cylinder retraction automatically occurs whenever the three-way valve is returned to the normal position.

When the circuit is pressurized, the cylinder is returned to the retracted position by low-pressure air controlled by *Valve A*. Air from the blind end of the cylinder is exhausted through *Valve B*. When *Valve B* is shifted, the blind end of the cylinder is exposed to maximum circuit pressure. The force generated by this pressure acting on the blind side of the piston is greater than the force generated by the lower pressure on the rod end. Therefore, the cylinder will extend. The low-pressure air from the rod end is exhausted through the vent of *Valve A*.

Two limiting factors must be considered with this design. First, the effective force is reduced due to the opposing pressures in the cylinder. Second, the venting capacity of the pressure regulator (*Valve A*) must be adequate to exhaust air from the rod end of the cylinder.

19.3.3 Allowing for Remote Pressure Adjustment

Figure 19-5 shows a circuit that allows *remote pressure adjustment*. A pilot-operated pressure regulator is used to change the setting of the system operating pressure. A pilot-operated pressure regulator uses air pressure in a chamber above the regulator diaphragm, instead of a spring, to establish the operating pressure of the unit. The design and operation of pilot-operated regulators is discussed in detail in Chapter 16, *Conditioning and Distribution of Compressed Air.*

Valve A is the main pressure regulator for the circuit. The pressure setting of this pilot-operated regulator

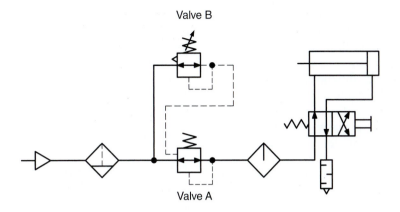

Valve B

Valve A

Goodheart-Willcox Publisher

Figure 19-5. Pilot-operated pressure regulators allow the adjustment of circuit pressure from a remote location. This allows the primary regulator to be placed close to other circuit components, while an adjustment valve is placed in a more convenient or safer location.

is established by the second, small-capacity regulator identified as *Valve B*. This design allows *Valve B* to be placed in a convenient location. Also, the primary pressure regulator can be placed close to the operating components. This results in minimum line lengths, which increases system response and efficiency.

> **NOTE**
>
> The circuit shown in **Figure 19-5** demonstrates how the filtration, pressure regulation, and lubrication functions can be located at separate points in a circuit, rather than combined in the FRL unit at one location.

19.4 Controlling System Airflow Rate

The rate at which compressed air is allowed to flow into a pneumatic cylinder or motor determines the operating speed of the actuator. The preferred method of flow control is the meter-out circuit. However, both meter-in and meter-out circuits can be found in pneumatic systems.

In pneumatic systems, flow control valves should be placed close to the actuator. The use of meter-out circuits and the close placement of flow control valves is suggested because of the compressibility of air. Meter-out circuits pressurize both sides of the cylinder piston. This provides more stable actuator movement. Short delivery lines contain less air volume, which minimizes the amount of compression. Both approaches produce more accurate control over actuator speed.

This section illustrates several basic flow control circuits that can be used to control actuator operating speeds. These basic circuits cover using single- and double-acting cylinders; four- and five-port directional control valves; and cam- or solenoid-operated, two-way valves.

19.4.1 Speed Control for Single- and Double-Acting Cylinders

Figure 19-6 illustrates a circuit that controls both the extension and retraction of a single-acting cylinder.

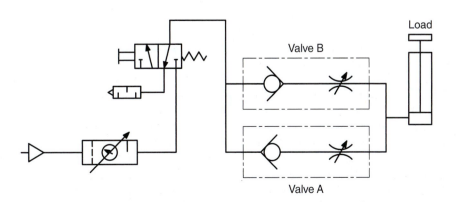

Load

Valve B

Valve A

Goodheart-Willcox Publisher

Figure 19-6. Operating speed is controlled by regulating the rate of airflow entering or exiting an actuator. This circuit controls both the extension and retraction speeds of a single-acting cylinder.

Valves A and *B* are needle valves containing integral check valves. They block the flow of air in one direction while allowing metered flow in the other direction. When the three-way directional control valve is shifted, airflow is blocked by the check valve in *Valve B*. Air is metered through *Valve A* to control the extension rate of the cylinder. When the directional control valve is returned to the normal position, the system load forces the retraction of the cylinder. During this phase, *Valve A* blocks airflow while *Valve B* meters the rate of cylinder exhaust. This controls the retraction speed of the cylinder.

The circuit in **Figure 19-7** shows a method used to control the extension and retraction speeds of a double-acting cylinder. A four-port directional control valve is used to shift the cylinder. The airflow rate is established by meter-out circuits for both extension and retraction. The design provides the preferred meter-out flow control. It allows placement of the flow controls close to the actuator, as recommended for effective speed control.

In this circuit, the cylinder is held in the retracted position when the four-way directional control valve is in the normal position. When the directional control valve is shifted to extend the cylinder, airflow passes through the check valve in *Valve A* to pressurize the blind end of the cylinder. The check valve in *Valve B* blocks the flow of the exhaust air. This forces the exhaust air to be metered through the needle valve. The extension speed is set with this needle valve. During retraction, the air freely passes through the check valve of *Valve B*

to pressurize the rod end of the cylinder. Exhaust air is metered through the needle valve of *Valve A* to control retraction speed.

The meter-out design used in both phases of this circuit causes both sides of the cylinder piston to be pressurized. This produces more stable cylinder movement under the varying load conditions encountered during system operation.

19.4.2 Speed Control Using a Five-Port, Four-Way Directional Control Valve

Figure 19-8 shows a cylinder speed-control circuit using a five-port, four-way directional control valve. This circuit provides meter-out flow control similar to the previous circuit. However, it uses a different layout and different types of valves. The primary differences between this circuit and the previous one are:

- Design of the directional control valve.
- Use of needle valves without bypass check valves.
- Placement of the flow control components in the exhaust lines of the directional control valve.

When this circuit is idle, the normal position of the directional control valve holds the cylinder in the retracted position. When the directional control valve is shifted to extend the cylinder, pressurized air goes directly to the blind end of the cylinder. As the cylinder begins to extend, the exhausted air passes through the directional control

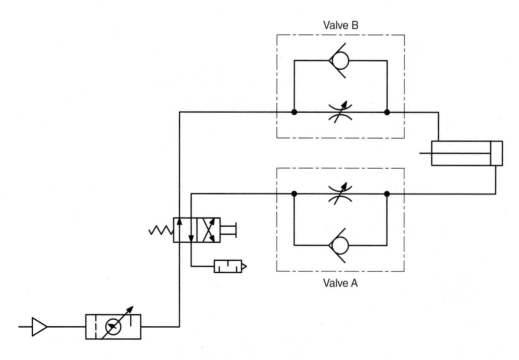

Figure 19-7. The meter-out flow control design is preferred for controlling the operating speed of pneumatic cylinders. This circuit shows the placement of flow control valves in a circuit using a four-port directional control valve.

Chapter 19 Applying Pneumatic Power   **473**

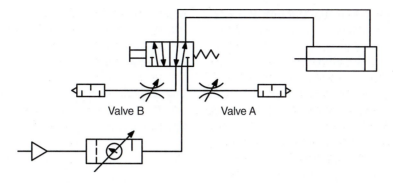

Figure 19-8. Flow control valves are often placed in the exhaust ports of five-port directional control valves. This arrangement eliminates the need for bypass check valves to allow reverse airflow.

valve and on to the needle valve of *Valve A*. This valve meters the exhaust air, which indirectly controls the volume of air that can enter the cylinder. The end result is control of cylinder extension speed. During retraction, the directional control valve passes pressurized air directly to the rod end of the cylinder. Exhaust air from the blind end passes through the directional control valve and on to the needle valve of *Valve B*. This results in the control of cylinder retraction speed.

Both this circuit and the previous circuit allow independent adjustment of extension and retraction speeds. An advantage of this design over the previous circuit is elimination of the integral check valves used to bypass the needle valves. What is lost is the close proximity of the flow control valves to the actuator.

19.4.3 One Retraction and Two Extension Speeds

The circuit shown in **Figure 19-9** provides two adjustable extension speeds and one adjustable retraction speed. The design allows rapid advance of a machine tool to a location where a lower speed is needed for a forming or machining operation. The retraction speed of the cylinder is also adjustable.

The low extension speed is activated by a mechanically operated, normally open, two-way valve. A solenoid valve can also be used. When the system is initially pressurized, the cylinder moves to the retracted position. Shifting the five-port, four-way directional control valve directs airflow to the blind end of the cylinder.

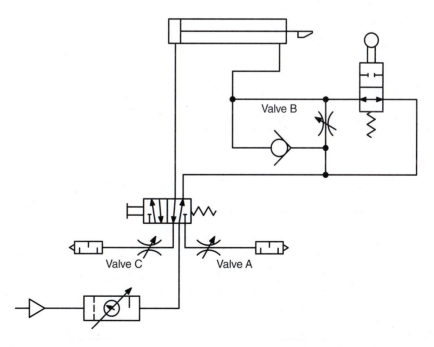

Figure 19-9. Multiple operating speeds can be achieved during cylinder extension or retraction using a mechanically or solenoid-operated, two-way valve. This circuit provides two different cylinder rod extension speeds.

Air from the rod end is exhausted through the normally open, two-way valve; to the four-way directional control valve; and on to needle valve of *Valve A*. The rate of exhaust through this needle valve establishes the higher speed of cylinder extension.

When the ramp on the cylinder rod shifts the mechanically operated, two-way valve, air is forced through the needle valve of *Valve B*. This valve reduces the airflow to slow the cylinder speed for the forming or machining operation. Air exhausted from *Valve B* passes to the atmosphere through the four-way directional control valve and *Valve A* with little restriction.

When the four-way directional control valve is shifted to retract the cylinder, pressurized air passes though the check valve. This allows reverse flow around *Valve B*. When the ramp moves off of the mechanically operated, two-way valve, the valve is shifted by the internal spring. The valve then provides an unrestricted route for the air to enter the rod end to aid in cylinder retraction. The air exhausted from the blind end is routed through the four-way directional control valve to *Valve C*. *Valve C* is an adjustable needle valve that controls cylinder retraction speed.

19.5 Controlling Actuator Motion

Pneumatic systems must provide a wide variety of motion-control functions. These functions range from simple manual controls for extending and retracting a cylinder to complex circuits that produce multiple movements. Often these multiple movements are automatically performed after a machine operator manually shifts a valve to start the sequence.

This section examines two circuits that provide automatic reciprocation of a cylinder. The principles involved in these simple designs can be expanded to more complex circuits to provide the sequencing of two or more actuators. The circuits shown use

mechanically operated, three-way valves controlling pilot-operated directional controls. A combination of limit switches and solenoid-operated valves is also commonly used to control these circuits.

19.5.1 Single-Cycle Operation

Figure 19-10 shows a circuit that provides for *automatic return* of the cylinder after it extends. To initiate the sequence, the operator momentarily depresses a three-way directional control valve (*Valve A*). This action directs pilot pressure to the five-port, four-way directional control valve (*Valve B*). As a result, *Valve B* shifts to extend the cylinder. When the ramp on the cylinder rod shifts *Valve C*, pilot pressure is supplied to shift the four-way directional control to the opposite position. This returns the cylinder to the original, retracted position. The cylinder is held in the retracted position until the machine operator again depresses *Valve A* to repeat the cycle.

19.5.2 Continuous Operation

The circuit shown in **Figure 19-11** provides continuous, *automatic reciprocation* of a cylinder. This continues as long as the system is pressurized and the operator holds the manually operated valve in the shifted position. The diagram shows the circuit before it is pressurized. The cylinder is in the fully retracted position with the extension phase of the cycle about to begin.

Depressing *Valve A* pressurizes the system and starts the sequence. Air passes through *Valves A* and *B*. *Valve B* provides pressure to the *Extension pilot* of the five-port, four-way directional control valve. The four-way valve shifts to direct air to the blind end of the cylinder. This air begins to extend the cylinder. As the cylinder extends, the ramp on the rod moves off the activation mechanism on *Valve B*. This allows the valve to shift and vent the *Extension pilot* line to the atmosphere. The

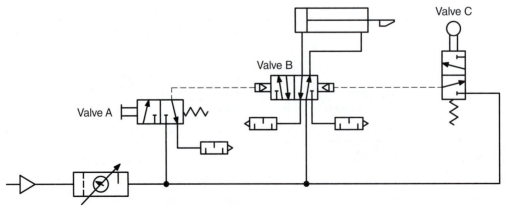

Figure 19-10. The automatic operation of portions of a circuit cycle often increases the efficiency of a machine. This circuit extends and retracts a cylinder after the control valve is activated.

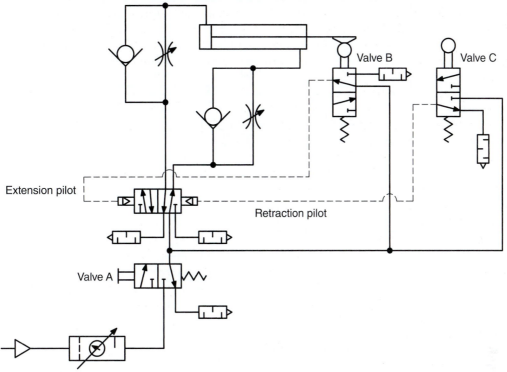

Extension pilot

Retraction pilot

Valve A

Valve B

Valve C

Figure 19-11. Automatic cylinder reciprocation is possible with this circuit. The cylinder continuously extends and retracts as long as the system is pressurized and valve A is activated.

cylinder continues to extend until the ramp depresses the activation mechanism on *Valve C*.

When *Valve C* is shifted, air is directed to the *Retraction pilot* of the four-way directional control valve. The pilot shifts the valve to direct air to the rod end of the cylinder. The cylinder begins to retract. The ramp moves off *Valve C*, which allows the *Retraction pilot* line to vent. The cylinder continues to retract until the ramp depresses the activation mechanism on *Valve B*. This allows system air pressure to again activate the *Extension pilot*, beginning another sequence. The cylinder continues to reciprocate as long as *Valve A* is held in the shifted position to keep the system pressurized.

19.6 Circuits for Other System Functions

A large number of additional basic circuits are often illustrated in materials prepared by component manufacturers, equipment manufacturers, trade magazines, and professional associations. The circuits shown in this section are included to illustrate the variety of specialized components and circuit segments available to support the operation of pneumatic equipment.

19.6.1 Booster Circuit

A pressure higher than that produced by the system compressor is sometimes needed for the operation of a section of a pneumatic circuit. This higher pressure can be provided by specialized air-to-air or air-to-oil **boosters**. Many boosters are basically small-volume, air-powered reciprocating pumps. They produce intensified pressure for use in a small section of a system.

Figure 19-12 illustrates the circuit and operating concepts involved in an air-to-air booster. The circuit shows components in position to begin the process. When the system is pressurized, air flows through *Valve A*, holding the five-port, four-way directional control valve in the position to retract the control rod and extend the piston of the booster pump. As the control rod retracts, *Valve A* shifts and vents the pilot pressure. As the piston of the booster pump extends, air moves out of the pump cylinder. This closes check valve in *Valve C* and opens check valve in *Valve D*. The air moved by the booster pump increases the pressure in the line leading to the circuit section requiring high pressure.

When the power cylinder approaches the end of its extension stroke, the ramp on the control rod shifts *Valve B*. Pilot pressure is then routed through that

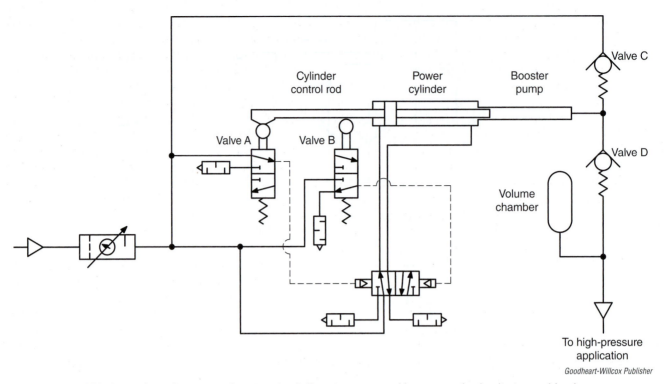

Figure 19-12. This is an air-to-air pressure booster circuit. Boosters are used in pneumatic circuits to provide air pressures higher than the pressure available from the distribution line.

valve to shift the four-way directional control valve to the opposite position. This causes the control rod to extend and the booster pump piston to retract. System air enters the pump through *Valve C. Valve D* closes because of the higher pressure in the high-pressure section of the circuit.

As the control rod extends, *Valve B* shifts and vents the pilot pressure. As the rod approaches the end of its extension stroke, the ramp shifts *Valve A*. Air flowing through this valve shifts the pilot-operated directional control valve to automatically repeat the cycle.

19.6.2 Quick-Exhaust Valve Circuit

A quick-exhaust valve is used to reduce back pressure during cylinder extension or retraction to allow maximum operating speeds. The valve should be close fitted to the ports of the actuator to obtain maximum performance.

Figure 19-13 shows a basic cylinder circuit using a quick-exhaust valve. The circuit is shown in the normal position with the cylinder retracted. System air is routed through the directional control valve to the

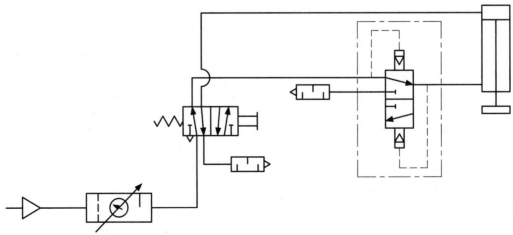

Goodheart-Willcox Publisher

Figure 19-13. Quick-exhaust valves reduce back pressure in the exhaust line of a cylinder by routing exhaust air directly to the atmosphere. Reducing back pressure allows the cylinder to reach maximum operating speed.

quick-exhaust valve. Pilot pressure shifts the valve to provide a route to the rod end of the cylinder. This forces the cylinder to fully retract. During this phase of operation, the quick-exhaust valve serves only as a route for the passage of air and does not contribute to circuit performance.

When the directional control valve is shifted to extend the cylinder, air flows from the directional control valve to the blind end of the cylinder. As the cylinder extends, the air exhausted from the rod end is routed to the quick-exhaust valve. The back pressure in the exhaust line caused by resistance to airflow through the line, the directional control valve, and the quick-exhaust valve itself causes the quick-exhaust valve to shift. This provides the exhausted air a short, direct route to the atmosphere.

19.6.3 Logic Function Circuits

Pneumatic components can be used to perform basic logic functions to provide system control. *Logic functions* include AND, OR, and NOR functions such as used in microprocessors. In pneumatic systems, the functions are completed by small moving parts in air-operated valves. These valves are not in the circuit to help produce motion and force. They are used to sense airflow and pressure to provide the desired functions. A number of manufacturers produce lines of miniature components that operate at this level.

Figure 19-14 illustrates a *memory circuit* that holds a valve position once the manually operated valve initiates the operation. The circuit uses three three-way directional control valves and one shuttle valve.

The control function is initiated by manually depressing *Valve A*. Air passes through that valve and shifts shuttle *Valve C*. This directs air to the pilot of the

system control valve (*Valve D*). That valve shifts to pass pressurized air to the system. It also passes air through *Valve B* to the second pressure port of the shuttle valve. When the start valve (*Valve A*) is released, the shuttle valve shifts, transferring the source of the pilot pressure for *Valve D* to the air line of the operating circuit (through *Valve B*).

The circuit holds *Valve D* open until the stop valve (*Valve B*) is shifted. When *Valve B* is shifted, the pilot pressure for *Valve D* is vented. This allows the internal spring in *Valve D* to shift the valve. Airflow and pressure to the operating circuit are blocked.

19.6.4 Time-Delay Circuits

Circuit actuators often need to be momentarily held at a point in their operation to allow the completion of a task. **Figure 19-15** shows a *time-delay circuit* that allows automatic timing of cylinder movement. This design uses simple needle valves, timing-volume reservoirs, and piloted-operated directional control valves to provide a desired sequence of operation and time delays.

Pilot-operated, five-port, four-way directional control valves are used to operate the two cylinders. The sequence in this circuit is extension of *Cylinder 1* followed by extension of *Cylinder 2* after a time delay. When the task performed by *Cylinder 2* is completed, that actuator can be immediately retracted, followed by *Cylinder 1* after another time delay.

When this circuit is pressurized, both *Cylinder 1* and *Cylinder 2* are moved to the retracted position. Circuit operation is initiated by the operator activating *Valve A*. This pressurizes the control portion of the circuit. Air flows through the check valve in *Valve B* and on to the timing reservoir and the pilot control of *Valve C*.

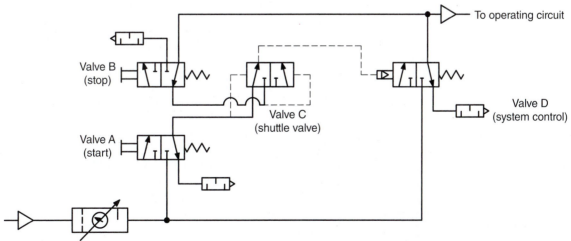

Goodheart-Willcox Publisher

Figure 19-14. Pneumatic control valves can be assembled into circuits that perform logic functions. This is a memory circuit that holds the system control valve in the activated position using only a momentary activation of a start valve.

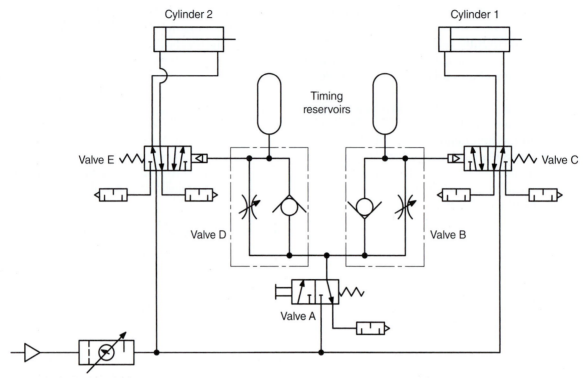

Figure 19-15. Timed delays in a sequence of cylinder movements can be produced using flow control valves coupled with simple timing reservoirs. These control segments time the activation of pilot-operated directional control valves.

The pressure quickly climbs to pressurize the timing chamber and activate the pilot. This shifts *Valve C* to extend *Cylinder 1*.

During the same period of time, the check valve in *Valve D* closes. This forces the control air to be metered through the needle valve. This metered air flows to the second timing reservoir, which slowly fills. After a period of time, the pressure in this reservoir increases to the point where the pilot mechanism shifts *Valve E*. Shifting *Valve E* extends *Cylinder 2*. This performs the second task of the operation.

When the tasks performed by the cylinders are completed, the machine operator allows *Valve A* to return to its normal position. This vents the control portion of the circuit to the atmosphere. The pilot air that shifted *Valve E* can easily vent through the check valve in *Valve D*. This allows the spring mechanism in *Valve E* to quickly shift the valve. As a result, *Cylinder 2* quickly returns to the retracted position. The retraction of *Cylinder 1* is delayed as the pilot air in the timing chamber must bleed through the needle valve in *Valve B*. When this pilot air has been vented to the atmosphere, *Valve C* shifts to return *Cylinder 1* to the retracted position.

The length of the time delays is set by adjusting the needle valves. Different timing-volume reservoirs are also available, which allows a large range of time delays.

19.6.5 Safety Circuits

The safe operation of machines is always a concern. Operators may be injured when they become distracted or take chances. Pneumatic control circuits are often used in an attempt to improve workplace safety.

Figure 19-16 shows a pneumatic safety circuit often used to reduce hand injuries. This circuit requires the use of both hands to activate the operation of a press or other machine that can cause serious hand injuries. A *two-hand safety circuit* generally prevents complete or partial machine operation if either of the hand controls is tied down in an attempt to override the system.

When this circuit is at rest, *Valve A* pressurizes *Pilot A* of the system valve (*Valve C*). This shifts *Valve C*, allowing the single-acting cylinder to vent to the atmosphere. The press platen in this example fully lowers.

Depressing only *Valve A* blocks system pressure and vents *Pilot A* to the atmosphere. Depressing only *Valve B* pressurizes *Pilot B*, which attempts to shift *Valve C*. However, the valve will not shift because both *Pilots A* and *B* are pressurized. This prevents movement of the valve control element. Simultaneously depressing *Valves A* and *B* vents *Pilot A* to the atmosphere and pressurizes *Pilot B*. This allows the system control valve (*Valve C*) to be shifted.

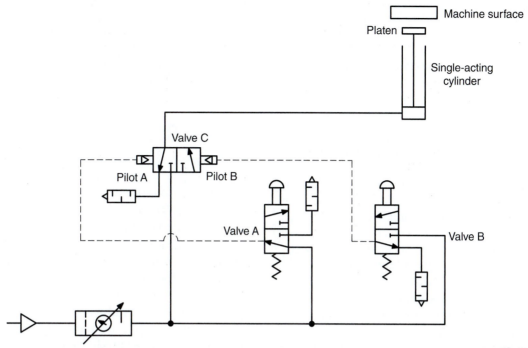

Figure 19-16. Machines often require both hands to activate the start cycle. These safety circuits will not function if an operator attempts to bypass the system by tying down either of the hand controls.

The system will not operate beyond a part of a cycle if a machine operator attempts to bypass the safety system by tying down one of the control valves. Tying down *Valve A* and attempting to operate the system by activating only *Valve B* results in the cylinder extending. However, there is no method to pressurize *Pilot A*, which is needed to shift *Valve C* to retract the cylinder. Tying down *Valve B* results in a similar situation. When *Valve A* is activated in this second situation, the air at *Pilot A* is vented. This allows the pressurized air coming from *Valve B* to shift *Valve C*. The cylinder extends, but cannot be lowered. *Valve B*, which is tied down, maintains a constant pressure on *Pilot B*. This prevents further shifting of *Valve C*.

Summary

- Pneumatic systems can be classified into basic circuit segments that are responsible for conditioning the air, controlling air pressure and flow rate, and producing a desired movement.

- Air preparation at the workstation of a pneumatic system includes air filtration, pressure reduction, and the addition of a lubricant if lubrication is required.

- The maximum operating pressure of a workstation circuit is determined by the setting of the pressure regulator located in the trio unit. Additional pressure regulators within segments of the workstation circuit can operate actuators at lower pressures. Pilot-operated regulators can be used when remote pressure adjustment is needed.

- Pneumatic systems use both meter-in and meter-out circuits for airflow control. However, the meter-out circuit is preferred, as the design pressurizes both sides of the internal elements of the actuator, providing more stable movement under varying loads.

- The most consistent control of actuator speed is obtained when the flow control valves are located as close to the actuator as practical and when the controlled air volume in the actuator and lines is as small as possible.

- When four-port, four-way directional control valves are used with a pneumatic circuit, flow control valves are typically placed between the actuator and the directional control valve. This arrangement requires the use of a flow control valve with an integral check valve to allow reverse airflow.

- When five-port, four-way directional control valves are used with a pneumatic circuit, flow control valves are placed in the exhaust port lines of the directional control valve. This arrangement eliminates the need for reverse flow check valves.

- Multiple flow rates producing a number of actuator speeds can be automatically produced in pneumatic circuits by using mechanical or solenoid activated valves to direct the route of airflow through a circuit.

- Mechanically, pilot-, and solenoid-operated valves can be used in circuit segments to produce a desired sequence of actuator movement.

- Pressure boosters can be used in pneumatic workstation circuits to produce pressures higher than those available from the compressor and compressed air distribution sections of a pneumatic system.

- Quick-exhaust valves provide a direct route to the atmosphere for the exhaust air, reducing back pressure in cylinder exhaust lines and increasing the operating speed of the cylinder.

- To reduce accidents and operator injuries, pneumatic-powered production equipment often includes a safety circuit that requires an operator to use both hands to initiate and maintain machine operation.

Internet Resources

The following are some useful resources available on the Internet. Enter a company or organization name into a search engine to access its website. Explore the various areas of the sites to discover useful fluid power resources.

Bosch Rexroth AG. The Rexroth media directory page. Use the search function to select the "Pneumatics" category (or a more specific topic of interest), and then view the extensive array of related technical articles and bulletins.

Fairchild Industrial Products Company. The page provides a number of pneumatic applications for the company's pneumatic products. The information is classified by the industries where the products are being used.

Hydraulics & Pneumatics Magazine. An extensive chapter on air logic controls. The descriptions of air logic elements include illustrations and circuit examples. Among the illustrations are logic symbols, standard pneumatic symbols, and graphic component cutaways.

International Fluid Power Society: Fluid Power Journal. A listing of professional papers, including some that support the use of pneumatic control of fluid power systems. This site is an example of materials provided by a professional organization.

Chapter Review

Answer the following questions using information in this chapter.

1. *True or False?* The challenge of understanding the operation of a pneumatic system can be reduced if it is considered to be groups of basic components and subcircuits.

2. The pneumatic workstation filter, regulator, and lubricator are often close fitted into a unit referred to as the _____.
   A. composite group
   B. trio unit
   C. multiple pressure
   D. logic function

3. The final preparation of air in a pneumatic system occurs at the _____.
   A. multiple-pressure circuit
   B. remote control planning
   C. workstation
   D. composite regulator

4. Whenever additional pressure regulators are used downstream from the air-preparation unit, they can regulate pressures to a level below the setting of the _____ pressure regulator.

5. A(n) _____-operated pressure regulator can be used when pneumatic system pressure must be controlled from a remote location.

6. In five-port, four-way directional control valves, flow control components are placed in the _____ lines.

7. *True or False?* Quick-exhaust valves are designed to increase the operating speed of cylinders.

8. _____ include AND, OR, and NOR functions such as used in microprocessors. They are used to provide system control.
   A. composite groups
   B. trio units
   C. multiple pressures
   D. logic functions

9. *True or False?* The purpose of a time-delay circuit is to manually time cylinder movement.

10. A two-hand safety circuit generally prevents complete or partial machine operation if _____.
    A. the circuit has no air supply
    B. either of the two hand controls is tied down
    C. either of the two hand controls is pressurized
    D. All of the above.

Apply and Analyze

1. Design and draw a circuit that allows you to control both the pressure and speed of cylinder retraction and extension independently.
   A. How is pressure regulated in your circuit?
   B. What directional control valves have you chosen? Why?
   C. How is actuator speed controlled? What other options could you choose?
   D. In constructing the circuit, what additional factors would you consider to produce the most stable and accurate actuator operating speeds?

2. Based on the circuit shown in **Figure 19-11**, complete the following exercises:
   A. Complete the chart below by estimating the relative air pressure expected at the pilot ports of the directional control valve. Describe the circuit action as the valves are activated and released. Assume instantaneous air exhaust.

Valve Operation	Relative Pressure at Pilot Ports		Circuit Action
	Retraction pilot	Extension pilot	
Valves A and B depressed.	Atmospheric pressure (Patm)	System pressure (Psystem)	Pilot-operated valve shifts allowing air to flow into blind end of the cylinder, cylinder begins to extend
Valve A depressed, B released.			
Valves A and C depressed.			
Valve A depressed, Valve C released.			
Valve A released.			

   B. What components affect the pressure at the blind and rod end of the cylinder during circuit operation?
   C. Modify the circuit so that air at lower pressure can be delivered to the pilot-operated directional control valve. Describe the impact of your modification on circuit operation.

3. Based on the time delay circuit shown in **Figure 19-15**, complete the following exercises:

A. Create a line graph that shows the change in relative pressure over time of operation for each of the following locations in the circuit: pilot side of *Valve C*, pilot side of *Valve E*, blind end of *Cylinder 1*, and blind end of *Cylinder 2*. Your graph should include a line for each of these five locations. Time starts when *Valve A* is depressed and ends when the system reaches equilibrium after *Valve A* is released. Indicate on your graph when *Valves C* and *E* shift and when the movement of both cylinders occurs.

B. What adjustments or modifications could you make if more time delay were required between movements of the cylinders? How would each of these changes affect your graph?

4. Design and draw a circuit for a basic drill-and-clamp-operation using two double-acting cylinders. Your circuit should allow for the following sequence of operations: extend clamp, advance drill, retract drill, and retract clamp.

A. What modifications would you make to allow the clamping force to be adjusted for use with different materials? Draw this modified circuit.

B. What modification would you make to automate the drilling operation? Draw this modified circuit.

Research and Development

1. Find and analyze a pneumatic circuit schematic for a machine. Create a poster that shows the pneumatic circuit, identifies the components used, and lists the steps that occur during operation.

2. Research pneumatic logic control devices. Prepare a report that provides examples of these devices, describes their operation, and explains when these devices might be used in a machine or pneumatic circuit.

3. Design, construct, test, and demonstrate a pneumatic circuit that could be used to complete one of the following tasks:

- Open a power door by pushing a button. The door needs to stay open for an adjustable amount of time before it automatically closes. Operators should also be able to close the door immediately or keep it open for extended periods of time if necessary.

- Safely cut potatoes into thick slices. Slices of potatoes must be removed from the cutting surface before a new potato is added.

- Lift and transfer parts to a separate bin for transport. The operator should be able to do this process automatically by pushing a button one time.

smereka/Shutterstock.com

Farming equipment makes great use of fluid power. This seed drill uses pneumatic power to plant seeds precisely and efficiently.

APPENDIX

Conversions

Area

To Convert	Into	Multiply By
square centimeters	square inches	.155
square feet	square centimeters	929.0
square inches	square centimeters	6.452
square inches	square millimeters	645.2
square meters	square feet	10.76
square meters	square yards	1.196

Distance

To Convert	Into	Multiply By
centimeters	feet	.0328
centimeters	inches	.3937
centimeters	meters	.01
centimeters	millimeters	10.0
feet	centimeters	30.48
feet	inches	12.0
feet	meters	.3048
feet	millimeters	304.8
inches	centimeters	2.54
inches	feet	.0833
inches	meters	.0254
inches	millimeters	25.4
meters	feet	3.281
meters	inches	39.37
millimeters	centimeters	.1
millimeters	inches	.0394
millimeters	feet	.00328
millimeters	meters	.001
yards	meters	.9144

Energy

To Convert	Into	Multiply By
kilowatt-hours	Btu	3415.0

Flow

To Convert	Into	Multiply By
cubic feet/minute	barrels/day	256.5
cubic feet/minute	cubic feet/hour	60.0
cubic feet/minute	cubic meters/hour	1.699
cubic feet/minute	gallons/minute	7.481
cubic feet/minute	liters/minute	28.32
cubic feet/hour	gallons/minute	.1247
cubic feet/hour	liters/hour	.472
cubic feet/hour	cubic feet/minute	.0167
cubic feet/hour	cubic meters/hour	.0283
cubic meters/hour	gallons/minute	4.403
cubic meters/hour	liters/minute	16.67
cubic meters/hour	cubic feet/minute	.5886
cubic meters/hour	cubic feet/hour	35.31
cubic meters/hour	barrels/day	150.9

gallons (US)/minute	cubic feet/minute	.133
gallons (US)/minute	cubic feet/hour	8.021
gallons (US)/minute	cubic inches/minute	231.0
gallons (US)/minute	cubic meters/hour	.227
gallons (US)/minute	liters/minute	3.785
liters/minute	gallons (US)/minute	.2642

Mass

To Convert	Into	Multiply By
kilograms	pounds	2.205
ounces	grams	28.35
pounds	kilograms	.4536
tons (metric)	pounds	2205.0
tons (short)	pounds	2000.0

Power

To Convert	Into	Multiply By
horsepower	foot-pounds/minute	33,000.0
horsepower	foot-pounds/second	550.0
horsepower	watts	745.7

Pressure

To Convert	Into	Multiply By
atmospheres	bar	1.013
atmospheres	inches of mercury	29.92
atmospheres	kilograms/square centimeter	1.033
atmospheres	kilopascals (kPa)	101.3
atmospheres	psi	14.7
bar	atmospheres	.9869
bar	feet of water column (wc)	33.45
bar	kilograms/square centimeter	1.020
bar	kilopascals (kPa)	100.0
bar	psi	14.5
inches of mercury	psi	.4912
inches of water (4°C)	psi	.0361
kilograms/cubic meter	pounds/cubic foot	.06243
millimeters of mercury	psi	.0194
pounds/cubic foot	grams/cubic centimeter	.016
pounds/square inch (psi)	atmospheres	.068
pounds/square inch (psi)	bar	.069
pounds/square inch (psi)	inches of mercury	2.036
pounds/square inch (psi)	inches of water (4°C)	27.7
pounds/square inch (psi)	kilopascals (kPa)	6.895
pounds/square inch (psi)	kilograms/square centimeter	.0703

Temperature

To Convert	Into	Multiply By
Celsius	Fahrenheit	$9/5 C° + 32°$
Celsius	Kelvin	$C° + 273°$
Fahrenheit	Celsius	$5/9 (F° - 32°)$
Fahrenheit	Rankine	$F° + 460°$

Volume

To Convert	Into	Multiply By
cubic centimeters	cubic inches	.061
cubic feet	cubic meters	.028317
cubic feet	liters	28.317
cubic inches	liters	.01639
cubic meters	cubic feet	35.31
gallons (US)	liters	3.785
liters	cubic feet	.0351
liters	cubic inches	61.02
liters	cubic meters	.001
liters	gallons (US)	.2642

Metric-Inch Equivalents

Millimeters	Decimal Inch	Fractional Inch	Millimeters	Decimal Inch	Fractional Inch
0.100	.0039		3.969	.1563	5/32
0.100	.0039		4.000	.1575	
0.200	.0079		4.100	.1614	
0.300	.0118		4.200	.1654	
0.397	.0156	1/64	4.300	.1693	
0.400	.0157		4.366	.1719	11/64
0.500	.0197		4.400	.1732	
0.600	.0236		4.500	.1772	
0.700	.0276		4.600	.1811	
0.794	.0313	1/32	4.700	.1850	
0.800	.0315		4.763	.1875	3/16
0.900	.0354		4.800	.1890	
1.000	.0394		4.900	.1929	
1.100	.0433		5.000	.1969	
1.191	.0469	3/64	5.100	.2008	
1.200	.0472		5.159	.2031	13/64
1.300	.0512		5.200	.2047	
1.400	.0551		5.300	.2087	
1.500	.0591		5.400	.2126	
1.588	.0625	1/16	5.500	.2165	
1.600	.0630		5.556	.2188	7/32
1.700	.0669		5.600	.2205	
1.800	.0709		5.700	.2244	
1.900	.0748		5.800	.2283	
1.984	.0781	5/64	5.900	.2323	
2.000	.0787		5.953	.2344	15/64
2.100	.0827		6.000	.2362	
2.200	.0866		6.100	.2402	
2.300	.0906		6.200	.2441	
2.381	.0938	3/32	6.300	.2480	
2.400	.0945		6.350	.2500	1/4
2.500	.0984		6.400	.2520	
2.600	.1024		6.500	.2559	
2.700	.1063		6.600	.2598	
2.778	.1094	7/64	6.700	.2638	
2.800	.1102		6.747	.2656	17/64
2.900	.1142		6.800	.2677	
3.000	.1181		6.900	.2717	
3.100	.1220		7.000	.2756	
3.175	.1250	1/8	7.100	.2795	
3.200	.1260		7.144	.2813	9/32
3.300	.1299		7.200	.2835	
3.400	.1339		7.300	.2874	
3.500	.1378		7.400	.2913	
3.572	.1406	9/64	7.500	.2953	
3.600	.1417		7.541	.2969	19/64
3.700	.1457		7.600	.2992	
3.800	.1496		7.700	.3031	
3.900	.1535		7.800	.3071	
			7.900	.3110	

Millimeters	Decimal Inch	Fractional Inch	Millimeters	Decimal Inch	Fractional Inch
7.938	.3125	5/16	12.303	.4844	31/64
8.000	.3150		12.400	.4882	
8.100	.3189		12.500	.4921	
8.200	.3228		12.600	.4961	
8.300	.3268		12.700	.5000	1/2
8.334	.3281	21/64	12.800	.5039	
8.400	.3307		12.900	.5079	
8.500	.3346		13.000	.5118	
8.600	.3386		13.097	.5156	33/64
8.700	.3425		13.100	.5157	
8.731	.3438	11/32	13.200	.5197	
8.800	.3465		13.300	.5236	
8.900	.3504		13.400	.5276	
9.000	.3543		13.494	.5313	17/32
9.100	.3583		13.500	.5315	
9.128	.3594	23/64	13.600	.5354	
9.200	.3622		13.700	.5394	
9.300	.3661		13.800	.5433	
9.400	.3701		13.891	.5469	35/64
9.500	.3740		13.900	.5472	
9.525	.3750	3/8	14.000	.5512	
9.600	.3780		14.100	.5551	
9.700	.3819		14.200	.5591	
9.800	.3858		14.288	.5625	9/16
9.900	.3898		14.300	.5630	
9.922	.3902	25/64	14.400	.5669	
10.000	.3937		14.500	.5709	
10.100	.3976		14.600	.5748	
10.200	.4016		14.684	.5781	37/64
10.300	.4055		14.700	.5787	
10.319	.4063	13/32	14.800	.5827	
10.400	.4094		14.900	.5866	
10.500	.4134		15.000	.5905	
10.600	.4173		15.081	.5938	19/32
10.700	.4213		15.100	.5945	
10.716	.4219	27/64	15.200	.5984	
10.800	.4252		15.300	.6024	
10.900	.4291		15.400	.6063	
11.000	.4331		15.478	.6094	39/64
11.100	.4370		15.500	.6102	
11.113	.4375	7/16	15.600	.6142	
11.200	.4409		15.700	.6181	
11.300	.4449		15.800	.6220	
11.400	.4488		15.875	.6250	5/8
11.500	.4528		15.900	.6260	
11.509	.4531	29/64	16.000	.6299	
11.600	.4567		16.100	.6339	
11.700	.4606		16.200	.6378	
11.800	.4646		16.272	.6406	41/64
11.900	.4685		16.300	.6417	
11.906	.4688	15/32	16.400	.6457	
12.000	.4724		16.500	.6496	
12.100	.4764		16.600	.6535	
12.200	.4803				
12.300	.4843				

Millimeters	Decimal Inch	Fractional Inch	Millimeters	Decimal Inch	Fractional Inch
16.669	.6563	21/32	21.034	.8282	53/64
16.700	.6575		21.100	.8307	
16.800	.6614		21.200	.8346	
16.900	.6654		21.300	.8386	
17.000	.6693		21.400	.8425	
17.066	.6719	43/64	21.431	.8438	27/32
17.100	.6732		21.500	.8465	
17.200	.6772		21.600	.8504	
17.300	.6811		21.700	.8543	
17.400	.6850		21.800	.8583	
17.463	.6875	11/16	21.828	.8594	55/64
17.500	.6890		21.900	.8622	
17.600	.6929		22.000	.8661	
17.700	.6968		22.100	.8701	
17.800	.7008		22.200	.8740	
17.859	.7031	45/64	22.225	.8750	7/8
17.900	.7047		22.300	.8780	
18.000	.7087		22.400	.8819	
18.100	.7126		22.500	.8858	
18.200	.7165		22.600	.8898	
18.256	.7188	23/32	22.622	.8906	57/64
18.300	.7205		22.700	.8937	
18.400	.7244		22.800	.8976	
18.500	.7283		22.900	.9016	
18.600	.7323		23.000	.9055	
18.653	.7344	47/64	23.019	.9063	29/32
18.700	.7362		23.100	.9094	
18.800	.7402		23.200	.9134	
18.900	.7441		23.300	.9173	
19.000	.7480		23.400	.9213	
19.050	.7500	3/4	23.416	.9219	59/64
19.100	.7520		23.500	.9252	
19.200	.7559		23.600	.9291	
19.300	.7598		23.700	.9331	
19.400	.7638		23.800	.9370	
19.447	.7656	49/64	23.813	.9375	15/16
19.500	.7677		23.900	.9409	
19.600	.7717		24.000	.9449	
19.700	.7756		24.100	.9488	
19.800	.7795		24.200	.9528	
19.844	.7813	25/32	24.209	.9531	61/64
19.900	.7835		24.300	.9567	
20.000	.7874		24.400	.9606	
20.100	.7913		24.500	.9646	
20.200	.7953		24.600	.9685	
20.241	.7969	51/64	24.606	.9688	31/32
20.300	.7992		24.700	.9724	
20.400	.8031		24.800	.9764	
20.500	.8071		24.900	.9803	
20.600	.8110		25.000	.9843	
20.638	.8125	13/16	25.003	.9844	63/64
20.700	.8150		25.100	.9882	
20.800	.8189		25.200	.9921	
20.900	.8228		25.300	.9961	
21.000	.8268		25.400	1.0000	1

Dimensional Data for Drilled Orifices, Hose, and Pipe

Diameter	Orifice	Internal Area (square inches)	
		Hose	Pipe (schedule 40)
1/32	.00077		
1/16	.00307		
3/32	.0069		
1/8	.01227	.01227	
5/32	.01917		
3/16	.02761		
7/32	.03758		
1/4	.04909		.104
9/32	.06213		
5/16	.0767		
11/32	.09281		
3/8	.1104	.11	.190
13/32	.1296		
7/16	.1503		
15/32	.1726		
1/2	.1963	.196	.304
17/32	.2217		
9/16	.2485		
19/32	.2769		
5/8	.3068	.307	
21/32	.3382		
11/16	.3712		
23/32	.4057		
3/4	.4418	.442	.533
13/16	.5185		
7/8	.6013		
15/16	.6903		
1	.7854	.785	.864
1 1/4	1.2272	1.227	1.496
1 1/2	1.767		2.036
2	3.1416	3.14	3.356
2 1/2	4.9088		4.788

Tubing Size

Outside Diameter (OD)		Wall Thickness		Inside Diameter (ID)	
Inches	Millimeters	Inches	Millimeters	Inches	Millimeters
1/8	3.18	.028	0.71	.069	1.76
		.032	0.81	.061	1.55
		.035	0.89	.055	1.40
3/16	4.76	.032	0.81	.123	3.13
		.035	0.89	.117	2.98
1/4	6.35	.035	0.89	.180	4.57
		.049	1.25	.250	6.35
		.065	1.65	.120	3.05
5/16	7.94	.035	0.89	.243	6.16
		.049	1.25	.215	5.45
		.065	1.65	.183	4.64
3/8	9.53	.035	0.89	.305	7.75
		.049	1.25	.277	7.04
		.065	1.65	.245	6.23
1/2	12.70	.035	0.89	.430	10.92
		.049	1.25	.402	10.21
		.065	1.65	.370	9.40
		.083	2.11	.334	8.48
5/8	15.88	.035	0.89	.555	14.10
		.049	1.25	.527	13.39
		.065	1.65	.495	12.58
		.083	2.11	.459	11.66
3/4	19.05	.049	1.25	.652	16.56
		.065	1.65	.620	15.75
		.083	2.11	.584	14.83
		.109	2.77	.532	13.51
7/8	22.23	.049	1.25	.777	19.74
		.065	1.65	.745	18.93
		.083	2.11	.709	18.01
		.109	2.77	.657	16.69
1	25.40	.049	1.25	.902	22.91
		.065	1.65	.870	22.10
		.083	2.11	.834	21.18
		.109	2.77	.782	19.86
1 1/4	31.75	.049	1.25	1.152	29.26
		.065	1.65	1.120	28.45
		.083	2.11	1.084	27.53
		.109	2.77	1.032	26.21
1 1/2	38.10	.065	1.65	1.370	34.80
		.083	2.11	1.334	33.88
		.109	2.77	1.282	32.56
1 3/4	44.45	.065	1.65	1.620	41.15
		.083	2.11	1.584	40.23
		.109	2.77	1.532	38.91
		.134	3.40	1.482	37.64
2	50.80	.065	1.65	1.870	47.50
		.083	2.11	1.834	46.58
		.109	2.77	1.782	45.26
		.134	3.40	1.732	43.99

Pipe Size

Nominal size	Outside Diameter	Wall Thickness		Inside Diameter		Flow Area	
		Schedule 40	Schedule 80	Schedule 40	Schedule 80	Schedule 40	Schedule 80
1/8	.41 in (1.03 cm)	.07 in (0.17 cm)	.10 in (0.24 cm)	.27 in (0.68 cm)	.22 in (0.55 cm)	.057 in^2 (0.366 cm^2)	.036 in^2 (0.234 cm^2)
1/4	.54 in (1.37 cm)	.09 in (0.22 cm)	.12 in (0.30 cm)	.36 in (0.92 cm)	.30 in (0.76 cm)	.104 in^2 (0.671 cm^2)	.072 in^2 (0.462 cm^2)
3/8	.68 in (1.71 cm)	.09 in (0.22 cm)	.13 in (0.32 cm)	.49 in (1.25 cm)	.42 in (1.07 cm)	.191 in^2 (1.231 cm^2)	.140 in^2 (0.906 cm^2)
1/2	.84 in (2.13 cm)	.11 in (0.28 cm)	.15 in (0.37 cm)	.62 in (1.58 cm)	.55 in (1.39 cm)	.304 in^2 (1.959 cm^2)	.234 in^2 (1.510 cm^2)
3/4	1.05 in (2.67 cm)	.11 in (0.28 cm)	.15 in (0.37 cm)	.82 in (2.09 cm)	.74 in (1.88 cm)	.533 in^2 (3.439 cm^2)	.432 in^2 (2.788 cm^2)
1	1.32 in (3.34 cm)	.13 in (0.34 cm)	.18 in (0.45 cm)	1.05 in (2.66 cm)	.96 in (2.43 cm)	.864 in^2 (5.573 cm^2)	.719 in^2 (4.638 cm^2)
1 1/4	1.66 in (4.22 cm)	.14 in (0.36 cm)	.19 in (0.49 cm)	1.38 in (3.51 cm)	1.28 in (3.25 cm)	1.495 in^2 (9.645 cm^2)	1.282 in^2 (8.272 cm^2)
1 1/2	1.90 in (4.83 cm)	.15 in (0.37 cm)	.20 in (0.51 cm)	1.61 in (4.09 cm)	1.50 in (3.81 cm)	2.035 in^2 (13.128 cm^2)	1.766 in^2 (11.395 cm^2)
2	2.38 in (6.03 cm)	.15 in (0.37 cm)	.22 in (0.55 cm)	2.07 in (5.25 cm)	1.94 in (4.93 cm)	3.354 in^2 (21.638 cm^2)	2.951 in^2 (19.041 cm^2)
2 1/2	2.88 in (7.30 cm)	.20 in (0.52 cm)	.28 in (0.70 cm)	2.47 in (6.27 cm)	2.32 in (5.90 cm)	4.785 in^2 (30.873 cm^2)	4.236 in^2 (27.330 cm^2)
3	3.50 in (8.89 cm)	.22 in (0.55 cm)	.30 in (0.76 cm)	3.07 in (7.79 cm)	2.90 in (7.37 cm)	7.389 in^2 (47.670 cm^2)	6.602 in^2 (42.592 cm^2)
3 1/2	4.00 in (10.16 cm)	.23 in (0.57 cm)	.32 in (0.81 cm)	3.55 in (9.01 cm)	3.36 in (8.54 cm)	9.882 in^2 (63.754 cm^2)	8.883 in^2 (57.312 cm^2)
4	4.50 in (11.43 cm)	.24 in (0.60 cm)	.34 in (0.86 cm)	4.03 in (10.23 cm)	3.83 in (9.72 cm)	12.724 in^2 (82.089 cm^2)	11.491 in^2 (74.136 cm^2)

Cylinder Force

The following table can be used to quickly determine the force, measured in pounds, that a cylinder will exert at a specific gauge pressure. Friction in the cylinder is disregarded in these calculations. Friction losses should be determined using information provided by the cylinder manufacturer.

Force during retraction of double-acting cylinders can be calculated by subtracting the force generated by the cross-sectional area of the cylinder rod from the appropriate figure in this chart.

Piston Diameter (inches)	Piston Area (square inches)	Operating Pressure (psig)									
		10	20	30	40	50	60	70	80	90	100
.500	.196	1.9	3.9	5.8	7.9	9.8	11.8	13.7	15.7	17.7	19.6
.625	.307	3.1	6.1	9.2	12.3	15.3	18.4	21.5	24.5	27.6	30.7
.750	.442	4.4	8.8	13.3	17.7	22.1	26.5	30.9	35.3	39.8	44.2
.875	.601	6.0	12.0	18.0	24.1	30.1	36.1	42.1	48.1	54.1	60.1
1.000	.785	7.8	15.7	23.6	31.4	39.3	47.1	54.9	62.8	70.7	78.5
1.125	.994	9.9	19.9	29.8	39.8	49.7	59.6	69.6	79.5	89.5	99.4
1.250	1.227	12.3	24.5	36.8	49.1	61.3	73.6	85.9	98.2	110.4	122.7
1.375	1.485	14.8	29.7	44.6	59.4	74.2	89.1	103.9	118.8	133.6	148.5
1.500	1.767	17.7	35.4	53.0	70.7	88.4	106.0	123.7	141.4	159.1	176.7
1.625	2.074	20.7	41.5	62.2	82.9	103.7	124.4	145.2	165.9	186.7	207.4
1.750	2.405	24.1	48.1	72.2	96.2	120.3	144.3	168.4	192.4	216.5	240.5
1.875	2.761	27.6	55.2	82.8	110.4	138.1	165.6	193.3	220.9	248.5	276.1
2.000	3.142	31.4	62.8	94.3	125.7	157.1	188.5	219.9	251.4	282.8	314.2
2.250	3.976	39.8	79.5	119.3	159.0	198.8	238.6	278.3	318.1	357.8	397.6
2.500	4.909	49.1	98.2	147.3	196.4	245.5	294.5	343.6	392.7	441.8	490.9
2.750	5.940	59.4	118.8	178.2	237.6	297.0	356.4	415.8	475.2	534.6	594.0
3.000	7.069	70.7	141.4	212.1	282.7	353.4	424.1	494.8	565.5	636.2	706.9
3.500	9.621	96.2	192.4	288.6	384.8	481.1	577.3	673.5	769.7	865.9	962.1
4.000	12.566	125.7	251.4	377.0	502.7	628.4	754.0	879.7	1005.4	1131.1	1257.0
4.500	15.904	159.0	318.1	477.1	636.2	795.2	954.2	1113.3	1272.3	1431.4	1590.4
5.000	19.635	196.4	392.7	589.1	785.4	981.8	1178.1	1374.5	1570.8	1767.2	1968.5
5.500	23.758	237.6	475.2	712.7	950.3	1188.0	1425.5	1663.1	1900.6	2138.2	2375.8
6.000	28.274	282.7	565.5	848.2	1130.9	1413.7	1696.4	1979.1	2261.9	2544.7	2827.4

Force (pounds)

Free Air Consumption

Use this table to determine the cubic feet of air consumed by a pneumatic cylinder. The values in this table indicate the cubic feet of air consumed for each inch of piston stroke. For single-acting cylinders, multiply the value in the table by the piston stroke (in inches). For retraction of double-acting cylinders, subtract the cross-sectional area of the rod from the piston area. Then, select the area closest to this figure in the area column to obtain an estimate of the free air required for retraction.

Piston Diameter (inches)	Piston Area (square inches)	Operating Pressure (psig)			
		50	60	70	80
.500	.196	.0005	.0006	.0007	.0007
.625	.307	.0008	.0009	.0010	.0011
.750	.442	.0011	.0013	.0015	.0016
.875	.601	.0015	.0018	.0020	.0022
1.000	.785	.0020	.0023	.0026	.0029
1.125	.994	.0025	.0029	.0033	.0037
1.250	1.227	.0031	.0036	.0041	.0046
1.375	1.485	.0038	.0044	.0050	.0055
1.500	1.767	.0045	.0052	.0059	.0066
1.625	2.074	.0053	.0061	.0069	.0077
1.750	2.405	.0061	.0071	.0080	.0090
1.875	2.761	.0070	.0081	.0092	.0103
2.000	3.142	.0080	.0092	.0105	.0117
2.250	3.976	.0101	.0117	.0133	.0148
2.500	4.909	.0125	.0144	.0164	.0183
2.750	5.940	.0151	.0175	.0198	.0221
3.000	7.069	.0190	.0208	.0236	.0264
3.500	9.621	.0245	.0283	.0321	.0359
4.000	12.566	.0320	.0370	.0419	.0469
4.500	15.904	.0405	.0468	.0530	.0593
5.000	19.635	.0500	.0578	.0655	.0732
5.500	23.758	.0605	.0699	.0792	.0886
6.000	28.274	.0720	.0831	.0943	.1054

Air Consumed (cfm)

Viscosity Conversions

Approximate Viscosities	
Material	**Viscosity (centipoise)**
Water	1
Oil, SAE 10	85–140
Oil, SAE 20	140–420
Oil, SAE 30	420–650
Oil, SAE 40	650–900

Viscosity Conversions (sp. gr. = 1)				
Centipoise (cPs)	**Poise (Ps)**	**Centistokes (cSt)**	**Stokes (St)**	**Saybolt Universal (SSU)**
1	.01	1	.01	31
20	.2	20	.2	100
30	.3	30	.3	160
40	.4	40	.4	210
60	.6	60	.6	320
80	.8	80	.8	430
100	1.1	100	1.0	530
120	1.2	120	1.2	580
140	1.4	140	1.4	690
160	1.6	160	1.6	790
180	1.8	180	1.8	900
200	2.0	200	2.0	1000
220	2.2	220	2.2	1100
240	2.4	240	2.4	1200
260	2.6	260	2.6	1280
280	2.8	280	2.8	1380
300	3.0	300	3.0	1475
320	3.2	320	3.2	1530
340	3.4	340	3.4	1630
360	3.6	360	3.6	1730
380	3.8	380	3.8	1850
400	4.0	400	4.0	1950
420	4.2	420	4.2	2050
440	4.4	440	4.4	2160
460	4.6	460	4.6	2270
480	4.8	480	4.8	2380
500	5.5	500	5.5	2480
600	6.0	600	6.0	2900
700	7.0	700	7.0	3380
800	8.0	800	8.0	3880
900	9.0	900	9.0	4300
1000	10.0	1000	10.0	4600
1100	11.0	1100	11.0	5200
1200	12.0	1200	12.0	5620

GLOSSARY

A

absolute pressure. A pressure scale where the zero point represents a perfect vacuum. (4)

absolute pressure ratio. A ratio used to calculate the volume of air consumed at system pressure. Equal to the operating system gauge pressure in pounds per square inch (psig) plus atmospheric pressure (14.7 psia) divided by atmospheric pressure (14.7 psia). (17)

absolute rating. A filter rating that states the largest pore opening, indicating that all particles larger than the stated size will be trapped. (12)

absolute viscosity. The resistance of a fluid to the relative movement of its molecules. Measured in pounds per foot-second in the US Customary system and by poise in the metric system. (6)

absolute zero. The temperature at which all molecular movement stops. (4)

absorbent filter. A filter that cleans a fluid using a medium to trap contaminants and then holds them by mechanical means. (12)

accumulator. A hydraulic component in which system liquid is stored under pressure using a mechanical device, compressed gas, or weight to maintain pressure. (11)

actuator. A component used to convert the energy in hydraulic fluid or compressed air into mechanical linear or rotary motion. (1)

adapter. Machine or system parts used to align components or change sizes of shafts and distribution lines. (2)

adapter fitting. A fitting used to connect components having dissimilar diameters or thread types. (8)

additive. A chemical compound added to a hydraulic fluid to modify its characteristics and improve system performance. (6)

adiabatic. The general term used to identify the process in which no heat is transferred when a change occurs in the volume of a gas. (11)

adiabatic compression. The process in which no heat is transferred as the volume of a gas is decreased. (14)

adiabatic expansion. The process in which no heat is transferred as the volume of a gas is increased. (14)

adsorbent filter. A filter that cleans a fluid using a medium that holds contaminants by molecular adhesion. (12)

aftercooler. A pneumatic system heat exchanger that cools compressed air as it leaves the compressor; used to both cool air and liquefy water vapors. (16)

air consumption. The volume of air used to operate a pneumatic actuator or system; usually measured in cubic feet per minute (cfm). (17)

air filter. A component designed to remove solid particles, moisture, and/or lubricant from pneumatic system air. (15)

air nozzle. A device located on the end of an air line and containing an outlet orifice designed to maximize kinetic energy from the controlled release of pressurized system air. Often used in a blowgun. (17)

air-distribution system. The network of lines used to distribute compressed air from the receiver to the various workstations in a pneumatic system. (16)

air-drying equipment. A device used to dry air after compression by cooling, centrifugal force, absorption, adsorption, or a combination of these methods. (16)

air-line respirator. A self-contained breathing device that provides a breathable atmosphere to workers in conditions where safe, clean air is not available. (3)

alternating current (AC). Electrical current that reverses direction on a regular cycle due to a change in voltage polarity. The common international standards are 50 or 60 cycles per second (or hertz). (4)

American National Standards Institute (ANSI). The primary standards coordinating group in the United States. The group consists of numerous professional, technical, trade, labor, consumer, and governmental organizations and agencies. It is also the United States member of the International Organization for Standardization (ISO). (5)

anti-wear agent. A chemical compound added to hydraulic fluid to help reduce wear on bearing surfaces during hydraulic system operation. (6)

API gravity. A comparison of the weight of a given volume of a substance to the weight of an equal volume of distilled water at a specified temperature. (6)

atmosphere. The pressure created by the weight of the atmosphere at sea level. The average atmospheric pressure is considered to be 14.7 pounds per square inch. (4) The layers of gases that surround earth between its surface and space. (14)

atom. The smallest parts of an element that can take part in a chemical reaction without changing the characteristics of the material. (4)

automatic reciprocation. A hydraulic or pneumatic circuit design that produces continuous extension and retraction movement of a linear actuator (as long as the system is activated). (19)

automatic return. A hydraulic or pneumatic circuit design that, when activated, produces a single extension and retraction stroke on a linear actuator. (19)

auxiliary-air receiver. A secondary air storage tank used in a pneumatic circuit to supply additional compressed air to a high-demand, intermittently operating actuator or other system function. (16)

axial-piston pump. A hydraulic pump using a design in which the reciprocating motion of the pistons is parallel to the power input shaft. (7)

B

back injury. An injury to the skeletal, muscular, or nervous-system elements of the back associated with lifting or other work-related activity. (3)

back-pressure check valve. A spring-loaded check valve used with some pilot-controlled hydraulic systems to maintain sufficient pressure during the idle portion of the cycle to shift directional control valves. (13)

baffle. A plate or other device placed in a hydraulic reservoir to direct oil flow between the system return lines and the pump inlet line. (8)

balanced-piston valve. A hydraulic valve that pressurizes both the control and system side of a primary control piston. Desired valve action is obtained using metering orifices and/or small, spring-loaded poppets to adjust pressure on the control side of the piston. (10)

balanced-poppet valve regulator. A pneumatic system regulator using a design in which the action of the valve control poppet is dampened by subjecting both ends of the poppet to valve outlet pressure. (16)

balanced-vane pump. A hydraulic vane pump using a design with two inlet and outlet chambers that balance the load forces on the bearing of the pump rotor shaft. (7)

ball valve. A valve primarily used as an on-off valve. The primary valve component is a ball with a bored passageway. The valve requires only 1/4 turn to go from fully open to fully closed. (10)

barrel. The tube portion of a hydraulic or pneumatic cylinder. (9)

bent-axis design. The axial-piston design in which the angle between the input shaft and the barrel determines the length of the piston stroke. (7)

biodegradable fluid. A hydraulic fluid formulated to degrade in nature to reduce environmental damage from spillage. (6)

bleed-off circuit. A hydraulic speed-control circuit in which the required flow rate to the actuator is set by returning unneeded pump output directly to the reservoir through a flow control valve. (13)

blowgun. A device containing an air nozzle and designed for cleaning parts and equipment with a high-velocity airstream. (17)

bonnet. The upper portion of a gate, globe, or needle valve that holds the adjustment mechanism and protects the valve from damage. (12)

booster. A hydraulic system component used to increase the pressure in a portion of the system to higher than the pump discharge pressure. Often called a pressure intensifier. (19)

Boyle's law. One of the basic gas laws that states the volume of a gas, held at a constant temperature, varies inversely with the pressure. (4)

brake valve. A pressure control valve used in some hydraulic motor circuits to slow the motor to a controlled stop against the rotational inertia of the external load. (10)

brazed-plate heat exchange. A heat exchanger formed from metal plates brazed together into a compact unit; used to cool or heat the fluid in a system. (12)

Btu. Abbreviation for British thermal unit, which is equal to the heat required to raise the temperature of one pound of water by one degree Fahrenheit. (4)

buoyancy. The force vertically exerted on a body by the fluid in which it is partially or wholly submerged. (4)

burst pressure. A rating for pressure vessels and conductors indicating the pressure at which the component failed during the prescribed test procedure. (8)

bypass control system. A pressure management system that bleeds excess air from the compressor to the atmosphere in order to maintain constant, maximum pressure in the distribution line. (18)

bypass flow control valve. A flow control valve that uses an integral control port to return excess fluid directly to the reservoir. (10)

bypass valve. A flow control valve that uses an integral control port to return excess fluid directly to the reservoir. (12)

C

cam-type radial-piston motor. A hydraulic motor in which the force generated by the pistons is applied against the slope of a cam located on the motor output shaft. This action produces motor rotation and torque. (9)

cap. The end closure of a hydraulic or pneumatic cylinder that is opposite of the end from which the rod extends. Often called the cylinder blind end. (9)

capacity-limiting system. A system used to control the maximum air pressure produced by the pneumatic system compressor. Compressor start-stop, inlet valve unloading, and other methods can control capacity. (15)

capillary viscometer. A test instrument containing a capillary tube calibrated to provide information adequate to determine the viscosity of fluid. (6)

catalyst. A substance that increases the rate of a chemical reaction without being consumed in the process. (6)

cavitation. The formation of gas within a liquid stream that occurs when pressure drops below the vapor pressure of the liquid; may result in excessive noise and pump damage. The condition in a hydraulic pump can result from restricted inlet flow or excessive oil viscosity. (7)

central air supply. An air-distribution system consisting of a single compressor station with compressed air delivered to workstations located throughout a facility. (15)

centralized grid. An air-distribution system consisting of a network of branch lines extending from a central trunk line. The branch lines supply workstations located throughout a facility. (16)

centrifugal pump. A pump using centrifugal force to create flow. A rotating impeller draws fluid into the pump where the fluid is moved by centrifugal force to the outer edges of the impeller and directed to the pump discharge. (7)

Charles' law. A basic gas law that states the volume of a gas, held at a constant pressure, varies directly with the absolute temperature. (4)

check valve. A valve that normally allows fluid flow in only one direction. (10, 18)

chipping hammer. A handheld, pneumatic, reciprocating tool designed to chip material from concrete, masonry, plaster, or other substances. (17)

circuit. A group of electrical, hydraulic, and/or pneumatic components assembled to produce a desired movement or sequence of movements. (4)

circuit diagram. A group of symbols combined to represent the components and interconnecting lines in a fluid power system. (5)

cleanout. A removable cover used in a hydraulic system reservoir to allow easy cleaning of the vessel. (12)

clevis mount. A cylinder rod and cap mounting configuration involving a C-shaped casting and a mounting pin that allows the cylinder to pivot during extension and retraction. (9)

closed-loop circuit. A circuit design in which pump output is returned directly to the pump inlet after passing through a motor. The design is compact and commonly used with hydrostatic drive systems. (9)

compact hydraulic unit. A hydraulic system design developed in the early twentieth century in a move away from large, centralized systems to compact hydraulic pumps that often serve a single machine or actuator. (1)

component group. A group of hydraulic or pneumatic components designed to perform specific system functions. (2)

composite symbol. A symbol that indicates a component containing more than one valve function. A simple example is a flow control valve with a built-in check valve. (19)

compound relief valve. A relief valve containing both a balancing piston and a small, direct-operated relief valve. The two designs work together to produce a valve that operates efficiently. (10)

compressed-air unit. Commonly used to designate a pneumatic compressor station that includes a prime mover, compressor, reservoir, and pressure control components. (15)

compression fitting. A fitting style used with both metal and plastic tubing that obtains a seal by tightening a component of the fitting, which tightly compresses around the tubing. (8)

compressor. The pneumatic system device that converts the energy input of the system prime mover into high-pressure compressed air. (2, 15)

compressor-capacity control. The method used to control compressor operation to obtain required air volume and pressure. Several methods may be used, depending on the demands placed on the pneumatic system. (15, 18)

conduction. A basic term used with both heat and electricity relating to the transfer of heat or electricity through the material. (4)

conductor. The element used in fluid power systems to allow the movement of hydraulic fluid or compressed air from component to component. Hose, pipe, and tubing are the most common fluid power conductors. (2, 8)

contaminant. Foreign materials carried in the liquid or air of a fluid power system: dirt, water, metallic particles, paint, and a variety of other materials that enter during system operation. (12)

control element. A valve component that functions to allow, direct, meter, and/or stop the flow of fluid in a control valve to obtain desired operation. (10)

control mechanism. A device used to control the operation of a control valve. This may include mechanically, pilot-, solenoid-, and manually operated devices. (5)

control valve. Any valve designed to control fluid pressure, flow rate, or actuator direction in a fluid power system. (2, 10)

convection. A basic method of heat transfer involving the movement of heated air caused by density differences of air of different temperatures. (4)

cooler. A pneumatic system dryer that uses a refrigeration element to lower the temperature of system air for the purpose of removing moisture. (15)

corrosion resistance. The ability of a material to resist damage from chemicals that may be formed when contaminants enter the fluid of a hydraulic system. (17)

counterbalance valve. A hydraulic system pressure control valve used as a safety device to prevent the movement of a loaded actuator without the application of additional pressure generated by the system power unit. (10)

coupler. A mechanical device used to directly connect the output shaft of the prime mover to a hydraulic pump or pneumatic system compressor. (7)

coupling. A general term used for devices that connect system components such as fluid conductors and power transmission shafts. (15)

cracking pressure. The pressure level required to produce the first fluid flow through a pressure control valve. (10)

crescent design. A low-pressure hydraulic pump using an internal gear design in which the pumping chamber is produced by a crescent-shaped element positioning the internal- and external-toothed gears. (7)

cumulative injuries. Injuries that occur from long-term exposure to unsafe environmental conditions. (3)

cup seal. A seal shaped like a cup and often used on cylinder pistons. Application of pressure to the lip edge of the seal ensures a tight seal. Initial development in the late nineteenth century was critical to the development of industrial hydraulics. (1)

current. The movement of electrons through a conductor. Current is measured in amperes. (4)

cushioning. A design feature in fluid power cylinders that reduces fluid flow near the end of the extension or retraction stroke to decelerate piston movement, which avoids both noise and component damage. (9)

cutaway symbol. A symbol that represents fluid power components in schematics using miniature cutaway drawings of the basic internal structure of valves and other parts. (5)

cylinder. An actuator used in fluid power systems to generate linear movement and force. (2, 9)

cylinder barrel. The tube-shaped portion of a cylinder (linear actuator) in which the piston and rod reciprocate. (7)

D

deceleration circuit. A speed-control circuit that involves controlled slowing of an actuator from a maximum speed to a selected feed rate. (13)

deceleration valve. A control valve specifically designed to allow maximum, deceleration, and feed flow rates. The valves often use mechanical activation and a check valve that allows an unrestricted return flow rate. (13)

decentralized grid. An air-distribution system consisting of multiple grids that serve small portions of a facility. These grid sections contain smaller compressors, but may be interconnected to provide air movement between grids to balance intermittent, high air demands. (16)

decompression valve. A valve designed to release pressurized hydraulic fluid in large presses or other installations where high pressure may be retained in the system. (13)

degree. A unit of measure typically associated with temperature scales. (4)

depth-type filter. A fluid filter that uses the depth of the filter material to provide many passageways to catch and hold contaminants. (12)

desiccant. A material in compressed-air dryers that uses the principle of adsorption to remove moisture from the air. (16)

detent. A mechanical device built into a directional control valve to hold the valve spool in a selected position. (10)

dew point. The temperature at which the water vapors in air will begin to condense (form dew). (14)

diaphragm. A flexible membrane used in various fluid power components to separate adjoining chambers that need to vary in volume. Often found in hydraulic accumulators, pneumatic pressure regulators, and low-capacity pumps and compressors. (18)

diaphragm-chamber regulator. A pneumatic system regulator that uses a flexible diaphragm between the outlet and control chambers of the valve. (16)

direct current (DC). An electrical current with a continuous, one-direction current flow. (4)

directional control valve. A hydraulic or pneumatic control valve that directs fluid flow to start, control direction of movement, and stop system actuators. (2, 10)

direct-operated regulator. A basic design of a pneumatic system pressure regulator in which the diaphragm is directly exposed to the pressure of the outlet chamber. (16)

direct-operated valve. A fluid power pressure control valve in which the operating pressure of the valve is directly dependent on physical force generated by a control spring. (10)

displacement. The volume of fluid displaced per unit of time. In fluid power applications, displacement is important in calculations involving cylinders, motors, pumps, and compressors. (7)

double-acting compressor. A compressor design in which air intake and compression are completed in chambers located on both the top and underside of the compressor pistons. These compressors were common in larger industrial installations, but are being replaced by current rotary designs. (15)

double-acting cylinder. A cylinder that can exert force on both the extension and retraction strokes. (9, 17)

double-rod-end cylinder. A cylinder that has rods attached to both sides of the piston, providing equal-displacement chambers that produce equal extension and retraction speeds, as well as two attachment points to the cylinder. (9)

draincock. A valve located at the lower end of a drop line in a pneumatic distribution system to allow removal of condensed water. A plug-type valve is often used for this application. (16)

drain return line. A line used to return to the reservoir fluid drained from externally-drained components. (8)

drop line. A pipe coming from the top of the air distribution lines to a system workstation. Tapping the line into the top of the distribution line prevents condensed water from entering the drop line. (16)

dry air. Pneumatic system air with a saturated vapor pressure well above the ambient temperature. (14)

dryer. The pneumatic system component designed to remove water vapor from the compressed air. The unit is usually located in the compressor station area and may use refrigeration, chemical, or mechanical means to reduce the water content of the air. (15)

dry filter. A pneumatic system filter that uses a dry element to remove dirt and debris from air at the compressor inlet and other system locations. (16)

drying. The process of removing moisture from pneumatic system air. (17)

Dryseal pipe thread. A type of pipe thread that provides a positive seal, preventing leakage of hydraulic oil through a threaded pipe connection. (8)

dual pump. A design in which two pumps are built into a single case. The pumps may be rated at different pressures and flow outputs to provide additional flexibility. (7)

dust mask. A personal protective device designed to remove dust and other airborne particles from the air an individual breathes, generally not suitable for removing fine particles and vapors. (3)

dynamic compressor. A device that compresses air or other gas using rotating vanes or impellers. These moving components increase pressure by converting the energy in the high-velocity air to pressure. (15)

dynamic seal. A seal between moving parts. These seal designs are subject to wear because the seal must contact the moving surfaces. (9)

E

earmuffs. Devices designed to protect the hearing of the user by reducing the decibel level of sound reaching the inner ear. The device is made from sound-deadening materials that snuggly fit over the ears. (3)

earplug. A device designed to protect the hearing of the user by reducing the decibel level of sound reaching the inner ear. The device is inserted into the outer ear canal. (3)

effective area. The area of a surface that contributes to the force generated by the pressure of a fluid acting on the surface. (9)

effective piston area. The area of a piston that contributes to the force generated by system pressure. For example, the effective area of a cylinder piston during retraction is the area of the piston minus the cross-sectional area of the piston rod. (9)

electrical control. Control of a variety of fluid power component and system functions ranging from simple manual on/off electrical switches to complex servomechanisms. (5)

electron. One of the three basic parts of an atom. These negatively-charged portions of the atom orbit the nucleus containing neutrons and protons. (4)

emergency shower. A shower bath located in a work area designed to allow a worker to immediately flood their body and clothing with water if exposed to corrosive or toxic chemicals. (3)

emulsion. A stabilized mixture of oil and water that typically has a milky appearance. An example is fire-resistant hydraulic fluid, which is classified as oil-in-water or water-in-oil. (6)

energy. The ability or potential to do work. (4)

energy conversion device. A device that has the ability to convert energy from potential to kinetic or kinetic to potential. (5)

entrained air. Air that has been churned into the fluid in a hydraulic system. Can shorten the life of the fluid by increasing oxidation or reduce system performance by producing spongy actuator operation. (7)

equipment maintenance. Service and repair work applied to equipment to ensure expected performance and service life. (3)

external force. A force applied to a component or system from an outside source. (10)

external-gear pump. A hydraulic pump design that uses gears with external teeth to move fluid through the pumping chamber of the unit. (7)

extreme-pressure agent. An additive blended into many hydraulic fluids to prevent direct contact of the surfaces of bearings and gear teeth when operated under heavy load conditions. (6)

eye protection device. Protective wear ranging from eyeglasses to full-face shields that are designed to prevent eye injuries from caustic fluids, flying debris, or harmful light rays. (3)

eyewash station. A small sink located in a work area and designed to allow workers to flood their eyes with water if exposed to corrosive or toxic materials. (3)

F

feedback control. A control system that uses information from system output to vary the input function to maximize performance. (5)

ferrule. The ring used on a hydraulic hose fitting to secure the hose to the threaded portion of the fitting. (8)

film. A thin layer or sheet of material. In fluid power systems, this may be a layer of oil or a solid sheet of material used to separate materials or protect the surface of component parts. (6)

filter. A fluid power component that is used to remove solid contaminants from both air and liquid. The term is often used to designate the complete component including the filter element, case, and mounting hardware. (2, 12)

filter-element-condition indicator. A device located on many filter designs that indicates the condition of the filter by measuring the pressure drop across the filter element. (12)

filtration system. Typically refers to the placement of the filter in the system. Filter placement includes pump inlet, return line, and working line. (12)

finned conductor. A simple form of heat exchanger consisting of a metal tube surrounded by thin metal disks that dispense heat to the surrounding air. Often used as an intercooler between the cylinders of two-stage compressors. (12)

fire point. The lowest temperature at which a volatile substance vaporizes rapidly enough to produce an air-vapor mixture that will continuously burn when ignited. (6)

first aid kit. A case containing a variety of basic first aid items suitable for emergency situations. These kits should be readily available in work areas and need to be maintained on a regular schedule. (3)

first law of thermodynamics. States that energy cannot be created or destroyed during a process, although it can be changed from one form to another. (4)

fitting. A small accessory part with standard dimensions used to assist the assembly of major component parts into systems. (2, 8)

five-port valve. A variation of the four-way directional control valve that provides a separate exhaust port for each valve position. (18)

fixed-centerline mount. A cylinder-mounting design in which the load carried by the cylinder rod and piston is supported at the centerline of the cylinder barrel, which is fixed to a machine member. (9)

fixed-delivery pump. Any hydraulic pump design that has a nonadjustable fluid delivery rate. (7)

fixed-displacement motor. Any hydraulic motor that does not allow the displacement per motor revolution to be adjusted. Varying the fluid flow rate through the motor controls the speed of these designs. (9)

fixed-noncenterline mount. A cylinder-mounting design in which the load carried by the cylinder rod and piston is transferred through the cylinder barrel to off-center mounts fixed to a machine member. (9)

fixed orifice. A hole of fixed size in a component part, allowing passage of a measured amount of fluid. (10, 18)

flared fitting. A fitting design used with metal tubing that involves flaring the end of the tube. A mechanically tight seal is obtained using a fitting with a matching angle and a compression nut. Fittings are available with 37° and 45° flares. (8)

flash point. The lowest temperature at which vapors from a volatile substance will momentarily ignite when exposed to an open flame. (6)

flexible conductor. A fluid transmission line in hydraulic and pneumatic systems that permits at least limited flexing or bending. These may range from flexible pipe joints that offer limited movement to hose and plastic tubing that offer almost unlimited movement. (16)

flexible hose. A hose used in both hydraulic and pneumatic systems that can operate under full system pressure while providing maximum freedom of movement. (16)

flow control valve. A component designed to provide control of the rate of flow of air or liquid in hydraulic and pneumatic systems. (2, 10)

fluid. A substance such as air, water, or oil that easily flows and tends to assume the shape of the container in which it is stored. (1)

fluid compressibility. The degree to which a set volume of fluid will be reduced when subjected to an outside force. Gases are considered compressible, while liquids are considered noncompressible. (1)

fluid conditioning. The methods used to provide a fluid that is clean and maintained at an acceptable operating temperature in a pneumatic or hydraulic system. (12)

fluid conductor. The lines used in a fluid power system to route the fluid from component to component in order to obtain the desired operation. (2)

fluid filter. The system component designed to remove contaminants from fluid power systems. The term typically refers to the filter element, case, and any mounting accessories. (7)

fluid flow weight. A factor involved in the hydraulic horsepower formula. Equal to the weight of the fluid output of the pump. (7)

fluid power. The transferring, controlling, and converting of energy using hydraulic and pneumatic systems. (1)

fluid turbulence. Fluid flow through a line in which the movement of the fluid is irregular, causing increased flow resistance. (8)

flux. The lines of force that surround a permanent magnet or electromagnet. (4)

force. An influence on a body that produces, changes, or stops the motion of a body. (4, 17)

Four-Ball Method. An ASTM test procedure designed to determine the wear-prevention characteristics of hydraulic fluids. The test involves rotating a steel ball under load against three stationary balls covered by the test fluid. (6)

four-way valve. A directional control valve providing four separate fluid-flow patterns, allowing a cylinder to be powered during both extension and retraction. (10, 18)

free air. Air displaying the characteristics of the atmospheric air at a specific location. (14)

friction. The force that retards or resists the movement of contacting surfaces. (4, 6)

FRL unit. A component group providing air filtration, pressure regulation, and lubrication (FRL) for a specific pneumatic system workstation. (16)

fulcrum. The support point of a lever about which the load and effort arms pivot. (4)

full-flow filtration. A hydraulic fluid filtration system that filters all pump output before returning it to the system reservoir. (12)

full-flow pressure. The pressure required to force all system fluid through a pressure control valve at a given setting. (10)

G

gas-charged accumulator. A component containing a sealed chamber pressurized with an inert gas and used to store hydraulic fluid under pressure. (11)

gate valve. A shutoff valve that provides straight-line flow between the valve inlet and outlet ports. Provides minimal pressure drop when fully open. (10)

gauge pressure. A pressure scale where the zero point is equal to atmospheric pressure. (4)

Gay-Lussac's law. Principles regarding the changing of a gas from one pressure and temperature condition to another while holding the volume constant. (4)

gear pump. A hydraulic pump design that uses gears to move fluid through the pumping chamber of the component. Gears may have internal or external teeth. (7)

general gas law. A general equation that expresses the relationships between pressure, volume, and temperature when applied to a gas ($P_1 \times V_1 \div T_1 = P_2 \times V_2 \div T_2$). (4)

general hydraulic horsepower. A computed figure used in hydraulic systems based on fluid flow and pressure. Equal to fluid flow weight in pounds per minute multiplied by the pressure in feet of head divided by 33,000 foot-pounds per minute. (7)

gerotor design. A common internal-gear pump or motor design. The pumping chamber of these units has two gear-shaped elements that form varying-size chambers as they rotate. (7)

globe valve. A shutoff valve that consists of a movable disk and a stationary seat. The passageway through the valve is offset, which causes fluid turbulence as the flow rate increases. (10)

graphic symbol. The most common type of symbol used to represent component features, system construction, and system operation in fluid power diagrams. Consists of a series of lines and shapes that represent specific components or features. (5)

gripper. An actuator commonly used with robots to handle parts in machining or assembly operations. (17)

H

head. The height of a column of water or other liquid necessary to develop a stated pressure. (4) The end closure of a hydraulic or pneumatic cylinder through which the cylinder rod passes. Often called the cylinder rod end. (9)

hearing-protection device. Protective wear, including earplugs and earmuffs, designed to prevent injury to the inner ear that could result in hearing loss. (3)

heat. Kinetic energy indicated by rapid molecular movement in a substance. (4)

heat exchanger. A component part in a fluid power system that adds heat to or subtracts heat from the system to keep the operating temperature within the desired range. (2, 12)

helical gear. A gear with teeth set at an angle to the axis of the gear. The design runs quietly, but has the disadvantage of producing a side thrust. (7)

herringbone gear. A gear tooth design used in some gear pumps that is the equivalent of two helical gears placed side by side with the teeth forming a V-pattern. The tooth pattern produces quiet operation with no side thrust. (7)

high-low pump circuit. A circuit using two fixed-displacement pumps. The pumps are connected, with appropriate pressure control valves, to provide a high-volume flow at low pressure and a low-volume flow when high pressure is required to operate a system. (13)

high-water-content fluids (HWCF). Hydraulic fluids that are primarily water with 2–5% soluble chemicals. Also known as 95/5 fluids. (6)

horsepower. A standard unit of power equal to 550 foot-pounds of work per second or 33,000 foot-pounds per minute. (4)

hose. A flexible conductor for carrying hydraulic fluid or compressed air to actuators. (2, 8)

hose-end fitting. Metal or plastic parts designed to provide a strong and leakproof connection between the end of a hose and a system component or rigid conductor. (8)

housekeeping. A general term used to indicate the process of keeping a workplace clean, orderly, and free from clutter. (3)

humidity. Atmospheric water vapor content expressed using several different measurement units. (16)

hydraulic accumulator. A hydraulic system component capable of storing liquid under pressure. Uses mechanical or compressed-gas elements to maintain system pressure. (1)

hydraulic intensifier. A component that increases the pressure in a selected section of the system over that of the system relief valve. The flow rate in the intensified pressure section is reduced in order to achieve the increased pressure. (1)

hydraulic pressure fuse. A device used to prevent excess pressure or uncontrolled fluid loss if a line should rupture. (10)

hydraulics. The study and technical application of liquids in motion, especially oils and other blended liquids used in industrial systems. (1)

hydrostatic drive. A fluid power drive system using a hydraulic pump and motor to transmit the power of a prime mover to the input of a machine. Available in either open- or closed-loop circuit designs. (9)

hydrostatic transmission. A hydrostatic drive in which the pump, motor, or both have variable displacement. Drives with these features can be manually or automatically adjusted to control torque, speed, and power. (9)

I

ideal gas laws. Mathematical equations that provide an approximation of how a gas reacts to pressure, temperature, and volume changes. These equations use absolute pressure and temperature. (4)

immersion heat exchanger. A component that heats or cools the fluid of a hydraulic system using water, electrical-resistance heaters, or steam. Usually located in the reservoir of the system. (12)

impact wrench. A portable hand tool, often operated by compressed air, used to install or remove bolts or nuts with a series of rapid-impact, rotary blows. (17)

impeller. The rotating component of a dynamic compressor or impeller pump. Energy is increased in the fluid as centrifugal force causes flow through the unit. (7)

inclined plane. One of the six basic machine types. The inclined plane is a slope used to reduce the force needed to lift an object. (4)

Industrial Revolution. A period extending from the 1700s through the 1800s in which rapid change occurred in industry, including the development of many fluid power concepts, components, and systems. (1)

inertia. The tendency of all objects to remain either at rest or moving in the same direction, unless acted on by an outside force. (4, 9)

inlet line. The conductor that provides system fluid to a component. (8)

inline design. An axial-piston design in which the pistons are parallel with the axis of the power input shaft. (7)

intake-line filter. A pneumatic system filter that removes dust and dirt from atmospheric air before air is allowed into the compressor inlet. These filters are available in two forms: sump strainers (course filtration) and suction filters (fine filtration). (16)

intercooler. A heat exchanger located between the stages of a two-stage pneumatic compressor. The intercooler reduces the temperature of the compressed air, thus increasing the performance of the compressor. (16)

internal force. The forces generated by internal valve components or by fluid pressure in the internal passageways of a valve. These forces are used to position valve elements to obtain a desired operation. (10)

internal-gear pump. A hydraulic pump design that uses an internal-tooth gear operated by an external-tooth gear to move fluid through the pumping chamber. Common internal-gear pumps use gerotor or crescent designs. (7)

International Organization for Standardization (ISO). An international agency for standardization located in Geneva, Switzerland. It includes the recognized national standards bodies of approximately 110 countries of the world. (5)

inverted emulsion. An emulsion in which oil surrounds finely divided water droplets. This principle is used in some forms of water-and-oil fire-resistant hydraulic fluids. (6)

ionosphere. The uppermost layer of the atmosphere, which extends from approximately 50–320 miles above earth's surface. (14)

isothermal. A process completed without a change in temperature. (11)

isothermal compression. The compression of gas in which any heat produced is removed, so that the temperature remains constant. The final pressure is based only on the decrease in volume. (14)

isothermal expansion. The expansion of a gas in which heat is added to maintain constant temperature. The final pressure is based only on the increase in volume. (14)

J

jet pump. A pump that uses a variation of Bernoulli's theorem to move fluid. The design forces a liquid through a nozzle placed in a venturi to produce a low-pressure area, which draws additional fluid from the primary fluid source. (7)

K

kinematic viscosity. A precise indicator of the viscosity of a liquid. The rating is based on the time required for a fixed amount of a fluid to flow through a calibrated viscometer under a fixed pressure and temperature. (6)

kinetic energy. The energy a body possesses because of its motion or velocity. (4)

L

laminar flow. Fluid flow arranged in distinct, thin layers. (8)

land. The full diameter portions of the control spool in a directional control valve. The annular clearance between lands directs fluid flow. The lands also block passageways through the valves to obtain desired operating characteristics. (10)

latent heat. The amount of heat absorbed or released by a substance without changing temperature during a change of state. (4)

lever. One of the simple machines. Consists of a rigid bar pivoted about a fulcrum. Used to multiply a force or movement. (4)

limited-rotation actuator. An actuator design that primarily produces rotational movement of one revolution or less. Various designs are available using a rack-and-pinion, vane, or helical shaft. Also known as a torque motor. (9)

linear actuator. A term often used to indicate a hydraulic or pneumatic cylinder. It converts fluid pressure and flow into linear mechanical force and movement. (17)

load factor. The estimated percentage of maximum actuator load delivered during operation. (16)

lobe pump. A hydraulic pump that uses specially-designed elements with lobes to provide a pumping chamber. Liquid moves through the unit very much like an external-gear pump design. (7)

lobe-type compressor. A compressor that usually provides only low-pressure air. Construction involves specially designed elements with lobes that provide a sealed pumping chamber. Air is swept from the compressor inlet to the outlet as the lobes rotate. Often referred to as a blower. (15)

lock-out device. A safety device designed so a machine cannot be accidentally turned on while maintenance work is being performed. (3)

logic function. A general term indicating several basic electronic or pneumatic circuits that perform problem-solving tasks. (19)

loop system. A hydraulic system design that allows continuous circulation of fluid through a motor and pump using only a low-volume system reservoir. (16)

lubricant. A substance used to reduce friction between moving surfaces, such as those found in bearings. (14)

lubrication. A method to reduce friction and minimize wear of parts moving in close proximity. Usually involves oil or other substance as a lubricant. (6)

lubricator. A device designed to apply a lubricant to a bearing surface. In pneumatic systems, the lubricator is a component of the FRL unit located at the workstation. (16)

lubricity. The ability of a lubricant to form a strong film between two bearing surfaces and adhere to those surfaces to reduce friction and wear. (6)

M

magnetic poles. The areas at the ends of a magnet that produce magnetic lines of force. The lines of force attract ferrous metals or may be used to generate electricity. (4)

magnetism. The phenomena produced by the magnetic lines of force surrounding natural or electromagnets. (4)

manifold. A fluid power component designed to provide branch connections to several conductors or valves to facilitate construction of circuits or systems. (8)

mass. The amount of matter in an object. (4)

material agitation. The use of compressed-air systems to mix or compact materials in process, packaging, or other industries. (17)

material transfer. The use of compressed-air systems to move powder or small-particle materials through tubes from one process or storage point to another. (17)

maximum operating pressure. The highest pressure recommended for operation of a hydraulic or pneumatic system. (13)

mechanical advantage. The ratio of the force produced by a machine to the applied input force. (4)

mechanical control. A mechanical mechanism that operates control valves, including levers and ramps (inclined planes). (5)

mechanical coupler. A mechanical device that connects the prime mover to the hydraulic pump or pneumatic compressor. (2)

mechanical efficiency. A ratio of the theoretical horsepower needed to operate a pump or compressor and the actual horsepower required. (4, 7)

memory circuit. A pneumatic control circuit that will hold an actuator in a selected position after only a momentary input signal. (19)

mesosphere. The portion of the atmosphere that extends from the stratosphere to the ionosphere. (14)

metal rolling. A metal-forming technique that is used by some manufacturers to attach the barrel to the cap and rod ends of the cylinder. (17)

meter-in circuit. A fluid power flow-control circuit in which fluid is metered into an actuator to control operating speed. (13, 18)

meter-out circuit. A fluid power flow-control circuit in which fluid is metered out of an actuator to control operating speed. (13, 18)

micron. A unit of measure equal to one millionth of a meter. Commonly used to designate the size of particles that filter media will remove. (12)

mill cylinder. A hydraulic cylinder constructed of heavy steel for use in industries such as foundries and steel mills. (9)

moisture separator. A device sometimes teamed with the compressor aftercooler to increase the effectiveness of the moisture-removal process during pneumatic system air compression. (16)

molecule. All substances are structured of molecules composed of atoms. Atoms are composed of electrons, neutrons, and protons. (4)

motor. An actuator that primarily produces continuous, rotary motion. These components may be powered by electricity, hydraulics, or pneumatics. (2, 9)

muffler. A device designed to reduce the level of noise produced at the air intake of a compressor or the exhaust of pneumatic control valves and motors. (18)

multiple maximum system operating pressures. A fluid power system designed to allow a number of maximum operating pressures by using remote pressure valves to control the operation of a primary hydraulic relief valve or pneumatic regulator. (13)

multiple-pressure circuit. A circuit that provides for selection of several different preset system pressure levels. (19)

muscular control. A valve control element that depends on manually applied force to shift the valve into a desired operating position. (5)

N

nail driver. A portable pneumatic tool used to drive nails. (17)

needle valve. A common flow control and shutoff valve used in both hydraulic and pneumatic circuits. The valve contains a fixed orifice and an adjustable screw with a tapered end that aligns with the center of the orifice. (10, 18)

neutron. One of the three basic parts of an atom. Neutrons have neither a negative nor a positive charge. Neutrons and protons form the nucleus of the atom. (4)

nominal rating. A rating that involves an approximation of performance rather than an exact rate. (12)

nominal sizing. A sizing system in which the actual dimension is approximately the same as the size stated. The measured dimension is uniform, but does not exactly match the stated size. (8)

nonpositive-displacement compressor. A compressor that uses an impellor or vanes operating at a high speed to compress air. (15)

nonpositive-displacement pump. A pump that does not have a variable-volume pumping chamber. An impeller or other device is used to move the fluid. The inertia of that fluid movement produces pressure when flow is resisted. (7)

normally closed. Refers to a pressure control valve in which the internal elements are closed when the system is shut down, blocking fluid passage through the valve. (10)

normally open. Refers to a pressure control valve in which the internal elements are open when the system is shut down, allowing fluid free passage through the valve. (10)

normal operating position. The position that the internal elements of a control valve assume when a fluid power system is shut down. (10)

nucleus. The center portion of an atom, which is made up of protons and neutrons. (4)

nutsetter. A specialized pneumatic impact wrench designed for assembly line work. The device installs nuts on bolts and torques them to a specified tightness. Available in single- and multiple-driver heads. (17)

O

Occupational Safety and Health Administration (OSHA). Federal agency that establishes requirements for safety, including safe and healthy working conditions and the encouragement of employees to follow safe working practices. (3)

off-line filtration. A hydraulic filtration method that continuously pumps fluid out of the reservoir, through a filter, and back to the reservoir. (12)

Ohm's law. Describes the fundamental relationship between voltage, current, and resistance. This is expressed as I (amps) = E (volts) ÷ R (ohms). (4)

oil-bath filter. An air-intake filter design used in pneumatic systems that involves a filter medium constantly bathed in oil. The oil catches the dust and dirt and transports it to a sump where it is removed during periodic service. (16)

oil-in-water emulsion. High-water-content, fire-resistant hydraulic fluids using approximately 5% soluble oil and 95% water. (6)

oil-wetted filter. An air-intake filter design using a filter medium made from a wire-screen form packed with oil-coated metal material that removes dust, dirt, and other debris. (16)

one-piece cylinder. A cylinder type that has permanently attached ends (head and cap). They are considered throw-away items as they cannot be disassembled for service. (9)

open-loop circuit. A circuit that uses the layout of a basic motor circuit with a directional control valve to control motor direction and a reservoir to hold surplus fluid. (9)

orbiting-gerotor motor. A variation of the gerotor motor that uses the internal-toothed gear of the gerotor set as a fixed gear. The external-toothed gear orbits following the internal-toothed gear. This produces higher torque and lower speed output. (9)

orifice. A precision-drilled hole in a component that is used to meter a desired flow rate through a flow control valve or the control section of other valve types. (10)

overall efficiency. A ratio of the theoretical energy produced by a system and the energy actually produced by the system. This ratio considers all energy losses that may occur in a system and is calculated by multiplying the efficiencies of individual subsystems or processes. (7)

overrunning load condition. An increased actuator speed resulting when an actuator uses a pulling action, thereby creating a vacuum that increases flow rate into the cylinder. (13)

oxidation. A chemical reaction that increases the oxygen content of a compound. (6)

oxidation inhibitor. An additive added to hydraulic fluids to slow the complex chemical reactions that oxidize the fluid. (6)

ozone layer. A layer of ozone that is located in the upper limits of the stratosphere. The layer acts as a filter to reduce the amount of ultraviolet energy that reaches earth's surface. (14)

P

packed-bore design. A design used with pneumatic directional control valves in which resilient seals are placed in annular grooves machined into the spool bore of the valve body. (18)

packed-slip joint. A fitting recommended for use around the pump inlet line as it passes through the cover of the hydraulic system reservoir. The fitting absorbs system vibration as well as reduces water and dirt entering the system. (8)

packed-spool design. A design used with pneumatic directional control valves in which resilient seals are placed in annular grooves machined into the land surfaces of the valve spool. (18)

parallel circuit. An electrical or fluid power circuit that simultaneously provides multiple paths for the current or fluid to follow as it moves through a circuit. (9)

Pascal's law. A basic scientific law that states pressure applied to a confined, nonflowing fluid is transmitted undiminished to all points in the fluid. (4)

paving breaker. A pneumatic percussion tool used for breaking pavement, compacted soil, and a variety of other materials. The unit may be handled by an individual operator or mounted on mobile equipment. (17)

phosphate ester. A synthetic hydraulic fluid. It has a high flash point and provides excellent fire resistance. (6)

pictorial symbol. A symbol often used to depict a component in a fluid power circuit. The symbols often represent the external shape of components, but no standardized shapes have been identified. (5)

pilot-operated regulator. A pneumatic system pressure regulator that is controlled by pilot pressure from a remote location. (16)

pilot pressure. The fluid pressure needed by pilot-control elements to obtain expected component performance. (10)

pinion gear. An external-toothed gear on the output shaft of a hydraulic limited-rotation actuator that uses a rack-and-pinion design. (9)

pintle. A fixed pin that acts as the bearing for the rotating cylinder block in a radial-piston pump. The inlet and outlet ports for the pump cylinders are also machined into the pintle pin. (7)

pipe. A rigid tube made from steel with adequate wall thickness to be threaded. Often used as a conductor in fluid power systems. (2, 8)

piston. A sliding cylinder that reciprocates in a bored circular hole in the body of a component. In a fluid power system, the piston is either moved by or moving against fluid pressure as it reciprocates. (9, 18)

piston motor. A fluid power rotary actuator using reciprocating pistons and motion-converting mechanisms to produce rotational movement and torque. (17)

piston pump. A fluid power pump that converts the power and rotary motion of a prime mover into fluid flow, using reciprocating pistons and a motion-converting mechanism. (7)

piston shoe. A component attached to the piston end in some piston pump and motor designs. The shoe is held in constant contact with the swash plate during operation. (7)

pivoting-centerline mount. A cylinder-mounting design that allows the cylinder to follow an arc as it powers a machine member. The load remains concentrated on the centerline of the cylinder. (9)

platen. A flat plate against which a load rests or is pressed. In a hydraulic press, the weight of the platen could contribute to the pressure in the system. (10)

pneumatic reciprocating motor. An air-powered motor that produces a repeated back-and-forth, linear motion. For example, this type of motor powers a jackhammer used in street maintenance. (17)

pneumatics. The study and technical application of air or other gas in motion. (1)

polyglycol. A fluid similar in chemical makeup to automotive antifreeze. Used in fire-resistant, water-glycol hydraulic fluid. (6)

poppet valve. One of several designs used in fluid power control components in which a face on the valve seals against a seat located in the body. (10, 18)

portable unit. A tool that is easily moved from one location to another. (15)

positive-displacement compressor. A compressor that confines air in a chamber that can be mechanically reduced in volume to increase pressure. (15)

positive-displacement pump. A pump that produces a fixed volume of fluid output for each revolution. Some designs are adjustable. (7)

potential. The difference in electrical charge between two points in an electrical circuit; measured in volts. This pressure causes electrons to move through a conductor. (4)

potential energy. Energy that is due to position and not motion. For example, crude oil, compressed springs, and charged hydraulic accumulators contain potential energy. (4)

pour point. The lowest temperature at which a fluid will flow as defined by a standardized test procedure. (6)

power. The rate at which work is performed or energy expended. Power = force × distance ÷ time. (4)

power unit. The unit in a fluid power system that provides energy for the system, moves fluid through the system, provides a safe maximum limit of system pressure, and maintains desired system temperature and fluid cleanliness. (2, 7)

precharging. The process of setting the pressure on the gas side of a gas-charged accumulator after the pressure of the hydraulic side has been reduced to 0 psi. (11)

pressure. Force per unit area, which is usually expressed in pounds per square inch (psi). (4)

pressure balancing. A design technique used to reduce or balance uneven loading of bearings and other pump components to reduce pump wear and increase efficiency. (7)

pressure booster. A fluid power component that can increase pressure in a selected section of a circuit above the general operating pressure of a system. A higher pressure is obtained by using two different-sized pistons with a low pressure acting on the larger piston. (18)

pressure compensation. An element in a flow control valve that maintains a constant pressure drop across the metering orifice of the valve, resulting in consistent flow rates through the valve even with varying system loads. (10)

pressure control. A component designed to limit maximum system pressure (relief), sequence actuator movement (sequence), restrain movement (counterbalance), unload pump output (unloading), and provide reduced pressure (pressure reducing). (5)

pressure control valve. A control valve in a fluid power circuit designed to protect the system from damage from excessive pressure, provide motion control, and protect operators from unexpected actuator movement. (2, 7, 10)

pressure override. The difference between the cracking pressure and the full-flow pressure of a valve. (10)

pressure rating. The maximum operating pressure recommended by the manufacturer for a fluid power component or system. (11)

pressure regulator. A pneumatic system component used to control air pressure at the system workstations. The valve maintains a constant, reduced pressure for use by tools and equipment operated at the station. (2, 16)

pressure safety valve. A valve used to protect a system from excessive pressure caused by valve failure or improper settings. Versions of these valves are used in both hydraulic and pneumatic systems. (3)

pressure surge. An unexpected increase in pressure in hydraulic systems caused by the inertia of fluid flow when valves close or loads suddenly increase or decrease. (11)

pressure-reducing valve. A hydraulic pressure control valve designed to maintain the pressure in a portion of a system at a lower level than the pressure set by the system relief valve. (10)

prime mover. The source that provides power to an actuator. Commonly used prime movers are electric motors and internal combustion engines. (1, 2, 7, 15)

prime mover horsepower. A rating used to indicate the power requirement of a fluid power system prime mover. Size calculations must include peak flow/pressure demands and the efficiency of the system pump. (7)

propeller pump. A nonpositive-displacement pump consisting of a rotating, propeller-shaped pumping element located in a fluid line. Used primarily as a fluid transfer pump in hydraulic systems. (7)

proportional filtration. A hydraulic filtration system that directs on a continuous basis a measured amount of system fluid to the filter and then returns it directly to the reservoir. (12)

proton. One of the three basic parts of an atom. Neutrons and protons form the nucleus of the atom. The proton carries a positive electrical charge. (4)

pulley. One of the simple machines, which allows the direction of a force to be changed. When used as a movable pulley or used in multiple sets, a mechanical advantage is provided. (4)

pump. A hydraulic component turned by the prime mover that produces fluid flow, transmitting energy through the system. (2, 7)

pumping chamber. The portion of a positive-displacement pump or compressor that creates the conditions necessary to draw in and then move fluid into the system. The chamber enlarges during the intake phase and reduces in size during the output portion of the cycle. (15)

pumping motion. The motion involved to change the volume of the pumping chamber, which acts to move fluid through a pump. Pumps are classified as either rotary or reciprocating. (15)

pump unloading control. A pump and unloading valve combination that allows the system pump to unload at near zero pressure when additional fluid is not needed to continue system operation. The combination reduces energy consumption without affecting system performance. (10)

Q

quick-disconnect coupling. A fitting used to facilitate rapid change of component units with minimum fluid loss. (8)

quick-exhaust valve. A pneumatic control valve used to increase the rate air can be exhausted from a cylinder. Increasing the exhaust rate allows faster extension or retraction speeds. (18)

R

rack gear. A bar containing gear teeth on one side that is meshed with a pinion gear. Used in some limited-rotation actuators to rotate a pinion gear attached to the output shaft of the actuator. (9)

radial-piston pump. A positive-displacement piston pump design in which the pistons are located in a cylinder block or housing that is perpendicular to the power input shaft. (7)

radiation. Energy transmitted by waves through space. Earth's atmosphere serves as protection from radiation coming from space. (4)

radiator. A heat exchanger used in both hydraulic and pneumatic systems to transfer heat from the system to atmospheric air. (12)

ram. A single-acting linear actuator in which the diameter of the rod is basically the same diameter as the bore of the actuator barrel. (9)

rammer. A pneumatic, portable, reciprocating tool used to compact sand in foundry molds and other applications. (17)

rapid-advance-to-work circuit. A fluid power circuit design that allows a tool or other machine element to rapidly move to a preset position, slow to a point where machining is to occur, and then move at an accurate cutting speed. May also allow rapid retraction of the work elements. (13)

rated capacity. The maximum capacity for which a component is designed. This rating may include details related to fluid flow, pressure, and other factors. (11)

receiver. A tank in a pneumatic system, located close to the compressor, that stores and assists in conditioning compressed air. (2, 15)

reciprocating compressor. A common compressor design using a cylinder, piston, crankshaft, and valves similar to an internal combustion engine. The reciprocating action of the piston brings air into the cylinder, where it is compressed and then moved into the system. (15)

reciprocating-plunger pump. An early reciprocating pump design used for both high-volume and high-pressure applications. They are currently used primarily for high-pressure, low-volume applications. (7)

reciprocating pump. One of a variety of pump designs used to move fluid. The drive mechanisms vary, but a reciprocating piston moves the fluid against system actuator resistance. (7)

reduced-pressure section. A section of a fluid power circuit that functions at a pressure lower than the primary relief valve or pressure regulator setting. (13)

regenerative-cylinder-advance circuit. A hydraulic circuit design in which fluid discharged from the rod end of a cylinder combines with pump delivery and is directed to the cylinder blind end. The design results in increased cylinder extension speed, but reduced force capability. (13)

relative humidity. The relationship between the actual amount of water in the air and the maximum amount the air could hold at a given temperature. (14)

relief valve. A pressure control valve that bypasses excess hydraulic system pump delivery to the reservoir in order to limit system pressure to a selected maximum level. (7, 10)

relieving-type regulator. A pneumatic system pressure regulator that contains a venting orifice. The vent prevents system pressure from exceeding the initial setting of the regulator, even when downstream conditions cause an increase in air pressure. (16)

remote location. A location for control components that is beyond the distance usually used for placement of components on machines. (13)

remote pressure adjustment. Adjustment of a pressure control valve from a location some distance from the valve installation. (19)

replenishment circuit. A circuit used with closed-loop hydraulic systems that provides makeup fluid to replace any fluid lost from leakage during system operation. (9)

reservoir. A component in a hydraulic system that holds the system fluid not currently in use in the pump, control components, actuators, and lines. (2, 8)

resilient material. A material that has the ability to return to its original shape after being compressed. This type of material is often used for the sealing surfaces on valve faces or seats in pneumatic components. (17)

resistance. Opposition to the movement of electron flow in wires, fluid flow in fluid conductors, or a load placed on a fluid power actuator. (4)

respirator face masks. A mask that covers only the nose and mouth or covers the whole face. The mask uses filters to remove toxins from the air. (3)

restriction check valve. A specialized hydraulic check valve that allows free flow in one direction and restricted flow in the opposite direction. (10)

restrictor flow control valve. A flow control valve design that meters the flow passing through the valve, while forcing excess pump output fluid to return to the reservoir through the system relief valve. (10)

return line. The hydraulic system fluid conductor that returns fluid from the operating system back to the reservoir. (8)

revolving-cylinder design. A radial hydraulic pump design in which a cylinder and piston assembly rotate as a unit. (7)

rigid conductor. A fluid power conductor that is not intended to be bent. Steel pipe and some forms of tubing and plastic pipe are examples. Various fittings and adaptors must be used to form these materials into required configurations. (16)

riveting hammer. A pneumatic, portable, reciprocating tool used to place and form the ends of rivets. (17)

rock drill. A pneumatic tool often used to make rock cuts in quarries, mines, and highway construction. May use a drill or a percussion tool for cutting. (17)

rod. A metal bar that is attached to one side of the cylinder piston and extends through the cylinder rod end (head). The rod transfers the force generated by the piston to the system load. (9)

rotary actuator. A fluid power actuator used to produce torque and rotary motion. Commonly called a hydraulic motor or pneumatic motor. (17)

rotary compressor. A compressor design that compresses air using a continuous process rather than the stop-and-go action of a reciprocating unit. Examples include positive-displacement screw and vane units and nonpositive-displacement centrifugal and axial-flow dynamic compressors. (15)

rotary pump. A pump design that moves liquid using a continuous process rather than the stop-and-go action of a reciprocating pump. Examples include positive-displacement gear and vane units and nonpositive-displacement centrifugal and propeller pumps. (7)

rotary screw compressor. A compressor unit that uses intermeshing screws to form chambers that linearly move air through the compressor. The design provides a continuous, positive displacement of the air. (15)

rotary sliding-vane compressor. A compressor unit using a rotor with chambers separated by sliding vanes. Turning the unit produces an almost pulsation-free stream of compressed air. (15)

rust inhibitor. An additive used in hydraulic oils that is designed to protect metal parts by neutralizing acids or forming a film on the metal surfaces to protect them from damage. (6)

S

safety helmet. Headgear designed as protection from the impact of falling or other moving objects. (3)

safety shoes. Shoes designed to protect a worker from specific dangers found in the work environment. Steel-toed and electrically insulated shoes and boots are examples. (3)

safety valve. A simple, direct-operated relief valve that opens to prevent damage resulting from excessive system pressure. (7, 10, 18)

saturation point. The condition when air, at a certain temperature, will not hold additional water vapor. (14)

Saybolt viscometer. A viscosity test apparatus that measures the number of seconds needed for a heated oil to drain through a calibrated orifice to fill a sample flask. No longer considered an accurate method of measuring viscosity. (6)

scaling hammer. A pneumatic, reciprocating hand tool used to remove flux and spatter from welds in fabricated products. (17)

schedule number. A standard of ANSI that rates the wall thickness of pipe. The standard uses schedule numbers from 10 to 160. Hydraulic systems typically use schedule 40, 80, or 160 pipe. (8)

scientific method. A systematic approach to the development of ideas. This includes accurate measurement, controlled testing, reproducibility of results, and systematic reporting of results. (1)

screw. One of the simple machines, which is an inclined plane wrapped around a rod to form a continuous spiral, or thread. The thread provides a high mechanical advantage. (4)

screw motor. A motor unit that uses intermeshing screws to form sealed chambers. Forcing compressed air or hydraulic oil into these chambers turns the motor, providing smooth, quiet, rotary power. (9)

screw pump. A pump unit that uses intermeshing screws to form chambers that linearly move hydraulic oil through the pump. The design provides a continuous, positive displacement of the oil. (7)

second law of thermodynamics. A basic scientific law that states heat flows between two bodies only when one body has a temperature that is higher than the other, with heat moving from the warmer to the colder body. (4)

self-contained breathing device. A device used where the conditions are such that a breathable atmosphere is not available to a worker. The device provides breathable air through a hose attached to a face mask. (3)

sensible heat. Heat that can be measured with a thermometer and will result in a feeling of warming or cooling as heat is applied or removed from a substance. Does not change the state of a substance. (4)

separator. A component, located in the compressor station area of a pneumatic system, that is designed to remove condensed water from the compressed air. The unit is often teamed with an aftercooler to optimize moisture removal. (2)

sequence valve. A hydraulic system pressure control valve designed to divert flow to a second portion of a circuit only after a preselected pressure has been applied to a primary actuator. (10)

sequencing. The control of valves and actuators to provide a specific order of movement in a fluid power circuit. (13)

series circuit. An electrical or fluid power circuit that provides only one path for the current or fluid to follow as it moves through the circuit. (9)

shell-and-tube heat exchanger. A component constructed of a bundle of tubes enclosed in a metal shell. The unit is used to either cool or warm hydraulic fluid by passing water through the tubes. Heat transfer occurs as the hydraulic fluid passes through the space between the tubes and the shell. (12)

shock pressure. A momentary high-pressure surge that exceeds the system relief valve setting. These pressures can occur in a hydraulic system when heavy loads are suddenly encountered or valves shift into closed positions. (8, 11)

shoe plate. A plate used in some axial-piston pumps to hold the piston shoes in contact with the swash plate. (7)

shutoff valve. A valve used in both hydraulic and pneumatic circuits to shut off the flow of fluid to a circuit. A variety of valve designs may be used for this function. (10, 18)

shuttle valve. A control valve that allows a circuit to operate from either of two fluid power sources. Typically, the source with the highest pressure shifts the valve and becomes the circuit's source of fluid. (18)

side loading. A situation in which a pump input shaft or the rod of a cylinder has more load applied from one side than the other. This can cause excessive wear, which shortens service life. (9)

single-acting compressor. A reciprocating-piston compressor in which intake and compression occur in the cylinder space above the piston during one rotation of the compressor crankshaft. The design may contain multiple cylinders, but the cylinders are not staged. (15)

single-acting cylinder. A cylinder design that exerts force only on extension or retraction and depends on some outside force to complete the second movement. (9, 17)

sliding plate. A control element used in one directional control valve design. The design involves two flat plates, held in close contact, that contain channels and ports. Moving the plates in relation to each other produces various fluid flow combinations. (18)

slippage. A condition that occurs in a nonpositive-displacement pump or compressor when flow resistance is met during operation. This condition exists because of the inherent clearances involved in the design of the units. (15)

sludge. An undesirable residue that forms in hydraulic system fluids, resulting from reactions of dirt, water, chemicals, and other substances that enter the system. (6)

specific gravity. The ratio of the weight of a given volume of a material to the weight of an equal volume of water at 4°C. (4)

specific heat. A ratio that compares the quantity of heat required to raise the temperature of one unit of a substance one degree in temperature to the heat needed to raise an equal weight of water 1°C. (4)

specific weight. The weight of a specific volume of a substance at a specific temperature and pressure. (4)

spiral clearance. The clearance that exists between the threads of a standard pipe thread even when they are tightened. This clearance is extensive enough to cause leakage when working with a pressurized fluid. (8)

spontaneous ignition. The temperature at which a hydraulic fluid will ignite and burn without the application of a spark or open flame. (6)

spool. A cylindrical, machined component precision-fit into a bore machined into the body of a control valve and used to control fluid flow and direction through the valve. (10, 18)

spool valve. A valve using a spool to control fluid flow. (10)

spraying. A process function involving the use of compressed air to apply paint, agricultural chemicals, or other materials. (17)

spring-loaded accumulator. An accumulator using a compressed spring to maintain pressure on system fluid. (11)

spur gear. A gear with radial teeth that are cut parallel to the axis of the gear. Commonly used in gear pumps and other fluid power components. (7)

staging. A process commonly used in the design of pneumatic compressors where the outlet of one compressor cylinder is connected to the intake of the next cylinder to obtain higher system pressures. (15)

standard. A set of specifications that define a process, product, or part. (5)

standard atmospheric pressure. The weight of the various gases that make up the atmosphere produces a pressure of approximately 14.7 pounds per square inch at sea level. (4)

static seal. A seal using materials compressed between two rigid, nonmoving parts. Paper gaskets or molded O-rings are common examples. (9)

stationary-cylinder design. A radial-piston hydraulic pump design in which a cam operates pistons located in a fixed cylinder block. (7)

stator. A stationary barrier built into the body of a vane-type, limited-rotation actuator. In conjunction with the rotor vane, the stator divides the body cavity of the actuator into two sealed chambers. (9)

steam engine. The first of the larger power-generating devices that contributed to the success of the Industrial Revolution and the application of hydraulic power. (1)

strainer. A coarse filter used on the end of a hydraulic pump inlet (suction) line submerged in the fluid of a reservoir. (7)

stratosphere. The part of the atmosphere that extends from 10 to 30 miles above earth's surface. (14)

suction filter. A filter located outside of the reservoir in the pump inlet (suction) line. (12)

sump strainer. The coarse filter located on the end of the pump inlet (suction) line. The unit is submerged in the fluid of the reservoir. Also called a strainer. (12)

surface-type filter. A single-surface filter medium containing numerous holes to allow the passage of fluid while trapping particles larger than the holes. (12)

swash plate. An inclined disk in a hydraulic axial-piston pump or motor that causes the pistons to reciprocate. (7)

symbol. A figure used on a diagram to represent fluid power components and features of those components. (5)

synchronization. The timing of one movement or operation with another, such as the identical movement of two actuators operating in a fluid power circuit. (13)

synthetic fluid. A refined product containing additives that produces high levels of certain factors desired by end users. (6)

system cycle time. The time required to complete the various steps involved in a repeating process. (13)

system function. The tasks completed by the operation of a fluid power system. (2)

T

tamper. A portable, pneumatic, reciprocating tool used to compact dirt at construction sites and other applications. (17)

telescoping cylinder. A linear actuator constructed of several nested tubes that can extend a distance equal to several times the actuator's retracted length. (9)

temperature. The degree of hotness or coldness of a substance. Temperature is determined by the rate of molecular movement within a substance. (4)

temperature compensation. A design feature of some flow control valves that adjusts the control orifice size to maintain a constant flow rate as viscosity changes due to changes in fluid temperature. (10)

thermodynamics. A science dealing with the relationships between the properties of matter affected by temperature and the conversion of energy from one form to another. (4)

threaded-end cylinder. A linear actuator design in which the cap and head are attached to the barrel of the cylinder by threads. (9)

three-way valve. A directional control valve that provides a means to extend a cylinder using system pressure while depending on an internal spring or external force for return of the cylinder. (10, 18)

tie-rod cylinder. A linear actuator design in which the cap and head components are secured to the barrel of the cylinder by external tie rods that run between those components. (9)

time-delay circuit. A fluid power circuit that provides an adjustable delay of time between various actuator movements. (19)

time factor. The percentage of time actuators function during system operation. (16)

timing-volume reservoir. A chamber used in the pilot lines of pneumatic circuits to allow variable delays in the shifting of pilot-operated directional control valves. (18)

Timken method. A laboratory test method for measuring the extreme pressure properties of lubricating fluids. (6)

torque. The turning or twisting force applied to a shaft. (4)

training. A learning experience characterized by specific instruction and practice. (3)

trio unit. The term sometimes used to designate the close-fitting unit containing the pneumatic workstation filter, regulator, and lubricator. Also called an FRL unit. (19)

troposphere. The layer of the atmosphere in which we live. This layer extends to 10 miles above earth's surface. (14)

trunnion mount. A cylinder mounting design that places fittings on the sides of cylinders, allowing the cylinder to pivot as it extends and retracts to move a machine member. (9)

tube: A long, cylindrical body with an open center. Tubes are used extensively in the fluid power field as fluid conductors and as component parts, such as the barrel of cylinders. (2)

tubing. A semirigid fluid conductor used in both hydraulic and pneumatic circuits and systems. Available in a number of different materials, ranging from plastic to stainless steel. (8)

turbine motor. Pneumatic motors in which pressurized air acts on a series of blades to produce rotary motion. Capable of producing high-speed operation. (17)

two-hand safety circuit. A fluid power circuit in which machine activation and continued operation requires both hands of the machine operator to continuously depress control valves. (13, 19)

U

unbalanced-vane pump. A hydraulic vane pump design using single inlet and outlet ports. The design produces side-loading pressure on the vane rotor and pump bearings. (7)

universal joint. A mechanical coupling used when shafts transmitting power are not in alignment. Used in bent axis piston pumps to drive the cylinder block. (9)

usable volume. The volume of hydraulic fluid an accumulator can supply in a specific system application. (11)

V

vacuum. A space in which the pressure is below normal atmospheric pressure. The concept applies to a variety of fluid power component and circuit applications, such as the intake lines of both hydraulic pumps and pneumatic compressors. (4)

valve body. The structural element of a valve made from standard metal stock or a special casting. The body contains bores, chambers, lines, ports, and often fittings for valve mounting. (10, 18)

valve plate. A valve design using a flat plate with a passageway or ports machined into the surface. Used in rotary directional control valves and some piston pump designs. (7)

vane motor. A hydraulic or pneumatic actuator using a rotor with slots containing movable vanes to generate rotary motion and torque. (17)

vane pump. A design of hydraulic pump that uses a rotor with slots containing movable vanes. Rotating the rotor moves fluid through the pump, producing fluid flow in a hydraulic system. (7)

vapor pressure. The pressure at which a liquid will begin to form a vapor. Vapor pressure varies with fluid type and formulation. (7)

variable-delivery pump. A pump design that permits changes in the size of the pumping chamber to produce a range of flow output. (7)

variable-displacement motor. A hydraulic motor design in which the displacement of the motor can be easily changed, allowing the speed to be changed without varying input fluid flow rate. (9)

varnish. An undesirable material that forms on internal surfaces of hydraulic components when hydraulic fluids oxidize as they are exposed to high operating temperatures. Contributes to corrosion of metal surfaces and promotes the formation of sludge. (12)

velocity. The speed at which fluid moves through a line or component element. Also the physical speed of a machine member. Expressed as distance per unit of time. (4)

vent connection. An external fitting on a compound relief valve connected to the control chamber of the valve. Allows remote adjustment of that pressure to vary the maximum operating pressure of a system. (13)

venturi. A shaped restriction in a component that increases velocity and lowers the pressure of the fluid passing through the opening. (7)

viscosity. The internal resistance to flow of the molecules of a liquid. (6)

viscosity grade. A rating indicating the viscosity of a liquid. Several different rating systems are available from standardizing groups. Each system provides consistent ratings within that system. (6)

viscosity index number. A number that expresses the relative change in viscosity that can be expected for a given change in temperature of a liquid. (6)

volumetric efficiency. A ratio of the volume of air admitted to a compressor cylinder at a given temperature and pressure to the full displacement of the cylinder. (7)

vortex. A whirlpool that can form around the inlet line if the pickup is too close to the fluid surface. (8)

W

water screw. An invention developed by Archimedes in the third century BCE, using a screw (simple machine) to raise water to a different elevation. It is still used today in some developing countries. (1)

water trap. A simple trap located at low points in pneumatic system distribution lines to collect water that has been condensed from the air in the lines. Water removal may be manual or automatic. (16)

water vapor. Water in its gaseous form. Note: fog or mist is not water vapor, but atomized liquid water. (14)

waterwheel. An early application of fluid power dating to ancient civilizations where moving water rotated a wheel to produce usable force. (1)

wear plate. A plate located on the end surfaces of gears and rotors in gear and vane pumps to promote sealing of the pumping chambers and to reduce wear on the components. (7)

wedge. One of the six basic machine types, consisting of two inclined planes that share a common base. (4)

weight. The result of the mass of a structure or body being acted on by gravity. (4)

weight-loaded accumulator. An accumulator design using physical weight acting on liquid in a vertical ram to store energy in a hydraulic system. The only accumulator design that provides a constant pressure as fluid is discharged. (11)

wheel and axle. One of the six basic machine types. Consists of a wheel attached to an axle. The common center of the wheel and axle is the fulcrum, allowing the device to act as a second- or third-class lever, depending where the effort is applied. (4)

windmill. An early application of fluid power that some people trace to boat sails. Machines resembling our concept of a windmill appeared in early history in the Middle East. Modern windmills are still in use throughout the world. (1)

wiper seal. A device that prevents dirt and other contaminants from entering a fluid power system as the cylinder rod is retracted. Often referred to as a scraper. (9)

work. The application of force through a distance. Work only results when the applied force moves the object. (4)

INDEX